Multimedia content is available for this book from http://iopscience.iop.org/book/978-0-7503-2678-0.

ISBN 978-0-7503-2678-0 (ebook)
ISBN 978-0-7503-2676-6 (print)
ISBN 978-0-7503-2679-7 (myPrint)
ISBN 978-0-7503-2677-3 (mobi)

DOI 10.1088/978-0-7503-2678-0

Version: 20200801

IOP ebooks

British Library Cataloguing-in-Publication Data: A catalogue record for this book is available from the British Library.

Published by IOP Publishing, wholly owned by The Institute of Physics, London

IOP Publishing, Temple Circus, Temple Way, Bristol, BS1 6HG, UK

US Office: IOP Publishing, Inc., 190 North Independence Mall West, Suite 601, Philadelphia, PA 19106, USA

Contents

Part II The modern physics behind experiments

Preface

The term 'modern physics' generally refers to the study of those facts and theories that emerged in the early decades of the twentieth century, when the classical methods and approaches encountered increasing difficulty in explaining experimental observations. These facts concern, on the one hand, the ultimate structure and interaction of matter and, on the other hand, the concepts of space and time. In terms of the basic knowledge of matter, all of modern physics is developed with reference to its elementary constituents, i.e. molecules, atoms and elementary particles. However, the atomicity of matter is not at all obvious and the student may well wonder how we can place so much confidence in our analysis, given that the observations of these particles are necessarily so indirect. Also, as far as the interplay between space and time is concerned, modern physics finds an answer, in the region of very high speeds, where classical mechanics breaks down.

This book first discusses the basic aspects of classical physics, from both the theoretical and experimental perspectives, and then starts dealing critically with the relevant aspects of modern physics. Thus, the first chapters are devoted to the teaching methods and technologies (chapters 1, 6 and 9) and the laboratory perspectives (chapters 2, 5 and 4) for learning classical physics. Taking advantage of the contents of these chapters, the same topics are subsequently analysed within a modern physics framework, exploiting critically the complexity of this 'new physics' compared to its classical counterpart. Then, the basic classical experiments that gave rise to the birth of modern physics are presented and discussed critically (chapter 7). In particular, they are illustrated considering what can be explained by means of classical physics and what cannot be justified and clarified using the classical 'beliefs'. By discussing empirical methods to explain these experiments, we provide arguments to support the claim that the difficulty in fully describing this phenomenology can be transformed into cultural challenges, underlining that new forms of complexity and reasoning are used.

Special relativity (chapter 3) and quantum mechanics (chapter 8) are thus presented in a way that undergraduate students can approach and fully understand. Special emphasis is then devoted to solid state physics and its relationship with 'modern physics', referring in particular to some experiments that manifest quantum macroscopic effects. We discuss some fundamental concepts of quantum optics and their applications in modern quantum technologies (chapter 12), the use of superconductors for quantum computers in superconducting materials (chapter 11) as well as spectroscopic methods that refer to the application of quantum mechanics paradigms (chapter 10), with the aim of providing examples of the manifestation of quantum effects in the realm of the 'classical' world.

Since teaching experiments can help students become more engaged and interested in the topic they are studying, in particular when used over time, the description of laboratory activities occupies an important position in physics teaching. Thus, we give the details of some classic experiments such as the Thomson experiment (chapter 13), the Millikan oil drop experiments (chapter 14), the Davisson and Germer experiment (chapter 15) and a method to measure Planck's constant (chapter 16). Moreover, taking advantage of modern technology, we also present computer-assisted experiments that can help to overcome some difficulties that occur in the traditional methods of teaching specific experiments, such as those concerning the notion of the dilatation of time intervals and contraction of lengths, as well as the quantum tunnelling effect (chapter 18).

Practice exercises are also included, and their solutions provided, in a clear and critical way (chapter 17).

The main purpose of this book is to create a bridge between classical and modern physics, filling the gap between the mainly descriptive treatment of phenomena, as given in elementary textbooks, and the extremely formal theoretical accounts which one may find in graduate-level textbooks. It is not encyclopaedic in its coverage; it aims to give the student a feeling for the principles of modern physics through the selection of topics which fit together and draw upon each other, in this way strengthening their interplay. To this end, only some parts of solid state physics building directly upon the fundamentals of quantum theory have been included, whereas topics such as band theory, which requires a digression from the main thrust of this book, are excluded.

This book was developed from lecture notes originally used in a specialized Master's course on 'New teaching methods in modern physics' devoted to advanced graduate students and high school teachers at Salerno University (Italy). The notes, which originally supplemented a number of books used in conjunction with the modern physics course, have been carefully revised and expanded by the course lecturers. More importantly, this material has been extensively class-tested with students having various mathematics and physics backgrounds and interests.

The content of this book is aimed at people willing to reflect on the complexities of thinking in physics. They are encouraged to understand productive syntheses of relativistic and quantum formalisms as forms of fruitful simplification, mainly based on an axiomatic formulation of the different types of mechanics, i.e. classical, relativistic and quantum. Furthermore, the comprehension of modern physics phenomenology is favoured, showing that these topics could be better interpreted using links to classical descriptions. In conclusion, we are confident that, using the adopted methodology, this book lays out a curriculum where students are progressively guided to manage more and more sophisticated forms of complexities up to those implemented in special relativity and quantum physics.

Acknowledgements

Editing this book has been more difficult than I anticipated, but more rewarding than I could have ever imagined. This book would not have been possible without the excellent organization that allowed me to develop, assemble and realize this editorial project over the last months.

None of this would have been possible without the editorial staff of IOP Publishing. I am deeply indebted to Caroline Mitchell and Robert Trevelyan for their wonderful editorial support and guidance.

I owe an enormous debt of gratitude to the contributors, the colleagues and high-school teachers, who wrote excellent chapters. They gave freely of their time to discuss the content and the text of their contributions, clarifying concepts, exploring particular features, and explaining complex concepts and topics simply.

I am also very grateful to Antonio Stabile, who also contributed as an author, for carefully reading the book from the early drafts. He was genuinely interested in this project, giving constructive critiques and suggestions.

I have greatly appreciated the wise counsel of my colleague and best friend Alfonso Romano during the final stages of completing this book, as well as the precious technical support given by Donato Noce to manage some parts of the book.

I would like to express my deepest appreciation for Vincenzo Di Marino who provided the possibility for me to complete this book. Indeed, he invested his full effort in achieving this goal, improving the presentation of all contributions, and managing technical problems in a clever and efficient way. Thank you very much Enzo.

Finally, my family: I want to thank my wife Rosangela—a lifelong partner who always makes the voyage worthwhile—for tolerating my disappearances into my home office, and my daughter Marialaura and my son Giuseppe for their continuous and unparalleled love, help and support, and for always cheering me up.

Editor biography

Canio Noce

Professor Canio Noce studied physics at the University of Salerno and earned a PhD in Physics at the University of Naples 'Federico II'. Since 1991 he has worked at the University of Salerno where he is currently Professor of Theoretical Physics, and the director of the specialized Master's course 'New teaching methods in modern physics'.

Contributors

Carmine Attanasio

Dipartimento di Fisica 'E R Caianiello', Università degli Studi di Salerno, Fisciano, Salerno, Italy;
Consiglio Nazionale delle Ricerche, Istituto SPIN, Fisciano, Salerno, Italy.

Professor Carmine Attanasio studied physics at the University of Salerno, where he earned a PhD in Physics. Since 1993 he has worked at the University of Salerno where he is currently Full Professor of Experimental Condensed Matter Physics.

Francesco Avitabile

Liceo Statale 'Ernesto Pascal', Pompei (NA), Italy.

Professor Avitabile Francesco studied physics at the University of Salerno, where he also earned a PhD in physics. He also obtained a II level university Master's course in Modern Physics at the University of Salerno. Since then, he has worked at Liceo Statale 'Ernesto Pasca', Pompei, where he is currently Professor of Mathematics and Physics.

Antonio Capolupo

Dipartimento di Fisica 'E R Caianiello', Università degli Studi di Salerno, Fisciano, Salerno, Italy;
Istituto Nazionale di Fisica Nucleare (INFN), Sezione di Napoli, Gruppo Collegato di Salerno, Fisciano, Salerno, Italy.

Professor Antonio Capolupo graduated in physics from the University of Salerno where he also earned a PhD in Physics. From 2014 to 2019 he was a researcher in the Department of Physics of the University of Salerno where he is currently Associate Professor of Theoretical Physics of Fundamental Interactions.

Mario Cuoco

Consiglio Nazionale delle Ricerche, Istituto SPIN, Fisciano, Salerno, Italy;
Dipartimento di Fisica 'E R Caianiello', Università degli Studi di Salerno, Fisciano, Salerno, Italy.

Dr Mario Cuoco studied physics at the University of Salerno, where he earned a PhD in Physics in 2000. He was a postdoctoral fellow at the University of Salerno, at the CNRS in Grenoble and at the National Institute of Condensed Matter. Since 2008 he has worked at CNR Institute SPIN where he is currently a researcher.

Roberto De Luca

Dipartimento di Fisica 'E R Caianiello', Università degli Studi di Salerno, Fisciano, Salerno, Italy.

Professor Roberto De Luca studied physics at the University of Salerno, where he graduated in 1986. In 1987 he obtained an MA in Physics at the University of Southern California. In 1992 he achieved a PhD in Physics at the Universities of Naples and Salerno. Since 1994 he has worked at the University of Salerno where he is currently Associate Professor of Physics Education.

Marco Di Mauro

Dipartimento di Fisica 'E R Caianiello', Università degli Studi di Salerno, Fisciano, Salerno, Italy;
Istituto Nazionale di Fisica Nucleare (INFN), Sezione di Napoli, Gruppo Collegato di Salerno, Fisciano, Salerno, Italy.

Dr Marco Di Mauro studied physics at the University of Salerno, where he earned a PhD in Physics. After a couple of years teaching mathematics and physics in secondary schools, since 2018 he has been working at the University of Salerno, where he is currently a Postdoctoral Fellow in the history and didactics of physics.

Marco Figliolia

Liceo Scientifico Statale 'P S Mancini', Avellino, Italy.

Professor Marco Figliolia studied physics at the University of Salerno (Bachelor's degree) and at University 'Federico II' of Naples, where he graduated. He also obtained a II level university Master's course in Modern Physics at the University of Salerno. Since 2017 he has worked at Liceo Scientifico 'P S Mancini' where he is currently Professor of Mathematics and Physics

Veronica Granata

Dipartimento di Fisica 'E R Caianiello', Università degli Studi di Salerno, Fisciano, Salerno, Italy;
Consiglio Nazionale delle Ricerche, Istituto SPIN, Fisciano, Salerno, Italy;
Istituto Nazionale di Fisica Nucleare (INFN), Sezione di Napoli, Gruppo Collegato di Salerno, Fisciano, Salerno, Italy.

Dr Veronica Granata studied physics at the University of Salerno, where she earned a PhD in Physics. She was a postdoctoral fellow at St Andrews University in Scotland, UK. Since 2010 she has worked at the University of Salerno where she is currently a postdoctoral fellow in Physics.

Delia Guerra

Liceo Scientifico 'Leonardo da Vinci', Vallo della Lucania (SA), Italy.

Professor Delia Guerra studied physics at the University of Salerno, where she graduated and earned a PhD in physics. She also obtained a II level university Master's course in Modern Physics at the University of Salerno. Since then, she has worked at Liceo Scientifico 'Leonardo da Vinci', where she is currently Professor of Mathematics and Physics.

Lazzaro Immediata

Liceo Scientifico 'Leonardo da Vinci', Vallo della Lucania (SA), Italy.

Professor Lazzaro Immediata studied physics at the University of Salerno, where he graduated. He also obtained a II level university Master's course in Modern Physics at the University of Salerno. Since then he has worked at Liceo Scientifico 'Leonardo da Vinci', where he is currently Professor of Mathematics and Physics.

Antonio Leo

Dipartimento di Fisica 'E R Caianiello', Università degli Studi di Salerno, Fisciano, Salerno, Italy;
Consiglio Nazionale delle Ricerche, Istituto SPIN, Fisciano, Salerno, Italy.

Dr Antonio Leo studied physics at the 'Federico II' University of Naples, where he graduated in 2005. In 2009, he earned a PhD in Physics at the University of Salerno, where he is currently a postdoctoral fellow.

Maria Teresa Mercaldo

Dipartimento di Fisica 'E R Caianiello', Università degli Studi di Salerno, Fisciano, Salerno, Italy.

Professor Maria Teresa Mercaldo studied physics at the University of Salerno, where in 2001 she earned a PhD in Physics. Since 2007 she has worked at the University of Salerno where she is currently Assistant Professor of Theoretical Condensed Matter Physics.

Martina Moccaldi

Istituto d'Istruzione Superiore 'Luigi Einaudi', Scafati (SA), Italy.

Professor Martina Moccaldi studied mathematics at the University of Salerno, where she graduated and earned a PhD in Mathematics. She also obtained a II level university Master's course in Modern Physics at the University of Salerno. Since then she has worked at Istituto 'Luigi Einaudi' where she is currently Professor of Mathematics and Physics.

Angela Nigro

Dipartimento di Fisica 'E R Caianiello', Università degli Studi di Salerno, Fisciano, Salerno, Italy;
Consiglio Nazionale delle Ricerche, Istituto SPIN, Fisciano, Salerno, Italy.

Professor Angela Nigro studied physics at the University of Salerno, where she graduated in physics. Since 1992 she has worked at the University of Salerno where she is currently Professor of Experimental Physics.

Canio Noce

Dipartimento di Fisica 'E R Caianiello', Università degli Studi di Salerno, Fisciano, Salerno, Italy;
Consiglio Nazionale delle Ricerche, Istituto SPIN, Fisciano, Salerno, Italy.

Professor Canio Noce studied physics at the University of Salerno and earned a PhD in Physics at the University of Naples 'Federico II'. Since 1991 he has worked at the University of Salerno where he is currently Professor of Theoretical Physics, and the director of the specialized Master's course of 'New teaching methods in modern physics'.

Sergio Pagano

Dipartimento di Fisica 'E R Caianiello', Università degli Studi di Salerno, Fisciano, Salerno, Italy;
Consiglio Nazionale delle Ricerche, Istituto SPIN, Fisciano, Salerno, Italy.
Professor Sergio Pagano studied physics at the University of Salerno, where he graduated in 1984. In 1987 he achieved a PhD in Physics at the Technical University of Denmark. From 1987 to 2005 he was a researcher at the Institute of Cybernetics of CNR, Italy. Since 2005 he has worked at the University of Salerno where he is currently Full Professor of Experimental Physics.

Ileana Rabuffo

Dipartimento di Fisica 'E R Caianiello', Università degli Studi di Salerno, Fisciano, Salerno, Italy.
Professor Ileana Rabuffo studied physics at the University of Salerno. Since 1982 she has worked at the University of Salerno where she is currently Professor of General Physics.

Alfonso Romano

Dipartimento di Fisica 'E R Caianiello', Università degli Studi di Salerno, Fisciano, Salerno, Italy;
Consiglio Nazionale delle Ricerche, Istituto SPIN, Fisciano, Salerno, Italy.
Professor Alfonso Romano studied physics at the University of Salerno and earned a PhD in Physics at the University of Naples 'Federico II'. Since 1992 he has worked at the University of Salerno where he is currently Professor of Theoretical Condensed Matter Physics.

Marcello Sette

Liceo Scientifico 'P S Mancini', Avellino (AV), Italy.
Professor Marcello Sette studied physics at the University of Naples, where he graduated. He also holds a PhD in Mathematics and Computer Science from the University of Naples and a II level university Master's in Modern Physics from the University of Salerno. Since 1995 he has taught in many high schools, the last of which is the 'P S Mancini' Scientific High School in Avellino, where he is currently Professor of Mathematics and Physics.

Alessandro Sorgente

Liceo 'G B Piranesi', Capaccio Paestum, (SA), Italy.
Professor Sorgente Alessandro studied physics at the University of Salerno, where he graduated and earned a PhD in Physics. He also obtained a II level university Master's course in Modern Physics at the University of Salerno. Since then he has worked at Liceo 'G B Piranesi' Capaccio Paestum, (SA), where he is currently Professor of Mathematics and Physics.

Antonio Stabile

Dipartimento di Fisica 'E R Caianiello', Università degli Studi di Salerno, Fisciano, Salerno, Italy;
Istituto Nazionale di Fisica Nucleare (INFN), Sezione di Napoli, Gruppo Collegato di Salerno, Fisciano, Salerno, Italy.
Dr Antonio Stabile studied physics at the University of Salerno, where he graduated in 2011 and earned a PhD in Physics in 2015. Since 2016 he has been a postdoctoral fellow at the University of Salerno in Theoretical Physics.

Antonio Vecchione

Consiglio Nazionale delle Ricerche, Istituto SPIN, Fisciano, Salerno, Italy;
Dipartimento di Fisica 'E R Caianiello', Università degli Studi di Salerno, Fisciano, Salerno, Italy.
Dr Antonio Vecchione studied physics at the University of Naples and earned a PhD in Physics at the University of Salerno in 1994. He was a postdoctoral fellow at the University of Salerno and an Associate Professor at Kyoto University, Japan. Since 2001 he has worked at CNR Institute SPIN where he is currently a researcher.

Part I

From classical to modern physics

Chapter 1

The basic concepts of classical physics as a useful path towards modern physics

Delia Guerra, Ileana Rabuffo and Alfonso Romano

Classical physics includes all the theories formulated after the seventeenth and before the twentieth century, characterized by the implementation of the scientific method proposed by Galileo Galilei [1]. From that moment, nature was investigated by scientists under a completely new perspective. The key feature of classical physics is undoubtedly the determinism which characterizes all its fields, from celestial mechanics to electrodynamics. According to Pierre Simon Laplace, 'an intelligence knowing all the forces acting in nature at a given instant, as well as the momentary positions of all things in the Universe, would be able to comprehend in one single formula the motions of the largest bodies as well as the lightest atoms in the world … nothing would be uncertain, the future as well as the past would be present to its eyes' [2].

The birth of quantum mechanics and the theory of relativity in the twentieth century undermined that vision of the Universe and created the crisis of classical physics, highlighting the need to adopt a totally different point of view (see chapters 3, 7 and 8). However, deep knowledge of the concepts of classical physics cannot be avoided if one is to understand modern physics. In fact, a physicist's mindset is shaped at the beginning when it approaches classic physics. In this chapter, our intent is to present some of the constitutive topics of classical physics, reworked in an educational manner; we propose alternative approaches to the canonical methods, in order to highlight their relevance to students and to prepare them for the study of modern physics. Moreover, the choice of subjects here has been made by considering each of them as preparatory and useful for the comprehension of the subsequent subjects.

In section 1.1 we introduce the three principles of dynamics with the purpose of underlining their conceptual value and their revolutionary importance. Section 1.2 deals with the concepts of work and energy, putting in evidence the strong correlation between these two quantities; in addition, profound knowledge of the

doi:10.1088/978-0-7503-2678-0ch1

classical concept of energy allows its quantization in quantum mechanics to be understood even more. This motivation, namely the ‘quantum jump’ which characterizes certain physical quantities, is also the reason why section 1.3 is devoted to the introduction of the momentum of a vector. In particular, angular momentum is generally a difficult concept for students; therefore we propose a pedagogical path in order to make it easier to understand. Section 1.4 is dedicated to a fundamental argument, which is hardly ever taken into account in basic physics courses, despite its unquestionable relevance: the link between symmetries and conservation laws, as investigated within the celebrated Noether theorem. Finally, sections 1.5 and 1.6 are devoted to the concept of the wave and to Maxwell’s equations, respectively, showing how it is possible to teach students that an undulatory behaviour for coupled electric and magnetic fields can, under suitable conditions, be predicted. We propose an alternative way to present this topic by introducing the concepts of divergence and curl of a vectorial field in a simplified but correct perspective.

It should be noted that in this context we mostly deal with mechanics rather than thermodynamics or electrodynamics. Actually, we believe that stressing these subjects is significant, since by learning to handle classical mechanics it is possible to acquire an universal method which turns out to be useful in every field of physics (and not only in physics) [3].

1.1 The Newton principles of dynamics

1.1.1 The principle of relativity and the first principle

According to Aristotle, the natural state of bodies is rest; thus, any body in movement tends to get slower and slower until it stops, unless it is pushed to continue its motion. Aristotle reached this conclusion, which is evidently wrong, based on his experiments. On the other hand, Galileo interpreted the same experiments in a completely different way: if the object tends to slow down then there is a force opposed to its motion, called the force of friction. He recognized (for the first time) the rank of ‘force’ for the phenomenon of friction. The difference between the two approaches is undoubtedly the scientific method used by Galileo; through it, the Aristotelian false belief about motion was corrected and the results can be summarized in the following statements:

1. If two laboratories move with uniform linear motion, no experiment will give different results in either of them.
2. A reference system moving with uniform linear motion is defined as inertial.
3. In an inertial reference system, a physical object preserves its rest or its uniform linear motion if it is not forced to change this state by the application of an external force.

The first point is the well-known principle of relativity. Point 3 is the principle of inertia, formulated by referring to the definition of an inertial reference system given in point 2. It should be noted that ‘force’ is intended here as a ‘real force’, i.e. an interaction due to something with physical relevance. A fictitious force, in contrast, is the product of the mass of the object through a non-inertial contribution to the acceleration that may emerge in the reference system in which it is located.

The first principle has revolutionary value: it states that the state of rest and the state of uniform linear motion are indistinguishable. This bold assertion is not simple to accept, since it is contrary to common intuition. The first principle also has a fundamental conceptual value, since it allows a special class of reference systems to be defined, that is, the inertial systems.

However, it also has a logical irregularity: according to its formulation, in order to know if one reference system is inertial we have to assure that no forces act on an object in it. Then, we have to verify the state of motion of the object: if it is at rest or moves with a constant speed, the reference system is inertial, otherwise, if it moves with a finite acceleration the reference system is not inertial. However, this demands certainty that no real force is acting on it, but this is formally impossible, since we do not have a quantitative definition of what we mean when we say that a force is zero. We can use the first principle 'back to front', namely as a principle allowing us to deduce whether a net force acts on an object. This, however, requires knowing that one is in an inertial system, so that one definitely ends up with a circular definition. In order to overcome this difficulty, it is assumed that an inertial reference system by definition exists, having its origin in the Sun and its axes oriented toward the fixed stars. As a consequence, all reference systems which are at rest or which move with constant speed with respect to this inertial system are inertial as well. In this way, once an inertial reference system is selected, one can immediately establish if some net forces act on a given object, by checking if it is at rest or in a state of uniform linear motion. It is clear that, under these conditions, the first principle can be used as a criterion to verify the presence or the absence of forces.

1.1.2 The second principle

The second principle deals with the concept of force. It was formulated by Isaac Newton who had in mind real forces only. There are several misconceptions about forces, which can be pointed out by asking students the following questions:

1. Do we need contact to exert a force?
2. Does the application of a force move an object?
3. If an object moves under the action of a force, does it always move in the same direction of the applied force?

Everyday experience leads to incorrect answers to these questions, in particular to the second one: students are tempted to answer 'yes' since they confuse the concept of force with the concept of energy: work always implies motion, whereas the same cannot be said for a force.

The first misconception can be immediately removed if we refer to well-known examples, such as the gravitational force or the electrostatic interaction, which demonstrate that contact is not necessary to exert a force on an object.

The answer to the second question can be suggested by pushing a desk and observing that it does not move if the applied force is not sufficient to overcome the maximum static friction force. Thus, it is evident to students that the application of a force does not always generate motion.

The third point is probably the most subtle, as one can understand on the basis of the following example: there is an attractive force between the Earth and the Moon and it is directed along the line which links their centres, however, the Moon does not move toward the centre of the Earth. Why does this happen? The answer is that the application of a force makes a body acquire an acceleration which is always parallel to the force, but this does not imply that its displacement, i.e. its velocity, is parallel to the force too.

The innovative significance of the second principle is that it establishes the real relation between the cause of the motion (a force $\vec{F}$ doing finite work) and its effect (the acceleration $\vec{a}$). Since it predicts what happens when causes are known, it introduces the profound concept of determinism, which is important also from a philosophical point of view: specifying position and speed at a given time, the equation $\vec{F} = m\vec{a}$ gives the possibility of describing the motion of an object completely. This point is crucial and characterizes classical mechanics, differently to what happens in quantum mechanics. The second principle has also an important conceptual significance, since it allows the mass of an object to be defined, beyond Newton's idea that it simply denotes the quantity of matter. The second principle refines this interpretation and defines the mass as the ratio between the modulus of the net force applied to an object and the modulus of the acceleration that it acquires. This ratio identifies the so-called inertial mass of an object, which is different to the gravitational mass, i.e. the property which we measure with a weight scale. Nevertheless, it is proved rigorously that for the same object the two values are the same.

In the following, we report three more widespread student beliefs that turn out to be false.

- Students often believe that gravity does not act in a vacuum; therefore, for example, in their opinion astronauts float in a spaceship because they are in a vacuum. This is of course not true: this actually happens because the spaceship is subject to gravity in exactly the same way as the astronauts are, in the sense that they are both in free fall. In order to clarify this concept, let us consider a lift and a student in it. For an inertial observer, we have that the projection of $\vec{F} = m\vec{a}$ along the vertical direction leads to

$$mg - N - T = ma, \tag{1.1}$$

where N is the normal force, T is the tension of the cable, a is the acceleration of the lift and $g = G\frac{M_T}{r^2}$ is the acceleration of gravity (G is the gravitational constant, M_T is the mass of Earth and r is the distance of the student from the centre of the Earth). Then, from equation (1.1) it follows that:

$$N = mg - ma - T = m(g - a) - T.$$

If the cable of the lift breaks ($T = 0$), then the lift is in free fall; this means that $a = g$ and, as a consequence, $N = 0$, i.e. the previous contact between the student's feet and the lift floor ceases to exist (see figure 1.1). The student, then, begins to float, since he is falling and the lift is falling in exactly the same way. This is what happens to the astronauts in the spaceship, namely the astronauts and the spaceship are together in free fall around the Earth.

- In order to point out a common mistake, sometimes reported in considerations of the physics of sport, one can consider the case of a ball hit by a racket; in the study of the motion of the ball, the force that the racket exerts on it must not be included in the equation $\vec{F} = m\vec{a}$, because this force only affects the initial conditions of the motion, determining the value of the initial velocity v_0. More precisely, we deal in this case with an impulsive force acting during a very short time interval Δt, which causes a variation of the quantity of motion in one dimension, given by

$$\Delta p = mv_0 - 0 = F\Delta t.$$

- Consider a pendulum and the projection of $\vec{F} = m\vec{a}$ in the direction of the cable; in some cases students are tempted to write

$$T - mg\cos\theta = 0$$

since there is no motion along the cable. Obviously, this is wrong since the normal component of the acceleration is clearly non-vanishing.

1.1.2.1 Most common errors

In this subsection, we discuss some common mistakes made by students in the application of the second principle of dynamics. This aspect should be taken into consideration: it is important to show students what should be done and what should

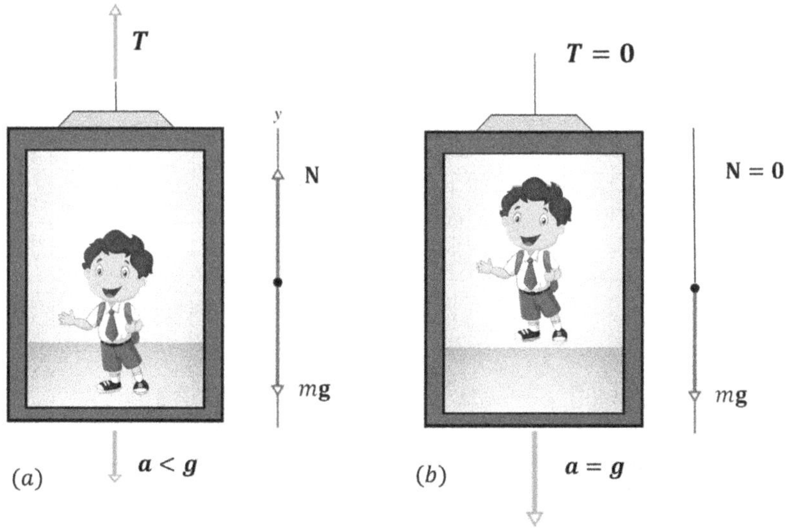

Figure 1.1. (a) A student in a lift going down and (b) the same system in free fall when the cable is broken.

not. In our opinion, students should be encouraged in particular when they get something wrong, since errors can be useful in order to obtain complete comprehension. In the left panel of figure 1.2, we show a mistake in writing the second law of dynamics, while the panels in the centre and on the right relate to two widespread errors with regard to the formal definition of the force of friction and the acceleration of an object on an inclined plane, respectively. Another misleading belief of students is that the so-called centrifugal force is an active force pushing an object out of its trajectory. This is obviously false, since it is a fictitious force, i.e. not due to a real interaction, that appears to act on an object when viewed in a rotating frame of reference. Some confusion may also arise with the concept of centripetal force, which students in some cases consider as a specific applied force. In contrast, the real physical quantity to refer to is the centripetal acceleration: when it is non-vanishing, then the product of mass for acceleration is the force, which sometimes is a tension, sometimes a reaction constraint, and so on.

Generally speaking, the most frequent errors are related to the use of vectors, which apparently often causes difficulties for students. In figures 1.3 and 1.4, we report other mistakes, in this case concerning the scalar and the vectorial products and the definition of pressure, respectively. In particular, the image on the right in figure 1.4 shows a double mistake: the first is, of course, that pressure is not a vector; the second one deals with the fact that, in vector algebra, the inverse of a vector is not defined (note that this mistake is also present in the image on the left of figure 1.2).

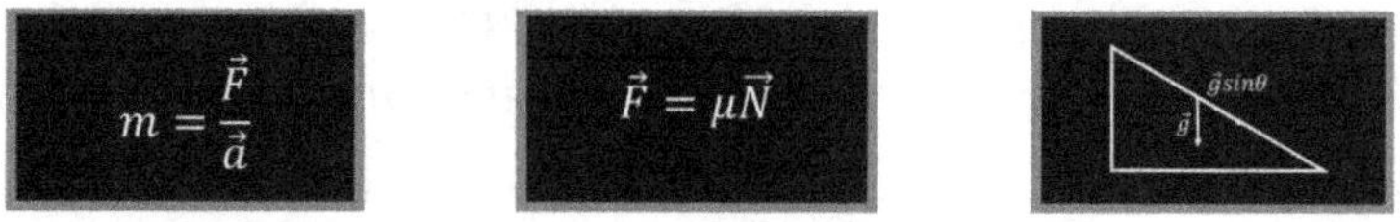

Figure 1.2. Serious mistakes in writing the second law of dynamics (left), the force of friction (centre) and the force of gravity on a inclined plane (right).

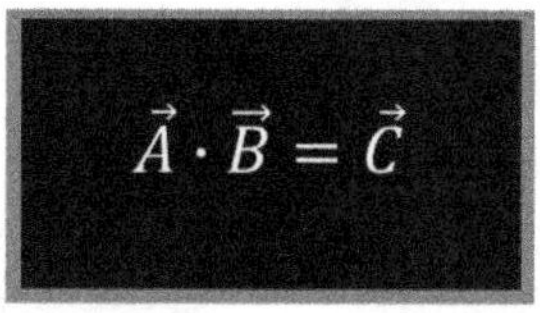

Figure 1.3. Two mistakes for the scalar and the vectorial product.

Figure 1.4. Incorrect relations defining pressure.

1.1.3 The third principle

An important aspect of the third principle is that opposite forces between two particles act along the same straight line. Without this clarification, in several cases students could be misled. For example, by considering an isolated system with two particles, the conservation of the linear momentum gives:

$$m_1\vec{v}_1 + m_2\vec{v}_2 = 0 \Rightarrow m_1\vec{a}_1 = -m_2\vec{a}_2 \Rightarrow \vec{F}_1 = -\vec{F}_2, \tag{1.2}$$

where $\vec{F}_1$ is the force that m_2 exerts on m_1 and $\vec{F}_2$ is the force that m_1 exerts on m_2. One could thus be tempted to conclude that the third principle is a consequence of the second one, but this is not true since equation (1.2) does not imply that $\vec{F}_1$ and $\vec{F}_2$ act along the same line. A possible suggestion to give to students is that every time we individuate a force acting on an object, there is always its partner, as required by the third principle; however, it is important to remember that the correct partner is not included in the equation of motion of the object we are interested in. A check on this point could be to ask a student: 'Consider an object lying on a table. Is the normal force exerted by the table on the object the partner of the weight on the object?'. Frequently the student answers 'yes', but this is the wrong answer, because the partner is the force exerted by the object on the Earth.

1.2 Work and energy

1.2.1 The concept of work

The concept of work is a central one, since it is strictly related to the concepts of kinetic and potential energy. This is of course of special relevance, since every form of energy (thermal, electric or nuclear) is always attributable to the forms of kinetic and potential energies of the elementary constituents. The importance of the concept of work is thus undoubted, thus it should be introduced in basic mechanics courses in a careful and appropriate way. The fundamental definition of work should be referred to the simplest situation, i.e. constant force and linear displacement; under these conditions, it is defined as the product between the force and the displacement which the force generates in the direction of the force itself. This means that all the forces giving rise to displacements orthogonal to their own direction do not produce work. This is what happens, for instance, in the case of a force acting on a particle which moves in a uniform circular motion. In contrast, if the force and the displacement are in the same direction, the work is maximum: this is what happens, for example, when a force pushes an object on a constraint in the direction parallel to the constraint. An elegant way to express this concept is the use of the scalar product:

$$\mathcal{L} = \vec{F} \cdot \Delta\vec{s} = F\Delta s \cos\theta. \tag{1.3}$$

Here θ is the angle formed by the direction of the force $\vec{F}$ and the direction of the displacement $\Delta\vec{s}$. This definition, which, as already stated, refers to the special case of constant force and linear displacement, can be immediately used to give the definition of work in the general case of variable forces and curvilinear trajectories.

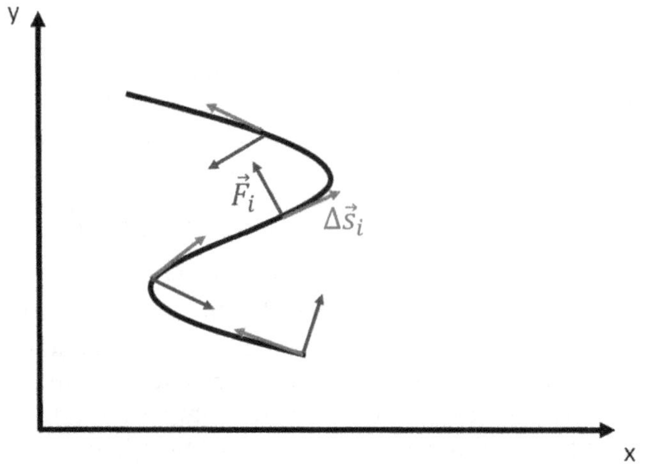

Figure 1.5. Illustration of the calculation of a line integral.

It is sufficient to divide a given trajectory C into a series of elements small enough that they can be considered linear and that the force along each of them can be assumed to be approximately constant. The general procedure is illustrated in figure 1.5, where the red vectors represent generic displacements $\Delta\vec{s}_i$ ($i = 1, 2, 3, \ldots$) along the elements into which the trajectory has been divided. Under the conditions specified above, the definition (1.3) can be applied to each linear element, so that one can introduce the sum

$$\vec{F}_1 \cdot \Delta\vec{s}_1 + \vec{F}_2 \cdot \Delta\vec{s}_2 + \cdots + \vec{F}_i \cdot \Delta\vec{s}_i + \cdots = \sum_i \vec{F}_i \cdot \Delta\vec{s}_i.$$

Work is then defined as the limit of this quantity when the number of elements into which the trajectory has been divided becomes arbitrarily large or, equivalently, the length of each them becomes arbitrarily small:

$$\mathcal{L} = \lim_{\Delta\vec{s}_i \to 0} \sum_i \vec{F}_i \cdot \Delta\vec{s}_i = \int_C \vec{F} \cdot d\vec{s}. \tag{1.4}$$

The integral in equation (1.4) refers to a calculation along a line; indeed, the subscript C denotes that the result of the integral is intimately connected to the path along which the calculation is performed. This is a point to stress with students, since the symbol $\int$ is also used to indicate ordinary integrals, defined, according to figure 1.6, as

$$\int_a^b F(x)dx = \lim_{\Delta x_i \to 0} \sum_i F(\xi_i)\Delta x_i. \tag{1.5}$$

It is evident that whereas the above quantity is only related to the dependence of F on x, in equation (1.4) the result is also dependent on the path C. Thus, equations (1.4) and (1.5) are profoundly different in their significance. Integrals such as the one

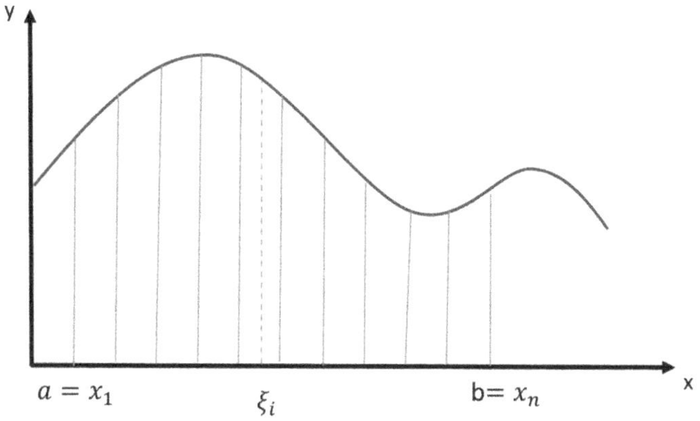

Figure 1.6. Illustration of the calculation of a definite integral.

in equation (1.4) are called line integrals to point out the importance of the path for their evaluation, while integrals such as that in equation (1.5) are the ordinary Riemann integrals. We also point out that when the line integral of a given vector is calculated around a loop, the integral is often called circulation of the vector.

In this context, it is worth mentioning a special case: the work done by a fluid as a consequence of variations of its volume, as usually considered in thermodynamics. The line integral, in this case, becomes an ordinary integral and the work reads

$$\mathcal{L} = \int_{V_A}^{V_B} P \mathrm{d}V,$$

where V_A and V_B are the values of V in the initial and final equilibrium states A and B, respectively. In the above expression, $P = F_\perp / A$ is the pressure of the fluid, with $F_\perp$ being the force orthogonal to its surface and A is the section of the recipient which contains the fluid, taken parallel to its surface. This integral is graphically represented in figure 1.7.

Line integrals are a widely used mathematical tool. The definitions of the electric potential difference and the Ampère circuital law, for example, are expressed in terms of line integrals:

$$V_B - V_A = -\int_A^B \vec{E} \cdot \mathrm{d}\vec{s}$$
$$\oint \vec{B} \cdot \mathrm{d}\vec{s} = \mu_0 i.$$

Going back to equation (1.4), it is evident that the value of a line integral is a function of the chosen path. However, there exists a special category of forces, called conservative forces, for which the line integral is independent of the path and only depends on its initial and final points. In these special cases, one introduces a function V which allows the result of the line integral to be written in the form

$$\int_C \vec{F} \cdot \mathrm{d}\vec{s} = V(f) - V(i), \tag{1.6}$$

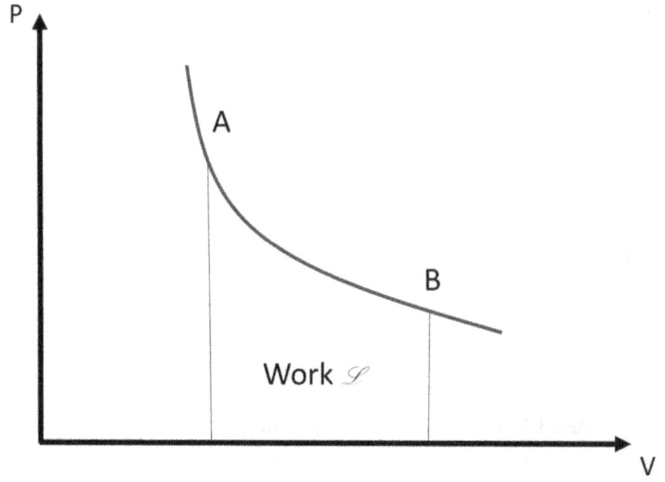

Figure 1.7. Representation of the thermodynamic work in the P–V plane.

whatever the line C connecting the initial and the final points i and f, such that the integral only depends on i and f. In addition, if the initial point i is kept fixed, equation (1.6) allows each point of the space to associate to a number, that is, the work done by the conservative force when a particle moves from i to that point (we deal with this argument in more detail in the next section). This also implies that if we choose a closed line, the value of the integral is always zero.

1.2.2 The concept of kinetic energy

Kinetic energy is defined as the energy of an object of mass m in movement with speed equal to v and it is expressed as $K = \frac{1}{2}mv^2$. It can be useful to motivate such a definition by showing students its origin and the important link with the concept of work illustrated in the previous section.

Let us consider a particle of mass m and all the forces acting on it. If we define $\vec{F}$ as the sum of all these forces, the second principle of dynamics asserts that $\vec{F} = m\vec{a}$, where $\vec{a}$ is the acceleration of the particle. Then, we can calculate the work $\mathcal{L}$ done by $\vec{F}$ relative to a displacement from point A to point B:

$$\begin{aligned} \mathcal{L} &= \int_{\mathrm{A}}^{\mathrm{B}} \vec{F} \cdot \mathrm{d}\vec{s} = \int_{\mathrm{A}}^{\mathrm{B}} m\vec{a} \cdot \mathrm{d}\vec{s} = m \int_{\mathrm{A}}^{\mathrm{B}} \frac{\mathrm{d}\vec{v}}{\mathrm{d}t} \cdot \mathrm{d}\vec{s} = m \int_{\mathrm{A}}^{\mathrm{B}} \mathrm{d}\vec{v} \cdot \vec{v} \\ &= m \int_{\mathrm{A}}^{\mathrm{B}} \frac{\mathrm{d}(v^2)}{2} = \frac{1}{2}mv^2 \bigg|_{\mathrm{A}}^{\mathrm{B}} = \frac{1}{2}mv_{\mathrm{B}}^2 - \frac{1}{2}mv_{\mathrm{A}}^2. \end{aligned} \tag{1.7}$$

Here, starting from $\vec{v} = \frac{\mathrm{d}\vec{s}}{\mathrm{d}t}$, we have used $\mathrm{d}(v^2) = \mathrm{d}(\vec{v} \cdot \vec{v}) = \vec{v} \cdot \mathrm{d}\vec{v} + \mathrm{d}\vec{v} \cdot \vec{v}$, so that $\mathrm{d}(v^2) = 2\vec{v} \cdot \mathrm{d}\vec{v}$. The calculation in equation (1.7), which expresses the so-called work–energy theorem, shows that the work done by all forces acting on a particle equals the variation of a quantity, $K = \frac{1}{2}mv^2$, depending on the modulus of the velocity, defined as the kinetic energy of the particle. The above result is also valid in special relativity [4]. Writing the force as

$$\vec{F} = \frac{d\vec{p}}{dt} = \frac{d}{dt}\frac{m\vec{v}}{\sqrt{1 - \frac{v^2}{c^2}}}$$

and calculating the derivative explicitly, we obtain

$$\vec{F} = \frac{m}{\sqrt{1 - \frac{v^2}{c^2}}}\frac{d\vec{v}}{dt} + \frac{m}{c^2}\frac{1}{\left(1 - \frac{v^2}{c^2}\right)^{\frac{3}{2}}}\left(\vec{v} \cdot \frac{d\vec{v}}{dt}\right)\vec{v}.$$

Given this expression of the relativistic force $\vec{F}$, we can evaluate the elementary work $d\mathcal{L}$. After some simple algebra, we obtain

$$d\mathcal{L} = \vec{F} \cdot d\vec{r} = \vec{F} \cdot \vec{v}dt = \frac{m}{\left(1 - \frac{v^2}{c^2}\right)^{\frac{3}{2}}}\vec{v} \cdot d\vec{v}.$$

This expression can also be written in the form

$$d\mathcal{L} = d\left(\frac{mc^2}{\sqrt{1 - \frac{v^2}{c^2}}}\right). \tag{1.8}$$

We can now evaluate the finite work $\mathcal{L}$ done by $\vec{F}$ when a particle moves from A to B by integrating equation (1.8):

$$\mathcal{L} = \int_A^B d\mathcal{L} = \int_A^B d\left(\frac{mc^2}{\sqrt{1 - \frac{v^2}{c^2}}}\right) = \frac{mc^2}{\sqrt{1 - \frac{v_B^2}{c^2}}} - \frac{mc^2}{\sqrt{1 - \frac{v_A^2}{c^2}}}.$$

This result allows the relativistic kinetic energy to be defined as

$$K = \frac{mc^2}{\sqrt{1 - \frac{v^2}{c^2}}} + K_0,$$

where K_0 is a constant. By imposing the condition $K(v = 0) = 0$ one immediately finds $K_0 = -mc^2$, so that the final expression of the relativistic kinetic energy is

Figure 1.8. A curved trajectory from point A to point B (blue line) and the corresponding linear displacement (red line).

$$K = \frac{mc^2}{\sqrt{1 - \frac{v^2}{c^2}}} - mc^2.$$

It is worth noting that, expanding this expression in Taylor series for $v/c \ll 1$ and retaining terms up to second order, one recovers the classical expression $\frac{1}{2}mv^2$. The above arguments thus prove that the concept of kinetic energy is intimately related to the concept of work also in a relativistic frame.

We now show that the arguments treated above can also be presented without the use of the line integral. Of course, in what follows there are simplifying hypotheses about the trajectories and the way in which an object travels along them; nonetheless this approach is worth discussing, since the line integral is a mathematical tool that students generally are not familiar with.

Suppose that a particle of mass m moves from A to B along a curved trajectory (see figure 1.8) under the action of a resulting force $\vec{F}$; for simplicity, we will consider a linear displacement $\Delta\vec{s}$ from A to B and we will suppose that $\vec{F}$ remains constant along this linear path. Then, the work done by $\vec{F}$ is given by

$$\mathcal{L} = \vec{F} \cdot \Delta\vec{s} = m\frac{\Delta\vec{v}}{\Delta t} \cdot \Delta\vec{s} = m\vec{v} \cdot \Delta\vec{v}.$$

Since we are considering a linear displacement, vectors $\Delta\vec{v}$ and $\vec{v}$ are parallel; then, $\vec{v} \cdot \Delta\vec{v} = v\Delta v$ and we write

$$\mathcal{L} = m\vec{v} \cdot \Delta\vec{v} = mv\,(v_B - v_A).$$

In this equation, the speed v is intended as the average speed $\frac{v_B + v_A}{2}$, then

$$\mathcal{L} = m\,\frac{v_B + v_A}{2}\,(v_B - v_A) = \frac{1}{2}m(v_B^2 - v_A^2) = \frac{1}{2}mv_B^2 - \frac{1}{2}mv_A^2. \tag{1.9}$$

Equation (1.9) is exactly the same theorem of kinetic energy obtained in the previous section (see equation (1.7)). The difference here is that we have obtained the same result with a formally easier mathematical approach.

1.2.3 The concept of potential energy and the principle of conservation of mechanical energy

We observe that in the most general case the resultant force includes two different categories of forces, conservative and non-conservative forces. As already mentioned in the previous section, the work done by a conservative force does not depend on the trajectory along which a particle moves. Looking at figure 1.8, this means that for the determination of its value, it is not important if the displacement of the particle from A to B is along a straight line or along a curve—conservative forces do the same work in the two cases. This implies, as already mentioned in the previous section, that if we keep point A fixed, we can associate a number with the point B, $V(B)$, equal to the work done by the conservative force when the particle moves from A to B. Generalizing this procedure, we can associate with each point of the space (C, D, E ...) a number (V(C), V(D), V(E) ...) representing the work done by the conservative force when the particle moves from A to that point. In this way, we have obtained a mapping of the space that associates with each point a number having the dimensions of an energy (as such measured in joules). Reversing the sign of this quantity, we obtain the potential energy U that a particle has when it is at that point. It is called 'potential' since it refers to an energy intended 'to be spent'. In this way conservative forces give to the points of the space a kind of energetic classification. Therefore, the work done by a conservative force when a particle moves from point A to point B can be written as

$$\mathcal{L} = -[U(B) - U(A)]. \tag{1.10}$$

Positive work done by the force thus implies a displacement accompanied by a decrease of the potential energy.

We have not justified so far the use of the term 'conservative'. What is conserved? We can rewrite the work–energy theorem separating in $\mathcal{L}$ the work $\mathcal{L}_{\text{cons}}$ done by conservative forces from the work $\mathcal{L}_{\text{non-cons}}$ done by non-conservative forces:

$$\mathcal{L}_{\text{cons}} + \mathcal{L}_{\text{non-cons}} = \frac{1}{2}mv_B^2 - \frac{1}{2}mv_A^2.$$

Then, using equation (1.10) one has

$$-[U(\text{B}) - U(\text{A})] + \mathcal{L}_{\text{non-cons}} = \frac{1}{2}mv_B^2 - \frac{1}{2}mv_A^2.$$

Rewriting this equation in the form

$$\mathcal{L}_{\text{non-cons}} = \left[\frac{1}{2}mv_B^2 + U(\text{B})\right] - \left[\frac{1}{2}mv_A^2 + U(\text{A})\right], \tag{1.11}$$

one obtains an expression for the work done by non-conservative forces. Note that in the absence of non-conservative forces, the quantity

$$E(P) = \frac{1}{2}mv_P^2 + U(P), \tag{1.12}$$

defined as the mechanical energy of the particle at a given point P, remains unchanged when the particle moves from A to B. In this case, the kinetic energy and potential energy both vary in time along the trajectory, but always in such a way that their sum stays constant. This is the reason why forces doing work not dependent on the trajectory are called conservative—when they are the only kind of forces acting on a particle, the mechanical energy defined in equation (1.12) is a conserved quantity. This conservation law is of great importance in mechanics since it entails a constraint on the behaviour of the particle. During its movement it can change speed (i.e. the kinetic energy) and position (i.e. the potential energy), but cannot change their sum. If conservative and non-conservative forces are simultaneously present, then equation (1.11) can be seen as an energy balance equation, since it states that the energy loss, which is the work done by non-conservative forces, is the difference between the final and the initial values of the mechanical energy. Obviously, this difference is always negative.

1.3 Angular momentum

It is well known that, in order to represent a vectorial quantity, we use arrows. In many cases an arrow can be moved in space without changing its magnitude and direction since this operation of 'parallel transport' does not change its meaning. Vectors of this type are, for example, force, velocity, acceleration and so on. For these quantities, a single arrow represents a class of equivalence. However, other physical quantities are associated with vectors which cannot be translated in space without consequences, since their application at a point has a different effect with respect to their application at another point. For this reason, the definition of these vectors is always given in combination with a position vector $\vec{r}$ specifying their point of application in space. It follows that in order to define the action of an applied vector $\vec{A}$, we have to assign a second vector $\vec{r}$, specifying the point of application P with respect to an origin O (figure 1.9). In particular, one can combine these two vectors by means of the operation of the vectorial product, in this way defining a fundamental quantity in physics, which is the momentum of $\vec{A}$ with respect to the point O:

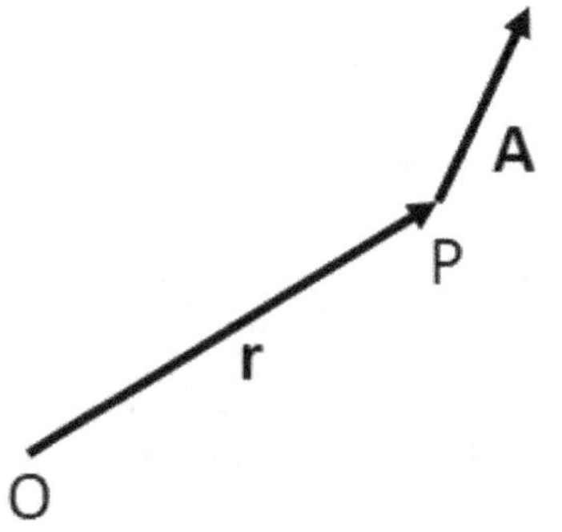

Figure 1.9. Schematic illustration of the applied vector $\vec{A}$ and the position vector $\vec{r}$ specifying the point of application P of $\vec{A}$ with respect to point O.

$$\vec{M} = \vec{r} \times \vec{A}. \tag{1.13}$$

It is orthogonal to the plane which contains $\vec{r}$ and $\vec{A}$ and retains memory of the relative position and orientation of the two vectors. The momentum has two relevant properties:

1. It depends on the position from which we see it—by changing O the vectorial product equation (1.13) changes correspondingly.
2. It does not change when $\vec{A}$ is moved rigidly along the straight line corresponding to its direction.

There are two fundamental quantities which are represented by vectors of this kind: the momentum of a force, or torque, and the momentum of the quantity of motion, or angular momentum.

The momentum of a force, called torque, is defined as $\vec{M} = \vec{r} \times \vec{F}$. In order to understand its role in mechanics, we can consider the case in which $\vec{F}$ changes its application point in space and $\vec{r}$, which follows it, does not change in magnitude. As is evident from figure 1.10, in this case the point of application of $\vec{F}$ moves on a circumference. Then, if $\vec{F}$ is applied to a mass m, regardless of whether we are considering a particle or a rigid body, the effect produced by $\vec{F}$ is a rotation.

The same arguments can be used for the definition of angular momentum, given by

$$\vec{L} = \vec{r} \times m\vec{v}.$$

Here we face the conceptual difficulty that $\vec{L}$ is a physical quantity not related to the causes of motion. Rather, it is intrinsic to the behaviour of the particle and thus its significance is not immediately evident. Assuming, for simplicity, that $\vec{r}$ does not change in magnitude, the particle describes arcs of circumference, so that the magnitude of its quantity of motion is equal to $mv = m\frac{ds}{dt} = mr\frac{d\theta}{dt} = mr\omega$, where ω is its angular velocity. In addition, the vector $\vec{v}$ has to remain orthogonal to $\vec{r}$ and then the magnitude of the angular momentum is

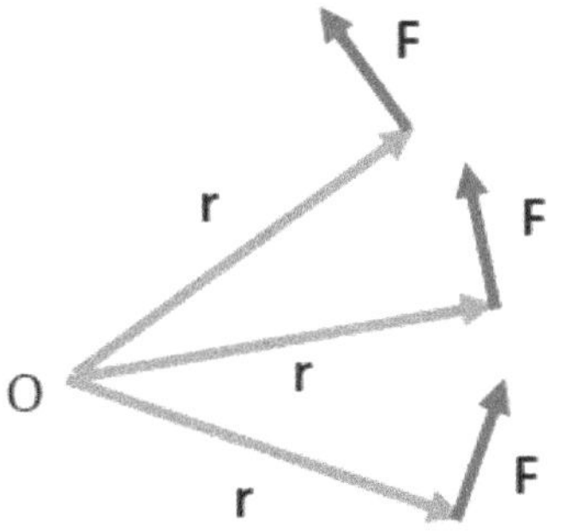

Figure 1.10. Schematic illustration of a vector $\vec{F}$ (red arrow) whose application point changes. The vector $\vec{r}$ (blue arrow) giving the position with respect to O of the application point is assumed to have a constant modulus.

$$|\vec{r} \times m\vec{v}| = mr^2\omega. \tag{1.14}$$

The direction of $\vec{L}$ is orthogonal to the plane which contains $\vec{r}$ and $\vec{v}$, with its orientation directly related, via the right-hand rule, to the direction of rotation. In this way, the angular momentum keeps track of everything dealing with rotation.

More interesting is the extension of the above considerations to the case of a system of particles and, in particular, to a rigid body. Suppose we have a symmetrical rigid body, treated for simplicity as a discrete set of particles, and we put it in rotation around one axis of symmetry. In this case, schematically shown in figure 1.11, a point of the rigid body with mass m_i and located at a distance r_i from the axis rotates on a circumference of radius r_i with an angular velocity ω which is the same for all particles in the body. Then, for every selected point in the body there is another point located symmetrically with respect to the rotation axis, so that the related contributions to the total angular moment $\vec{L}$ have components that cancel out in the direction perpendicular to the axis and sum up along the axis itself. Then, the magnitude of the angular momentum is given by the sum of many contributions, each having the form of equation (1.14). Defining the angular velocity as a vector $\vec{\omega}$ of magnitude ω and the direction of the rotation axis, oriented in such a way as to see the rotation taking place counter-clockwise, we can say that the direction of $\vec{L}$ is in this case the same as that of $\vec{\omega}$. Thus we can write

$$\vec{L} = \left(\sum_i m_i r_i^2\right)\vec{\omega} = I\vec{\omega}. \tag{1.15}$$

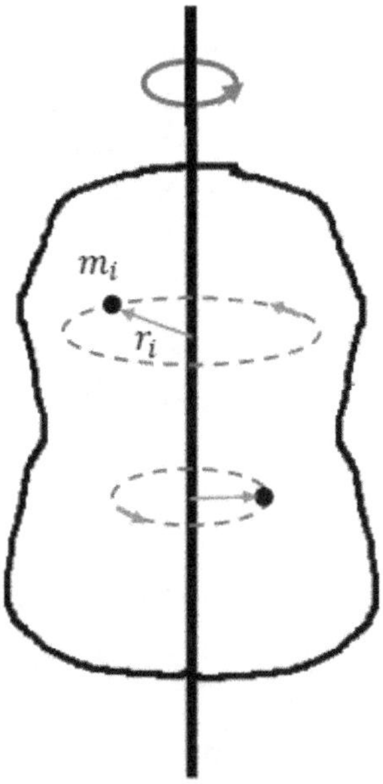

Figure 1.11. A symmetric rigid body rotating around its symmetry axis.

The scalar quantity $I = \sum_i m_i r_i^2$, defined as 'the moment of inertia', gives information not only on the mass, but also on the way it is distributed around the axis of rotation, which is a relevant point. From equation (1.15) one can easily understand why $\vec{L}$ is also called the momentum of the quantity of motion—the analogy between this equation and the definition of the quantity of motion $\vec{p} = m\vec{v}$ is evident and allows us to guess that the quantity I is a measure of the inertia that a rigid body offers to rotations, exactly in the same way as the mass is a measure of the inertia that a particle offers to translations. However, it should be noted that equation (1.15) is valid only if the rotation takes place around a symmetry axis of the rigid body. When this condition is not realized, $\vec{L}$ is not in the direction of $\vec{\omega}$, but nonetheless for its axial component $L_{\parallel}$ one can write a scalar relation reminiscent of equation (1.15):

$$L_{\parallel} = I\omega.$$

The above considerations are of great relevance, since they allow students to be shown an interesting analogy between the translational motion, governed by $\vec{F} = m\vec{a}$, and the rotational one, governed by the momentum of the forces through an equation which can be immediately obtained by deriving equation (1.15) with respect to time (again considering rotations around a symmetry axis):

$$\vec{M} = I\vec{\alpha}. \tag{1.16}$$

Here $\vec{M}$ is the resultant momentum of the external forces with respect to a pole lying on the rotation axis, I is the moment of inertia with respect to the same axis and $\vec{\alpha}$ is the angular acceleration taking into account variations of the angular velocity. These quantities can be seen as the rotational analogue of the total force $\vec{F}$, the mass m and the linear acceleration $\vec{a}$, respectively. In this context, it is important to note that rotational dynamics is affected not only by the value of the mass of the rotating body, but also by its distribution around the rotation axis.

When the angular momentum is not parallel to $\vec{\omega}$, namely it is not along the axis of rotation, but has also a transverse component $\vec{L}_{\perp}$ ($\vec{L}_{\perp} = L_x\hat{x} + L_y\hat{y}$, if the z-axis coincides with the axis of rotation) then an additional term appears in the equation of the rotational motion:

$$\vec{M} = I\vec{\alpha} + \frac{\mathrm{d}\vec{L}_{\perp}}{\mathrm{d}t}.$$

Separating axial and transverse contributions according to

$$\vec{M}_{\parallel} = I\vec{\alpha} \qquad \vec{M}_{\perp} = \frac{\mathrm{d}\vec{L}_{\perp}}{\mathrm{d}t},$$

we see that $\vec{M}_{\parallel}$ and $\vec{M}_{\perp}$ are responsible for variations of the angular momentum along the axis of rotation and orthogonally to this axis, respectively. It is evident that when $\vec{M}_{\perp} = 0$ we go back to equation (1.16). In this case a motion with constant angular velocity can take place with no need to apply external momenta. Equivalently, we can say that the most efficient way to put a rigid body in rotation is to make this happen around an axis of symmetry.

The analysis can be made more general, observing that for every rigid body it is always possible to find three orthogonal axes such that in a rotation around one of them, the direction of the angular momentum would coincide with the one individuated by that particular axis. We can thus say that if $\hat{a}$, $\hat{b}$ and $\hat{c}$ are the unit vectors of these 'special' axes, then in analogy with equation (1.16) the total angular momentum is $\vec{L}_a = I_a\omega\, \hat{a}$ if the body rotates around the $\hat{a}$-axis, is $\vec{L}_b = I_b\omega\, \hat{b}$ if it rotates around the $\hat{b}$-axis and is $\vec{L}_c = I_c\omega\, \hat{c}$ if it rotates around the $\hat{c}$-axis. The scalar quantities I_a, I_b and I_c give information about the distribution of the mass with respect to the three axes and are called central moments of inertia. Each of them has the form $\sum_i m_i r_i^2$, where r_i are the distances from the specific axis we are considering of the particles making up the rigid body (in a discretised description). When the rotation is around a generic axis, the relation connecting the angular momentum to the angular velocity is more complicated and takes the matrix form

$$\vec{L} = \mathcal{I}\vec{\omega},$$

where $\mathcal{I}$ is a 3×3 matrix, known as the tensor of inertia. We thus see that different to the case of rotation around a symmetry axis, each component of $\vec{L}$ is now expressed as a linear combination of all three components of $\vec{\omega}$.

1.4 Symmetries and conservation laws

Conservation laws are of great relevance in physics since they capture some general regularities of Nature. Even though their deduction is based on the second and the third principles of dynamics, such that they have the same 'information content' as the Newtonian laws, they allow a deeper understanding of the behaviour of a dynamic system, at the same time providing a powerful tool for the solution of specific problems.

The three fundamental conservation laws of mechanics, which concern mechanical energy, quantity of motion and angular momentum, have never been violated by any experiment or theory. They hold in quantum mechanics as well, where their role is even more crucial than in classical mechanics or relativity, given that they often represent a unique instrument allowing us to understand microscopic phenomena. Therefore, conservation laws are universal and control all physical processes directly—if they do not respect these laws, they are forbidden. Conversely, if a predicted event respects conservation laws, this could certainly be verified by means of a suitable experiment. In this sense, we could say that conservation laws delimit what can happen from what cannot.

The reason why conservation laws are inviolable lies in their direct link to symmetry principles, as established by Noether's theorem. In general, if something is symmetrical it means that it shows an invariance with respect to a change of the point of view, that is, it does not change under certain space transformations. For example, the object in figure 1.12(a) is invariant if we rotate it by 120° with respect to the centre (discrete symmetry), whereas the object of figure 1.12(b) is invariant under

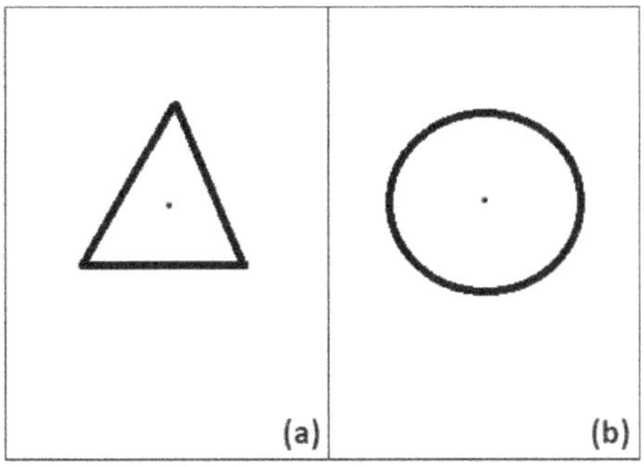

Figure 1.12. Schematic representation of objects characterized by certain symmetries.

rotation by every possible angle (continuous symmetry). For a given symmetry, there is a transformation of coordinates leaving unchanged the equations which describe the dynamics of the system under consideration. The technical definition of symmetry is indeed invariance under certain transformations. Among all the transformations which can be applied to the equations of motion in any context, three of them play a special role:

1. Translation of the axis of time.
2. Translation of the origin (implying a change of position).
3. Rotation of the axes (implying a change orientations).

Symmetry means that these transformations leave the equations of motion unchanged, this being associated, respectively, with:

1. Homogeneity of time.
2. Homogeneity of space.
3. Isotropy of space.

Emmy Noether demonstrated that a conservation law stems from each of these fundamental symmetries, where the concept of symmetry is intended as invariance with respect to a continuous coordinate transformation. The physical quantities conserved correspondingly are:

1. Mechanical energy.
2. Quantity of motion.
3. Angular momentum.

1.5 A brief description of waves

1.5.1 General remarks

A wave is a disturbance which travels in space and time. More specifically, wave motion is a perturbation on a microscopic scale which moves on a macroscopic scale. Waves carry energy and linear momentum, but not matter. This implies that the material through which they move is stationary. A stone thrown into a lake can intuitively give the idea of the propagation of a wave: a series of concentric rings is generated, and they travel away from the point at which the stone fell into the water.

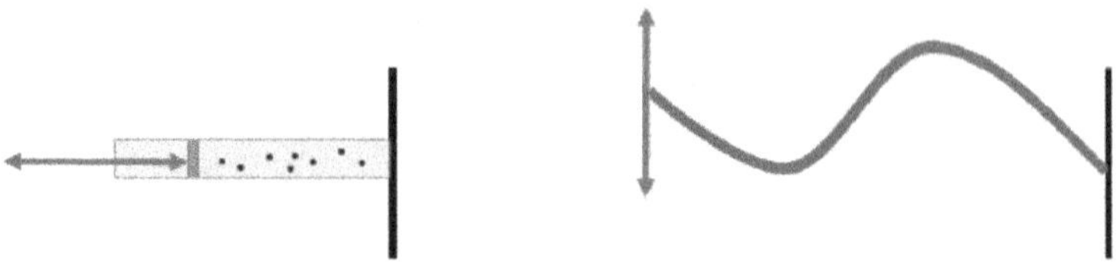

Figure 1.13. Schematic illustration of how a longitudinal wave in a gas and a transverse wave on a rope can be generated. In the left panel the source moves left and right, while in the right panel the source moves up and down.

In this case, the physical quantity which is perturbed is the position of the particles of the liquid which oscillate around their equilibrium position. We can consider the wave as the propagation of a state of motion in matter. This is the way, for instance, in which sound or an earthquake propagate (but there is also a kind of wave which does not need matter to propagate, the electromagnetic wave). If the material is isotropic and there are no obstacles, the wave front is spheric and the corresponding wave is a spheric wave. If we observe the wave at a point which is far away from the source that generates it, the portion of the wave front is assimilable to a plane—in this case, we deal with a plane wave. The evolution of a wave depends on the physical properties of the material in which the wave propagates, in particular on the nature of the boosting forces acting on the particles of the perturbed fluid. The simplest case is when the force is an elastic one, then the wave is an elastic wave.

We can classify waves on the basis of the direction of movement of the individual particles of the medium relative to the direction of propagation of the wave itself. According to this, in an elastic medium there are two kinds of waves: longitudinal waves, which are characterized by the fact that the particles move in the same direction as the propagation of the wave, and transverse waves, in which particles move in a direction which is orthogonal to the direction of propagation. For example, sound waves are longitudinal waves, while electromagnetic waves are an example of transverse waves (although, as mentioned above, in this case wave propagation does not involve matter). Figure 1.13 presents a schematic illustration of the propagation of a longitudinal wave in a gas and of a transverse wave on a rope.

1.5.2 Mathematical description

According to general considerations, we can give a mathematical description of a wave starting from the following expression:

$$\alpha = f(x \pm vt). \tag{1.17}$$

Here α is the physical quantity whose perturbation propagates along the x-axis. It is mathematically described by a function depending on space and time through the combination $x \pm vt$, where v is the speed of the wave. This feature is the hallmark of undulatory behaviour. We distinguish between progressive waves, i.e. waves propagating in the positive direction of the x-axis, and waves propagating in the opposite direction. They are described by the functions $f(x - vt)$ and $f(x + vt)$, respectively.

Common sense always associates a sinusoidal curve with the concept of a wave, however, this is not necessarily so. Only when the source of the wave is an oscillating disturbance, the correspondent wave is a harmonic one, described by a sinusoidal function:

$$y(x, t) = A \sin k(x - vt). \tag{1.18}$$

A classical example of a propagating wave refers to a string with an oscillating extremity, the other end being fixed. The disturbance propagates along the string with transverse oscillation and every point located at x is at time t at a height y given by equation (1.18). A graphic of y as a function of t for fixed x, and as a function of x for fixed t, is presented in figure 1.14. Looking at equation (1.18), we see that the constant A is the maximum value which y can assume and thus is called the amplitude of the wave. The fact that A does not depend on x and t means that there is no dissipation, namely the wave does not decay in its propagation. The quantity k is inserted as a constant making the argument of the sine in equation (1.18) adimensional, but it also has a relevant physical meaning. By writing equation (1.18) in the form

$$y(x, t) = A \sin(kx - kvt), \tag{1.19}$$

it is easy to see that, for a given value of t, when $kx = 2\pi n$ with n being an integer, the function in equation (1.19) repeats itself in space. This means that all points $x = \frac{2\pi}{k}n$ distant from each other are in 'spatial phase', and thus the quantity

$$\lambda = \frac{2\pi}{k}, \tag{1.20}$$

called the wavelength, represents the spatial period of the wave. As a consequence, $k = \frac{2\pi}{\lambda}$ is the spatial frequency of a wave, i.e. the number of wavelengths per unit distance, for this reason k is called the wave number.

On the other hand, kv is linked to the temporal periodicity of the wave. For a fixed value of x, one has that when $kvt = 2\pi n$, n being again an integer, the function (1.19) repeats itself in time. Then, waiting for a time equal to an integer multiple of

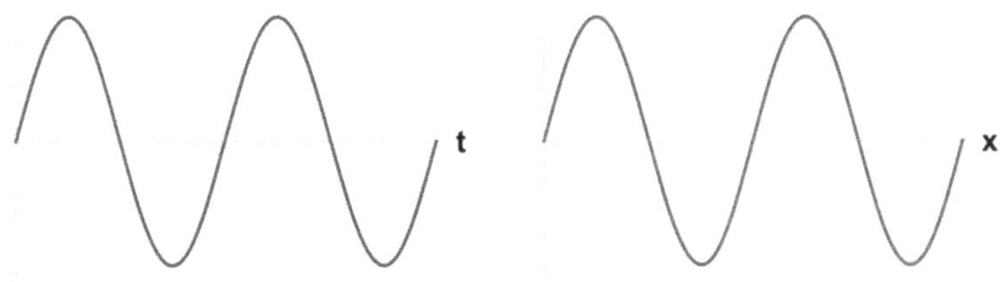

Figure 1.14. Evolution in time (left) and space (right) of a wave.

$$T = \frac{2\pi}{kv}, \tag{1.21}$$

the wave resumes its value at the chosen point x. Equation (1.21) defines the temporal period T of the wave. One also defines the frequency ν of the wave as

$$\nu = \frac{1}{T} = \frac{kv}{2\pi}. \tag{1.22}$$

This can be interpreted as the number of waves that pass through a given point per unit time. It is also useful to introduce the angular frequency ω, defined as

$$\omega = 2\pi\nu.$$

Then, taking into account equations (1.22) and (1.20), we have

$$\omega = 2\pi\frac{1}{T} = kv = \frac{2\pi}{\lambda}v.$$

From the previous equations, we also obtain

$$v = \frac{\lambda}{T}.$$

This relation is important since it connects the spatial and the temporal periodicity of the wave via its speed, this being at the origin of the undulatory behaviour. The characteristic parameters defined above are represented in figure 1.15, where a wave is plotted as a function of the space variable x at a fixed time (top panel), and as a function of time for a given value x (bottom panel). We also remark that when two waves move simultaneously in a medium, they give rise to a new wave, according to the so-called superposition principle. On the basis of the above considerations, the

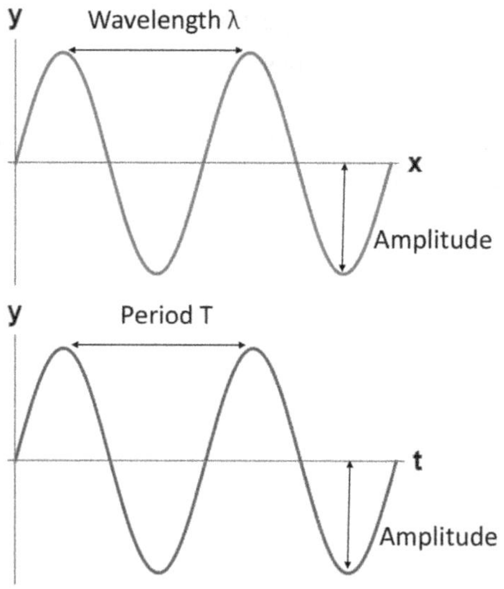

Figure 1.15. Spatial (top) and time (bottom) evolution of a wave with its characteristic parameters.

equation that the function (1.17) must obey in order to correctly describe a wave has to take into account the following points:

1. A wave is always described in terms of a function of space and time variables appearing in the special combination $(x \pm vt)$.
2. When two waves move simultaneously in a medium, they form a new wave, according to the so-called superposition principle.

This implies that the equation we are looking for has to treat the coordinate x and the product vt (point 1) on an equal footing and it has to be linear (point 2).

We propose a heuristic approach to derive this equation, limiting ourselves for simplicity to waves propagating in one dimension with speed equal to v. First of all, the variation of the function f with respect to x is expected to appear in the equation in the same form as the variation of f with respect to vt does. A starting point may be

$$\frac{\Delta f}{\Delta x} = \frac{\Delta f}{\Delta (vt)}. \tag{1.23}$$

Equation (1.23) is relative to finite variations. In order to extend equation (1.23) to a continuum, it has to be written using derivatives. In particular, since f is a function of more variables, we have to introduce partial derivatives and thus we could tentatively write

$$\frac{\partial f}{\partial x} = \frac{1}{v}\frac{\partial f}{\partial t}. \tag{1.24}$$

Here the symbol ∂ denotes a partial derivative, performed with respect to a given variable and considering the others as if they were constant quantities. Equation (1.24) is, however, not correct, since it does not treat the dependence on $+vt$ and $-vt$ on the same footing as it should. A possible way to overcome this difficulty could be to square equation (1.24), but in this way we would lose the linearity of the equation. The problem is settled if we consider an equation of the form (1.23), but involving second derivatives rather than first derivatives:

$$\frac{\partial^2 f}{\partial x^2} = \frac{1}{v^2}\frac{\partial^2 f}{\partial t^2}. \tag{1.25}$$

This is exactly the equation of waves known as the d'Alembert equation, which is a second-order linear partial differential equation. It is easy to verify that a solution of equation (1.25) is given by the sinusoidal function defined in equation (1.18). Solutions of equation (1.25) satisfy the superposition principle, according to which a linear combination of solutions of the equation is again a solution. Even though the derivation presented above strictly refers to waves which propagate in one dimension, nevertheless a similar description holds when waves in more dimensions are considered.

1.5.3 Interference and diffraction

The superposition principle has fundamental implications as far as wave propagation is concerned. Suppose we have two waves with the same amplitude A, the same wavelength λ and the same frequency ν, described by the functions

$$y_1(x, t) = A \sin(kx - \omega t + \varphi) \tag{1.26}$$

$$y_2(x, t) = A \sin(kx - \omega t), \tag{1.27}$$

with φ being the so-called phase constant. A finite value of φ gives rise to the behaviour represented in figure 1.16, where one of the two waves is shifted with respect to the other. By applying the superposition principle, the wave emerging from the overlap of equations (1.26) and (1.27) is

$$y(x, t) = y_1(x, t) + y_2(x, t) = 2A \cos\frac{\varphi}{2} \sin\left(kx - \omega t + \frac{\varphi}{2}\right), \tag{1.28}$$

where we have used the prostapheresis formula. Equation (1.28) describes a wave which has the same wavelength and frequency as the original waves and has an amplitude which is not the sum of the amplitudes of the wave equations (1.26) and (1.27). A common misconception! This is just the peculiar phenomenon called interference. In particular, we observe that the amplitude of the wave described by equation (1.28),

$$A_0 = 2A \cos\frac{\varphi}{2},$$

is equal to the sum of the amplitudes of equations (1.26) and (1.27) only in the particular case in which $\varphi = 0$ or, in general, when $\varphi = 2n\pi$, n being an integer. This is the case of constructive interference. In contrast, if $\varphi = \pi/2$ or, more generally, $\varphi = 2\pi(n + \frac{1}{2})$, the total amplitude is zero and we have destructive interference. This phenomenon is of great relevance and characterizes the wave behaviour in a unique way—we generally refer to it to recognize undulatory propagation in nature.

In addition, interference gives rise to another very relevant phenomenon, typical of wave propagation—diffraction. It occurs when a wave encounters an obstacle or

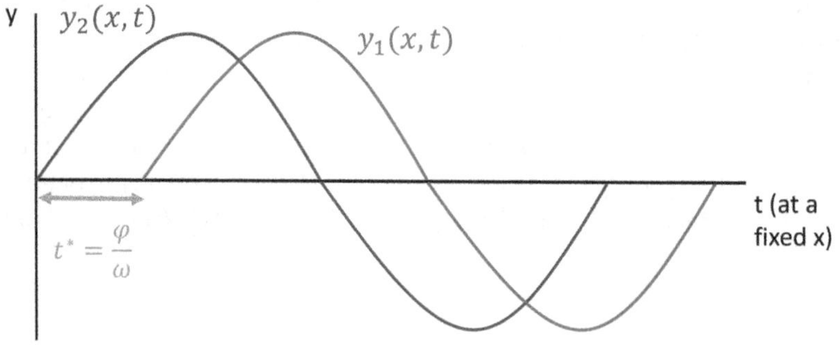

Figure 1.16. Waves out of phase.

a slit, beyond which secondary waves are generated in different directions with respect to the original incidence wave. Since these waves are characterized by path differences, we have interference between them that at some points is constructive and at other points destructive. This behaviour generates the characteristic patterns on a screen, called diffraction patterns, such as those shown in figure 1.17. Their distinctive feature is the alternation of bright and dark zones corresponding to constructive and destructive interference, respectively. Note that diffraction is an interference of a wave with itself which develops only if $\lambda \sim d$, where λ is the wavelength of the wave and d represents the dimension of the slit. Conversely, when $\lambda \gg d$, the wave moves undisturbed, as if the obstacle was not there, whereas if $\lambda \ll d$, undulatory behaviour does not emerge and the wave behaves like a ray. This is the reason why for many years it was not understood if light was a wave or if it was made of particles.

1.6 Maxwell's equations and electromagnetic waves

In this section, our purpose is to present Maxwell's equations in integral form and to deduce from them their differential form [5, 6]. Since the mathematical tools needed for that are usually not included in basic calculus courses, we provide students with qualitative arguments only. In particular, we will apply our approach by referring to a simple arrangement of electric and magnetic fields, which will turn out to be useful to show that these fields must satisfy the d'Alembert equation.

1.6.1 The integral and the differential forms of Maxwell's equations

Maxwell's equations in integral form are expressed in terms of two fundamental integrated quantities associated with a given vector, that is, its line integral around a loop and its flux through a closed surface. The line integral is defined accordingly

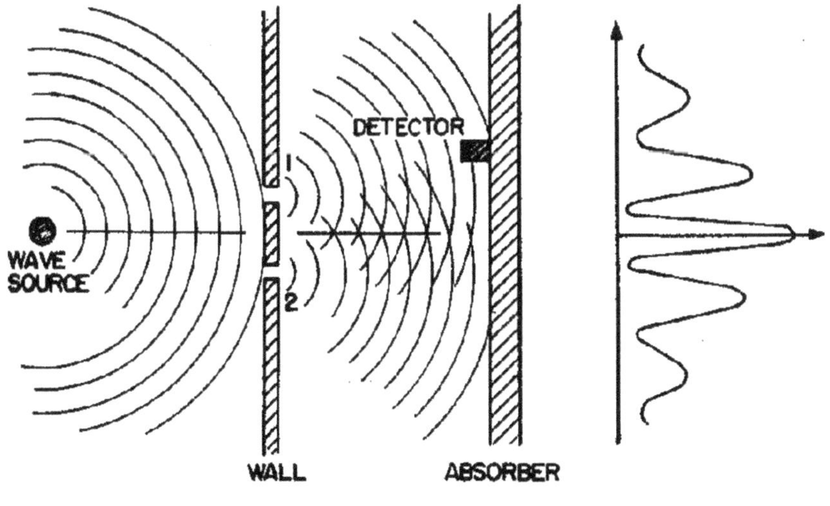

Figure 1.17. Diffraction pattern (reproduced from https://cronodon.com/Atomic/Photon.html).

using the procedure illustrated in section 1.2 (see equation (1.4)). Concerning the flux $\Phi(\vec{A})$ of a vector $\vec{A}$, for its definition one first divides a closed surface S into many elements of area ΔS_i (the index i denotes the ith element), small enough that they can be considered planar and such that the value $\vec{A}_i$ of $\vec{A}$ can be assumed to be constant on each of them. Introducing the vector $\Delta\vec{S}_i = \Delta S\ \hat{n}_i$, where the unit vector $\hat{n}_i$ denotes the outward normal to ΔS_i, one can define the sum

$$\vec{A}_1 \cdot \Delta\vec{S}_1 + \vec{A}_2 \cdot \Delta\vec{S}_2 + \cdots + \vec{A}_i \cdot \Delta\vec{A}_i + \cdots = \sum_i \vec{A}_i \cdot \Delta\vec{S}_i.$$

The flux of $\vec{A}$ through the closed surface S is then defined as the limit of this quantity when $\Delta S_i \to 0$:

$$\Phi(\vec{A}) = \lim_{\Delta S_i \to 0} \sum_i \vec{A}_i \cdot \Delta\vec{S}_i = \oiint_S \vec{F} \cdot \mathrm{d}\vec{S}.$$

The flux $\Phi(\vec{A})$ is proportional to the density of the field lines and to the area through which it is calculated. In addition, it varies according to the direction of the flow with respect to the surface. We also note that the flux of a given vector keeps track of how many times that vector 'pierces' the surface, in the sense that a non-vanishing value of Φ signals the presence of sources or sinks inside the surface. In contrast, when the flux vanishes, for every field line entering the surface there is another one going out from it.

In terms of the above defined quantities, Maxwell's equations in integral form are the following:

$$\oiint_S \vec{E} \cdot \mathrm{d}\vec{S} = \frac{\sum_i Q_i^{\mathrm{int}}}{\varepsilon_0} \tag{1.29}$$

$$\oint_\Gamma \vec{E} \cdot \mathrm{d}\vec{l} = -\frac{\mathrm{d}\Phi(\vec{B})}{\mathrm{d}t} \tag{1.30}$$

$$\oiint_S \vec{B} \cdot \mathrm{d}\vec{S} = 0 \tag{1.31}$$

$$\oint_\Gamma \vec{B} \cdot \mathrm{d}\vec{l} = \mu_0 \sum_j i_j^{\mathrm{int}} + \varepsilon_0\mu_0 \frac{\mathrm{d}\Phi(\vec{E})}{\mathrm{d}t}. \tag{1.32}$$

Equations (1.29) and (1.31) are the Gauss theorem for the electric and magnetic fields $\vec{E}$ and $\vec{B}$, respectively, with $\sum_i Q_i^{\mathrm{int}}$ being the total charge inside the closed surface S. Equation (1.30) is the Faraday–Neumann law relating the line integral of $\vec{E}$ around the loop Γ to the flux Φ of $\vec{B}$ through a surface having Γ as a boundary. Finally, equation (1.32) is the generalized Ampère theorem (also called the Ampère–Maxwell theorem), relating the line integral of $\vec{B}$ around the loop Γ to the algebraic sum $\sum_j i_j^{\mathrm{int}}$ of the electric currents passing through Γ and to the so-called displacement current expressed in terms of the flux of $\vec{E}$ enclosed by Γ. The constants ε_0 and μ_0 are the vacuum permittivity and the vacuum permeability, respectively.

Two fundamental theorems of mathematical analysis allow us to obtain Maxwell's equations in differential form. They require the introduction of two quantities, known as the divergence of a vector and the curl of a vector, constructed as special combinations of the derivatives of the components of that vector. How can the related definitions be proposed to students? The first obstacle is to make them familiar with the differential operator nabla $\vec{\nabla}$. For our purposes, it can be introduced simply as a mathematical tool defined as a strange vector without numerical coordinates but such that each component gives rise to the corresponding derivative when applied to a given function:

$$\vec{\nabla} = \left(\frac{\partial}{\partial x}, \frac{\partial}{\partial y}, \frac{\partial}{\partial z}\right).$$

In this way, if we have a vector $\vec{A} = (A_x, A_y, A_z)$, we can formally introduce the scalar product

$$\vec{\nabla} \cdot \vec{A} = \frac{\partial A_x}{\partial x} + \frac{\partial A_y}{\partial y} + \frac{\partial A_z}{\partial z},$$

which defines the divergence of $\vec{A}$, as well as the vectorial product

$$\begin{aligned}\vec{\nabla} \times \vec{A} &= \text{Det}\begin{pmatrix} \hat{i} & \hat{j} & \hat{k} \\ \frac{\partial}{\partial x} & \frac{\partial}{\partial y} & \frac{\partial}{\partial z} \\ A_x & A_y & A_z \end{pmatrix} \\ &= \left(\frac{\partial A_z}{\partial y} - \frac{\partial A_y}{\partial z}\right)\hat{i} + \left(\frac{\partial A_x}{\partial z} - \frac{\partial A_z}{\partial x}\right)\hat{j} + \left(\frac{\partial A_y}{\partial x} - \frac{\partial A_x}{\partial y}\right)\hat{k},\end{aligned}$$

which defines the curl of $\vec{A}$.

In terms of these quantities, the two above-mentioned theorems, known as the divergence and the curl theorems, respectively, are expressed as

$$\oiint_S \vec{A} \cdot \mathrm{d}\vec{S} = \int_V (\vec{\nabla} \cdot \vec{A})\mathrm{d}V \tag{1.33}$$

$$\oint_\Gamma \vec{A} \cdot \mathrm{d}\vec{l} = \int_S (\vec{\nabla} \times \vec{A}) \cdot \mathrm{d}\vec{S}. \tag{1.34}$$

The divergence theorem, equation (1.33), allows a surface integral defined on a closed surface S to be transformed into a volume integral over the volume V enclosed by S. Using the curl theorem, equation (1.34), a line integral over a closed curve Γ can be transformed into a surface integral defined on a generic surface S having Γ as a boundary.

We now provide arguments to show how a flux through a closed surface is linked to a divergence (theorem 1.33) and how a circulation is linked to a curl (theorem 1.34).

This discussion closely follows that presented in [7]. Starting with the case of the divergence theorem, let us consider a small cube whose sides are given by Δx, Δy and Δz ($\Delta x = \Delta y = \Delta z$), as represented in figure 1.18. We want to find the flux of $\vec{A}$ through the surface of the cube summing the fluxes through each of the six faces. First consider the two faces, numbered 1 and 2 in the figure, perpendicular to the y-axis. Assuming that the cube is small enough that the value of $\vec{A}$ can be considered constant on each face, we have that the outward flux on the face 1 at $y \neq 0$ is $A_y(1)\Delta x\Delta z$, while the outward flux at the opposite face is $-A_y(2)\Delta x\Delta z$ (in this case we have to consider the negative of the y-component of $\vec{A}$). The fact that the cube is very small also implies that the values of $A_y(1)$ and $A_y(2)$ are very close to each other, so that we can write

$$A_y(1) \simeq A_y(2) + \frac{\partial A_y}{\partial y}\Delta y.$$

Summing the fluxes $\Phi(1)$ and $\Phi(2)$ through faces 1 and 2, respectively, we thus have

$$\begin{aligned}\Phi(1) + \Phi(2) &= [A_y(1) - A_y(2)]\Delta x\Delta z\\ &= \frac{\partial A_y}{\partial y}\Delta x\Delta y\Delta z.\end{aligned}$$

In the same way, we have for the flux through faces 3 and 4

$$\Phi(3) + \Phi(4) = \frac{\partial A_y}{\partial y}\Delta x\Delta y\Delta z$$

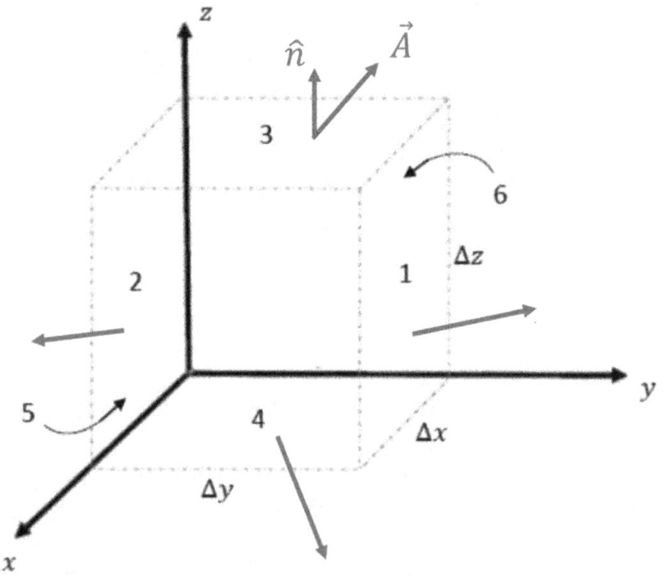

Figure 1.18. A cube with sides Δx, Δy and Δz ($\Delta x = \Delta y = \Delta z$). The faces orthogonal to the axes x, y and z are numbered 5–6, 1–2 and 3–4, respectively.

and for the flux through faces 5 and 6

$$\Phi(5) + \Phi(6) = \frac{\partial A_y}{\partial y}\Delta x \Delta y \Delta z.$$

We thus have that the total flux Φ through the six faces of the cube is

$$\Phi = \left(\frac{\partial A_x}{\partial x} + \frac{\partial A_y}{\partial y} + \frac{\partial A_z}{\partial z}\right)\Delta x \Delta y \Delta z. \tag{1.35}$$

This result tells us that the outward flux of a vector $\vec{A}$ from the surface of a very small cube is equal to the divergence of $\vec{A}$ multiplied by the volume of the cube. Going to a finite volume, we can use the fact that the total flux out of it is the sum of the fluxes out of each of the parts into which we imagine dividing that volume. This means that if we integrate equation (1.35) over the entire volume, we obtain the divergence theorem given by equation (1.33).

Similarly, let us show how, according to equation (1.34), the flux of $\vec{\nabla} \times \vec{A}$ is connected to the circulation of $\vec{A}$. To this purpose, we evaluate the line integral of a given vector $\vec{A}$ around a small square such as the one in figure 1.19. In particular, we assume that $\vec{A}$ does not change much on each side of the square—of course, the smaller the square the better this assumption is. Moving from the lower left corner in the direction indicated by the arrow, we have that along the side marked 1 (see the figure), the line integral is

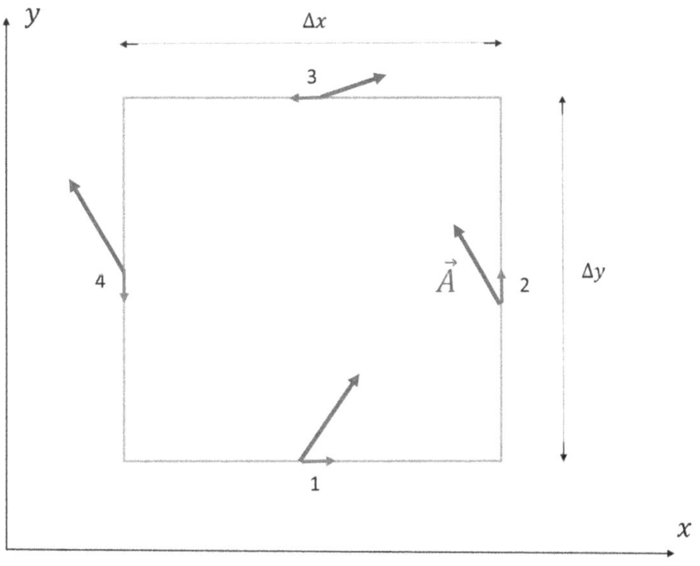

Figure 1.19. A square with sides Δx and Δy ($\Delta x = \Delta y$) around which the circulation of a vector $\vec{A}$ is calculated. The sides parallel to the x-axis are numbered 1 and 3, while the sides parallel to the y-axis are numbered 2 and 4.

$$\Gamma_1 = \vec{A}(1) \cdot \Delta\vec{x} = A_x(1)\Delta x.$$

Along the side marked 3 one has analogously

$$\Gamma_3 = \vec{A} \cdot \Delta\vec{y} = -A_x(3)\Delta x,$$

where the minus sign takes into account that the line direction on side 3 is opposite to the one on side 1. Summing up these two contributions we have

$$\Gamma_1 + \Gamma_3 = [A_x(1) - A_x(3)]\Delta x. \tag{1.36}$$

Since the square is very small, we can assume that the values $A_x(1)$ and $A_x(3)$ are very close. We can thus write

$$A_x(3) \simeq A_x(1) + \frac{\partial A_y}{\partial y}\Delta y$$

so that equation (1.36) becomes

$$\Gamma_1 + \Gamma_3 = -\frac{\partial A_x}{\partial y}\Delta x\Delta y.$$

Repeating the calculation for sides 2 and 4, one similarly has

$$\Gamma_2 + \Gamma_4 = \frac{\partial A_y}{\partial x}\Delta x\Delta y$$

and thus the line integral Γ around the square takes the form

$$\Gamma = \sum_{i=1}^{4}\Gamma_i = \left(\frac{\partial A_y}{\partial x} - \frac{\partial A_x}{\partial y}\right)\Delta x\Delta y. \tag{1.37}$$

We can see that the quantity in the parentheses is exactly the z-component of the curl of $\vec{A}$, while the product $\Delta x\Delta y$ is the area ΔS of the square. Considering that the z-component of $\vec{A}$ is also the component normal to the square, we have that equation (1.37) can be put in the form

$$\Gamma = (\vec{\nabla} \times \vec{A}) \cdot \hat{n}\Delta S, \tag{1.38}$$

where $\hat{\vec{n}}$ is the unit vector in the direction normal to the square. When we consider the circulation of a vector around any loop C, we can consider any surface having C as a boundary and divide it into a set of very small surface elements, small enough that they can be considered flat and very nearly a square. It is then possible to show that the sum of the circulations around the closed boundary of these very small squares equals the circulation around the original loop C, due to the cancellation of the contributions coming from the adjacent portions of the small squares. In this way, applying equation (1.38) to each of the small squares and summing up all the corresponding contributions, one recovers the curl theorem (1.34).

The application of the divergence theorem to the surface integrals appearing in Maxwell's equations (1.29) and (1.31) allows these equations to be expressed in differential form. Introducing the charge density ρ associated with the charge in volume V,

$$\sum_i Q_i^{\text{int}} = \int_V \rho \mathrm{d}V,$$

equation (1.29) becomes

$$\int_V (\vec{\nabla} \cdot \vec{E}) \mathrm{d}V = \frac{1}{\varepsilon_0} \int_V \rho \mathrm{d}V$$

and thus, V being a generic volume,

$$\vec{\nabla} \cdot \vec{E} = \frac{\rho}{\varepsilon_0}.$$

Similarly one obtains for $\vec{B}$

$$\vec{\nabla} \cdot \vec{B} = 0.$$

In the same way, the application of the curl theorem to the line integrals of $\vec{E}$ and $\vec{B}$ in equations (1.30) and (1.32), respectively, leads to

$$\vec{\nabla} \times \vec{E} = -\frac{\partial \vec{B}}{\partial t} \tag{1.39}$$

$$\vec{\nabla} \times \vec{B} = \mu_0 \vec{j} + \varepsilon_0 \mu_0 \frac{\partial \vec{E}}{\partial t}. \tag{1.40}$$

These two equations imply that variations in time of one of the two fields produce 'rotational' configurations of the other in planes which are orthogonal to the field which generates them.

1.6.2 Electromagnetic waves

We can now show that Maxwell's equations establish that the dependence on time and space of the electric and the magnetic fields is wave-like [5]. Without loss of generality, we consider a simple case where the direction of propagation and the two fields $\vec{E}$ and $\vec{B}$ are mutually orthogonal:

$$\vec{v} = v\hat{i} \qquad \vec{E} = E\hat{j} \qquad \vec{B} = B\hat{k}. \tag{1.41}$$

According to this choice, equation (1.39) gives

$$\frac{\partial E}{\partial x} = -\frac{\partial B}{\partial t}. \tag{1.42}$$

In the same way, from equation (1.40) one obtains

$$\frac{\partial B}{\partial x} = -\varepsilon_0 \mu_0 \frac{\partial E}{\partial t}. \tag{1.43}$$

Equations (1.41), (1.42) and (1.43) suggest that if one of the two fields varies in space the other varies in time, with the two fields remaining orthogonal to each other. The undulatory behaviour of the two fields can be verified by deriving equation (1.42) with respect to the spatial variable x,

$$\frac{\partial^2 E}{\partial x^2} = -\frac{\partial}{\partial x}\frac{\partial B}{\partial t} = -\frac{\partial}{\partial t}\frac{\partial B}{\partial x}, \tag{1.44}$$

where one has taken into account that the order in which the derivatives with respect to different variables are performed is irrelevant. By substituting equation (1.43) in equation (1.44) we have

$$\frac{\partial^2 E}{\partial x^2} = \varepsilon_0\mu_0\frac{\partial^2 E}{\partial t^2}. \tag{1.45}$$

This is exactly the wave equation (1.25), with the constant factor $\varepsilon_0\mu_0$ correctly having the dimension of the inverse of the square of a velocity. We can thus put

$$v = \frac{1}{\sqrt{\varepsilon_0\mu_0}} \tag{1.46}$$

so that equation (1.45) takes the form

$$\frac{\partial^2 E}{\partial x^2} = \frac{1}{v^2}\frac{\partial^2 E}{\partial t^2}. \tag{1.47}$$

On the other hand, combining the two equations obtained by deriving equation (1.43) with respect to x and equation (1.42) with respect to t, we similarly obtain for the magnetic field

$$\frac{\partial^2 B}{\partial x^2} = \frac{1}{v^2}\frac{\partial^2 B}{\partial t^2}. \tag{1.48}$$

Equations (1.47) and (1.48) imply that the solutions E and B of Maxwell's equations both satisfy the D'Alembert equation. It is thus confirmed that the two fields have wave-like behaviour. Together they constitute the so-called electromagnetic wave, propagating in a vacuum with a speed which does not depend on its frequency and wavelength. Its magnitude can be calculated by substituting the numerical values of the constants ε_0 and μ_0 in equation (1.46). We obtain

$$v = \frac{1}{\sqrt{\left(8.85\times10^{-12}\frac{\mathrm{C}}{\mathrm{N\,m^2}}\right)\left(4\pi\times10^{-7}\frac{\mathrm{T\,m}}{\mathrm{A}}\right)}} \cong 3\times10^{8}\ \mathrm{ms^{-1}}. \tag{1.49}$$

It is significant that the velocity of propagation of the electromagnetic waves is expressed in terms of the two fundamental constants, ε_0 and μ_0, appearing in the description of electric and magnetic phenomena, respectively. Its numerical value, as given in equation (1.49), is exactly equal to the speed of light in a vacuum, thus confirming that the light is an electromagnetic wave.

References

[1] Galilei G 2001 *Dialogue Concerning the Two Chief World Systems, Ptolemaic and Copernican* (New York: Modern Library)
[2] Laplace P S 1995 *A Philosophical Essay On Probabilities* (New York: Dover)
[3] For a thorough presentation of the laws and applications of mechanics see, for example, Resnick R, Halliday D and Krane K S 2001 *Physics* vol 1 (New York: Wiley)
[4] Resnick R 1968 *Introduction to Special Relativity* (New York: Wiley)
[5] Halliday D, Resnick R and Krane K S 2001 *Physics* vol 2 (New York: Wiley)
[6] Jackson J D 1998 *Classical Electrodynamics* (New York: Wiley)
[7] Feynman R, Leighton R and Sands M 2011 *The Feynman Lectures on Physics* vol II (New York: Basic Books)

Chapter 2

Transition from classical physics to quantum physics: the role of interference

Lazzaro Immediata and Sergio Pagano

2.1 Introduction

The study of the nature of light has appeared throughout the history of physics. The interpretations have been highly varied, but those with more success are the corpuscular, undulatory and quantum interpretations. The study of the behaviour of light and the theories regarding its nature present a perfect example of the evolutionary history of physics. It is also thanks to studies carried out on the nature of light that modern physics, based on two fundamental pillars, the theory of relativity and quantum mechanics, was born. In particular, we will discuss the Michelson and Morley experiment, a historical experiment that ideally connects the principles introduced by Einstein in his Special Theory of Relativity with the most recent gravitational wave detectors, the Ligo and Virgo gravitational interferometers.

2.2 Light

When we talk about light we refer to that part of the electromagnetic spectrum that is visible to the human eye ranging from about 400 nm to about 700 nm in wavelength, i.e. with a frequency between 790 and 435 THz. As it is a type of electromagnetic wave, light interacts with matter. Light, therefore, presents phenomena such as absorption, dispersion, reflection, refraction and diffraction. Reflection is the main phenomenon that allows the objects around us to be seen. A classic example of diffusion is that in the terrestrial atmosphere which makes the sky bright and blue. Light has a dual nature, as a wave and as a flow of particles, called photons. For the description of the wave interpretation, classical electromagnetism allows an almost complete treatment, while quantum mechanics describes light also through its particle behaviour and allows purely quantum phenomena to be explained such as, for example, the photoelectric effect.

doi:10.1088/978-0-7503-2678-0ch2

2.2.1 Corpuscular theory

The first complete treatment of the corpuscular theory of light was the work by Isaac Newton in the seventeenth century. According to this model, light is composed of material particles (corpuscles) emitted, from a source, in all directions. The mathematical treatment of this theory is simpler than that of the wave theory and explains some characteristics of light propagation easily. In fact, using the laws of classical mechanics we can predict propagation in a straight line and a very high, but finite, speed. Reflection can also be explained by classical mechanics based on the dynamics of impacts. Refraction, even if it presents more difficulty, is also explained in the context of classical mechanics. The treatment of diffraction and interference is not possible in the classical context.

2.2.2 Wave theory

In the formulation of Christiaan Huygens in 1678, light is treated as a wave that propagates in a medium, called the ether, in a way that is very similar to mechanical waves. The ether would be a particular substance pervading the whole Universe, formed by particles that give it particular properties that would allow the propagation of light. The wave theory explained all the phenomena known at the time of its formulation. Thomas Young demonstrated, in 1801, how diffraction and interference could be explained using the wave theory. Again thanks to the wave theory, Joseph von Fraunhofer and, subsequently, Kirchhoff and Bunsen, succeeded in explaining and interpreting the solar spectrum. Moreover, assuming that light has a microscopic wavelength, it is possible to explain the fact that although it is propagating in a straight line, the light is able to bypass an obstacle. Finally, the wave theory predicts that the speed of light in a medium is less than that in a vacuum.

2.2.3 Classic electromagnetic theory

Proposed by James Clerk Maxwell towards the end of the nineteenth century, this theory suggests that light is of an electromagnetic nature. For most practical applications this interpretation is still valid today. In electromagnetic theory, light is part of the electromagnetic spectrum. With Maxwell's equations, the electrical, magnetic and optical phenomena were completely unified. In its original version, the electromagnetic theory still included the presence of ether. Only later was it shown that light can propagate even in a vacuum, without a medium being necessary.

2.2.4 Quantum theory

When, at the beginning of the twentieth century, Max Planck posed the problem of dealing with black body radiation, he introduced a mathematical tool that avoided the interpretative problems that arose in the classical treatment. This approach consisted of introducing a quantity that derived from a discrete, and not continuous, nature of the energies involved. Finally, when Albert Einstein gave an explanation of the photoelectric effect, what appeared to be a simple mathematical tool turned out to be a new interpretation of Nature and its way of revealing itself: it was the birth of quantum mechanics.

2.3 Light as a wave

2.3.1 What is a wave?

The propagation in space and time of a perturbation, possibly periodic, generated by a physical quantity is called a wave. A wave carries energy and momentum, but not matter. It cannot exist in a single point of space, but must extend from one point to another. Light and sound are both forms of energy that propagate through space in the form of waves. The sinusoid is the graphic representation of a wave of a single frequency. The characteristics of a wave are shown in figure 2.1.

The period T is the time in which a complete oscillation takes place. The unit of measurement is that of time (seconds). The wavelength λ is the distance between two recurring points of the oscillation. The unit of measurement is that of lengths (metres). The frequency ν is the number of complete oscillations in a unit of time. The unit of measurement is cycles per second or hertz. Amplitude A is related to the energy carried by the wave as it propagates. Frequency and period are linked by the relationships

$$\nu = \frac{1}{T},$$

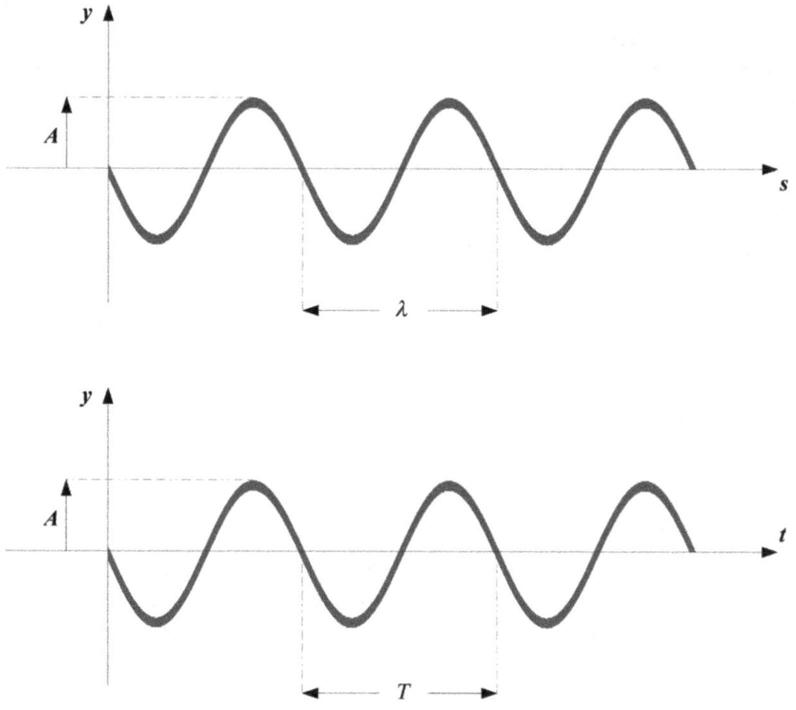

Figure 2.1. Properties of a generic wave.

$$T = \frac{1}{\nu}.$$

The wave propagation speed v is given by

$$v = \nu\lambda.$$

The intensity or power density (energy per unit area and unit time) carried by the wave is proportional to A^2

$$W \propto A^2.$$

A characteristic of mechanical waves is also that of being able to transport energy associated with the vibrations of the particles of the vehicle crossed. The waves can be:

- *Transversal*, if the oscillation is perpendicular to the direction of wave propagation. The waves that are established in the stretched strings of musical instruments and on the surfaces of liquids, as well as electromagnetic waves are of this type.
- *Longitudinal*, if the oscillation occurs along the direction of wave propagation (e.g. sound waves).

Multiple waves can simultaneously coexist in the same space—the effect can be a reinforcement, an attenuation or a simultaneous elision. When the crest of a wave overlaps that of another their individual effects are added together, producing a wave of greater amplitude (constructive interference), and when the crest of a wave overlaps the valley of another the effects are cancelled out (destructive interference). If we tie a rope to a wall and shake the free end up and down, we produce a wave; the wall is too rigid to shake so the wave is reflected backwards. By shaking the rope appropriately, you can make the initial incident wave and the reflected wave form a stationary wave in which parts of the string, called nodes, remain at rest (this can be produced for both longitudinal waves and transverse waves). The positions of a stationary wave in which the amplitude is maximum are called bells (they are produced halfway between the nodes). Stationary waves are due to interference: when two waves with the same amplitude and the same wavelength propagate along the same path, but in opposite directions, the waves are always in phase opposition at the nodes. The nodes are stable regions of destructive interference.

2.3.2 Electromagnetic waves

Electromagnetic fields are produced by the oscillation of electric charges. The oscillation of a charge determines the variation of the flow of the electric field. Due to this variation a magnetic field will be generated (the fourth Maxwell equation, see below) which in turn will determine the variation of the electric field (the third Maxwell equation, see below) and so on. Once the electromagnetic field has been produced by the oscillation of a charge, it will have an autonomous existence. As this is a characteristic of waves we speak of electromagnetic waves. Due to their undulatory nature, electromagnetic waves will carry energy but, unlike mechanical waves, they do not need a means to propagate.

A monochromatic electromagnetic wave (i.e. with a well-defined frequency and wavelength) consists of an electric field $\vec{E}$ and a mutually perpendicular magnetic field $\vec{B}$ that oscillate in phase with each other perpendicularly to the direction of propagation (see figure 2.2). So an electromagnetic wave actually consists of two coupled components: an electric one and a magnetic one. A wave of this type is called a flat polarized wave, and the plane of polarization is the plane in which the electric field oscillates. In energy terms, an electromagnetic wave can be thought of as a flow of energy, which in a vacuum propagates at the speed of light, in the form of electric and magnetic fields. Each of the two components of the electromagnetic wave, electric and magnetic, carries the same amount of energy. Electromagnetic waves are transverse waves. The electric field and the magnetic field are, in fact, orthogonal to each other, and are always perpendicular to the direction of propagation. The definitions of wavelength, period and frequency and the relationships between these magnitudes extend to electromagnetic waves. In the particular case of a planar electromagnetic wave that is linearly polarized and sinusoidal, the energy density $u_{\text{em.wave}}$ is proportional to E^2 and to B^2 (where E is the amplitude of the electric field and B is that of the magnetic field):

$$u_{\text{em.wave}} = \frac{1}{2}\varepsilon_0 E^2 + \frac{1}{2\mu_0} B^2,$$

where $\varepsilon_0 = 8.854^{-12}$ F m^{-1} is the dielectric constant of the vacuum and $\mu_0 = 4\pi \times 10^{-7}$ H m^{-1} is the magnetic permeability of the vacuum.

2.3.3 Classification of electromagnetic waves

By the electromagnetic spectrum we mean the set of frequency bands of electromagnetic radiation (see figure 2.3). The frequencies vary from a few kHz to 10^{22} Hz. The corresponding wavelengths vary from hundreds of kilometres to dimensions of the order of the atomic nucleus (10^{-13} m). The electromagnetic spectrum varies continuously across the various parts into which it is customarily divided, which are all connected together. The names of the different types of radiation traditionally represent frequency bands for which there is a common detection and production technique. Since the energy of photons varies proportionally with the frequency, different parts of the electromagnetic spectrum have very different properties. For

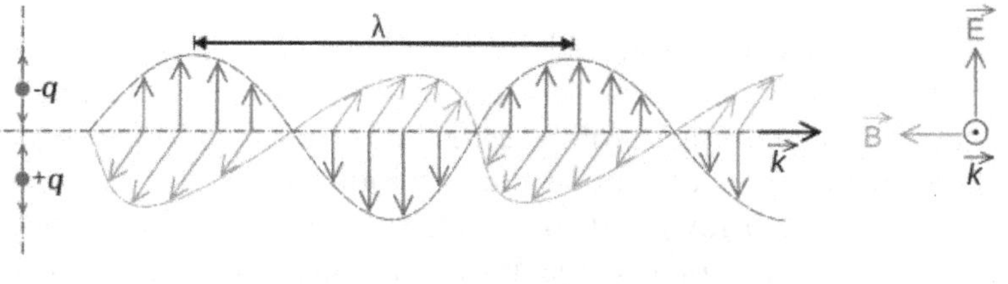

Figure 2.2. Schematic view of the propagation of an electromagnetic wave. k is the propagation vector. Author: SuperManu.

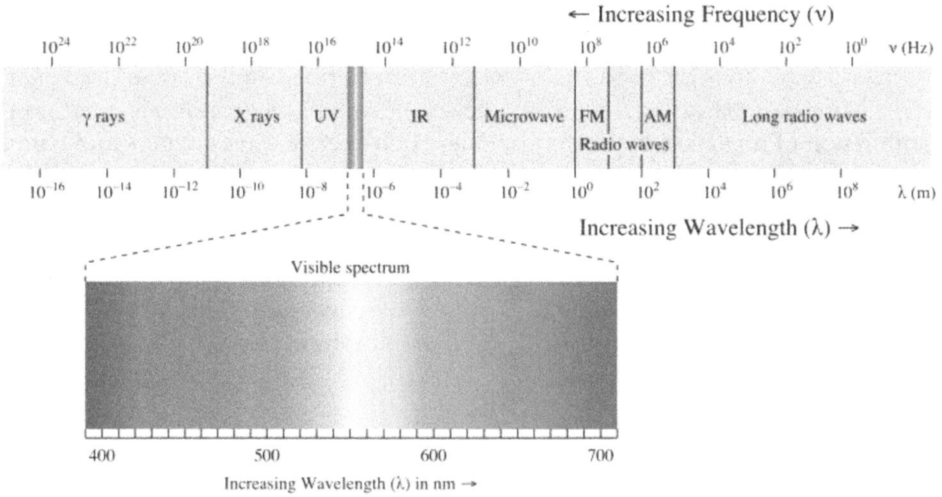

Figure 2.3. The electromagnetic spectrum. User: Zedh.

example, infrared, visible and ultraviolet radiation, which transport energy in the order of the energy characteristics of electrons in atoms, interact strongly with atoms and molecules. Radio waves (large wavelengths) generally interact weakly with matter. They are used, in particular, for radio broadcasts. Due to their long wavelength they have the property of overcoming even considerable obstacles. For some types of radio waves it is possible to exploit the reflective properties of the ionosphere, the ionized layer that surrounds the Earth beyond the atmosphere, to increase the transmission distance.

Microwaves are the radiation between radio waves and infrared. They are generally obtained by means of special electronic tubes (e.g. klystrons, magnetrons), which exploit the interactions between the electromagnetic wave emitted by a generator and an electron beam. Microwaves have applications in directional radio transmissions.

Infrared waves are electromagnetic waves with wavelengths between those of microwaves and the red side of visible light (0.8 μm). They are used in some medical therapies and, among other applications, to photograph hot objects in dark environments.

Visible light is the very limited part of the electromagnetic spectrum accessible to the human eye, located between infrared and ultraviolet waves in the electromagnetic spectrum, or between 0.4 and 0.7 μm. The visible light perceived by humans is emitted, for example, by incandescent bodies.

Ultraviolet radiation lies between visible light and x-rays. It is invisible to the human eye but can register on photographic plates and make some substances fluorescent. They are used in the sterilization of environments, as chemical catalysts, and in various applications in therapeutic medicine and biology.

X-rays and *γ-rays*. X-rays have a wavelength of approximately between 10 nm and 1 pm. They are generally divided into soft x-rays, with a wavelength greater than

0.1 nm, and hard x-rays with a wavelength greater than 0.1 nm. Hard x-rays are similar to γ-rays, being more energetic, and are distinguished from them by their origin: x-ray photons are produced by electron transitions on their quantized energy levels, while gamma rays derive from nuclear or subatomic transitions, or from the annihilation of a particle by its antiparticle. High-energy waves, x-rays and gamma rays, whose wavelengths are comparable to the size of atomic nuclei, are able to interact at the nuclear level. X-rays are used in diagnostic medicine (x-ray, γ-ray), in chemistry for x-ray spectrophotometry (XRF), in the analysis of the structure of materials and in absorption spectroscopy. γ-rays, due to their ability to kill bacteria, are used to sterilize medical equipment and food packaging. They are also used in diagnostic medicine (e.g. proton emission tomography)

2.4 Electromagnetism

2.4.1 History

The story of electromagnetism developed over the nineteenth century, starting from the observation of the correlations between electrical and magnetic phenomena, which, until then, had been treated separately. We owe to James Clerk Maxwell the first modern theory of electromagnetism (1864), which summarizes all electromagnetism in a single theory. Some of the electrical properties of matter have been known since antiquity. In the eighteenth and nineteenth century there was a considerable increase in interest in studies on the electric properties of matter. The fundamental stages of this period can be summarized as follows:

18th century and 19th century

- Pieter van Musschenbroek invents the capacitor (1745)
- Benjamin Franklin introduces the concept of electric charge (1746)
- Charles-Augustin de Coulomb discovers the law of force among electric charges (1785)
- Alessandro Volta invents the electric battery: it allows to generate a constant current flow to be genereted (1799)
- Hans Christian Ørsted discovers the magnetic effect of the current (1820)
- André-Marie Ampère discovers the force between currents (1826)
- Michael Faraday discovers the law of electromagnetic induction (1831)
- James Clerk Maxwell unifies the electromagnetism and light (1864)
- Röntgen discovers the x-rays (1896)
- Joseph John Thomson discovers the electron and determines its charge (1897)

We recall the work of Stephen Gray on conduction (1726), Charles François Dufay on electric duality (1730), Cavendish on electrostatic phenomena (1772) and Galvani on animal electricity (1791). However, it was only through the work of Charles-Augustin de Coulomb, still in the seventeenth century, that a mathematical organic formulation of the forces between the two states of matter (known as positive and negative electrical charges) was formulated. This formulation is known as Coulomb's law and is given by

$$\vec{F} = \frac{1}{4\pi\varepsilon}\frac{q_1 q_2}{r^2}\hat{r},$$

where ε is the dielectric constant of the medium in which the charges are immersed. The forces that develop between electrical charges are, as is known, attractive or repulsive according to whether the charges are in different or equal states. Starting from Coulomb's law it is possible to deduce that an electrically charged body produces an electric field in the space given by

$$\vec{E} = \frac{\vec{F}}{q_2} = \frac{1}{4\pi\varepsilon}\frac{q_1}{r^2}\hat{r}.$$

As far as natural magnetism is concerned, this was already known to the ancient Greeks in the fifth to sixth centuries BCE, and probably also in ancient China where, it is thought, the compass was already in use. However, the first true treatise on magnetism was produced by William Gilbert, with his *De Magnete* (1600), but we had to wait until the end of the eighteenth century and the beginning of the nineteenth century to obtain a quantitative formulation of magnetism and associated phenomena, with the works of the French scientists Biot and Savart and, subsequently, Ampère. A correlation between electricity and magnetism was hypothesized by the Danish physicist Hans Christian Ørsted, who in his experiments showed that a wire carrying an electric current generates a magnetic field.

The British scientist Michael Faraday conducted experiments with which he demonstrated that a current-carrying conductor immersed in a magnetic field was affected by a force. Biot and Savart found the formula for the magnetic field $\vec{B}$ generated by an indefinite wire carrying electric current i:

$$\vec{B} = \frac{\mu}{2\pi}\frac{i}{r}\hat{i} \times \hat{r},$$

where r is the distance from the wire, $\hat{i}$ is direction of the current in the wire, $\hat{r}$ is the r unit vector and μ is the magnetic permeability of the medium in which the wire is immersed. The mathematical formulation of the force exerted by a magnetic field on the electric current is due to André-Marie Ampère, who through the Ampère experiment concluded that between two wires of length l separated by the distance r, carrying currents of intensities i_1 and i_2, respectively, a force $\vec{F}$ is exerted whose form is

$$\vec{F} = \frac{\mu}{2\pi}\frac{i_1 i_2 \cdot l}{r}\hat{r}.$$

Electrical and magnetic phenomena were studied more rigorously from the beginning of the nineteenth century and research and discoveries followed, which gradually brought to light new phenomena. Experiments and the formulation of laws then followed which led to the first connections between electricity and magnetism, considered until then as separate fields. We owe to James Clerk Maxwell the synthesis of the knowledge acquired until then, which he converged in *A Treatise on Electricity and Magnetism* and in the electromagnetic theory. Examining the experimental results and the laws formulated until then, Maxwell provided a formally symmetrical organization of the electric and magnetic phenomena in which the interdependence between magnetism and electricity is evident—that is, the two fields do not exist separately but are the expression of a single electromagnetic field. Finally, the twentieth century was a period of veracious revolutions for physics and for technological applications deriving from these discoveries, which took place at dizzying speeds, and whose steps can be summarized as follows.

Twentieth century:

- Atomic theory (Einstein, Bohr).
- Quantum energy (Planck).
- Theory of the relativity (Einstein).
- Quantum mechanics (Bohr, Heisenberg, Schrödinger, Pauli, Dirac).
- Quantum electrodynamics (Feynman).

2.4.2 Maxwell's equations

The importance of Maxwell's work lies not only in the great synthesis he carried out, but also in the thrust he gave to modern physics, starting with the crisis in Galilean relativity in favour of Einstein's relativity. Precisely to maintain the invariance of Maxwell's equations for inertial reference systems, Lorentz deduced his famous eponymous transformations, taken up and reinterpreted by Einstein. Einstein's Theory of Special Relativity is one of the emblems of modern physics that revolutionized the way we see and think of space and time, also profoundly influencing philosophical thought in the 1900s and, not least, the collective imagination. Maxwell's equations are (with the corresponding physical law in parentheses):

$$\vec{\nabla} \cdot \vec{E} = \frac{\rho}{\varepsilon_0} \text{ (Gauss law)},$$

$$\vec{\nabla} \cdot \vec{B} = 0 \text{ (the magnetic field force lines are closed)}, \tag{2.1}$$

$$\vec{\nabla} \times \vec{E} = -\frac{\partial \vec{B}}{\partial t} \text{ (Faraday induction law)}, \tag{2.2}$$

$$\vec{\nabla} \times \vec{B} = \mu_0 \vec{J} + \mu_0 \varepsilon_0 \frac{\partial \vec{E}}{\partial t} \text{ (Ampère circuital law)}, \tag{2.3}$$

where $\vec{J}$ is the current density and ρ the charge density. Let us now see a very important consequence of Maxwell's fourth equation. In the case of a charged particle that oscillates in a vacuum ($i = 0$ because of the absence of circuits) it will become

$$\vec{\nabla} \times \vec{B} = \mu_0 \varepsilon_0 \frac{\partial \vec{E}}{\partial t}.$$

The product $\mu_0 \varepsilon_0$ has the dimensions of the square of the inverse of a speed. Replacing the values of the two constants μ_0 and ε_0,

$$\frac{1}{\sqrt{\mu_0 \varepsilon_0}} \cong 3 \times 10^8 \text{ m s}^{-1}. \tag{2.4}$$

With similar procedures Maxwell showed that the ratio considered in equation (2.4) has the dimensions of a velocity and that the value thus obtained corresponds with good approximation to that found by Armand Hippolyte Louis Fizeau and Jean Bernard Léon Foucault for the speed of light c. This evidence, together with other considerations, led Maxwell to hypothesize the electromagnetic nature of light. From this reasoning the Scottish genius deduced that light is made up of electromagnetic waves. It can be shown that the electromagnetic wave equations descend from Maxwell's equations. As already mentioned, a charged particle oscillating in space will produce an electromagnetic wave that propagates in every direction. The fields generated, $\vec{E}$ and $\vec{B}$, are in phase in a plane orthogonal to the propagation direction. The amplitudes of the two fields have direct proportionality, with a proportionality factor given by the speed of the electromagnetic wave. In the case of a wave propagating in a vacuum,

$$|\vec{E}| = c \cdot |\vec{B}|,$$

where $c \cong 3 \times 10^8$ m s^{-1}. At a generic point at distance x from the source (placed at the origin of the axes) and in the generic instant t, the values of the intensities of the fields $\vec{E}$ and $\vec{B}$ are given by

$$\begin{aligned} E(x, t) &= E_{\text{MAX}} \sin\left[2\pi\left(\frac{t}{T} - \frac{x}{\lambda}\right)\right], \\ B(x, t) &= B_{\text{MAX}} \sin\left[2\pi\left(\frac{t}{T} - \frac{x}{\lambda}\right)\right], \end{aligned} \tag{2.5}$$

where E_{MAX} and B_{MAX} are the amplitudes of the oscillations. The electric field $\vec{E}$ and magnetic $\vec{B}$ field are in phase. This means that the two fields assume, at the same time, either the maximum value, or the minimum value, or the zero value. In the case of electromagnetic waves, remember that

$$\lambda = c \cdot T.$$

The propagation of an electromagnetic wave also involves the propagation, with the same speed, of energy. The amount of energy per unit of time passing through the surface unit orthogonal to the propagation direction is given by the Poynting vector $\vec{I}$ defined in a vacuum as

$$\vec{I} = \frac{1}{\mu_0}\vec{E} \times \vec{B},$$

and having units of W m^{-2}. The direction of the Poynting vector is the same as the speed of propagation, therefore this vector can be used to represent the direction of propagation of an electromagnetic wave.

One of the most interesting things that can be deduced from Maxwell's laws is that the propagation speed of electromagnetic waves is independent of any reference system. In fact, the speed of propagation assumes a typical value without there being any need to refer to a particular reference system. This has fundamental importance, because, for the first time in the history of physics, it was shown that, at least as far as the speed of light is concerned, the Galilean principle of speed composition in inertial reference systems is lost. This is where Einstein started from for his reasoning underlying special relativity.

2.5 Interference

The phenomenon of interference is due to the superposition, at a point in space, of two or more waves. What is observed is that the intensity of the resulting wave at that point can be different from the sum of the intensities associated with each single initial wave. Specifically, it can vary between a minimum, at which no undulatory phenomenon is observed, and a maximum given by the sum of all the intensities. In general, it is said that interference is constructive when the resulting intensity is the sum of each individual original intensity. The term is usually used when considering two coherent waves, usually coming from the same source. Phenomena of interference that can be observed in day-to-day life are, for example, those relating to the ripples that form on the surface of water or to the beats of sound waves. Interference has played a fundamental role in the evolution from classical to modern physics. In this section, we will make a brief journey through some experiments that, using the phenomenon of interference, have left their mark in the history of physics and humanity.

2.6 The Michelson and Morley experiment

The hypothesis of the wave nature of light was verified using various experiments in the mid-nineteenth century, although only with Maxwell's equations was it possible to prove that light was a phenomenon attributable to electromagnetic waves. Until then, therefore, it was not clear what kind of waves the luminous ones were, and it was hypothesized that they propagated in a particular material medium, called luminiferous ether, which was thought to permeate the entire Universe. The motion of the Earth and the planets, crossing the ether, gave rise to the so-called 'wind of ether'. Given the validity of Maxwell's equations, and to avoid contradictions with these, it was assumed that they were applicable to light waves only in a reference system in which the ether was at rest. It was clear, therefore, that in order to study

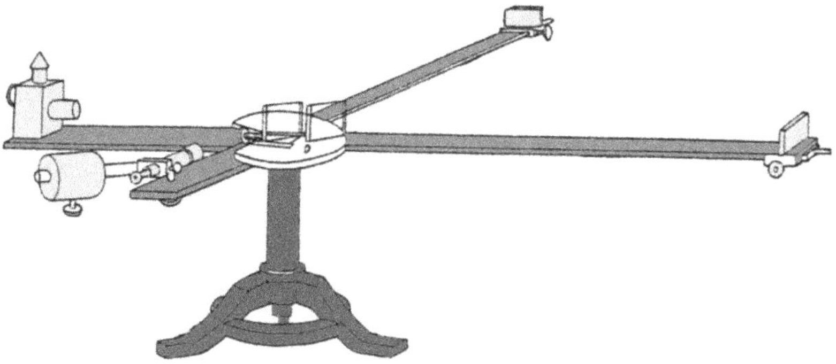

Figure 2.4. The Michelson interferometer.

the behaviour and properties of light, it was necessary to take into account the 'ether wind' caused by the motion of the Earth through the ether itself. The experiments that Michelson and Morley carried out towards the end of the 1800s were aimed precisely at verifying the existence of the ether and understanding the behaviour of light when crossing it. The Michelson and Morley experiment, performed in 1887, finds justification in the field of ether theories, according to which the speed of light is equal to c only in reference systems at rest with respect to the ether.

According to classic (Galilean) physics, in an inertial system in relative motion with respect to the ether, the speed of light is calculated with Galileo's transformation laws. Michelson performed the experiment in an attempt to measure the speed of the Earth relative to the ether. The instrument to highlight such motion was an interferometer of his own invention (figure 2.4). Thanks to his work, Michelson won the Nobel Prize in 1907. The Michelson interferometer scheme is illustrated in figures 2.5 and 2.6. A monochromatic light source S emits a ray of monochromatic light, with wavelength L, in the $\bar{SP}$ direction. P is a semitransparent mirror, placed at an angle of 45° with respect to the light beam. A part r_1 of this beam is reflected towards the mirror S_1 and a part r_2 is transmitted towards the mirror S_2. For the first ray r_1, reflected by S_1, one part crosses P and is collected by the telescope C, and part of the second ray r_2, reflected by S_2, is also reflected by P towards the telescope C which thus receives two coherent light rays (i.e. with the same amplitude and wavelength) [1, 2].

Moving along the screen the optical paths of the two beams change and light zones (constructive interference) and dark zones (destructive interference) alternate. Michelson and Morley mounted their interferometer on a heavy square slab of granite with sides equal to 15 cm and with a thickness of 5 cm and, for stability reasons, they made the device float in a tank of liquid mercury, so that it could be gently rotated around a central pivot passing through P. To make the path of light as long as possible, they placed a series of mirrors on the slab so as to make the rays, with successive changes, travel eight round trips. The two light beams have the two sections $\bar{SP}$ and $\bar{PC}$ in common. The overall paths, however, differ because the first

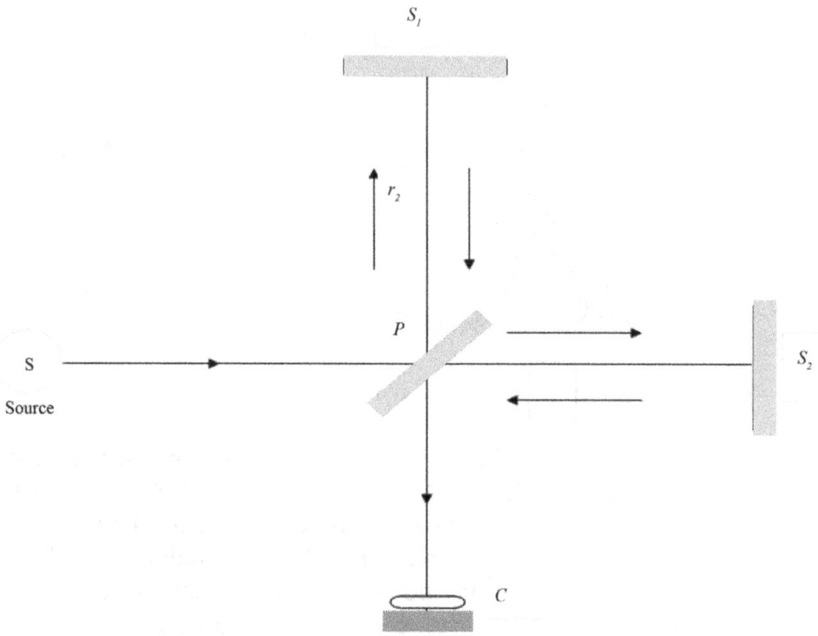

Figure 2.5. The Michelson interferometer schematic.

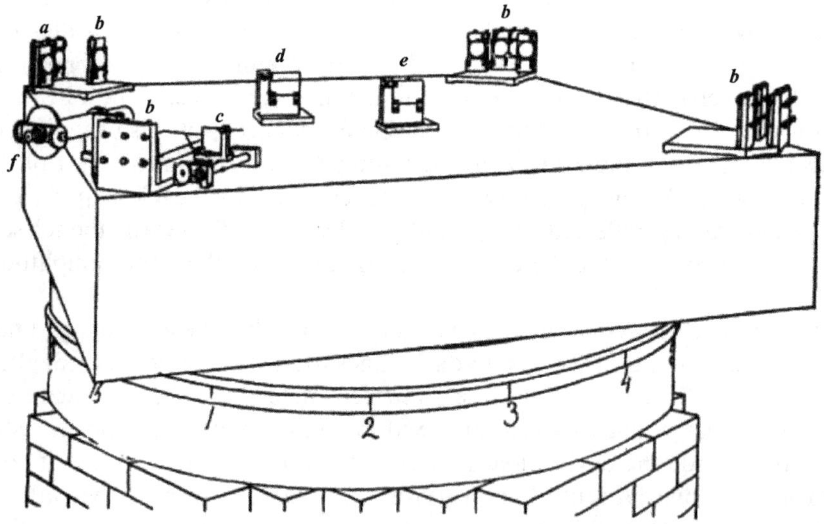

Figure 2.6. A more detailed version of Michelson's experiment: a = source; b = mirror; c = rotating mirror; d = mirror; e = semitransparent mirror; and f = telescope [1].

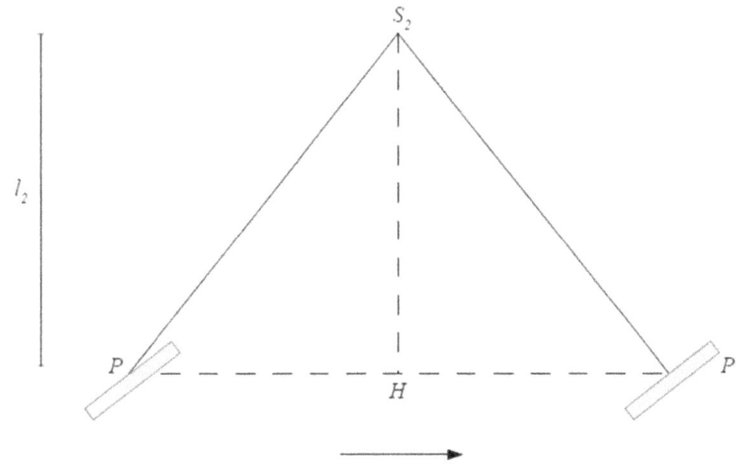

Figure 2.7. The path of a light beam between two reflecting mirrors, considering the motion of the Earth.

travels back and forth along the $\bar{PS_1}$ section (of length l_1) while the second travels back and forth along the $\bar{PS_2}$ section (of length l_2, equal to l_1).

The $\bar{PS_1} = \bar{PS_2}$ section was longer than one metre, but with a system of multiple mirrors each beam travelled back and forth eight times, so that the optical path was $l_1 = l_2 = 11$ m. Suppose that initially the experimental apparatus is oriented so that the $\bar{PS_1}$ arm is parallel to the velocity v of the Earth. Compared to the apparatus, the radius r_1 moves, in the hypothesis of the ether, with speed $c + v$ while it travels from P to S_1 and with speed $c - v$ while it travels from S_1 to P. The time t_1 to make the outward and return journey is

$$t_1 = \frac{l_1}{c+v} + \frac{l_1}{c-v} = l_1\left(\frac{2c}{c^2 - v^2}\right) = \frac{2l_1}{c\left(1 - \frac{v^2}{c^2}\right)}.$$

It is more difficult to determine the speed of the radius r_2 which moves in a direction orthogonal to the speed v. The situation is complicated by the fact that, if the Earth is in motion, the actual path in the ether, supposedly still, of the ray of light is what appears in figure 2.7. In it the semi-silver mirror is represented twice, at the instant $t = 0$ and at the instant $t = t'$.

Since the $\bar{PS_2}$ and $\bar{S_2P}$ segments have the same length, it will be enough to calculate the time taken by the light to cover one of the two, and double the result. Indicating with Δt the time taken by the light to travel the $\bar{PS_2}$ stretch, and taking into account the fact that this stretch is travelled at speed c, we obtain

$$c\Delta t = \bar{PS_2} = \sqrt{\bar{PH}^2 + \bar{HS_2}^2} = \sqrt{(v\Delta t)^2 + l_2^2}$$

$$c^2\Delta t^2 = (v\Delta t)^2 + l_2^2 \tag{2.6}$$

$$t_2 = \frac{2l_2}{\sqrt{c^2 - v^2}} = \frac{2l_2}{c\sqrt{1 - \frac{v^2}{c^2}}}. \tag{2.7}$$

We observe that t_1 and t_2 are different, i.e. the rays that are collected in the telescope C have a certain difference in optical path and should give rise to interference fringes, such as those visible in figure 2.8. The difference in optical path between the two beams will be

$$\Delta x = c(t_1 - t_2)$$

from which, by replacing the values t_1 and t_2, we obtain

$$\Delta x = \frac{2}{\sqrt{1 - \frac{v^2}{c^2}}}\left(\frac{l_1}{\sqrt{1 - \frac{v^2}{c^2}}} - l_2\right). \tag{2.8}$$

If $\Delta x = nL$, where L is the wavelength of the light used (590 nm with a sodium source), then constructive interference will occur. If $\Delta x = (n + 1/2)L$ then there will be destructive interference. As a consequence, with the telescope we see a sequence of interference fringes. However, if we rotate the interferometer by 90°, instead of the horizontal radius, the orbital velocity of the Earth will be added to the vertical radius, and therefore the difference in optical path between the two rays will vary, obtaining

$$\Delta x' = \frac{2}{\sqrt{1 - \frac{v^2}{c^2}}}\left(l_1 - \frac{l_2}{\sqrt{1 - \frac{v^2}{c^2}}}\right).$$

Figure 2.8. Interference pattern.

You must therefore have a relative displacement in the interference fringes equal to

$$\Delta N = \frac{\Delta x - \Delta x'}{\lambda} \approx \frac{l_1 + l_2}{\lambda}\frac{v^2}{c^2}.$$

Using the experimental values $l = 590\,\text{nm}$, $l_1 = l_2 = 11$ m, $v = 30\,\text{m s}^{-1}$ and $c = 3 \times 10^5$ km s^{-1}, one obtains $\Delta N = 0.37$.

The displacement thus obtained should be about half a wavelength of yellow light, and therefore should be such as to bring the dark fringes onto the light fringes and vice versa, just as shown in figure 2.9. The observation did not show any displacement of the fringes and the accuracy was estimated to be compatible with a speed of the laboratory v_L with respect to the ether

$$v_\text{L} = 10\text{ m s}^{-1} \approx 1/3 v_\text{Earth}.$$

If, instead of Galilean transformation, we take the Lorentz transformations, then we will have a contraction of the lengths in the direction of the motion, obtaining

$$l \to l' = l\sqrt{1 - \frac{v^2}{c^2}}.$$

From which

$$\Delta x = \frac{2}{\sqrt{1-\frac{v^2}{c^2}}}\left(\frac{l_1\sqrt{1-\frac{v^2}{c^2}}}{\sqrt{1-\frac{v^2}{c^2}}} - l_2\right) = \frac{2}{\sqrt{1-\frac{v^2}{c^2}}}(l_1 - l_2)$$
$$\Delta x = \frac{2}{\sqrt{1-\frac{v^2}{c^2}}}\left(l_1 - \frac{l_2\sqrt{1-\frac{v^2}{c^2}}}{\sqrt{1-\frac{v^2}{c^2}}}\right) = \frac{2}{\sqrt{1-\frac{v^2}{c^2}}}(l_1 - l_2) \tag{2.9}$$

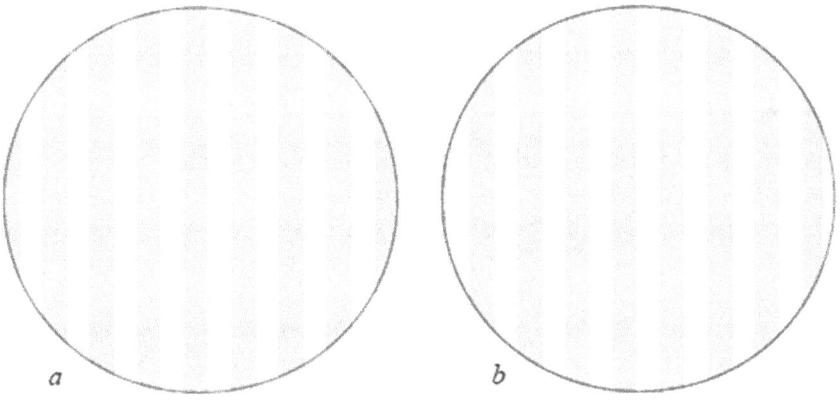

Figure 2.9. Expected interference pattern in Michelson and Morley experiment (a) before and (b) after rotation.

Observer	*Year*	*Place*	*D [cm]*	*2D/l (v/c)2 [fringe]*	*A [fringe]*	*Ratio*
Michelson *(a)*	1881	Potsdam	120	0.04	0.01	2
Michelson & Morley *(b)*	1887	Cleveland	1100	0.40	0.005	40
Morley & Miller *(c)*	1902 – 04	Cleveland	3220	1.13	0.0073	80
Miller *(d)*	1921	Mt. Wilson	3200	1.12	0.04	15
Miller *(e)*	1923 – 1924	Cleveland	3200	1.12	0.015	40
Miller (sunlight) *(f)*	1924	Cleveland	3200	1.12	0.007	80
Tomaschek (starlight) *(g)*	1924	Heidelberg	860	0.3	0.01	15
Miller *(h)*	1925 – 1926	Mt. Wilson	3200	1.12	0.044	13
Kennedy *(i)*	1926	Pasadena and Mt. Wilson	200	0.07	0.001	35
Illingworth *(j)*	1927	Pasadena	200	0.07	0.0002	175
Piccard & Stahel *(k)*	1927	Mt. Rigi	280	0.13	0.003	20
Michelson, *et al. (l)*	1929	Mt. Wilson	2590	0.9	0.005	90
Joos *(m)*	1930	Jena	2100	0.75	0.001	375

(a) A.A.Michelson, Am. J. Sci. 22, 120 (1881); Phil Mag.; 3, 236 (1882)
(b) A.A.Michelson and E.W.Morley, Am. J. Sci. 34, 333 (1887); Phil Mag.; 24, 449 (1887)
(c) E.W.Morley and D.C.Miller, Phil Mag.; 9, 680 (1905); Proc. Am. Acad. Arts Sci. 41, 321 (1905)
(d) D.C. Miller, Data sheets of Observations December 9 to 11, 1921 (unpublished)
(e) D.C. Miller, Observations, August 23 to September 4, 1923; June 27 to July 26, 1924 (unpublished)
(f) D.C. Miller, Observations with Sunlight on July 8 to 9, 1924, Proc. Natl. Acad. Sci. 11, 311 (1925)
(g) R.Tomaschek, Ann d Physik 73, 105 (1924)
(h) D.C. Miller, Revs. Modern Phys. 5, 203 (1933)
(i) R.J. Kennedy, Proc. Natl. Acad. Sci. 12, 621 (1926; Astrophys. J. 68, 367 (1928)
(j) K.K. Illingworth, Phys. Rev. 30, 692 (1927)
(k) A. Piccard and E.Stahel, Compt. Rend. 183, 420 (1926); 184, 152, 451 (1927); 185, 1198 (1927); J.Phys. Radium 8, 56 (1927)
(l) Michelson, Pease and Pearson, Nature 123, 88 (1929); J. Opt. Soc. Am. 18, 181 (1929)
(m) G. Joos, Ann. Physik 7, 385 (1930); Naturwiss. 38, 784 (1931)

Figure 2.10. Testing of the Michelson–Morley experiment [3].

which will therefore result in a displacement in the interference fringes equal to

$$\Delta N = \frac{\Delta x - \Delta x'}{\lambda} = 0.$$

The fringes were observed during the continuous rotation of the device and a displacement of 1/100 of the fringe should have been highlighted. Observations were made during the day and at night and during all seasons of the year, but no fringe movement was detected. In fact, the experimental conclusion was that there was no fringe shift. This null result was such a blow to the ether hypothesis that the experiment was repeated by many experimenters for the next 50 years (see figure 2.10). The null result was confidently confirmed and constituted a great stimulus for theoretical and experimental research.

2.6.1 Conclusions

The experiment of Michelson and Morley was conceived to show that light can have different speeds for different observers in relative motion with respect to the ether, through the demonstration of the existence of a sort of 'wind of ether' and relative motion with respect to it of the Earth along its orbit, on the basis of the presumed validity of the Galilean composition of velocities. The fact that the experiment failed dramatically could not help but deny the initial assumptions, showing once and for all that light always has the same speed for all observers and that evidently Galileo's transformations are not valid for all reference systems in motion relative to each other. Some of the possible explanations are as follows:

- The Earth is integral to the ether (it is the simplest explanation but it must be discarded for obvious reasons).
- The Earth partially drags the ether, as it pulls the air.
- Maxwell's equations are incorrect. (This possibility was already considered to be unacceptable at the end of the last century, above all due to the tests on the existence of electromagnetic waves that took place precisely in the years of the Michelson–Morley experiment.)
- Light is not a wave, but is made up of corpuscles that follow Newton's usual laws of dynamics.
- The result of the experiment should be taken for what it is and we need to review the physical concepts we used previously and 'come to terms with the fact that they do not add up'.

The last one is, as is known, the right solution, proposed by Einstein in 1905, but already anticipated by some ideas by Lorentz and Poincaré, in particular. It is clear that this idea should have led, as indeed actually happened, to the revision of Newtonian mechanics and its laws. To adapt to a newborn theory, the mechanics of Newton, which had so much success, particularly in explaining the motion of heavenly bodies, had to be revised from the ground up. For his invention of the interferometer and for his many optical experiments, Michelson received the Nobel Prize for Physics in 1907.

2.7 Gravitational interferometers

Einstein's theory of general relativity suggests that the Earth should be continuously impacted by energetic radiation of gravitational origin produced by the interactions of massive stars, even from far away. The energy emitted by a cosmic perturbation (the explosion of a supernova or a collapse between two massive bodies, such as a pair of neutron stars or black holes) travels at the speed of light in the form of gravitational waves which, in their propagation, should cause distortions in the regions of space crossed. These waves could be measured by the distortion effects on space caused by their passage using interferometric methods that follow, ideally, the approach taken by Michelson and Morley in their experiment. To understand the effect of a gravitational wave on space and matter, consider a long cylinder placed transversely to the direction of propagation of a gravitational wave. Propagating along the transverse direction of the cylinder, the wave distorts the section as it passes, and then lets it return to its original shape once it has passed through. If, instead, we consider masses spaced apart from each other on a plane and imagine a gravitational wave propagating perpendicular to the plane itself, we will first see the masses distancing themselves in a direction perpendicular to the direction of wave propagation, for example North–South, and, at the same time, approaching in a direction perpendicular to it, that is along the East–West direction, to then see the inverted effect with respect to the two directions mentioned. Such variations, although infinitesimal, can be seen if the magnitude of the gravitational wave that generated them is great enough. During the passage of the gravitational wave, the stretching and compression of the space between the masses continue and invert periodically. The generation of gravitational waves is, in some respects, similar to that of electromagnetic waves in the sense that, while an electrically charged body radiates electromagnetic waves with an amplitude proportional to its charge and acceleration, gravitational waves depend on the mass of the body that generates them. However, due to the principle of conservation of momentum, in an isolated system the acceleration produced is zero: for every action there is an equal and opposite reaction and the effect is annulled. This means that most of the waves emitted by a mass that accelerates are cancelled by the waves emitted by a second mass of the system. The cancellation, however, is not perfect because the positions of the two masses cannot coincide. This means that there is a slight phase shift between the waves generated by the two masses that would allow the waves to be detected. In the collapse of very massive bodies, the effect is amplified by the inhomogeneity and distortion of the masses of the system that is formed. Intense gravitational waves are emitted when the quadrupole moment (which is a measure of the inhomogeneity of the mass distribution in a body or a system) of a large mass body undergoes deformations, such as the case of massive bodies that orbit around each other or in a gravitational collapse between massive bodies (figure 2.11).

2.7.1 The LIGO interferometer

Several interferometers are currently operating on the Earth's surface. At present, the most sensitive is LIGO (figure 2.12), an acronym for Laser Interferometer Gravitational Wave Observatory. LIGO has three detectors: one in Livingston, Louisiana, the other two (in the same vacuum tube) at the Hanford Site and Richland in the state of Washington. Like its older sibling the Michelson–Morley interferometer, LIGO is made up of two long arms, from 2 to 4 km long, placed at 90° angles to each other, containing vacuum tubes 1 m in diameter, within which light travels. The passage of a gravitational wave will slightly lengthen one and shorten the other, causing diffraction patterns that the interferometer reveals with extreme sensitivity. Gravitational waves produced by a binary system of neutron stars, for example, should generate deformations in an appropriate detector placed on the Earth of one part on 10^{20} of the length of the arms.

But even with such long arms, the strongest gravitational waves will change the distance between the ends by a maximum of approximately 10^{-18} m. LIGO is designed to detect gravitational waves capable of generating displacements of the order of 10^{-22} m. The improvements implemented for LIGO and other detectors

Figure 2.11. Binary system neutron stars collapsing. Courtesy: NASA/Goddard Space Flight Center.

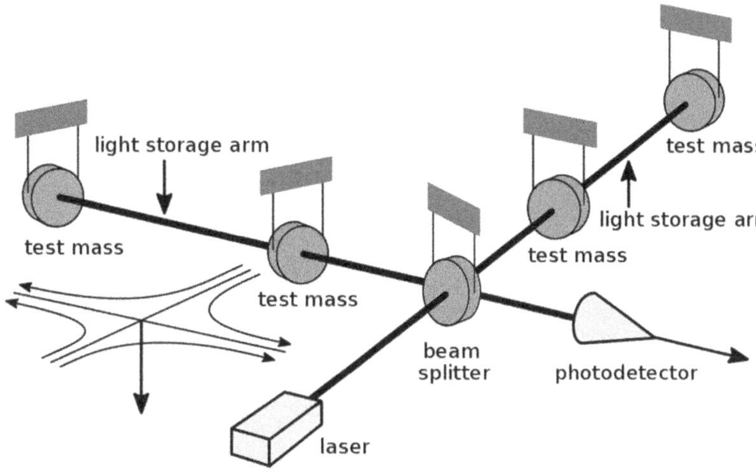

Figure 2.12. LIGO schematic.

such as VIRGO, GEO 600 and TAMA 300 would further increase the sensitivity. Interferometric detectors are limited to high frequencies due to shot noise, which happens due to the fact that light has a quantized structure (it is a stream of photons). An analogy for this refers to rain: the rate of rainfall, like the intensity of the laser, is measurable, but raindrops, like photons, fall at randomized times, causing fluctuations around the average value. This leads to disruption in the output of the detector, similar to radio static. Furthermore, due to the sufficiently high power of the laser, the random momentum transferred in the test masses by the laser photons shakes the mirrors, masking the signals at low frequencies. Thermal disturbance—for example Brownian motion—is another limitation on sensitivity. In addition to these 'stationary' (constant) disturbance sources, all terrestrial detectors are also limited to low frequencies by seismic disturbance, other forms of environmental vibration and other 'non-stationary' disturbance sources: deformations in mechanical structures, lightning or other large electrical disturbances, etc. They can also create disturbances by masking an event or even imitating it. All of this (what physicists call 'noise') must be taken into consideration and excluded from the analysis before a study can be considered a true event of gravitational waves.

On 24 September 2015 at 09:50:45 UTC (universal time) a signal attributable to the passage of a gravitational wave was unmistakably detected for the first time in history (figure 2.13). This signal was detected simultaneously by the two LIGO gravitational interferometers. The wave was produced by the fusion of two supermassive black holes with a mass of 36 and 29 solar masses [4, 5].

2.7.2 The VIRGO interferometer

Another currently active gravitational interferometer is VIRGO (figure 2.14 and 2.15), located in Cascina (Pisa), Italy. Virgo is composed of two arms 3 km in length

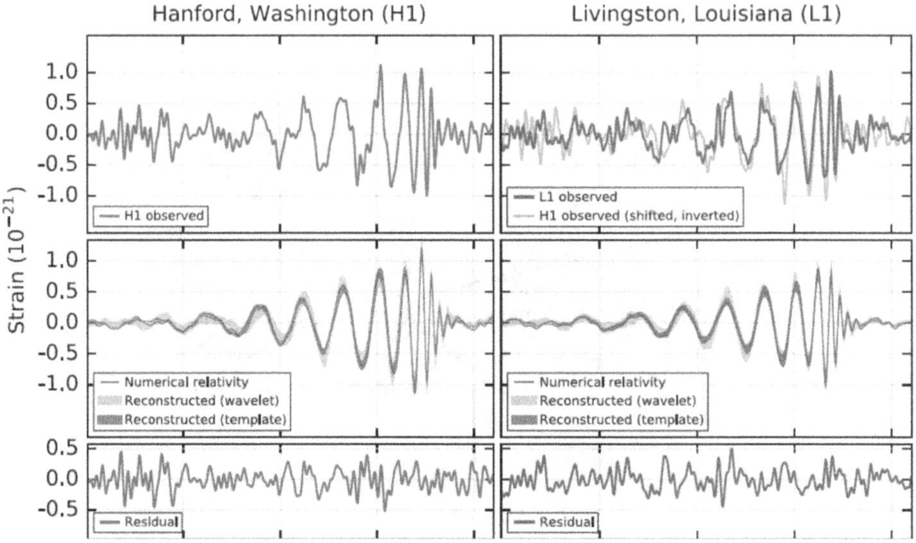

Figure 2.13. The signal detected by LIGO on 24 September 2015 at 09:50:45 UTC. Courtesy Caltech/MIT/LIGO Laboratory.

Figure 2.14. The VIRGO interferometer. Source: The Virgo collaboration.

positioned perpendicular to each other. One arm extends in an approximately South–North direction and the other in an approximately East–West direction.

At the entrance of the Virgo detector, a laser beam impinges on a semitransparent mirror, inclined at 45°, called a beam splitter (BS), and is divided into two beams which are sent inside two straight 3 km vacuum tubes, housed in the perpendicular arms of Virgo. The Virgo laser emits a beam of infrared light that is extremely

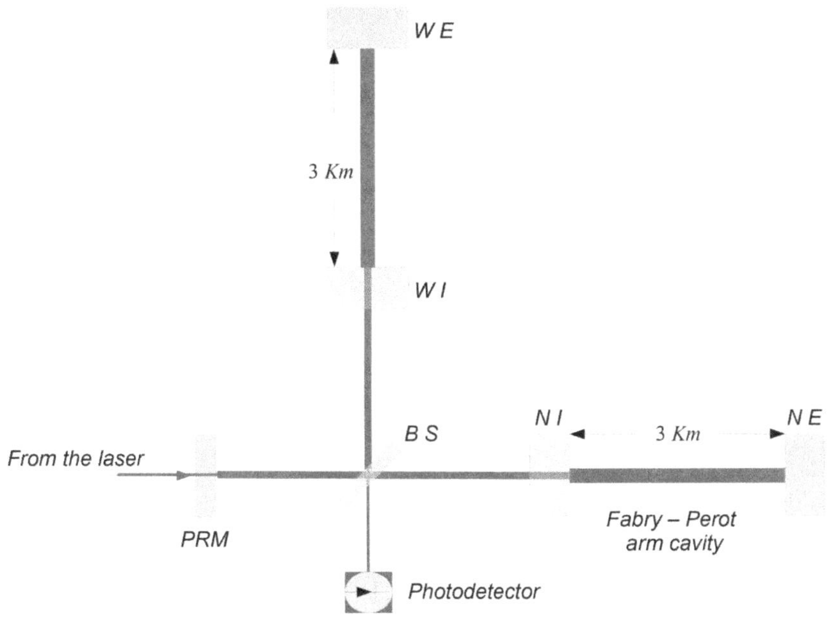

Figure 2.15. The VIRGO interferometer scheme.

stable in frequency and power, with a wavelength of 1064 nm. The first version of Virgo used a 20 W laser that provided a 10 W beam at the interferometer input. In Advanced Virgo the laser will have a maximum power of 200 W, and will provide a 125 W beam at the interferometer input. In Virgo the laser beams go back and forth hundreds of times along the arms, before recombining and interfering.

When a gravitational wave passes near the detector, space and thus the distances of the terminal mirrors lengthen and shorten alternately: the fringes begin to fluctuate at the frequency of the gravitational wave. These fluctuations, recorded by means of a photocell, are accurately analysed with appropriate algorithms using a bank of computers. As mentioned above, the main difficulty consists in distinguishing the genuine but very weak signals of gravitational waves from those produced by spurious effects. Both in the extreme west (WE) and in the extreme north (NE) of the tubes there is a mirror that reflects the beam back. Each mirror weighs 40 kg and is perfectly smooth, reflecting 99.999% of the incident light. The reflected beams recombine on the mirror at 45°. There, the interference pattern produced (the interference fringes) is observed by a photocell, a transducer that converts the incident light into an electric current, which is subsequently amplified, recorded and analysed. To avoid the propagation of seismic and acoustic vibrations, all the mirrors are suspended in a vacuum using up to eight cascade pendulum stages. With an attenuation factor of one hundred times for each stage, the seismic noise is reduced by up to 1016 times. Advanced Virgo will have the ability to detect gravitational waves with frequencies between 10 Hz and 10 kHz.

On 16 October 2017 the LIGO and Virgo collaborations announced the first ever detection of gravitational waves originating from the coalescence of a binary

Figure 2.16. The NGC 4993 galaxy and the GRB 170817A gamma-ray burst, associated with the GW170817 event, as captured by Hubble. Hubble Space Telescope, NASA and ESA.

neutron star system. The observation of the GW170817 transient, which occurred on 17 August 2017, allowed the masses of the neutron stars involved to be constrained between 0.86 and 2.26 solar masses. Further analysis allowed a greater restriction of the mass values to the range of 1.17–1.60 solar masses, with the total system mass measured to be 2.73–2.78 solar masses. The inclusion of the Virgo detector in the observation effort allowed for an improvement of the localization of the source by a factor of ten. This in turn facilitated the electromagnetic follow-up of the event. In contrast to the case of binary black hole mergers, binary neutron star mergers were expected to yield an electromagnetic counterpart, that is, a light signal associated with the event. A gamma-ray burst (GRB 170817A) was detected by the Fermi Gamma-ray Space Telescope, occurring 1.7 s after the gravitational wave transient (figure 2.16). The signal, originating near the Galaxy NGC 4993, was associated with the neutron star merger. This was corroborated by the electromagnetic follow-up of the event (AT 2017gfo), involving 70 telescopes and observatories and yielding observations over a large region of the electromagnetic spectrum which further confirmed the neutron star nature of the merged objects and the associated kilonova [4, 5].

With the discovery of the optical counterpart for the gravitational wave GW170817 a new form of astronomy was born. The gravitational interferometers, used simultaneously with the highly classical astronomical observation instruments, will allow phenomena to be understood better and more intimately, even exotic phenomena, which will be able to provide fundamental data on the nature of the Universe, on its origin and its future.

2.7.3 The future of gravitational interferometers

The next generation of instruments (Advanced LIGO and Advanced Virgo) will be more than ten times more sensitive than current ones. Another highly sensitive interferometer (LCGT) is currently being designed. A key point is that in a ten-fold increase in sensitivity ('range') the volume of space accessible to the instrument is increased by a thousand times. This increases the percentage of detectable signals

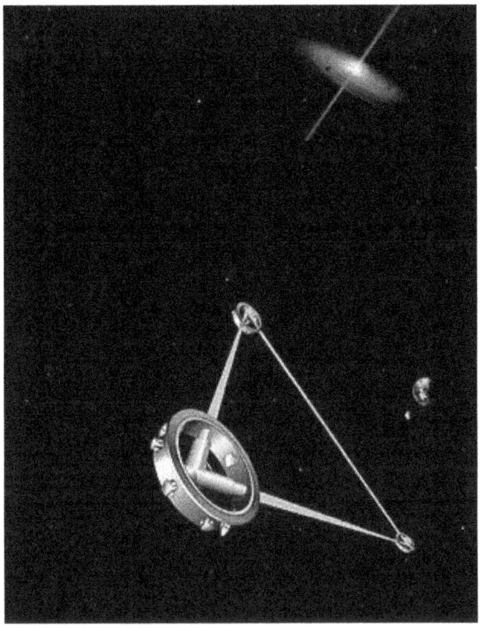

Figure 2.17. The LISA project. Image courtesy of NASA.

that could be seen from one in tens of years of observation, to tens per year. Spatial interferometers, such as LISA and DECIGO, have also been developed. The LISA project calls for three test masses forming an equilateral triangle, with lasers from each spacecraft to the others, forming two independent interferometers. LISA is designed to occupy a solar orbit following the Earth, with each arm of the triangle being five million kilometres long (figure 2.17). This puts the detector in an optimal vacuum condition away from terrestrial disturbance sources, although it is still susceptible to disturbing discharges, as well as to artefacts caused by cosmic rays and solar wind.

2.7.4 Another use of interferometers

When surrounded by a transparent emission region, black holes are expected to reveal a dark shadow caused by gravitational light bending and photon capture at the event horizon. To image and study this phenomenon, the Event Horizon Telescope (EHT) was assembled, a global very-long-baseline interferometry (VLBI) array observing at a wavelength of 1.3 mm. This allows event-horizon-scale images to be reconstructed of the supermassive black hole candidate in the centre of the giant elliptical galaxy M87. The central compact radio source was resolved as an asymmetric bright emission ring with an angular diameter of 42 μas. The EHT is a VLBI array that comprises millimetre- and submillimetre-wavelength telescopes separated by distances comparable to the diameter of the Earth. At a nominal operating wavelength of ~1.3 mm, the angular resolution (λ/D) of EHT is ~25 μas.

References

[1] Resnick R 1968 *Introduction to Special Relativity* (New York: Wiley)

[2] Michelson A A and Morley E 1887 On relative motion of the Earth and the luminiferous Ether *Am. J. Sci.* **34** 333

[3] Shankland R S *et al* 1955 New analysis of the interferometer observations of Dayton C Miller *Rev. Mod. Phys.* **27** 167

[4] Abbott B P *et al* 2016 Observation of gravitational waves from a binary black hole merger *Phys. Rev. Lett.* **116** 061102

[5] Abbott B P *et al* 2016 GW151226: observation of gravitational waves from a 22 solar mass binary black hole coalescence *Phys. Rev. Lett.* **116** 211103

IOP Publishing

Modern Physics
A critical approach
Canio Noce

Chapter 3

Special relativity: an introduction

Roberto De Luca, Marcello Sette and Alessandro Sorgente

This chapter is divided into two main sections. In the first we will summarize the salient points of the birth of the special theory of relativity, limiting ourselves to kinematic considerations. This part is well suited to undergraduate science students since calculus is used only when strictly necessary. The second part is mainly devoted to the proof of the invariance of the Maxwell equations under Lorentz transformations and to the deduction of the new relativistic form of Newton's laws.

3.1 Kinematics and dynamics

3.1.1 Reference systems and events

We will consider only inertial systems in relative motion with speed v, directed for convenience along the common axis x, x'. Non-inertial systems (i.e. with reciprocal accelerations) are the subject of general relativity, and are beyond the aim of this book. We will call the four-dimensional vector $(x,\ y,\ z,\ t)$ an *event* and we will study the relationship between events measured in the two reference systems.

3.1.2 Transformations and principles of relativity

Galileo transformations
These are transformations of the vector coordinates, defined by

$$\begin{cases} x' = x - vt \\ y' = y \\ z' = z \\ t' = t \end{cases}. \tag{3.1}$$

These transformations are invertible, such that

$$\begin{cases} x = x' + vt' \\ y = y' \\ z = z' \\ t = t' \end{cases}.$$

If we call $(u_x, u_y, u_z) = \left(\frac{dx}{dt}, \frac{dy}{dt}, \frac{dz}{dt}\right)$ and $(u_x', u_y', u_z') = \left(\frac{dx'}{dt'}, \frac{dy'}{dt'}, \frac{dz'}{dt'}\right)$ the components of the velocity measured in the two reference systems, taking the derivative with respect to time of equation (3.1), we obtain the classical relation of the addition of velocities:

$$\begin{cases} u_x' = u_x - v \\ u_y' = u_y \\ u_z' = u_z \end{cases}.$$

By taking the second derivative, we then obtain equal accelerations in the two reference systems.

Classical mechanics and Newtonian relativity

In addition to acceleration, even the mass is not affected by the relative motion of the two reference systems. Therefore, Newton's second law and the equations of motion of a particle also have the same form in all inertial reference systems. Furthermore, since in mechanics it can be shown that the conservation principles (e.g. energy, momentum and angular momentum) are all consequences of Newton's laws, then we can also say that all the laws of mechanics have the same form in all inertial reference systems [1].

No inertial reference system is privileged over others and all inertial systems are equivalent with respect to mechanics.

Consequently:

1. No mechanical experiment, performed entirely in a single inertial reference system, can tell the observer what the motion of that reference is with respect to another inertial reference system.
2. The relative velocities can be described in the two reference systems only by observing the two reference systems from the outside, but there is no way to determine the absolute speed of an inertial reference system by means of observations limited to that single reference.

The laws of transformation, in general, will change some quantities but will leave others unchanged. These unchanged quantities are called *invariants of the transformation.* A statement concerning the invariant quantities under given transformations is called a *principle of relativity*. Therefore Newton's principle of relativity can be expressed in this way: the acceleration, the mass and Newton's laws are invariant under Galilean transformations[1].

[1] Note here that a principle of relativity must always explicitly express what the transformations are that generate the invariants.

Electromagnetism

Let us now examine the situation from the electrodynamics point of view, asking ourselves if Maxwell's laws are invariant under Galilean transformations. If this were true, Newton's principle of relativity would be valid not only for mechanics, but for all physics, so that we might generalize it as follows:

- No inertial reference would be privileged over others.
- No physics experiment (not just mechanics experiment) performed in a single reference system would be capable of measuring the absolute speed of that reference system.

However, since the speed of light $c = \frac{1}{\sqrt{\epsilon_0 \mu_0}}$ is not invariant under Galilean transformations, there would be one and only one reference system where the speed of light is exactly c. We would therefore have a physical way of determining an absolute (and at rest) reference system, named the *ether reference system*, in which the speed of light is exactly c and determining, with optical experiments performed exclusively within another reference system, the absolute speed of that reference system (by means of the difference in the measurement of the speed of light in that system and the absolute value c). We must therefore admit that Newton's principle of relativity applies to mechanics but not to electromagnetism, so only one of the following alternatives is the correct one:

1. There is a principle of relativity (the Newton one, dependent on Galilean transformations) for mechanics, but not for electrodynamics (and so there exists the ether reference system).
2. There is a principle of relativity (the Newton one, dependent on Galilean transformations) both for mechanics and for electrodynamics, but Maxwell's laws are not correct (we should then correct them in order to make them invariant under Galilean transformations).
3. There is a principle of relativity for both mechanics and for electrodynamics (different to the Newton one, therefore not based on Galilean transformations); in this case the new transformations should be rewritten so as to make both Maxwell's laws and Newton's laws invariant (and perhaps the latter would be further adapted to them).

Attempts to identify an absolute reference system failed and the Michelson and Morley experiment (1887), verifying that the speed of light is c in any inertial system, definitively proved the falsity of the first option. Later, some tried to preserve the concept of the privileged reference of the ether—the contraction hypothesis of Lorentz and Fitzgerald (1892) was a complicated and artificial theory that did not confirm other experimental results. The hypothesis of a trailing ether would explain the results of the Michelson–Morley experiment, but would not then explain other effects such as stellar aberration and Fizeau's convection coefficient (on the propagation of electromagnetic waves in moving media). The second option was attempted by other researchers using electrodynamics theories, so-called emissive theories, in which the speed of light is c with respect to the source and which is independent of the medium that transmits the light. However, all the emissive theories are contradicted by two types of experiments, one which consists of the observations of the double (binary) stars and another which, in a similar way to the Michelson–Morley experiment, exploits extraterrestrial light sources. Almost no one believed it initially, but the last alternative proved to be the correct one [2].

3.1.3 Einstein's relativity

Basic postulates and hypothesis

1. The laws of physics are the same in all inertial reference systems. There is no privileged inertial reference system (the principle of relativity).
   This principle has two immediate and important consequences:
   a. Space and time are homogeneous (all points of space–time are equivalent).
   b. Space is isotropic (it has the same properties in all directions).
2. The speed of light in empty space has the same value c in all inertial systems (the principle of the invariance of the speed of light).

Lorentz transformations

Let S and S' be the two reference systems under study. The space isotropy hypothesis allows us to choose the axes x–x' so that they are coincident and the other axes y–y' and z–z' are parallel to each other. The hypothesis of homogeneity of space and time requires that the equations of transformations be linear:

$$\begin{cases} x' = a_{1\,1}x + a_{1\,2}y + a_{1\,3}z + a_{1\,4}t \\ y' = a_{2\,1}x + a_{2\,2}y + a_{2\,3}z + a_{2\,4}t \\ z' = a_{3\,1}x + a_{3\,2}y + a_{3\,3}z + a_{3\,4}t \\ t' = a_{4\,1}x + a_{4\,2}y + a_{4\,3}z + a_{4\,4}t \end{cases} . \tag{3.2}$$

If this were not the case, an object of length l in the unaccented system would be transformed into objects of different length depending on theri position in space. For example, if we had $x' = a_{1\,1}x^2$, a bar of unitary length, placed between $x_1 = 1$ and $x_2 = 2$, in the accented system would have length $l' = 3a_{1\,1}$, while, placed between $x_1 = 2$ and $x_2 = 3$, in the accented system it would have length $l' = 5a_{1\,1}$. Similarly, we can exclude any dependence on t that is not linear, since the time interval of an event must not depend on the numerical indication of the hands of the clock. We now can go on to determine the values of the 16 parameters of the transformation.

1. The x-axis is always the same as the x'-axis. So the equations $y = z = 0$ (of the x-axis) should be transformed into $y' = z' = 0$. This means that $a_{2\,1} = a_{2\,4} = a_{3\,1} = a_{3\,4} = 0$. Then:

$$\begin{cases} y' = a_{2\,2}y + a_{2\,3}z \\ z' = a_{3\,2}y + a_{3\,3}z \end{cases} .$$

2. In a similar way the x–y plane (with equation $z = 0$) should transform into the x'–y' plane (with equation $z' = 0$) and the x–z plane (with equation $y = 0$) should transform into the x'–z' plane (with equation $y' = 0$). This means that $a_{2\,3} = a_{3\,2} = 0$, and

$$\begin{cases} y' = a_{2\,2}y \\ z' = a_{3\,3}z \end{cases}.$$

3. These last two constant coefficients could be determined using the relativity postulate. We show how to do this for finding the value of $a_{2\,2}$. Let a bar of unitary length, with respect to the system S, be fixed on the y-axis, with one end on the origin. When he measures its length the S' observer would find a value $l' = a_{2\,2}l = a_{2\,2}$. Let us suppose now the same bar is fixed in the S' system, in the same position. This time the S observer would measure a length $l = \frac{1}{a_{2\,2}}l' = \frac{1}{a_{2\,2}}$. As the two reference systems should be equivalent, we have $a_{2\,2} = 1$. An analogous argumentation gives $a_{3\,3} = 1$.

   To summarize:

$$\begin{cases} y' = y \\ z' = z \end{cases}.$$

4. Using a symmetry based argumentation we can say that t' should not depend on either y or z. If this was not the case, two synchronous clocks, located in a symmetrical position with respect to the plane x–z or to the plane x–y (e.g. at the points $(0, y, 0)$ and $(0, -y, 0)$ or at points $(0, 0, z)$ and $(0, 0, -z)$), would become asynchronous when observed by S', a contradiction with respect to the space isotropy[2]. Then it should be $a_{4\,2} = a_{4\,3} = 0$.
5. Let us consider an object at rest in the origin of S', its law of motion is constantly $x' = 0$, while the S observer should measure a law of motion $x = vt$. This means that $a_{1,4} = -va_{1,1}$ and the equations (3.2) reduce to

$$\begin{cases} x' = a_{1\,1}(x - vt) \\ y' = y \\ z' = z \\ t' = a_{4\,1}x + a_{4\,4}t \end{cases}. \tag{3.3}$$

6. Let us put two further constraints on the coefficients, namely $a_{1\,1} > 0$ and $a_{4\,4} > 0$. This corresponds to choosing the same orientation on the x and x' axes and to letting time increase in both reference systems.
7. The last three coefficients will be found using the principle of constancy of the speed of light. To this end, let us suppose the two systems of reference are

[2] Otherwise we can identify one privileged direction on the y-axis or on the z-axis.

coincident at $t' = t = 0$ and at the same time a spherical electromagnetic wave starts from the common origin. Both observers should register a propagation described by the equation of a sphere centred on the origin (0, 0, 0) and a radius time proportional to, respectively, the values ct and ct' (the speed of light is the same in the two systems of reference):

$$x^2 + y^2 + z^2 = c^2t^2 \tag{3.4}$$

$$x'^2 + y'^2 + z'^2 = c^2t'^2. \tag{3.5}$$

Substituting equation (3.3) into (3.5), we obtain

$$a_{11}^2(x - vt)^2 + y^2 + z^2 = c^2(a_{41}x + a_{44}t)^2$$

and factorizing the common terms

$$(a_{11}^2 - c^2a_{41}^2)x^2 + y^2 + z^2 - 2(va_{11}^2 + c^2a_{41}a_{44})xt = (c^2a_{44}^2 - v^2a_{11}^2)t^2.$$

Requiring the previous equation be the same as the equation (3.4), it should be

$$\begin{cases} c^2a_{44}^2 - v^2a_{11}^2 &= c^2 \\ a_{11}^2 - c^2a_{41}^2 &= 1 \\ va_{11}^2 + c^2a_{41}a_{44} &= 0 \end{cases}.$$

Calling $\beta = \frac{v}{c}$ and $\gamma = \frac{1}{\sqrt{1-\beta^2}}$ ($\beta < 1$ and $\gamma > 1$), the coefficients a_{11}, a_{14} and a_{44} can be rewritten as follows:

$$\begin{cases} a_{44} = \dfrac{1}{\sqrt{1 - \frac{v^2}{c^2}}} = \gamma \\ a_{11} = \dfrac{1}{\sqrt{1 - \frac{v^2}{c^2}}} = \gamma \\ a_{41} = -\dfrac{\frac{v}{c^2}}{\sqrt{1 - \frac{v^2}{c^2}}} = -\frac{\beta}{c}\gamma \end{cases}.$$

Finally, the transformation rules (3.3) become

$$\begin{cases} x' = \gamma(x - \beta ct) \\ y' = y \\ z' = z \\ t' = \gamma\left(t - \dfrac{\beta}{c}x\right) \end{cases}, \tag{3.6}$$

commonly called *the Lorentz transformation.*

They are invertible and their inverse, using simple calculation, are (as we can reasonably expect):

$$\begin{cases} x = \gamma(x' + \beta ct') \\ y = y' \\ z = z' \\ t = \gamma\left(t' + \dfrac{\beta}{c}x'\right) \end{cases}.$$

Furthermore, for small relative velocity ($\beta \ll 1$ and $\gamma \longrightarrow 1$), they become

$$\begin{cases} x' = x - vt \\ y' = y \\ z' = z \\ t' = t \end{cases},$$

returning the Galilean transformations.

3.1.4 Some important implications

Length contraction

Let us suppose we would like to compare in the two reference systems the length of a bar put on the common x–x' axes but at rest with respect to S'. We say that the *proper length* L_0 is the length which is measured in the reference system in which the object is at rest (in this case S').

Let L be the length in the S system. To obtain a value for L we have to subtract the positions of the bar ends, but as the object is moving with respect to S those positions are to be registered at the same time[3] (e.g. $t = t_1$). Disregarding the y and z coordinates, the two measured events in S are $A = (x_1, t_1)$ and $B(x_2, t_1)$. In this way $L = |x_2 - x_1|$. The proper length is, instead,

$$L_0 = |x_2' - x_1'| = |\gamma(x_2 - \beta ct_1) - \gamma(x_1 - \beta ct_1)| = \gamma|x_2 - x_1| = \gamma L.$$

This result can be stated as follows:

> The length of an object is greater when it is at rest in a reference system (it has a proper length L_0). More precisely, when it moves with a velocity v with respect to an observer, its measured length is contracted in the direction of movement by a factor $1/\gamma$, while its dimensions in any direction orthogonal to the movement direction are not affected in any way.

Time dilation

In a single reference system, measuring a time interval means to evaluate the difference in time of the occurrence of two events, independent of their position

[3] An observer on S' could also measure those positions at different times, as the bar is at rest with respect to him.

(because in that same reference system all the clocks are synchronous). We call a time interval T_0 *proper* when it separates two events happening in the same position. Let A and B be two events in the same position in the reference S: $A(x_1, t_1)$ and $B(x_1, t_2)$. Then $T_0 = |t_2 - t_1|$. In the reference S', instead,

$$T = |t_2' - t_1'| = \left| \gamma\left(-\frac{\beta}{c}x_1 + t_2\right) - \gamma\left(-\frac{\beta}{c}x_1 + t_1\right) \right| = \gamma|t_2 - t_1| = \gamma T_0.$$

This result can be stated as:

A watch appears to go slower when it is at rest with an observer (and it measures a shorter proper time interval T_0). More precisely, when the watch moves with respect to the observer, it appears to go faster by a factor of γ (and it measures a longer time interval T).

Transformation of velocities

If an object is moving with speed $\vec{u} = (u_x, u_y)$ with respect to S and $\vec{u}' = (u_x', u_y')$ with respect to S', then

$$u_x' = \frac{\Delta x'}{\Delta t'} = \frac{\gamma(\Delta x - \beta c \Delta t)}{\gamma\left(-\frac{\beta}{c}\Delta x + \Delta t\right)} = \frac{\frac{\Delta x}{\Delta t} - \beta c}{-\frac{\beta}{c}\frac{\Delta x}{\Delta t} + 1} = \frac{u_x - v}{1 - \frac{u_x v}{c^2}},$$

$$u_y' = \frac{\Delta y'}{\Delta t'} = \frac{\Delta y}{\gamma\left(-\frac{\beta}{c}\Delta x + \Delta t\right)} = \frac{\frac{\Delta y}{\Delta t}}{\gamma\left(-\frac{\beta}{c}\frac{\Delta x}{\Delta t} + 1\right)} = \frac{u_y\sqrt{1 - \frac{v^2}{c^2}}}{1 - \frac{u_x v}{c^2}}.$$

Interval invariance

Let us define the *interval* between two events:

$$(\Delta s)^2 = c^2(\Delta t)^2 - (\Delta x)^2.$$

Technically we should also subtract $(\Delta y)^2$ and $(\Delta z)^2$, but we know that no interesting things happen in the transverse directions, so we ignore those contributions. Using the Lorentz transforms, we can rewrite $(\Delta s)^2$ in the S' coordinates, namely $\Delta x'$ and $\Delta t'$:

$$\begin{aligned}(\Delta s)^2 &= c^2\Delta t^2 - \Delta x^2 = \gamma^2 c^2(\Delta t' + v\Delta x'/c^2)^2 - \gamma^2(\Delta x' + v\Delta t')^2 \\ &= \gamma^2[\Delta t'^2(c^2 - v^2) - \Delta x'^2(1 - v^2/c^2)] \\ &= c^2\Delta t'^2 - \Delta x'^2 = (\Delta s')^2.\end{aligned}$$

We can then say:

The interval value does not depend on the reference system—it is an invariant.

Beware that an interval is a distance between space–time events. It is not an Euclidean distance, because it can also assume negative values. Let us examine the three possible alternatives.

- **Case 1.** $\Delta s^2 > 0$ (*time interval*).
  As $(c\Delta t)^2 > \Delta x^2$ then there exists some reference system S' which is moving with speed $v = \left|\frac{\Delta x}{\Delta t}\right| < c$ and for which, using the Lorentz transforms,

$$\Delta x' = \gamma(\Delta x - v\Delta t) = 0.$$

  The events appear to happen in the same position. In that same reference $\Delta t'$ was called earlier the proper time interval T_0.
  We can also say that there exists some reference system S' in which

$$\Delta s^2 = c^2\Delta t^2 - \Delta x^2 = c^2\Delta t'^2 = c^2 T_0^2.$$

- **Case 2.** $\Delta s^2 < 0$ (*space interval*).
  As $(c\Delta t)^2 < \Delta x^2$ then there exists some reference system S' which is moving with speed $v = c^2\left|\frac{\Delta t}{\Delta x}\right| < c$ and for which, using again the Lorentz transforms,

$$\Delta t' = \gamma\left(-\frac{v}{c^2}\Delta x + \Delta t\right) = 0.$$

  The two events appears to happen simultaneously.
  We can also say that there exists some reference system S' in which

$$\Delta s^2 = c^2\Delta t^2 - \Delta x^2 = -\Delta x'^2 = -L_0^2.$$

  Where L_0 is the proper length.
- **Case 3.** $\Delta s^2 = 0$ (*light interval*). This is an interval between two events which occur on the trajectory of a photon. There exists no reference system S' in which the two events could happen in the same position or at the same time, because such a system should move at speed c.

3.1.5 Further work

To conclude the path we began in section 1.2.3, we should verify the invariance of Maxwell's laws under Lorentz transforms. This task is beyond our aim here of remaining simple, so we will discuss it in the next section [3]. In addition, analogous verification would prove that Newton's laws are not invariant under the Lorentz transform [4]. However, again, in the next section we allow them to be invariant as well (but a little uglier).

3.2 Relativistic field transformations

3.2.1 Fields transformations in special relativity

Force transformations
Newton's *second* law in the form

$$\vec{F} = \frac{\mathrm{d}\vec{p}}{\mathrm{d}t} \tag{3.7}$$

retains its validity in relativistic mechanics, provided we use the *relativistic momentum*:

$$\vec{p} = \frac{m\vec{v}}{\sqrt{1-\beta^2}}.$$

In another inertial system S', moving with respect to the first with speed V, the law of motion must retain the same form. Therefore, we may write

$$\vec{F'} = \frac{\mathrm{d}\vec{p'}}{\mathrm{d}t'}. \tag{3.8}$$

It is known that one can introduce a four-vector $(E/c, \vec{p})$, where E is the energy of the particle, that behaves and transforms as $(ct, \vec{x})$ [4]; in this way, we have

$$\begin{cases} p_x' = \dfrac{p_x - \beta\frac{E}{c}}{\sqrt{1-\beta^2}} \\ p_y' = p_y \\ p_z' = p_z \end{cases}. \tag{3.9}$$

Differentiating both components of equation (3.9) one has

$$\begin{cases} \mathrm{d}p_x' = \dfrac{\mathrm{d}p_x - \beta\frac{\mathrm{d}E}{c}}{\sqrt{1-\beta^2}} \\ \mathrm{d}p_y' = \mathrm{d}p_y \\ \mathrm{d}p_z' = \mathrm{d}p_z \end{cases}. \tag{3.10}$$

Furthermore, by differentiating the Lorentz transformation for the time t we obtain

$$\mathrm{d}t' = \frac{\mathrm{d}t - \beta\frac{\mathrm{d}x}{c}}{\sqrt{1-\beta^2}}, \tag{3.11}$$

and we can finally rewrite the components of the forces by placing equations (3.10) and (3.11) into equation (3.8):

$$\begin{cases} F'_x = \dfrac{\mathrm{d}p'_x}{\mathrm{d}t'} = \dfrac{\frac{\mathrm{d}p_x}{\mathrm{d}t} - \frac{\beta}{c}\frac{\mathrm{d}E}{\mathrm{d}t}}{1 - \frac{\beta v_x}{c}} \\ F'_y = \dfrac{\mathrm{d}p'_y}{\mathrm{d}t'} = \dfrac{\frac{\mathrm{d}p_y}{\mathrm{d}t}\sqrt{1-\beta^2}}{1 - \frac{\beta v_x}{c}} \\ F'_z = \dfrac{\mathrm{d}p'_z}{\mathrm{d}t'} = \dfrac{\frac{\mathrm{d}p_z}{\mathrm{d}t}\sqrt{1-\beta^2}}{1 - \frac{\beta v_x}{c}} \end{cases}. \tag{3.12}$$

By recalling that[4]

$$F \cdot v = \frac{\mathrm{d}E}{\mathrm{d}t},$$

and by equation (3.7), equation (3.12) can be written as follows:

$$\begin{cases} F'_x = \dfrac{F_x - \frac{\beta}{c}\vec{F} \cdot \vec{v}}{1 - \beta v_x / c} \\ F'_y = \dfrac{F_y\sqrt{1-\beta^2}}{1 - \beta v_x / c} \\ F'_z = \dfrac{F_z\sqrt{1-\beta^2}}{1 - \beta v_x / c} \end{cases}. \tag{3.13}$$

---

[4] This formula can be obtained as follows. Work is the line integral of the force

$$W = \int F \cdot \mathrm{d}l.$$

The *work–energy theorem* holds relativistically:

$$W = \int \frac{\mathrm{d}\vec{p}}{\mathrm{d}t} \cdot \mathrm{d}\vec{l} = \int \frac{\mathrm{d}\vec{p}}{\mathrm{d}t} \cdot \frac{\mathrm{d}\vec{l}}{\mathrm{d}t}\mathrm{d}t = \int \frac{\mathrm{d}\vec{p}}{\mathrm{d}t} \cdot \vec{u}\mathrm{d}t.$$

On the other hand, we have

$$\frac{\mathrm{d}\vec{p}}{\mathrm{d}t} \cdot \vec{u} = \frac{\mathrm{d}}{\mathrm{d}t}\left[\frac{m\vec{u}}{\sqrt{1 - \frac{u^2}{c^2}}}\right] \cdot \vec{u} = \frac{m}{\left(1 - \frac{u^2}{c^2}\right)^{3/2}} \frac{\mathrm{d}\vec{u}}{\mathrm{d}t} \cdot \vec{u}.$$

It is easy to recognize that the last expression can be written as

$$\frac{\mathrm{d}}{\mathrm{d}t}\left[\frac{mc^2}{\sqrt{1 - \frac{v^2}{c^2}}}\right] = \frac{\mathrm{d}E}{\mathrm{d}t}.$$

If the particle is at rest in S, so that $\vec{v} = 0$, we have

$$\begin{cases} F_x' = F_x \\ F_y' = F_y\sqrt{1-\beta^2} = \dfrac{1}{\gamma}F_y \\ F_z' = F_z\sqrt{1-\beta^2} = \dfrac{1}{\gamma}F_z \end{cases}.$$

That is, the component of $\vec{F}$ parallel to the motion of S is unchanged, whereas the perpendicular components are divided by $\gamma = \frac{1}{\sqrt{1-\frac{v^2}{c^2}}}$ or, concisely,

$$\begin{aligned} \vec{F}_{\parallel}' &= \vec{F}_{\parallel}, \\ \vec{F}_{\perp}' &= \frac{1}{\gamma}\vec{F}_{\perp}. \end{aligned} \tag{3.14}$$

Therefore, while equation (3.13) represents the way the force $\vec{F}$ transforms when we go from an inertial reference frame S to a second reference frame S', moving at velocity $\vec{v} = v\hat{x}$ with respect to S, the simple relation equation (3.14) applies to cases where the particle is seen at rest in S.

Transformation laws for the fields $\vec{E}$ and $\vec{B}$

In deriving the transformation laws for the fields $\vec{E}$ and $\vec{B}$, we start from considering the Lorentz force $\vec{F}$ on the charge test particle q, moving with velocity $\vec{v}$ given by

$$\vec{F} = q(\vec{E} + \vec{v} \times \vec{B}),$$

where $\vec{E}$ and $\vec{B}$ are the electric and the magnetic fields (with magnitudes E and B, respectively).

Let us rewrite, for convenience, the laws of the transformation of forces (3.13) and their inverse

$$\begin{cases} F_x' = \dfrac{F_x - \frac{\beta}{c}\vec{F}\cdot\vec{v}}{1-\beta v_x/c} \\ F_y' = \dfrac{F_y\sqrt{1-\beta^2}}{1-\beta v_x/c} \\ F_z' = \dfrac{F_z\sqrt{1-\beta^2}}{1-\beta v_x/c} \end{cases}.$$

By the substitution $\beta \longrightarrow -\beta$, we obtain the inverse transformations

$$\begin{cases} F_x = \dfrac{F'_x + \frac{\beta}{c}\vec{F}'\cdot\vec{v}'}{1 + \beta v'_x/c} \\ F_y = \dfrac{F'_y\sqrt{1-\beta^2}}{1 + \beta v'_x/c} \\ F_z = \dfrac{F'_z\sqrt{1-\beta^2}}{1 + \beta v'_x/c} \end{cases}. \tag{3.15}$$

To derive the field transformation equations, let us start from a simple case. Consider a fixed charge q in the system S' moving with velocity $\vec{v} = v\hat{x}$ with respect to S. In S' then we have $\vec{v}' = 0$ and equation (3.15) therefore becomes

$$\begin{cases} F_x = F'_x \\ F_y = F'_y\sqrt{1-\beta^2} \\ F_z = F'_z\sqrt{1-\beta^2} \end{cases}.$$

In S' only an electric field is present, so that the Lorentz force will be

$$\vec{F}' = q\vec{E}',$$

while in S there will also be the contribution of the magnetic force because the observer in S sees the charge moving with velocity $\vec{v}$:

$$\begin{cases} F_x = [q(\vec{E} + \vec{v}\times\vec{B})]_x = qE'_x \\ F_y = [q(\vec{E} + \vec{v}\times\vec{B})]_y = qE'_y\sqrt{1-\beta^2} \\ F_z = [q(\vec{E} + \vec{v}\times\vec{B})]_z = qE'_z\sqrt{1-\beta^2} \end{cases}. \tag{3.16}$$

Calculating the cross product

$$\vec{v}\times\vec{B} = \begin{vmatrix} \hat{x} & \hat{y} & \hat{z} \\ u & 0 & 0 \\ B_x & B_y & B_z \end{vmatrix} = -uB_z\hat{y} + uB_y\hat{z} \qquad \vec{v} = u\hat{x};$$

here we are assuming that the particle is at rest in a frame moving with velocity v. We can explicitly rewrite equation (3.16) in the following form:

$$\begin{cases} E_x = E'_x \\ E_y - uB_z = E'_y\sqrt{1-\beta^2} \\ E_z + uB_y = E'_z\sqrt{1-\beta^2} \end{cases}, \tag{3.17}$$

and finally

$$\begin{cases} E'_x = E_x \\ E'_y = \dfrac{(E_y - uB_z)}{\sqrt{1-\beta^2}} \\ E'_z = \dfrac{(E_z + uB_y)}{\sqrt{1-\beta^2}} \end{cases}. \tag{3.18}$$

From equation (3.17) we can see that the components of the electric field in the direction of motion remain unchanged, while the transverse components of the electric field contain the magnetic field.

We now propose determining the laws of transformation of magnetic fields starting again from a simple physical model. Suppose we have a charge q that moves in the system S' with velocity $\vec{v}' = v'_y\hat{y}$. The Lorentz force felt by the charge in S' is

$$\vec{F}' = q(\vec{E}' + \vec{v}' \times \vec{B}').$$

Projecting equation (3.18) on the coordinate axes, we have

$$\begin{cases} F'_x = q\left(E'_x + v'_y B'_z\right) \\ F'_y = qE'_y \\ F'_z = q\left(E'_z - v'_y B'_x\right) \end{cases}.$$

The velocity transformation law allows us to express the velocities for the present case in the following way:

$$\begin{cases} v_x = u \\ v_y = v'_y\sqrt{1-\beta^2} \\ v_z = 0 \end{cases}. \tag{3.19}$$

The Lorentz force in S is

$$\begin{cases} F_x = q(E_x + v_y B_z - v_z B_y) \\ F_y = q(E_y + v_z B_x - v_x B_z) \\ F_z = q(E_z + v_x B_y - v_y B_x) \end{cases}. \tag{3.20}$$

By substituting equation (3.19) into equation (3.20) we obtain

$$\begin{cases} F_x = q\left(E_x + v'_y B_z\sqrt{1-\beta^2}\right) \\ F_y = q(E_y - uB_z) \\ F_z = q\left(E_z + uB_y - v'_y B_x\sqrt{1-\beta^2}\right) \end{cases}. \tag{3.21}$$

The transformation of forces equation (3.13), applied to our particular case, for which $\vec{v}'_y = v'_y\hat{y}$, can be written in the following form, because of equation (3.20):

$$\begin{cases} F_x = F'_x + \dfrac{\beta}{c}F'_y v'_y = q\left(E'_x + v'_y B'_z + \dfrac{\beta}{c}E'_y v'_y\right) \\ F_y = F'_y\sqrt{1-\beta^2} = qE'_y\sqrt{1-\beta^2} \\ F_z = F'_y\sqrt{1-\beta^2} = q\left(E'_z - v'_y B'_x\right)\sqrt{1-\beta^2} \end{cases}.$$

Comparing equations (3.20) and (3.21) we obtain the following system of equations

$$\begin{cases} E_x + v'_y B_z\sqrt{1-\beta^2}\,x = E'_x + v'_y B'_z + \dfrac{\beta}{c}E'_y v'_y \\ E_y - uB_z = E'_y\sqrt{1-\beta^2} \\ E_z + uB_y - v'_y B_x\sqrt{1-\beta^2} = \left(E'_z - v'_y B'_x\right)\sqrt{1-\beta^2} \end{cases}. \tag{3.22}$$

We observe that equation (3.22) contains the transformation laws for the electric fields given in equation (3.18). Placing equation (3.18) into equation (3.22) we obtain

$$\begin{cases} B'_x = B_x \\ B'_y = \text{undetermined} \\ B'_z = \dfrac{B_z - \frac{\beta}{c}E_y}{\sqrt{1-\beta^2}} \end{cases}. \tag{3.23}$$

The second term of equation (3.23) cannot be determined by means of equation (3.22) because it does not appear in the latter. To obtain the transformation law for B'_y we repeat the previous exercise, this time taking $\vec{v}' = v'_z\hat{z}$. In this new situation the components of the Lorentz force in the S system are

$$\begin{cases} F_x = q(E_x - B_y v_z) = q\left(E_x - B_y v'_z\sqrt{1-\beta^2}\right) \\ F_y = q(E_y - B_x v_z - uB_z) = q\left(E_y - B_x v'_z\sqrt{1-\beta^2}\,z - uB_z\right) \\ F_z = q(E_z + uB_y) \end{cases}.$$

The latter were obtained by exploiting the speed composition laws:

$$\begin{cases} v_x = u \\ v_y = 0 \\ v_z = v'_z\sqrt{1-\beta^2} \end{cases}.$$

In S' the Lorentz force has the following components:

$$\begin{cases} F'_x = q\left(E'_x - B'_y v'_z\right) \\ F'_y = q\left(E'_y - B'_x v'_z - uB'_z\right). \\ F'_z = q\left(E'_z + uB'_y\right) \end{cases} \tag{3.24}$$

To derive B'_y just work on the first equation of (3.24). Recalling the law of the transformation of forces, let us consider the x-component

$$F_x = F'_x + \frac{\beta}{c} F'_z v'_z,$$

we have

$$q\left(E_x - B_y v'_z \sqrt{1 - \beta^2}\right) = F'_x + \frac{\beta}{c} F'_z v'_z = q\left(E'_x - B'_y v'_z\right) + \frac{\beta}{c} q\left(E'_z + uB'_y\right) v'_z.$$

By means of equation (3.18), the above equation can be rearranged as follows:

$$B'_y = B_y \sqrt{1 - \beta^2} + \frac{\beta}{c} E'_z = B_y \sqrt{1 - \beta^2} + \frac{\beta}{c} \frac{(E_z + uB_y)}{\sqrt{1 - \beta^2}}.$$

Simplifying, finally we obtain

$$B'_y = \frac{\frac{\beta}{c} E_z + B_y}{\sqrt{1 - \beta^2}}.$$

Of course, we can always switch from the accented to the unaccented system by changing $\beta \to -\beta$. Below is a summary table of the transformation of electric and magnetic fields. Note how the components of $\vec{E}$ and $\vec{B}$ remain unchanged in the direction of boost $\vec{u} = u\hat{x}$.

3.2.2 Applications

Aberration of light and relativistic Doppler effect

From the transformation laws of fields, as reported in table 3.1, we may derive the equations that describe the Doppler effect for electromagnetic waves, as well as the phenomenon of light aberration.

Suppose that the electric field in S' is described by a monochromatic plane wave

$$E'_x = A' \sin(\vec{k}' \cdot \vec{r}' - \omega' t').$$

In S the electric field is described by the same equation because a plane wave remains such due to the linearity of the Lorentz transformations:

$$E_x = A \sin(\vec{k} \cdot \vec{r} - \omega t).$$

Table 3.1. Field transformations.

Relativistic transformations of the fields $\vec{E}$ and $\vec{B}$		
$E'_x = E_x$	$E'_y = \frac{(E_y - uB_z)}{\sqrt{1-\beta^2}}$	$E'_z = \frac{(E_z - uB_y)}{\sqrt{1-\beta^2}}$
$B'_x = B_x$	$B'_y = \frac{\frac{\beta}{c}E_z + B_y}{\sqrt{1-\beta^2}}$	$B'_z = \frac{B_z - \frac{\beta}{c}E_y}{\sqrt{1-\beta^2}}$

Because of transformation laws of fields in table 3.1, we have $E_x = E'_x$. In this way we write

$$\begin{gathered} A' = A \\ \vec{k}' \cdot \vec{r}' - \omega' t' = \vec{k} \cdot \vec{r} - \omega t. \end{gathered} \tag{3.25}$$

The wave fronts are parallel planes in the x–y plane and the wave vector $\vec{k}' = (k'_x, k'_y, 0)$, where

$$k'_x = k' \cos\theta' = \frac{2\pi}{\lambda'} \cos\theta'$$

and

$$k'_y = k' \sin\theta' = \frac{2\pi}{\lambda'} \sin\theta'.$$

From equation (3.25) we obtain

$$k'_x x' + k'_y y' - \omega' t' = k_x x + k_y y - \omega t,$$

which we can rewrite as (see figure 3.1)

$$\left[\frac{x'\cos\theta' + y'\sin\theta'}{\lambda'} - \nu' t'\right] = \left[\frac{x\cos\theta + y\sin\theta}{\lambda} - \nu t\right].$$

We may now replace x', y' and z' by their respective Lorentz-transformed expressions, such that

$$\begin{cases} x' = \dfrac{x - ut}{\sqrt{1-\beta^2}} \\ y' = y \\ z' = z \\ t' = \dfrac{t - \frac{\beta}{c}x}{\sqrt{1-\beta^2}} \end{cases},$$

and, by replacing into the above equation, we obtain

$$\left[\frac{\frac{x-ut}{\sqrt{1-\beta^2}}\cos\theta' + y\sin\theta'}{\lambda'} - \nu' \frac{t - \frac{\beta}{c}x}{\sqrt{1-\beta^2}}\right] = \left[\frac{x\cos\theta + y\sin\theta}{\lambda} - \nu t\right].$$

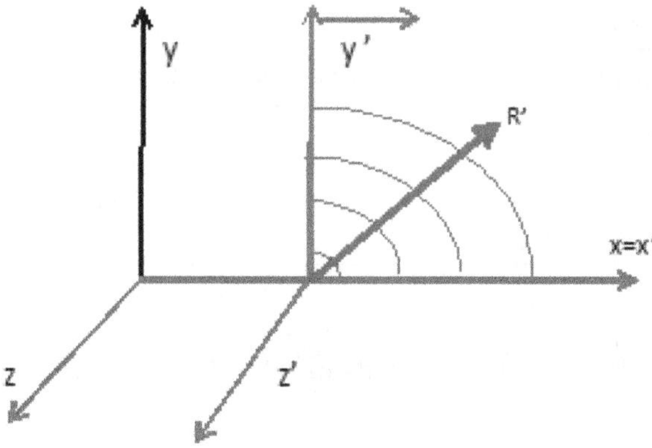

Figure 3.1. Plane wave generated in O', with a wave vector $\vec{k}'$. The direction of propagation forms an angle θ' with x'.

Therefore, by equating the coefficient of the variables x, y, z and t, we obtain

$$\frac{\cos\theta'}{\lambda'\sqrt{1-\beta^2}} + \frac{\frac{\beta}{c}\nu'}{\sqrt{1-\beta^2}} = \frac{\cos\theta}{\lambda}$$
$$\frac{\sin\theta'}{\lambda'} = \frac{\sin\theta}{\lambda} \tag{3.26}$$

$$\frac{u\cos\theta'}{\lambda'\sqrt{1-\beta^2}} + \frac{\nu'}{\sqrt{1-\beta^2}} = \nu. \tag{3.27}$$

Remembering that $\lambda'\nu' = \lambda\nu = c$, from equation (3.27), we have

$$\nu = \frac{\nu'}{\sqrt{1-\beta^2}}(1+\beta\cos\theta'), \tag{3.28}$$

$$\nu' = \frac{\nu}{\sqrt{1-\beta^2}}(1-\beta\cos\theta). \tag{3.29}$$

Equations (3.29) were obtained from equation (3.28) by the substitution $\beta \to -\beta$. Equations (3.28) and (3.29) represent the famous relativistic *Doppler effect* formula. Considering the ratio between (3.26) and (3.27), we obtain

$$\tan\theta = \frac{\sin\theta'\sqrt{1-\beta^2}}{\beta+\cos\theta'}$$
$$\tan\theta' = \frac{\sin\theta\sqrt{1-\beta^2}}{\cos\theta-\beta}. \tag{3.30}$$

Equation (3.30) represents the relativistic formula for the *aberration* of light. We observe that both the relativistic Doppler effect and the aberration of light are effects which contain first-order and second-order terms in β; for small speeds compared to c, i.e. $\beta \ll 1$, by neglecting β^2 with respect to 1, we have

$$\nu = \nu'(1 + \beta \cos\theta'),$$
$$\nu' = \nu(1 - \beta \cos\theta).$$

These are the formulas of the classic Doppler effect which, as we can see, represent a first-order effect in $\beta = u/c$. We may consider the following special cases:

1. If $\theta' = 0$, $\nu = \nu'(1 + u/c)$ we have a stationary source and the observer is moving towards the source.
2. If $\theta' = \pi$, $\nu = \nu'(1 - u/c)$ we have a stationary source and the observer moving away from the source.
3. If $\theta' = \pi/2$, $\nu = \nu'$, classically no effects are observed if the source and the observer move perpendicularly with respect to each other.

Let us now return to the relativistic case where the condition $\beta \ll 1$ is removed. We therefore consider an electromagnetic wave that moves in the x'-direction, with $\theta = 0$, we then consider an electromagnetic wave that moves in the x-direction, with $\theta = 0$ (the observer moves towards the source), then equation (3.28) becomes

$$\nu = \frac{\nu'}{\sqrt{1 - \beta^2}}(1 + \beta) = \nu'\sqrt{\frac{1 + \beta}{1 - \beta}}. \tag{3.31}$$

From system S', the detected frequency is

$$\nu' = \nu\sqrt{\frac{1 - \beta}{1 + \beta}}.$$

Similarly to the case already seen, let us now consider that the source and observer move away, $\theta = \pi$. In this case, by repeating the calculations, we will have

$$\nu = \nu'\sqrt{\frac{1 - \beta}{1 + \beta}} \tag{3.32}$$

or

$$\nu' = \nu\sqrt{\frac{1 + \beta}{1 - \beta}}. \tag{3.33}$$

Equations (3.30) and (3.32) or equivalently equations (3.31) and (3.33) are often referred to as a *longitudinal Doppler effect*. Placing $\theta = \pi/2$ or $\theta = 3\pi/2$ in equation (3.28), we find an effect that has no classic analogue and derives only from Lorentz transformations:

$$\nu = \nu'\sqrt{1 - \beta^2}. \tag{3.34}$$

The expression in equation (3.34) is referred to as a *transverse Doppler effect*.

Fields $\vec{E}$ and $\vec{B}$ for a charged particle in uniform motion
Here we propose calculating the electric field $\vec{E}$ and the magnetic field $\vec{B}$ generated by a charge q in motion with constant speed.

Let us suppose that the charge q is placed at the origin of the reference S' which moves with velocity $\vec{v} = u\hat{x}$. The fields $\vec{E}$ and $\vec{B}$, measured by an observer solidarly moving with S', are

$$\vec{E}' = \frac{1}{4\pi\varepsilon_0}\frac{q}{r^3}\vec{r}; \qquad \vec{B}' = 0. \tag{3.35}$$

Applying the transformations of the fields according to table 3.1 and considering equation (3.35) we have

$$\begin{cases} E_x = E'_x \\ E_y = \gamma E'_y \\ E_z = \gamma E'_z \end{cases} \quad \begin{cases} B_x = B'_x = 0 \\ B_y = -\gamma \dfrac{\beta}{c} E'_z \\ B_z = \gamma \dfrac{\beta}{c} E'_y \end{cases}, \tag{3.36}$$

where $\gamma = \frac{1}{\sqrt{1 - \frac{u^2}{c^2}}}$. The magnetic field depends directly on the components of the electric field, so it is sufficient to find the components of the electric field. Let us express the electric fields in S' according to Coulomb's law and then, by equation (3.36), let us write

$$\begin{cases} E_x = \dfrac{1}{4\pi\varepsilon_0}\dfrac{q\gamma(x - ut)}{(\gamma(x - ut)^2 + y^2 + z^2)^{3/2}} \\ E_y = \dfrac{1}{4\pi\varepsilon_0}\dfrac{\gamma q}{(\gamma(x - ut)^2 + y^2 + z^2)^{3/2}} y. \\ E_z = \dfrac{1}{4\pi\varepsilon_0}\dfrac{\gamma q}{(\gamma(x - ut)^2 + y^2 + z^2)^{3/2}} z \end{cases} \tag{3.37}$$

Equation (3.37) can be summarized in a single equation in vector form:

$$\begin{aligned} \vec{E} &= \hat{x}E_x + \hat{y}E_y + \hat{z}E_z \\ &= \frac{1}{4\pi\varepsilon_0}\frac{\gamma q}{(\gamma(x - ut)^2 + y^2 + z^2)^{3/2}}(\hat{x}(x - ut) + \hat{y}y + \hat{z}z). \end{aligned} \tag{3.38}$$

Setting $x - ut = X$ and taking θ as the angle between $\vec{R}$ and the x-axis, see figure 3.2, we can write $X = R\cos\theta$ and $\sqrt{y^2 + z^2} = R\sin\theta$ (see figure 3.3). With these geometric considerations equation (3.38), which defines the electric field $\vec{E} = \vec{E}(\vec{R})$, is transformed into a theta-dependent field:

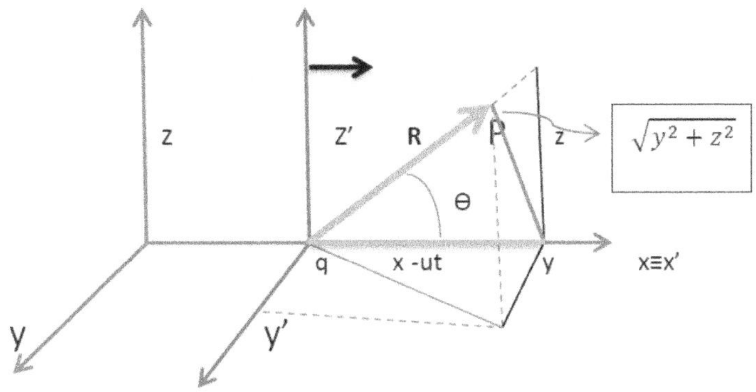

Figure 3.2. The electric field E, in frame S, has the direction R. θ is the angle between R and the x-axis.

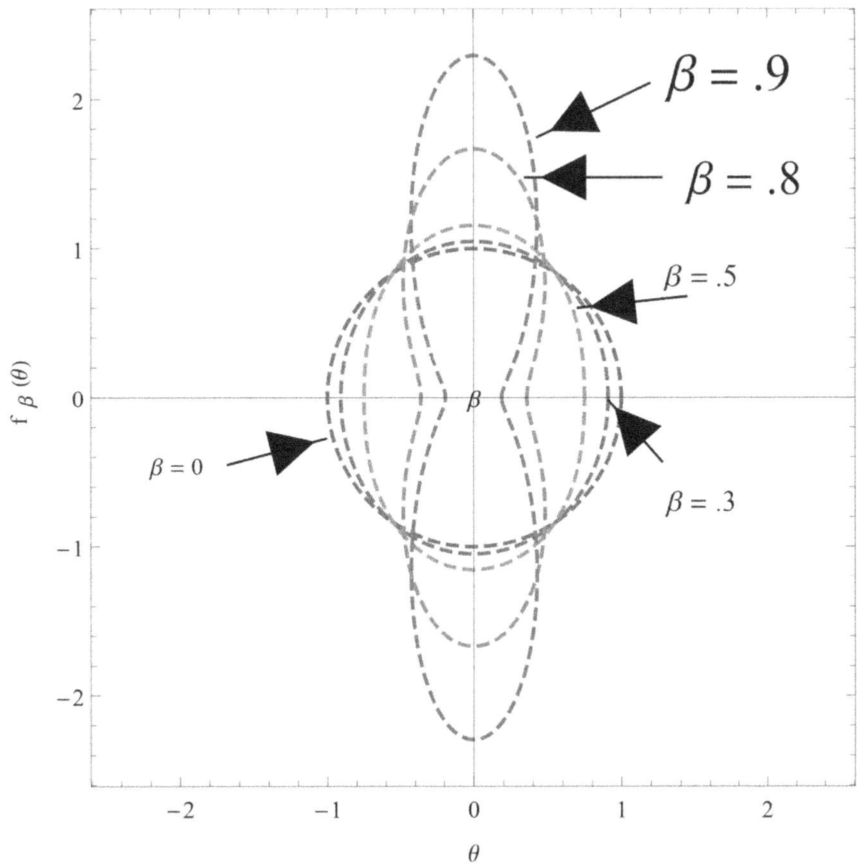

Figure 3.3. Representation of the electric field in the function of angle θ for $\beta = (0, 0.3, 0.5, 0.8, 0.9)$.

$$\vec{E} = \frac{1}{4\pi\varepsilon_0}\frac{\gamma q\vec{R}}{(\gamma^2 R^2 \cos\theta^2 + R^2 \sin^2\theta)^{3/2}}$$
$$= \frac{1}{4\pi\varepsilon_0}\frac{\gamma q\vec{R}}{R^3(\gamma^2 \cos^2\theta + \sin\theta)^{3/2}}$$
$$= \frac{1}{4\pi\varepsilon_0}\frac{\gamma q\vec{R}}{R^3((\gamma^2 - 1)\cos^2\theta + 1)^{3/2}} \tag{3.39}$$
$$= \frac{1}{4\pi\varepsilon_0}\frac{\gamma(1 - \beta^2)^{3/2} q\vec{R}}{R^3(1 - \beta^2 \sin^2\theta)^{3/2}}$$
$$= \frac{(1 - \beta^2)}{(-\beta^2 \sin\theta^2 + 1)^{3/2}}\frac{q\vec{R}}{4\pi\varepsilon_0 R^3}.$$

Equation (3.39) indicates that the electric field of a charge in motion with constant speed, seen by an observer S, is equal to the field of a fixed charge multiplied by a form factor $f(\beta, \theta) = \frac{(1-\beta^2)}{(-\beta^2 \sin\theta^2 + 1)^{3/2}}$ that depends on the angle theta.

From the graph of figure 3.3, we observe that as the velocity of the charge increases the field $\vec{E}$, seen from S, intensifies in the direction perpendicular to the motion of the charge, losing its spherical symmetry. From the same graph we can also see that the field reaches high peaks for $\theta = \pi/2$ and for $\theta = 3\pi/2$, that is in the direction perpendicular to the motion of the charge. Moreover, the field amplitude in these directions increases with β (see figure 3.4).

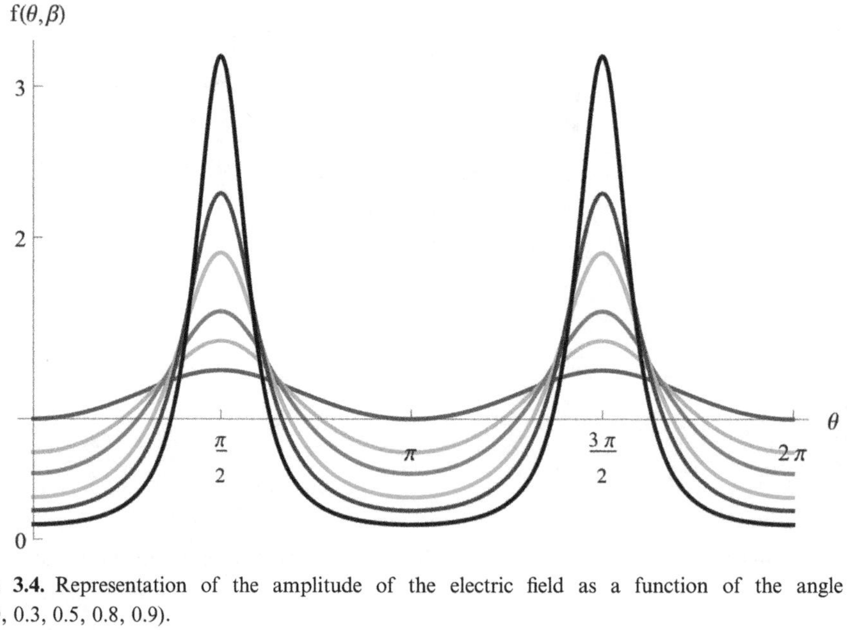

Figure 3.4. Representation of the amplitude of the electric field as a function of the angle θ for $\beta = (0, 0.3, 0.5, 0.8, 0.9)$.

Finally, we report a spectral representation of the electric field in S and in S'. Also in this representation we note the thickening of the field vectors around the vertical axis (see figure 3.5).

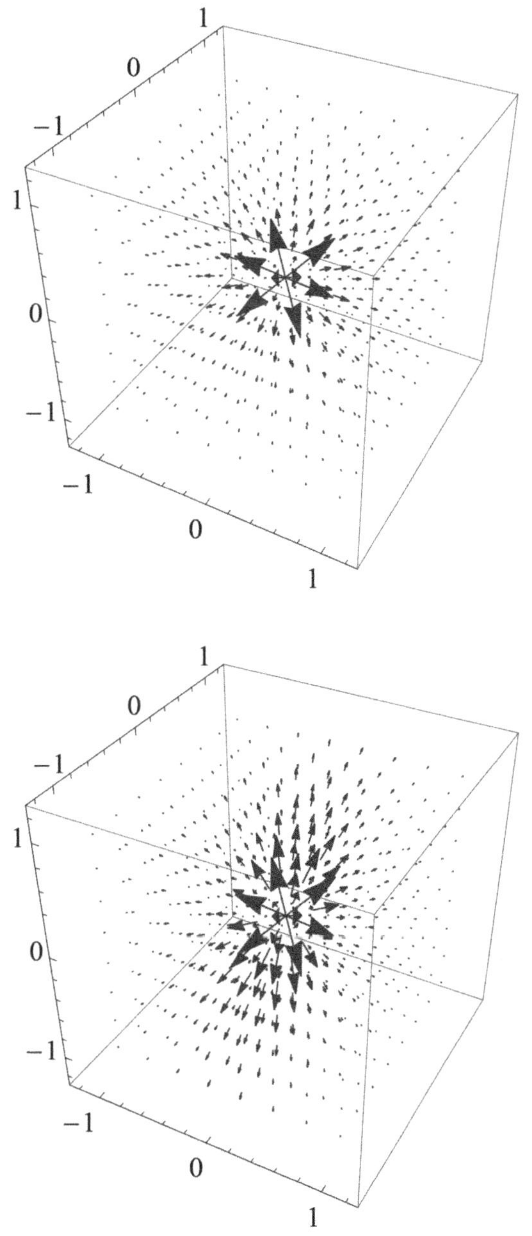

Figure 3.5. Spectral representation of the electric field. The top panel represents the field in S' (spherical symmetry) and the bottom panel shows the field in S.

Fields $\vec{E}$ and $\vec{B}$ near a current line

Consider a charge q, see figure 3.6, at rest in S, located near a wire in which a current I flows. The current is due to the motion of the electrons in the wire whose average velocity is $\vec{u}$. The positive ions are fixed in the wire and their linear charge density λ_+ is equal to that of the electrons in motion λ_-, with opposite sign. The current I in the reference frame S is $I = \lambda u$, having the opposite direction to the average velocity of electrons. The total electric charge in the wire is zero because $\lambda = \lambda_+ + \lambda_- = 0$, so there is no electric field: $\vec{E} = 0$ (if there were a density on the wire the electric field would be $E = \frac{\lambda}{2\pi\varepsilon_0 r}$ for the Gauss theorem), but there is a magnetic field $B = \frac{\mu_0 I}{2\pi r} = \frac{\mu_0 \lambda_- u}{2\pi r}$ (the Biot–Savart law). We would like to determine the total force on the charge q. Therefore, a test charge on the y-axis at a distance d from the wire will not feel any force; indeed

$$\begin{cases} E_x = 0 \\ E_y = 0 \\ E_z = 0 \end{cases} \begin{cases} B_x = 0 \\ B_y = 0 \\ B_z = \dfrac{\mu_0 \lambda_- u}{2\pi r} \end{cases}. \tag{3.40}$$

We would thus like to determine the fields in S'. From the law of the transformation of fields, considering equation (3.40), we have

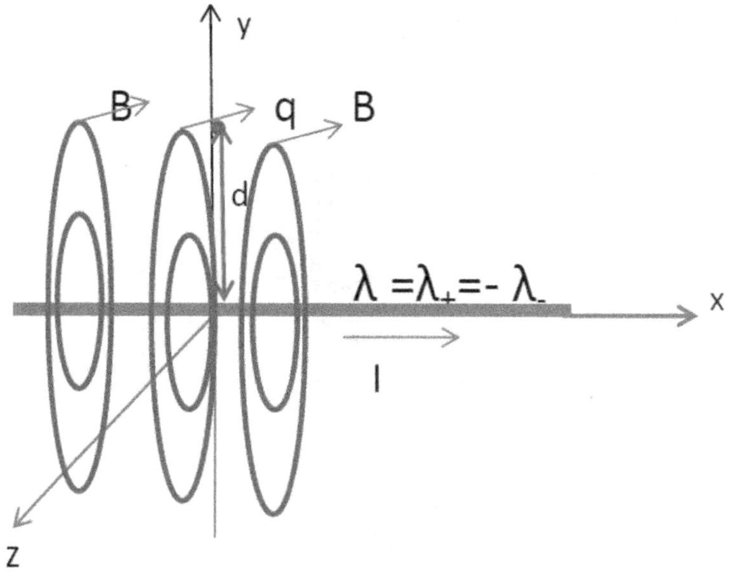

Figure 3.6. A current in a wire with zero charge density and charge q stops on the y-axis at a distance d from the wire.

$$\begin{cases} E_x' = 0 \\ E_y' = -u\gamma B_z \\ E_z' = 0 \end{cases} \quad \begin{cases} B_x = 0 \\ B_y = 0 \\ B_z' = \gamma B_z \end{cases}.$$

The Lorentz force will therefore be

$$F' = q(E_y'\hat{y} + \vec{v}' \times \vec{B}') = q(-u\gamma B_z\hat{y} - u\gamma B_z\hat{y}) = 0,$$

where have set $v' = -u$.

We have shown that a fixed charge does not experience any force from a neutral current-carrying wire if the force is measured either by a fixed observer, or by an observer in uniform motion with respect to a fixed observer.

Forces between electrical charges in motion

We now return to the problem of the interaction between two fixed charges in S'; in particular we study the specific case of two charges of the same sign moving in the direction of positive x (figure 3.7) in the reference system S. In S' the charges are at rest, so the field generated can be expressed in its components as follows:

$$E_x' = 0; \qquad E_{y'}' = \frac{1}{4\pi\varepsilon_0}\frac{q}{d^2}\hat{y}; \qquad E_z' = 0;$$

$$B_x' = B_y' = B_z' = 0.$$

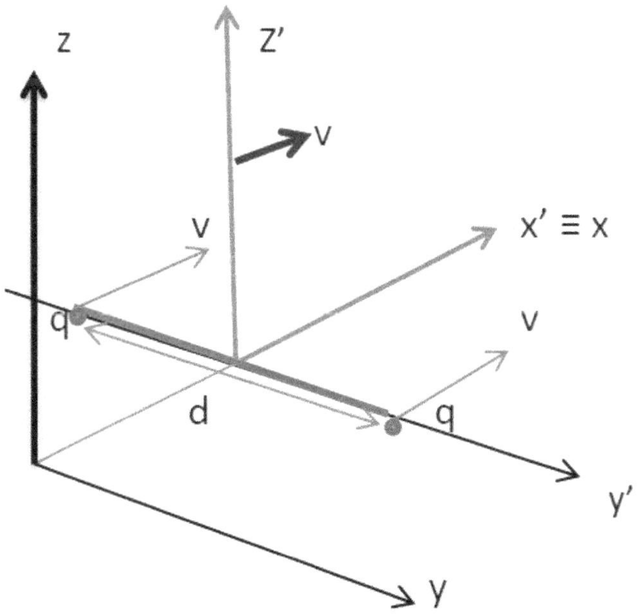

Figure 3.7. Forces between charges in uniform motion.

The fields measured in S must be transformed according to the field transformation rules (see table 3.1). In the present case, therefore, we have

$$\begin{cases} E_x = E'_x = 0 \\ E_y = \gamma E'_y \\ E_z = \gamma E'_z = 0 \end{cases} \begin{cases} B_x = B'_x = 0 \\ B_y = 0 \\ B_z = \gamma \dfrac{\beta}{c} E'_y \end{cases} . \tag{3.41}$$

The force between moving charges in the S system is

$$\begin{aligned} F &= q(\vec{E} + \vec{v} \times \vec{B}) = q\left(\gamma E'_y \hat{y} + v(\hat{x} \times \hat{z})\gamma \frac{\beta}{c} E'_y\right) \\ &= q\left(\gamma E'_y \hat{y} - \gamma \frac{v^2}{c^2} E'_y \hat{y}\right) \\ &= \frac{1}{\gamma} \frac{1}{4\pi\varepsilon_0} \frac{q^2}{d^2} \hat{y} = \left(\sqrt{1 - \frac{v^2}{c^2}}\right) \frac{1}{4\pi\varepsilon_0} \frac{q^2}{d^2} \hat{y} = \sqrt{1 - \frac{v^2}{c^2}} F'. \end{aligned} \tag{3.42}$$

The last equation says that the electromagnetic interaction between moving charges is smaller than the electromagnetic interaction between fixed charges. This result, on the other hand, can be obtained directly by applying the transformation law of forces in equation (3.15).

Appendix

A.1 Relativistic invariance of Maxwell's equations

According to the principle of relativity, all the laws of physics must have the same form in inertial reference systems. Thus Maxwell's equations must be covariant in form under the group of Lorentz transformations. In other words, we want to show that the form of Maxwell's equations is preserved when passing from one reference system to another in uniform motion with respect to the first.

Maxwell's equations

The four complete Maxwell equations, written in a fixed system, are

$$\begin{cases} \vec{\nabla} \cdot E = \rho/\varepsilon_0 \\ \vec{\nabla} \cdot B = 0 \\ \vec{\nabla} \times E = -\dfrac{\partial B}{\partial t} \\ \vec{\nabla} \times B = \mu_0\left(j + \varepsilon_0 \dfrac{\partial E}{\partial t}\right) \end{cases} . \tag{A.1}$$

The operator $\vec{\nabla}$ (nabla) is a vector operator defined as

$$\vec{\nabla} = \left(\frac{\partial}{\partial x}, \frac{\partial}{\partial y}, \frac{\partial}{\partial z}\right).$$

If $\vec{A}$ is a vector field, define the *divergence* of $\vec{A}$ as follows:

$$\vec{\nabla} \cdot \vec{A} \equiv \frac{\partial A_x}{\partial x} + \frac{\partial A_y}{\partial y} + \frac{\partial A_z}{\partial z} = \sum_{i=1}^{3} \frac{\partial A_i}{\partial x_i}.$$

We also define the *rotor* of $\vec{A}$ as

$$\vec{\nabla} \times \vec{A} \equiv \begin{vmatrix} \hat{x} & \hat{y} & \hat{z} \\ \frac{\partial}{\partial x} & \frac{\partial}{\partial y} & \frac{\partial}{\partial z} \\ A_x & A_y & A_z \end{vmatrix} = \hat{x}\left(\frac{\partial A_z}{\partial y} - \frac{\partial A_y}{\partial z}\right) + \hat{y}\left(\frac{\partial A_x}{\partial z} - \frac{\partial A_z}{\partial x}\right) + \hat{z}\left(\frac{\partial A_y}{\partial x} - \frac{\partial A_x}{\partial y}\right).$$

Let be $\Phi(x, y, z)$ a scalar field, then the vector

$$\vec{\nabla}\Phi(x, y, z) \equiv \left(\frac{\partial \Phi}{\partial x}, \frac{\partial \Phi}{\partial y}, \frac{\partial \Phi}{\partial z}\right)$$

is called the *gradient* of $\Phi(x, y, z)$. Later we will often use the notation $(\frac{\partial}{\partial x} = \partial_x)$ and so on for the other variables. Let $F = F(x, y, z, t)$ be a real function of the variables x, y, z and t:

$$\begin{cases} x = x(x', y', z', t') \\ y = y(x', y', z', t') \\ z = z(x', y', z', t') \\ t = t'(x', y', z', t') \end{cases}.$$

In this way, $F(x, y, z, t) = F(x(x', y', z', t'), y(x', y', z', t'), z(x', y', z', t'), t(x', y', z', t'))$, and

$$\begin{cases} \partial_x F = \partial_x(x')\partial_{x'}(F) + \partial_x(y')\partial_{y'}(F) + \partial_x(z')\partial_{z'}(F) + \partial_x(t')\partial_{t'}(F) \\ \partial_y F = \partial_y(x')\partial_{x'}(F) + \partial_y(y')\partial_{y'}(F) + \partial_y(z')\partial_{z'}(F) + \partial_y(t')\partial_{t'}(F) \\ \partial_z F = \partial_z(x')\partial_{x'}(F) + \partial_z(y')\partial_{y'}(F) + \partial_z(z')\partial_{z'}(F) + \partial_z(t')\partial_{t'}(F) \\ \partial_t F = \partial_t(x')\partial_{x'}(F) + \partial_t(y')\partial_{y'}(F) + \partial_t(z')\partial_{z'}(F) + \partial_t(t')\partial_{t'}(F) \end{cases}. \tag{A.2}$$

From equation (A.2) we can deduce that the representation of the partial derivation operator in terms of the accented variables is the following:

$$\begin{cases} \dfrac{\partial}{\partial x} = \dfrac{\partial x'}{\partial x}\dfrac{\partial}{\partial x'} + \dfrac{\partial y'}{\partial x}\dfrac{\partial}{\partial y'} + \dfrac{\partial z'}{\partial x}\dfrac{\partial}{\partial z'} + \dfrac{\partial t'}{\partial x}\dfrac{\partial}{\partial t'} \\ \dfrac{\partial}{\partial y} = \dfrac{\partial x'}{\partial y}\dfrac{\partial}{\partial x'} + \dfrac{\partial y'}{\partial y}\dfrac{\partial}{\partial y'} + \dfrac{\partial z'}{\partial y}\dfrac{\partial}{\partial z'} + \dfrac{\partial t'}{\partial y}\dfrac{\partial}{\partial t'} \\ \dfrac{\partial}{\partial z} = \dfrac{\partial x'}{\partial z}\dfrac{\partial}{\partial x'} + \dfrac{\partial y'}{\partial z}\dfrac{\partial}{\partial y'} + \dfrac{\partial z'}{\partial z}\dfrac{\partial}{\partial z'} + \dfrac{\partial t'}{\partial z}\dfrac{\partial}{\partial t'} \\ \dfrac{\partial}{\partial t} = \dfrac{\partial x'}{\partial t}\dfrac{\partial}{\partial x'} + \dfrac{\partial y'}{\partial t}\dfrac{\partial}{\partial y'} + \dfrac{\partial z'}{\partial t}\dfrac{\partial}{\partial z'} + \dfrac{\partial t'}{\partial t}\dfrac{\partial}{\partial t'} \end{cases}.$$

In our specific case equations (3.41) are the Lorentz transformations, such that

$$\begin{cases} \dfrac{\partial}{\partial x} = \gamma\dfrac{\partial}{\partial x'} - \gamma\dfrac{\beta}{c}\dfrac{\partial}{\partial t'} \\ \dfrac{\partial}{\partial y} = \dfrac{\partial}{\partial y'} \\ \dfrac{\partial}{\partial z} = \dfrac{\partial}{\partial z'} \\ \dfrac{\partial}{\partial t} = -\gamma v\dfrac{\partial}{\partial x'} + \gamma\dfrac{\partial}{\partial t'} \end{cases}. \tag{A.3}$$

We can rewrite equation (A.3) in matrix form

$$\begin{pmatrix} \partial_t \\ \partial_x \\ \partial_y \\ \partial_z \end{pmatrix} = \begin{pmatrix} \gamma & -v\gamma & 0 & 0 \\ -\beta\gamma/c & \gamma & 0 & 0 \\ 0 & 0 & 1 & 0 \\ 0 & 0 & 0 & 1 \end{pmatrix}\begin{pmatrix} \partial_{t'} \\ \partial_{x'} \\ \partial_{y'} \\ \partial_{z'} \end{pmatrix}.$$

Invariance of Maxwell's equations under Lorentz transformations

We would like to show that the set of Maxwell's equations (A.1) preserves its form under Lorentz transformations; i.e. we would like to prove that in S' Maxwell's equations can be written as follows:

$$\begin{cases} \vec{\nabla}' \cdot \vec{E}' = \rho'/\varepsilon_0 \\ \vec{\nabla}' \cdot \vec{B}' = 0 \\ \vec{\nabla}' \times \vec{E}' = -\dfrac{\partial \vec{B}'}{\partial t'} \\ \vec{\nabla}' \times \vec{B}' = \mu_0\left(\vec{j}' + \varepsilon_0\dfrac{\partial \vec{E}'}{\partial t'}\right) \end{cases}.$$

We first prove the invariance of the two Maxwell equations that do not contain field sources (charges and currents): $\vec{\nabla} \cdot \vec{B} = 0$ and $\vec{\nabla} \times \vec{E} = -\frac{\partial \vec{B}}{\partial t}$. Writing these equations explicitly we have

$$\begin{cases} \partial_x B_x + \partial_y B_y + \partial_z B_z = 0 \\ \partial_y E_z - \partial_z E_y + \partial_t B_x = 0 \\ \partial_z E_x - \partial_x E_z + \partial_t B_y = 0 \\ \partial_x E_y - \partial_y E_x + \partial_t B_z = 0 \end{cases}. \tag{A.4}$$

Let us develop them individually. Because of equation (A.3) and of the field transformation rules in table 3.1 the first becomes

$$\begin{aligned} &\gamma\left(\partial_{x'} - \frac{\beta}{c}\partial_{t'}\right)B'_x + \gamma\partial_{y'}\left(B'_y - \frac{\beta}{c}E'_z\right) + \gamma\partial_{z'}\left(B'_z + \frac{\beta}{c}E'_y\right) = \\ &\nabla' \cdot B' - \frac{\beta}{c}((\nabla' \times E')_{x'} + \partial_{t'}B'_{x'}) = 0. \end{aligned} \tag{A.5}$$

Let us now transform the second equation in (A.4). By means of equation (A.3) and the field transformation rules in table 3.1, we have

$$\begin{aligned} &\gamma\left(\partial_{y'}\left(E'_z - vB'_y\right)\right) - \left(\partial_{z'}\left(E'_y + vB'_z\right)\right) + \gamma(\partial_{t'} - v\partial_{x'}B'_x) \\ &= -v\nabla' \cdot B' + (\nabla' \times E')x' + \partial_{t'}B'_{x'} = 0. \end{aligned} \tag{A.6}$$

By transforming the third equation in (A.4), we obtain

$$\begin{aligned} &(\partial_{z'}(E'_x)) - \gamma^2\left(\partial_{x'} - \frac{\beta}{c}\partial_{t'}\right)(E'_z - vB'_y) + \gamma^2(\partial_{t'} - v\partial_{x'})\left(B'_y - \frac{\beta}{c}E'_z\right) \\ &= \partial_{z'}(E'_x) - \partial_{x'}(E'_z) + \partial_{t'}(B'_y) = (\nabla' \times E')y' + \partial_{t'}B'_y = 0. \end{aligned} \tag{A.7}$$

Finally we transform the last equation of (A.4):

$$\begin{aligned} &\gamma^2\left(\partial_{x'} - \frac{\beta}{c}\partial_{t'}\right)\left(E'_y + vB'_z\right) - \left(\partial_{y'}\left(E'_y + \gamma^2\right)\right)(\partial_{t'} - v\partial_{x'})\left(B'_z + \frac{\beta}{c}E'_x\right) \\ &= (\nabla' \times E')_{z'} + \partial_{t'}B'_z = 0. \end{aligned} \tag{A.8}$$

We can note that the components y and z in equations (A.7) and (A.8) directly satisfy the two Maxwell's equations we have examined. In order to obtain a complete proof of the invariance of these equations, the component x must also satisfy the rotor equation and also the solenoidality of $\vec{B}'$. In this regard we rewrite equations (A.5) and (A.6) as follows:

$$\begin{cases} \nabla' \cdot B' - \frac{\beta}{c}((\nabla' \times E')_{x'} + \partial_{t'}B'_{x'}) = 0 \\ -v\nabla' \cdot B' + (\nabla' \times E')x' + \partial_{t'}B'_{x'} = 0 \end{cases}. \tag{A.9}$$

The homogeneous system (A.9), in the unknowns $\nabla' \cdot B'$ and $(\nabla' \times E')_{x'} + \partial_{t'}B'_{x'}$, admits null solutions if and only if the determinant of the coefficients is not null.

In this case the determinant of the coefficients is $(1 - \frac{v^2}{c^2})$, i.e. always different from zero.

It is therefore proved that

$$\begin{cases} \nabla' \times E'_{x'} + \partial_{t'}B'_{x'} = 0 \\ \nabla' \cdot \vec{B}' = 0 \end{cases}. \tag{A.10}$$

Based on equations (A.7) and (A.10) we complete the demonstration of the invariance:

$$\vec{\nabla}' \cdot \vec{B}' = 0$$

and

$$\vec{\nabla}' \times \vec{E}' + \partial_{t'}\vec{B}' = 0.$$

With the procedure used previously, the invariance of the other two Maxwell equations in the absence of charges and currents can be demonstrated. To demonstrate the invariance of these equations in the presence of charges and current it must be considered that one can define a four-vector current j^μ that transforms according to the laws of equation (3.6).

References

[1] Noce C 2018 *Introduzione alla Fisica Moderna* (Rome: Aracne)
[2] Resnick R 1968 *Introduction to Special Relativity* (New York: Wiley)
[3] Jackson J D 1998 *Classical Electrodynamics* (New York: Wiley)
[4] Rindler W 1991 *Introduction to Special Relativity* (Oxford: Oxford Science)

IOP Publishing

Modern Physics
A critical approach
Canio Noce

Chapter 4

What happens to light when it passes through a prism? The early history of spectroscopy

Francesco Avitabile and Angela Nigro

At the end of the nineteenth century, many discoveries, both experimental and theoretical, showed how classical physics was inadequate for understanding the nature of radiation and matter and for the analysis of phenomena on an atomic scale. Indeed, surprising phenomena were discovered when studying the radiation coming from matter. The study of the internal structure of atoms, molecules and solids through radiation is called spectroscopy and it has been crucial in the development of the most fundamental theories in physics, including quantum mechanics, the special and general theories of relativity, and quantum electrodynamics. The type of spectroscopy that studies the visible part of the spectrum of electromagnetic radiation coming from matter is called optical spectroscopy. The measuring instrument for observing the visible spectrum is the prism spectrometer, which uses the property of prisms to divide white light into its fundamental components. In this chapter, after a historical introduction to optical spectroscopy and after illustrating the theoretical foundations describing the interaction between radiation and matter [1–4], an experimental activity will be described based on the use of a prism spectrometer, commonly used for teaching, in secondary schools.

4.1 Spectroscopy

Spectroscopy is the study of the absorption and emission of light by matter as a function of radiation frequency (or energy, or wavelength). The type of spectroscopy that studies visible light is called optical spectroscopy. However, spectroscopy is a method of investigation that is not limited only to the visible field of the radiation spectrum. In fact, there are other absorption and emission mechanisms that allow us to study the entire spectrum. In particular, below the frequencies of light (the infrared and radio frequencies), spectroscopy exploits the physical phenomena related to vibrations and deformations of atomic and molecular bonds, while above

doi:10.1088/978-0-7503-2678-0ch4

the frequencies of light (x-rays), it studies the electronic transitions from the innermost atomic orbitals and the energies involved are very high. We can find spectroscopy in several technical fields of science and technology. For example, the medical technique called magnetic resonance imaging (MRI), used to analyse the internal soft tissues of the body, is based on radio-frequency radiation spectroscopy. High-frequency radiation produced by high-energy particle accelerators are used in spectroscopic techniques to study the internal structure of the proton and neutron. Optical, radio and x-ray spectroscopy are used to determine the chemical composition of matter both in routine laboratory analysis and in determining the constituents of distant stars. The frequencies of the electromagnetic spectrum range from the radio wave frequencies of about 10^5 Hz to the gamma wave frequencies greater than 10^{19} Hz. As we shall see in section 4.1.4, the fundamental mechanism of interaction between light and matter links the energy exchanges between light and matter to the frequency of radiation ν through the relation $E = h\nu$, where $h = 6.62\ldots\times10^{-34}$ J s is the Planck constant. The absorption or emission of an electromagnetic wave by an atom is due to the transition of an electron from one energy level to another.

4.1.1 The origin and development of optical spectroscopy

Spectroscopy has always been available for human activity. In fact, the simple distinction of colours is a rough analysis of visible light. A typical human eye will respond to wavelengths from 390–700 nm (see figure 4.1). In terms of frequency this corresponds to the range of 430–790 THz.

However, despite natural phenomena like rainbows having always been known by humanity, and the fact that the Romans were already aware that rainbow-like phenomena could be generated by a prism, until the seventeenth century light was

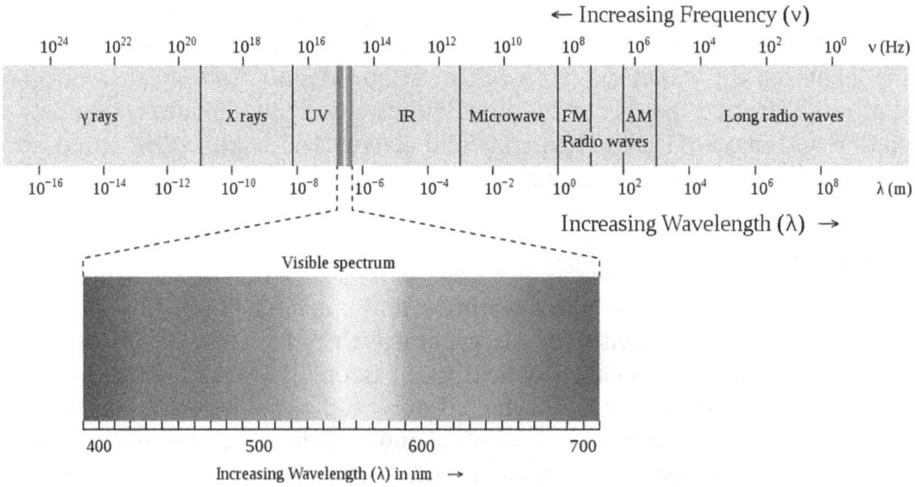

Figure 4.1. The spectrum of electromagnetic waves with frequencies and wavelengths [5]. Author: Philip Ronan

considered colourless and the colours were considered to be emanations of coloured bodies, or mixtures of white and black.

The first person to provide a modern interpretation of the behaviour of light through methodical study was the great physicist and astronomer Isaac Newton (1642–1727). In his work, titled *Opticks*, he argues that light is made up of corpuscles that move in a straight line and explains the phenomena of refraction, reflection and dispersion. In particular, refraction is the basis of spectroscopic observations because it allows the decomposition of light into its fundamental coloured components by means of a prism (as we can see in figure 4.2). The sources of white light used by Newton included flames and stars, and the Sun. However, the corpuscular theory hypothesized by Newton was inadequate to explain some phenomena, such as interference and diffraction.

At the end of the seventeenth century, Christiaan Huygens (1629–1695) introduced a wave-like model of light that was developed by Auguste-Jean Fresnel (1788–1827) and Thomas Young (1773–1829) in the following century. In 1802, Young carried out the first measurement of the wavelength of light. This wave model of light, analogous to sound which does not propagate in a vacuum, was thought to require a medium, called 'ether', which was hard, elastic, undetectable, and capable of supporting and transmitting light waves. Thus, the two elementary theories, the corpuscular and the wave-like models, were incompatible, however, both explained physical phenomena related to the behaviour of light.

In the early 19th century, thanks to improvements in the manufacture of spectroscopes (see figure 4.3), dark lines were observed in the solar spectrum for the first-time by William Hyde Wollaston (1766–1828) and later by Joseph von Fraunhofer (1787–1826). However, they were unable to explain the causes of these absorption lines. An explanation of the dark lines was provided by Gustav Robert Kirchhoff (1824–1887), who attributed the cause to the selective absorption of light by the elements of the solar atmosphere. Kirchhoff, together with his fellow chemist Robert Bunsen (1811–1899), systematically studied the nature of light generated by

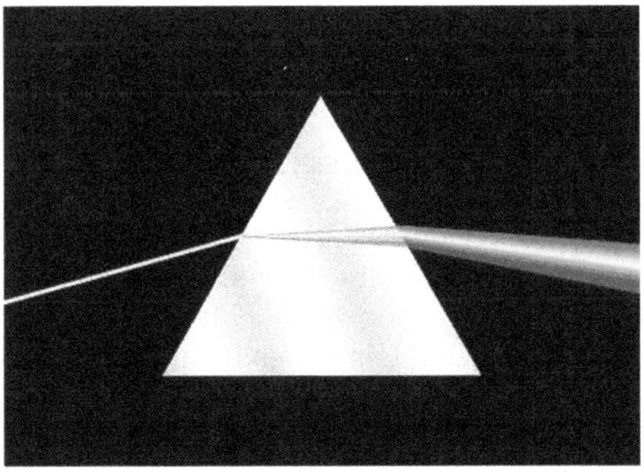

Figure 4.2. Decomposition of white light by a prism into its constituent colours. Creative Commons license CC SA 1.0 [6].

Figure 4.3. Kirchhoff's first spectroscope. Source: Historisk Fysik bind I - Author: Poul la Cour & Jacob Appel [7].

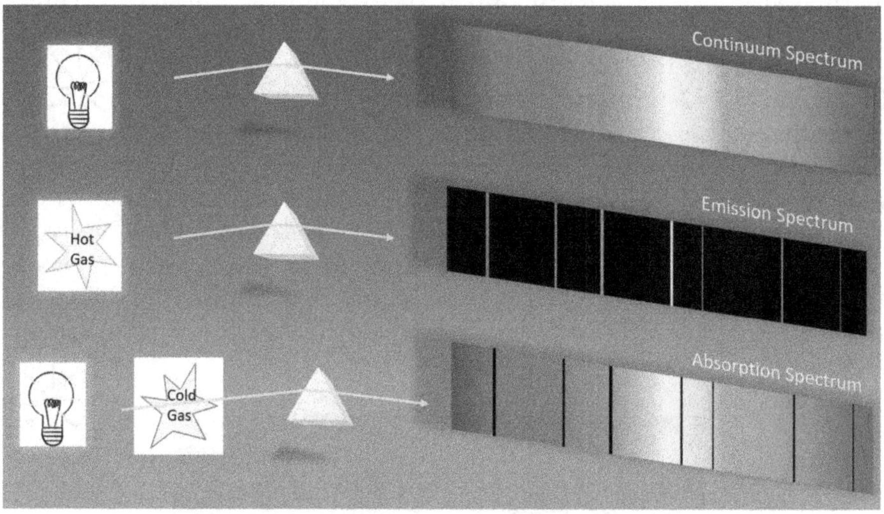

Figure 4.4. Summary sketch of Kirchhoff's laws.

sources in different physical conditions. This study, carried out thanks to improvements in spectroscopic devices, led to the formulation of the famous empirical laws on emission and absorption (see figure 4.4):

1. An incandescent solid, liquid or gas under high pressure emits a continuous spectrum.
2. A hot gas under low pressure emits a 'bright-line' or emission-line spectrum.
3. A continuous spectrum source viewed through a cool, low-density gas produces an absorption-line spectrum.

Then, the emission spectra of materials heated by flames or by electric discharges were studied. By studying various pure elements, Kirchhoff noticed that they had a characteristic spectrum of absorption and emission lines, and thus he gave rise to the technique for the spectroscopic analysis of the chemical compositions of substances.

In 1859, James Clerk Maxwell (1831–1879), elaborated the first modern theory of electromagnetism, unifying the previous knowledge of the so-called Maxwell equations, where light is explained in terms of the mutual transformation of oscillating electric fields in oscillating magnetic fields, and vice versa. The famous experiment of Albert Michelson (1852–1931) and Edward W Morley (1838–1923) in 1887 undermined the foundations of Galilean relativity and the concept of ether in which electromagnetic waves were thought to have propagated. The results of this experiment were later (1905) explained by Albert Einstein (1879–1955), with his Theory of Special Relativity and the revolutionary hypothesis according to which the speed of light is a universal constant for all inertial reference systems.

In 1890, Max Planck (1858–1947) created a model that explains the mechanisms of absorption and emission of radiation, overcoming the contradictions between experiments and the theory of radiation emitted by a black body. A black body is an object capable of absorbing all the electromagnetic radiation that affects it. The absorbed radiation heats the body which thus emits electromagnetic radiation whose energy depends only on its temperature (see figure 4.5). According to the equations of Maxwell, each body, at any temperature, would be able to emit an infinite amount of energy in the form of radiation for wavelengths λ close to zero (the ultraviolet catastrophe). The experimental data, however, refuted such a theoretical result, showing that for $\lambda \longrightarrow 0$ the emitted power was cancelled. The solution to the problem of the ultraviolet catastrophe, thanks to Planck (which he considered a mere mathematical convenience) provided that within a black body the electromagnetic radiation could be absorbed and emitted only in discrete quantities, called

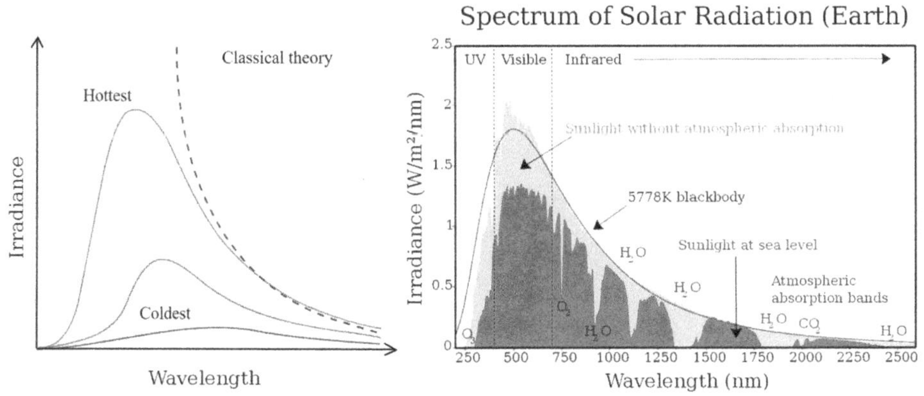

Figure 4.5. Left: a comparison between the experimental spectra of emission of a black body at different temperatures and the theoretical curve calculated according to classical theories. Source: helix84 [8]. Right: superposition of the experimental data obtained for the emission spectrum of the Sun (yellow zone) and of the curve obtained through the Planck model for a black body at temperature $T = 5778$ K (full dark grey line) [9].

quanta. Subsequently, Einstein applied the Planck model to light, demonstrating, in the photoelectric effect, that it is quantized in particles of light, called photons.

Finally, in 1927, Louis de Broglie (1892–1987) put an end to the debate on the nature of light by associating photons, like any other particle, with a wave-like as well as a corpuscular nature that reveals itself according to the circumstances. Practically, they are vibrations of the electromagnetic field and propagate like waves. However, each wave carries an energy packet depending on the wave frequency and can be seen as corpuscular.

4.1.2 Refraction and dispersion

As is well known, a beam of light hitting the surface of a transparent material is partly reflected and partly refracted. Generally, the refracted radius undergoes a deviation with respect to the direction of propagation of the incident ray. Specifically, in the passage from a vacuum to materials such as air or water, the beam of transmitted light forms a smaller angle with the surface normal than the incident beam. In contrast, if it passes into a vacuum from a medium like air or water, the transmitted beam forms a greater angle. This phenomenon is called refraction.

Refraction is described by the wave model of light in terms of the refractive index. The refractive index n is the ratio

$$n = \frac{c}{v},$$

where c is the speed of light in a vacuum and v is the velocity in the considered medium. More precisely, two laws apply to refraction:

1. The incident ray and the refracted ray have the same frequency. Moreover, they belong to the same plane together with the perpendicular line to the surface passing through the incidence point.
2. If n_1 and n_2 are the refractive indices of the two media and θ_1 and θ_2 are the incidence and refraction angles, respectively, we have the following relation, called Snell's law:

$$\frac{\sin\theta_1}{\sin\theta_2} = \frac{n_2}{n_1}.$$

The ratio n_2/n_1 is called the relative refraction index n_{21} of the second medium with respect to the first. From Snell's law we see that if $n_2 > n_1$ ($n_{21} > 1$), i.e. if the incident wave comes from a less refractive medium, then $\theta_1 > \theta_2$ and the refracted ray approaches the normal to the interface plane. The other way around, if $n_2 < n_1$ ($n_{21} < 1$), i.e. if the incident wave comes from a more refractive medium, then $\theta_1 < \theta_2$ and the refracted ray moves away from the normal to the interface plane. In this second case, if the angle θ_1 is such that $\sin\theta_1 = n_{21}$, we have that $\sin\theta_2 = 1$ and therefore $\theta_2 = \frac{\pi}{2}$. In this case θ_1 is the limit angle beyond which the phenomenon of refraction would no longer be present and total reflection would occur.

The first law, moreover, imposes the equality of the frequency of the incident wave and of the refracted wave, therefore we have

$$f_1 = \frac{v_1}{\lambda_1}$$
$$f_2 = \frac{v_2}{\lambda_2}$$

from which

$$\frac{v_1}{\lambda_1} = \frac{v_2}{\lambda_2}$$

and therefore, if the first medium is the vacuum, $n_1 = 1$, $\lambda_1 = \lambda_0$ (the wavelength in a vacuum) and the refractive index for any medium becomes

$$n = \frac{\lambda_0}{\lambda}, \tag{4.1}$$

which means that shorter the wavelengths, the greater is the deviation from θ_1 (see also figure 4.6).

However, the dependence of n on the wavelength expressed by equation (4.1) applies only if the medium is perfectly transparent and absorption by the medium is neglected. Indeed, it is never possible to obtain a perfectly transparent medium and if considering absorption the expression of $n(\lambda)$ becomes much more complicated (see figure 4.7). Limited to the visible spectrum, it can be approximated well by the empirical expression of Cauchy:

$$n = A + \frac{B}{\lambda^2}.$$

The coefficients A and B change for each material, for example for dense flint glass $A = 1.728\,0$ and $B = 1.342 \times 10^{-8}\ \text{m}^2$.

In general, the light that we usually see is a mixture of waves that have different wavelengths. For example, sunlight, which is white, has all the wavelengths of the visible spectrum. Each of the colours of the visible spectrum is composed of waves of

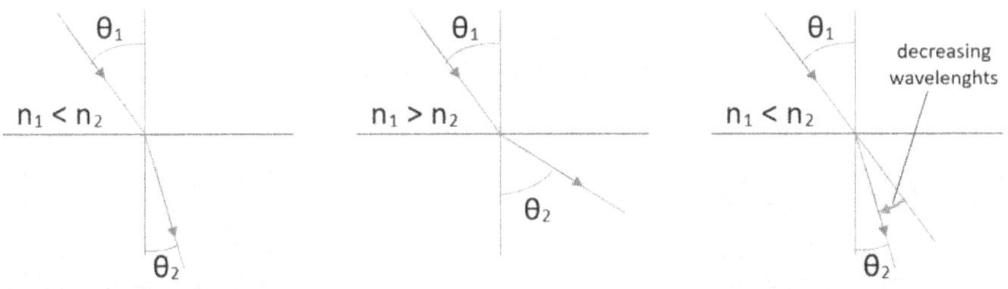

Figure 4.6. In the left and middle panels we can see how an incident wave is refracted at an interface between two media with different refractive indices. The right panel illustrates how the refraction angle varies as wavelength varies.

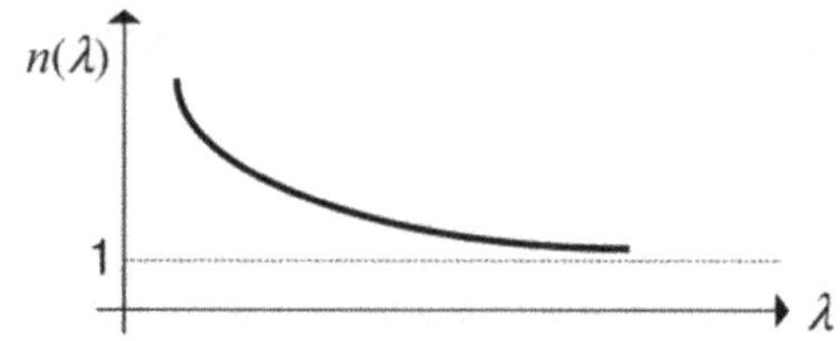

Figure 4.7. The qualitative trend in the absence of absorption of the refractive index as a function of the wavelength. Creative Commons license CC BY-SA 3.0 [10]. Source: Claudio Oleari. Schede di OTTICA GEOMETRICA.

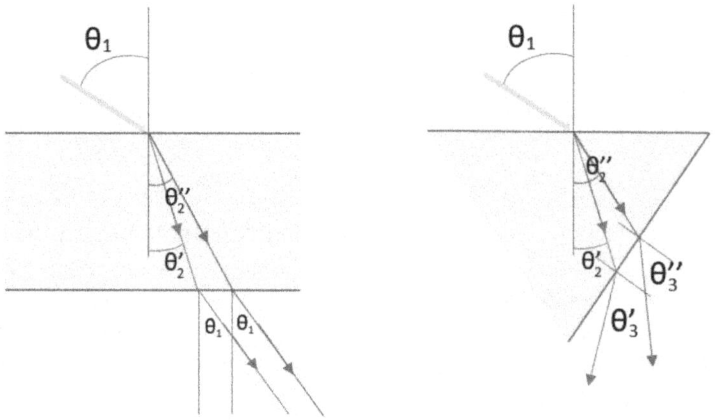

Figure 4.8. Comparison of the white light radiation refracted from a medium to parallel faces (left) and from a prism (right).

the same wavelength, therefore, the radiation is called monochromatic. If non-monochromatic radiation hits the interface between two media with a certain angle, as the refractive index depends on the wavelength of the incident wave, each component of the radiation is deviated with a different refraction angle. This phenomenon is called dispersion (see figure 4.8).

To measure the refraction angle, the refracted ray needs to emerge again from the medium towards the air. In the case of a second surface parallel to the first, the emerging rays will have the same direction as the incident rays and the radiation will be reconstructed. If the second interface surface is not parallel to the first, as in a prism, each wave is refracted at a different angle from the others and, in the passage from the prism to the less refractive air, is further deviated with respect to the other components such that the waves with different wavelengths emerge as more divided.

4.1.3 The hydrogen atom spectrum

The spectra of the simplest atoms played a key role in the development of a fundamentally new physical theory. Thanks to the increased accuracy in the measurement of spectra, in 1885 the Swiss physicist Johann Balmer (1825–1898) empirically discovered that the numerical relation that provided the frequencies corresponding to the lines of the visible spectrum of monoatomic hydrogen was

$$f = k\left(\frac{1}{4} - \frac{1}{n^2}\right),$$

where k is a proportionality constant and n is a positive integer number greater than 2. Varying n, all lines of the so-called Balmer spectral series are obtained.

With the discovery of new spectroscopic techniques, in the early 1900s it was possible to observe other spectral series in the ultraviolet and infrared spectra. Johannes Rydberg (1854–1919) subsequently restated and expanded Balmer's result in the Rydberg equation:

$$f = cR_{\mathrm{H}}\left(\frac{1}{m^2} - \frac{1}{n^2}\right), \tag{4.2}$$

where $R_{\mathrm{H}} = 1.097 \times 10^7\ \mathrm{m}^{-1}$ is a proportionality constant, called the Rydberg constant. The terms m and n are two positive integers, with $n > m$. The Balmer series is obtained for $m = 2$. The frequencies belonging to the ultraviolet are obtained for $m = 1$ and the obtained spectral series is known as the Lyman series. For $m = 3$ and $m = 4$ we obtain the Paschen series and the Brackett series, respectively, both in the infrared spectrum. Two other series have specific names—the Pfund series and the Humphreys series with $m = 5$ and $m = 6$, respectively.

At the time of these observations it was not clear why every chemical element should have a characteristic emission spectrum. Subsequent discoveries in the field of the structure of matter made the reasons for these results clear.

4.1.4 Atomic theory

By the end of the nineteenth century it had been established that matter was made up of atoms because this model could explain a large amount of experimental data in the field of chemistry and thermodynamics. With the discovery of cathode rays and related studies, Joseph John Thomson (1856–1940) formulated the so-called plum pudding model which posited that the atom is constituted by a positively charged sphere in which negative electrons are immersed (see figure 4.9).

In 1911, Ernest Rutherford (1871–1973) proposed a new theory of the deflections experienced by alpha particles passing through matter. According to this theory, the greatest part of the mass and the totality of the electric charge are not distributed in a region of the size of the atom but are confined in its centre in a very small region. The extension of this region was evaluated to be around 10^{-12} cm, while the negative charge is distributed on a sphere of radius comparable to the atomic radius 10^{-8} cm.

However, Rutherford's model presented theoretical difficulties and led to predictions that were in contrast to some important experimental results. First, according to classical electrodynamics, a particle in accelerated motion emits electromagnetic waves. As a result, the electrons in orbit around the nucleus have non-zero acceleration, so they should progressively lose energy until they fall onto the nucleus. This instability is contradicted by experience. Moreover, as it approaches the nucleus, the frequency of radiation emitted by an electron, which for electrodynamics is equal to the frequency of revolution, should grow

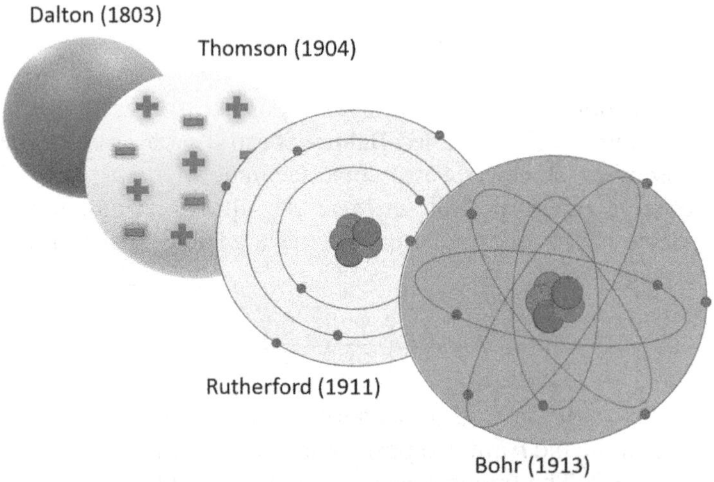

Figure 4.9. Evolution of the various atomic models. From left to right: Dalton atomic model, where the atoms of an element are all equal, of equal mass and not divisible; Thomson's plum pudding model, where negative charges are inserted in a diffused positive charge distribution; the Rutherford model, where the electrons gravitate as planets around the Sun and all orbits are possible; and, finally, the Bohr model, in which electrons can only occupy quantized energy levels.

continuously and therefore we should have a continuous frequency spectrum, but this does not occur. For these reasons, the physicist Niels Bohr (1885–1962) understood that the planetary atomic model proposed by Rutherford could not be correct.

To explain the stability of atoms, Bohr introduced the following postulates:

1. Atoms can be in a number of states in which there is no irradiation which are called stationary states. Therefore, for the electrons of the atom there exist discrete favoured orbits and consequently the energy values corresponding to them are discrete.
2. Each emission or absorption of radiation corresponds to a transition between two stationary states. According to the Planck hypothesis, during this process there is the emission or absorption of a single quantum. If the transition occurs between two energy orbits E_n and $E_{n'}$ (with $E_n > E_{n'}$) and if f is the frequency of the emitted (or absorbed) radiation, for the principle of conservation of energy we have the Planck relation

$$E_n - E_{n'} = hf,$$

   with h the Planck constant.
3. The dynamic equilibrium of the system is determined by classical mechanics which, however, cannot be used in the transition between two states.
4. The possible different stationary states of a system formed by an electron that rotates around a positive nucleus are those for which the orbits are circular and determined by the relation:

$$l = n\hbar,$$

where l is the electron angular momentum, $\hbar = h/2\pi$ is the reduced Planck constant, and n is an integer varying from 1 to ∞ and is called the quantum number. Therefore, the angular momentum is quantized.

According to the fourth postulate, the electron rotating around the nucleus has angular momentum:

$$m_e vr = n\hbar,$$

where m_e and v are the electron mass and velocity, respectively, while r is the orbital radius. Moreover, by third postulate we have the equality between the centripetal force and the Coulomb force

$$\frac{m_e v^2}{r} = \frac{k_0 Z e^2}{r^2},$$

where Z is the atomic number (the number of protons or electrons) and $k_0 = 8.99 \times 10^9$ N m^2 C^{-2} is the Coulomb constant. The last two equations provide the following quantities:

$$r_n = \frac{n^2\hbar^2}{k_0 m_e e^2 Z} \tag{4.3}$$

$$v_n = \frac{Ze^2 k_0}{n\hbar}, \tag{4.4}$$

where r_n and v_n represent the radius r and velocity v for a fixed steady state n, respectively.

Assuming $Z = 1$ and $n = 1$, which represents the hydrogen atom in the configuration in which the electron is closest to the nucleus, we obtain the fundamental radius of the hydrogen atom, i.e. the Bohr radius, which is

$$a_0 = \frac{\hbar^2}{k_0 m_e e^2} = 0.529 \times 10^{-10}\ \text{m}.$$

We can express equation (4.3) in terms of a:

$$r_n = \frac{n^2}{Z} a_0.$$

The electron energy is

$$E = T + U = \frac{1}{2} m_e v^2 - \frac{k_0 Z e^2}{r} + \text{constant}.$$

The constant is zero if the zero-energy state of the electron is chosen at an infinite distance from the nucleus and with zero velocity. Replacing the values r_n and v_n one has

$$E_n = -\frac{m_e Z^2 e^4 k_0^2}{2n^2 \hbar^2}. \tag{4.5}$$

The fundamental state energy (obtained for $n = 1$ and $Z = 1$) is

$$E_0 = \frac{m_e e^4 k_0^2}{2\hbar^2} = 13.6\,\text{eV}.$$

Then, the energy given by equation (4.5) becomes

$$E_n = -\frac{Z^2}{n^2} E_0.$$

According to Bohr's second postulate, a hydrogen atom radiates energy if an electron jumps from a more external stationary state to an inner one of lower energy. The difference between the two energies is emitted as a single quantum of electromagnetic radiation whose frequency is given by

$$f = \frac{E_n - E_{n'}}{h}, \tag{4.6}$$

where, if we insert the values of E_n and $E_{n'}$ given by equation (4.6), we obtain

$$f = \frac{2\pi^2 m_e e^4 k_0^2}{h^3}\left(\frac{1}{n^2} - \frac{1}{n'^2}\right).$$

The last expression has exactly the same form as the Rydberg experimental law expressed by equation (4.2) if we set

$$R_H = \frac{2\pi^2 m_e e^4 k_0^2}{ch^3}.$$

This theoretical value for the Rydberg constant is in perfect agreement with the experimental value and this determined the success of Bohr's theory.

The graphical representation of spectral terms and spectroscopic lines is obtained with a Grotrian diagram of energy levels (see figure 4.10). In this diagram, the energy values are shown on the ordinates. Moving upwards, the distance between the levels decreases, up to representing a continuous spectrum in the upper part of the diagram. At the limit value $n \to \infty$ corresponds to $E = 0$, i.e. the electron is separated from the nucleus. Energy levels are represented by horizontal lines while vertical lines are transitions from one level to another.

4.1.5 Optical spectroscopy analysis

Spectroscopy is a very powerful investigation tool because it can analyse the chemical and physical characteristics of a gas even at a great distance. For this reason it is also very important for the study of stars and galaxies. The position of the spectral lines becomes the signature of the analysed chemical elements. However, chemical composition is not the only thing that can be investigated. Indeed, from the characteristics of a spectrum we can obtain the following information:

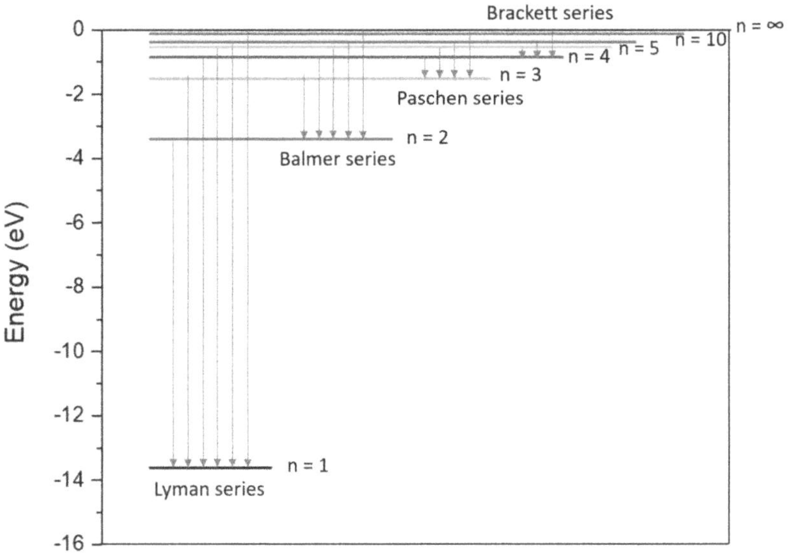

Figure 4.10. Grotrian diagram of the hydrogen atom energy levels.

- *Source temperature.* By identifying the maximum radiation spectrum distribution, the temperature of the source can be determined.
- *Quantitative analysis of the source chemical components.* This investigation can be performed through the intensity of the lines that characterize how much each element is present in the mixture of analysed gas.
- *The speed of the source.* The Doppler effect causes the movement of the spectral lines coming from the source towards red light (moving away) or blue (approaching). From the frequency shift the speed is obtained by using specific formulas [11].
- *The presence of magnetic fields.* In some cases, the spectral lines appear to be split into multiplets, this is due to the Zeeman effect that appears when the electrons are immersed in an external magnetic field. The larger the distance between the split lines, the greater the intensity of the magnetic field.
- *Density and interaction with other matter.* The lines are not perfectly monochromatic, they have a certain width that includes a range of wavelengths. Such an enlargement is due to several factors which are also related to the density, temperature and pressure of the gas.

4.2 Measuring the line spectra of inert gases and metal vapours using a prism spectrometer

4.2.1 General description of the experiment

If we excite inert gases or metal vapours using electromagnetic radiation, they emit spectral lines characteristic of the respective elements. The wavelengths of these spectral lines can give us information on the nature of the light source. As we saw in

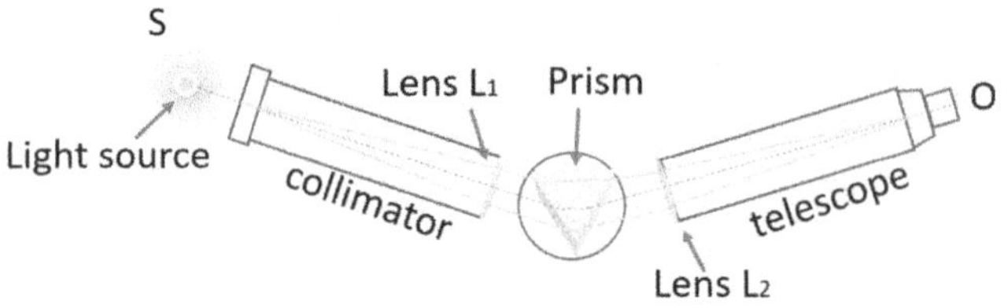

Figure 4.11. Sketch of the main parts of the prism spectrometer.

section 4.1.2, it is possible to use a prism to divide the various wavelengths of light, thanks to their dependence on the refractive index. Therefore, the measuring instrument used is a prism spectrometer.

With the help of figure 4.11, we can understand the operating principle of this apparatus. Starting from the left, we have the light source, consisting of a spectral lamp with helium or cadmium, inserted in a vertical cylinder and powered by a power supply. The emerging light propagates in all directions, however, a vertical slit adjustable in width and height allows only a limited beam to pass out. Then, the beam passes through an objective lens L_1 placed at a distance equal to its focal distance from the slit S. In this way, the system formed by the slit and the lens forms a collimator (see section 4.1.2). Behind the lens, the light falls on the prism P as a parallel beam, i.e. all rays are incident on the prism at the same angle. The flint glass prism refracts the light, and each wavelength is deviated at a different angle. Aligned with one of these emerging directions, there is the focal axis of a second objective lens L_2 which focuses all parallel rays of a specific wavelength to an image of slit S in the focal plane of the lens. In this way, the system formed by the L_2 lens and the ocular O forms an astronomical telescope. The deviation angle can be measured by a graduated circle and, thanks to a couple of magnifiers and the Vernier scale, it is therefore possible to measure the angle with the sensitivity of the minute of degrees. For each spectral line, the telescope angular position is marked and the corresponding wavelength is obtained by the comparison with the calibration curve. The calibration curve takes into account the non-linearity between the refractive index (therefore of the refraction angle) and the wavelength. However, many instruments are not calibrated and, therefore, the calibration is performed by a lamp for which the spectral lines are known and are also distributed over the whole spectrum. In this way it is then possible to trace the wavelengths of unidentified sources.

4.2.2 Carrying out the experiment

Step 1. Aligning the optical axis of the telescope, the collimator and the prism plane. From the first law of refraction we have that the incident ray, the refracted ray and the perpendicular to the surface between the two media at the point of incidence must be on the same plane. Therefore, the collimator, the prism and the telescope need to align on the same plane. This alignment is first done roughly freehand, then refined through precision screws.

Step 2. Calibrating the spectrometer using a He spectral lamp.
We perform the calibration through the wavelengths emitted by the helium lamp. Indeed, from the literature we know that the first spectral line meets for a wavelength of 706.5 nm (see all the spectrum lines observed in figure 4.12). Then we set our zero for

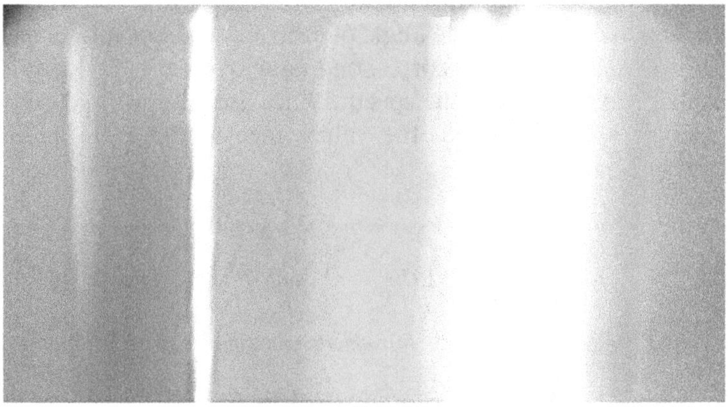

Figure 4.12. The spectral lines visible through the ocular lines (the lighter lines).

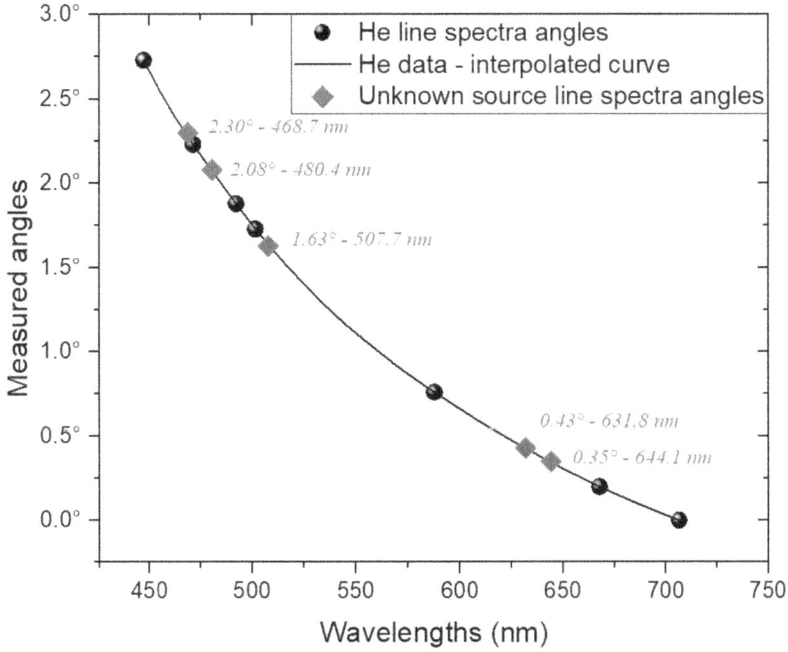

Figure 4.13. Calibration curve of the prism spectrometer. The black points indicate the measured values of α angles in correspondence with the He spectral lines, the continuous black line is the interpolated curve of these points, the red rhombuses indicate the unknown source measured points. The measured angles and the corresponding wavelengths for the unknown source line spectra are shown in red. From a comparison with the literature data we find that the unknown source is composed of cadmium. (Data from [12].)

the angles at this first value. Iterating this operation, the whole calibration is performed by noting the angles corresponding to the other spectral lines for which we know the wavelengths (see figure 4.13).

Plotting these data and interpolating them, we obtain the calibration curve.

Step 3. Measuring the line spectra of other light sources, e.g. spectral lamp Cd. After changing the helium lamp, the angle of each spectral line allows us to go back, using the calibration curve, to the corresponding wavelength. To identify unknown light sources, the wavelengths of the spectral lines are read off and compared with the literature values. As expected, the values measured for the unknown lamp correspond to cadmium.

References

[1] Ashcroft N W and Mermin N D 1976 *Solid State Physics* (Philadelphia, PA: Saunders College)

[2] Träbert E and Heckman P H 1986 *Introduction to Spectroscopy of Atoms* (Amsterdam: North Holland)

[3] Jackson J D 1998 *Classical Electrodynamics* (New York: Wiley)

[4] Bransden B H and Joachain C J 2003 *Physics of Atoms and Molecules* (London: Pearson Education)

[5] Ronan P 2008 EM spectrum *Wikimedia Commons* https://commons.wikimedia.org/wiki/File:EM_spectrum.svg

[6] Diagram of a dispersion prism *Wikimedia Commons* https://commons.wikimedia.org/wiki/File:Dispersion_prism.jpg

[7] la Cour P and Appel J 1896 Kirchhoff's first spectroscope *Wikimedia Commons* https://commons.wikimedia.org/wiki/File:Kirchhoffs_first_spectroscope.jpg

[8] Radiation of completely black body *Wikimedia Commons* helix84

[9] Solar spectrum Wikimedia Commons 2013 https://commons.wikimedia.org/wiki/File:Solar_spectrum_en.svg

[10] Oleari C Refraction index vs wavelength graph *Wikimedia Commons* https://commons.wikimedia.org/wiki/File:Refraction_index_vs_waveleght_graph.png

[11] Lambourne R 1997 The Doppler effect in astronomy *Phys. Educ.* **32** 34–40

[12] LD Physics Leaflets P5.7.1.1 - LD DIDACTIC Group https://www.ld-didactic.de/en/ld-didactic-group.html

Chapter 5

Electrical resistivity measurements reveal transport properties

Lazzaro Immediata and Carmine Attanasio

5.1 Introduction

In a metal, the atoms are arranged in a crystal-like configuration. The type of metal will determine how the bonds are arranged and how closely the atoms are grouped. Electrons can inhabit energy levels. Generally, only the 'outer' electrons in an atom interact to form bonds with other atoms. These outer electrons are held to the atom with a relatively small amount of energy. Normally, they inhabit an energy level we call the *valence* band. This is their *ground* state. The addition of energy can raise these electrons out of the valence band and into the *conduction* band. In the conduction band they are free to move about within the crystal structure. The application of an electric potential will influence them to move in a particular direction. In a metal the valence band is relatively close to the conduction band—that is, very little energy is necessary to cause electrons to jump from their valence state into the conduction band. In fact, we think of metals as having a large population of free electrons in the conduction band at all times and the application of an electric potential will cause them to move—a current flow. Thus metals generally have a relatively low (although not zero) resistance. In a material such as glass, there is a large energy gap between the valence and conduction bands. This means there are very few free electrons available for current flow, and it takes a large input of energy to raise any electrons into the conduction band.

Within a metal conductor, even though there are free electrons, there is still resistance to current flow. This can be described using simple models, but apparently only quantum electron theories accurately deal with the behaviour of metals under extreme conditions such as very low temperatures. Replacing the idea of electrons as particles with electrons as waves solves the problems of the simpler models. You can picture these electron waves oscillating through the metal lattice (which can also be pictured as a wave-like structure)—the interference of the lattice structure with the

doi:10.1088/978-0-7503-2678-0ch5

electrons causes resistance. This resistance is caused mainly by two things. One is impurities in the metal, which cause irregularities in the periodicity of the lattice. The other is the disturbance or *vibration* of the lattice caused by heat. Since some heat is always present (except at absolute zero) there is always some resistance from this source which prevents the electrons from moving freely through the metal.

Resistivity measurements are of paramount importance in the study of the electrical transport properties of both metallic and semiconducting materials. The electrical resistance of conducting materials is a physical quantity that is relatively easy to determine experimentally. It represents a very useful tool to study the physical mechanisms at play in most of the elements of the periodic table and it can shed light even on the structural properties of the investigated specimens. In this case, even though conventional diffraction analysis gives more precise information since resistivity measurements represent an average value of the scattering of the electrons in all directions, they are still of great importance because of their simplicity and availability. Resistivity measurements can immediately answer the question of whether you are dealing with a material that is a metal, a semiconductor (discriminating with the determination of the Hall coefficient if it is of n- or p-type) or a superconductor. Furthermore, such methods can provide useful knowledge about the crystalline nature of a conducting solid, its amorphousness, or the change of its properties when in the form of a thin film. In general, the determination of the resistivity value, temperature and magnetic field dependence are often the first tests of the most advanced theories of transport in solids. However, the concrete realization of an experimental set-up which returns reliable values for the resistivity is far from being trivial. As a rule, great care should be taken to perturb the specimen as little as possible using, for example, a low-level DC input. Completely different schemes should be used, for example, when measuring high- instead of low-resistance samples. This chapter, far from being an exhaustive treatment, will address briefly only the case of low-impedance measurements.

In the following, we will deal with methods to measure the resistivity and Hall coefficient of materials of different nature and shape. Resistivity and the concentrations of both charge carriers and impurities in materials are correlated and, for most materials, such correlations are well-known such that, in order to determine the above properties, it can be sufficient to know only one of the three values. In some cases, constraints are needed to obtain reliable results and, in general, it is more convenient to perform measurements on very small surfaces. Some of the methods which we address briefly here are only applicable in principle while only a few are effective in practice. In particular, the van der Pauw method, proposed in the late 1950s, is still valid for the measurement of the specific resistivity and the Hall coefficient of a flat sample of arbitrary shape.

5.2 General considerations

The methods that will be considered can be used for poly-crystalline or mono-crystalline materials and, in general, additional conditions or particular interpretations of the results are required. It is necessary to take into account, for example, the role played by the sample edge, in which inhomogeneity of the material can accumulate and anomalous behaviours may occur which could distort the measurement of resistivity. Generally, it is not always possible to predict whether the measured resistivity might be lower or greater than the true resistivity of the sample. Other difficulties arise when the sample has several phases with different resistivities or if, despite being present in a single phase, this phase is anisotropic. If we consider a multi-layer sample in which, for example, high and low resistivity can be alternated or if we have p–n anisotropic sections, the measurements made using currents parallel to the layers can be different from those made with currents perpendicular to the same layers. These samples can be analysed only with some of the methods that will be presented below. Surface effects in thin samples can induce resistivity measurements different from those obtained in bulk specimens. If we consider, for example, a layer of a given material and a probe system used for measurement, the probes may not pierce the layer but only measure the resistivity of the layer. If the probes cross the layer and come into contact with the underlying bulk, the layer and the bulk will give rise to a parallel circuit, with all that it can be achieved. Furthermore, some inaccurate measurements may occur due to abrasion of the sample surface. This is what happens, for example, to silicon and germanium samples with low resistivity—one might find a greater resistivity at room temperature due to micro-lesions present on the surface of the sample. If the abrasion is deep, approximately proportional to the diameter of the abrasive used, which is of the order of an appreciable fraction of the thickness of the sample, errors of various percentage points can arise. This problem also occurs when the surface is abraded intentionally on thin films to facilitate contact of the probes. In contrast, materials with extremely high resistivity may exhibit low resistivity measurements caused by effects resulting from defects due to accidentally induced damage. Moreover, resistivity is naturally influenced by the geometry of the sample and the conditions of the edges. As a consequence, a measurement of resistivity must be corrected for some empirical factors, which are usually tabulated in the literature. Most semiconductors require corrective resistivity factors at high temperatures. Therefore, if a precise measurement is required or if the environment has a significant influence on the measurements, these corrective factors must be taken into account. Since the correction factors can vary significantly in the presence of impurities, it is necessary to pay attention to the data obtained using these conditions.

5.3 Basic methods

5.3.1 The direct method

The oldest method still in use for the measurement of resistivity consists of using a rectangular sample of known size to measure its resistance R and, indirectly, the resistivity ρ by using the relationship

$$R = \frac{\rho L}{A},$$

where L is the length of the sample and A is the surface of the section. The disadvantage of this method is that R will also contain a contribution to its value due to the contacts, which can be appreciable for materials with very low resistivity [1].

5.3.2 The two-point probe method

In the case of semiconducting materials, if the cross section of the sample is relatively uniform, the contribution of the resistance of the contact points of the probes can be eliminated by using a two-point probe, as shown in figure 5.1. When using this method it is necessary that the current is kept low enough to prevent the sample from heating. Moreover, the voltmeter must have a high input impedance and the measurements must be made at points sufficiently far from the contacts to prevent any injected carrier minority from recombining. The requirements for contacts are not very restrictive. However, if the contact is very poor the equipotential lines will be distorted near the ends. To minimize this effect, it is necessary that the maximum size of the cross section A is not more than one third of the length L of the sample and that the measurement is taken at the midpoint of the bar. By taking several readings it is possible to trace $\mathrm{d}V/\mathrm{d}x$ and calculate the resistivity from

$$\rho = \frac{A}{i}\frac{\mathrm{d}V}{\mathrm{d}x},$$

where x is the distance along the direction of length L.

5.3.3 Linear four-point probes

The most common way to measure the resistivity of a metallic material is to use a four-point collinear probe. This technique involves the contact of four equidistant probes with a material of unknown resistance. The probe matrix is positioned in the centre of the material, as shown in figure 5.2. The two external probes supply current and the two internal probes are used to measure the resulting voltage drop on the surface of the sample. The volume resistivity is calculated as follows:

$$\rho = \frac{\pi}{\ln 2}\frac{V}{i}tk,$$

where ρ is volume resistivity (Ω cm), V is the measured voltage (volts), i is the source current (amperes), t is the sample thickness (centimeters) and k is a correction factor. For some materials, such as thin films and coatings, the sheet resistance, or surface

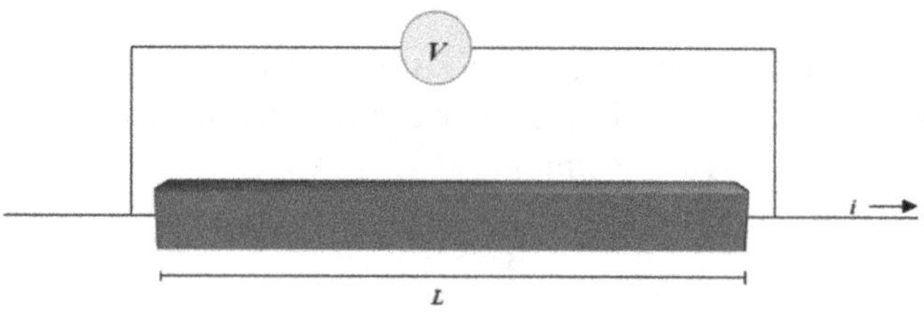

Figure 5.1. Schematic of the two-probe configuration.

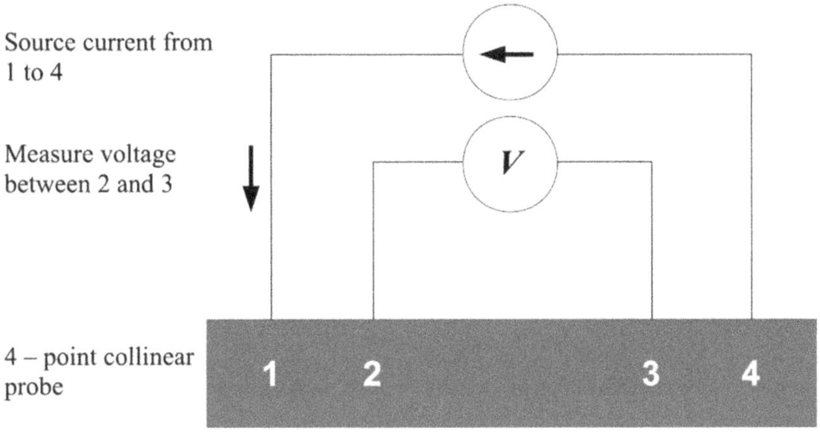

Figure 5.2. Schematic of the four-probe configuration.

resistivity, is determined instead, which does not take the thickness into account. The sheet resistance σ is calculated as follows:

$$\sigma = \frac{\pi}{\ln 2} \frac{V}{i} k. \tag{5.1}$$

The use of the four probes eliminates measurement errors due to the probe resistance, the spreading resistance under each probe, and the contact resistance between each metal probe and the material. Figure 5.3 is another representation of the four-point collinear probe set-up that shows some of the circuit resistances. The R_L terms represent the test lead resistance. R_C represents the contact resistance between the metal probe and the material under study. The contact resistance R_C can be several hundred to a thousand times higher than the resistance of the sample material, which is represented by R_S. Notice that the current flows through all the resistances in the first and fourth set of leads and probes, as well as through the metallic material. However, the voltage is only measured between probes 2 and 3. Since between probes 2 and 3 the current only flows through R_{S1}, only the voltage drop due to R_{S2} will be measured by the voltmeter. The lead R_L and contact R_C resistances will not be measured [1].

5.3.4 Non-collinear probe spacing

The probe array does not need to be linear and, in principle, can have any configuration. The most commonly used is a square configuration. However, several other possibilities which have advantages for specific applications are also taken in consideration.

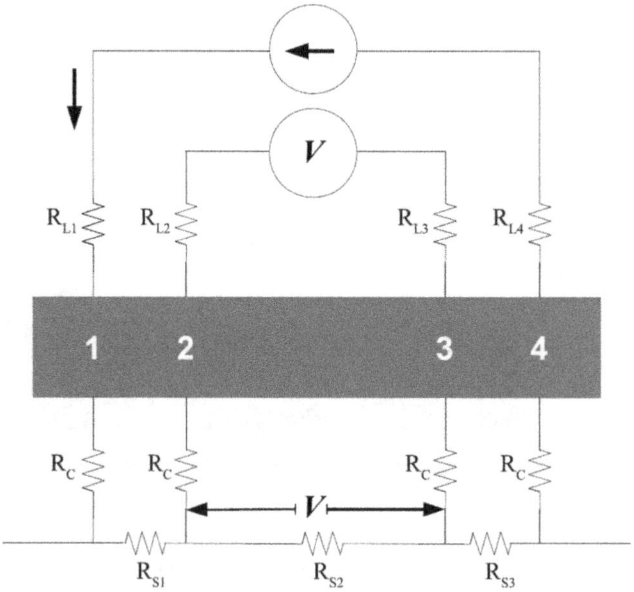

Figure 5.3. Schematic of the four-probe configuration: R_L is the lead resistance, R_C is the contact resistance, R_S is the sample resistance and V is the measured voltage between probes 2 and 3. Only the voltage drop due R_{S2} is measured by the voltmeter.

5.3.5 Square array

Various corrections have been developed for the square array. They are not as extensive as for the linear case but are still adequate for most circumstances. The basic equation for a square array resting on the surface of a semi-infinite medium is

$$\rho = \frac{2\pi S}{2 - \sqrt{2}} \frac{V}{i},$$

where S is the square-array side length.

5.3.6 The Delta four-point probe

The Delta four-point configuration and the following over–under configuration have been developed in an attempt to allow the direct measurement of high resistivity thin layers in direct contact with low resistivity substrates. In this case, the resistivity is a complex function of the distance between the probes, the thickness of the layer and the resistivity of the substrate, such that the thickness of the substrate and more correction factors are required. For example, for a layer thickness of 10 μm and a spacing of 24 μm, it is possible to measure a layer resistivity which is approximately 20 times the substrate resistivity.

5.3.7 The over–under probe

In this configuration, the current flows through probes 1 and 4 of figure 5.2 and the voltage is measured between probes 2 and 3. As in the previous configuration the solutions are complex. A possible result is that for a layer resistivity of the order of 0.1 Ω cm, the probe spacing must be less than 1 μm.

5.4 The van der Pauw method

The van der Pauw method is a technique used for measuring the resistivity and the Hall coefficient of a sample. Its usefulness is due to the possibility of measuring these properties in samples of arbitrary shape, provided that the sample is approximately two-dimensional (the thickness is much smaller than the width) and that the electrodes are placed on the perimeter.

For some particular cases, the 'bridge-type' configuration can be used, as illustrated in figure 5.4. The contacts for the measurement of potential and current in this configuration have a relatively large surface, so as to reduce the effects of the resistor due to the contact points. The properties that can be investigated are:

- The resistivity of the material.
- The type of doping in the case of semiconductors.
- The density of carriers per unit area of the majority charge carriers—from this measure the charge density and the doping level are calculated.
- The mobility of majority holders.

In order to use the van der Pauw method, as already mentioned, the sample thickness must be much smaller than the width and length of the sample itself. To reduce errors in the calculations a symmetrical shape is preferable. Furthermore, there must be no isolated holes in the sample (a simply connected surface). The measurement requires four ohmic contacts on the sample which must meet certain conditions. In particular:

- They must be on the edge of the sample (or as close to it as possible).
- They must be infinitely small, i.e. they must be as small as possible.

Each error given by their finite size is of the order of D/L, where D is the average diameter of the contact and L is the distance between the contacts. Furthermore, the electrical connections to the contacts should be made of the same type of material to

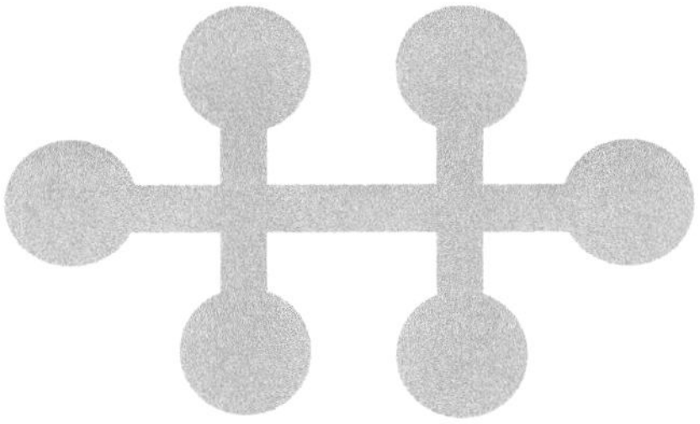

Figure 5.4. A bridge-type sample.

minimize the thermoelectric effect. Measurements are normally performed on thin sheets, which can be represented as in figure 5.5. The resistivity can be calculated directly from measuring the voltage drop and the distance between the contacts O and P, the current i and the size of the sheet.

In order to calculate the Hall coefficient, the lamina is subjected to a magnetic field B which acts perpendicularly to the direction of the current i and to the direction which contains O and Q, diametrically opposite points on the lamina [1].

5.4.1 Methods for measuring resistivity: the case of a flat sample of arbitrary shape

As mentioned, a thin, hole-free sheet must be considered. We must, therefore, create four contacts, which we denote here by M, N, O and P, on the periphery of the lamina itself (figure 5.6). We now apply a current i_{MN} entering at M and going out at N. We measure the potential difference $V_{\mathrm{P}} - V_{\mathrm{O}}$ and define

$$R_{\mathrm{MN,\,OP}} = \frac{V_{\mathrm{P}} - V_{\mathrm{O}}}{i_{\mathrm{MN}}}.$$

In a similar way we define

$$R_{\mathrm{NO,\,PM}} = \frac{V_{\mathrm{M}} - V_{\mathrm{P}}}{i_{\mathrm{NO}}}.$$

The method of measurement is based on the theorem that establishes between $R_{\mathrm{MN,\,OP}}$ and $R_{\mathrm{NO,\,PM}}$ the relation

$$\exp\left(-\frac{\pi d}{\rho}R_{\mathrm{MN,\,OP}}\right) + \exp\left(-\frac{\pi d}{\rho}R_{\mathrm{NO,\,PM}}\right) = 1, \tag{5.2}$$

where d is the thickness of the lamellae and ρ is the resistivity of the material. To demonstrate equation (5.2) let us consider the following factors. If we know d, $R_{\mathrm{MN,\,OP}}$ and $R_{\mathrm{NO,\,PM}}$, then equation (5.2) is an equation with only unknown ρ, which will therefore be easily computable. Measurements and calculations are particularly simple if the sample under examination shows an axial symmetry. In this case, M

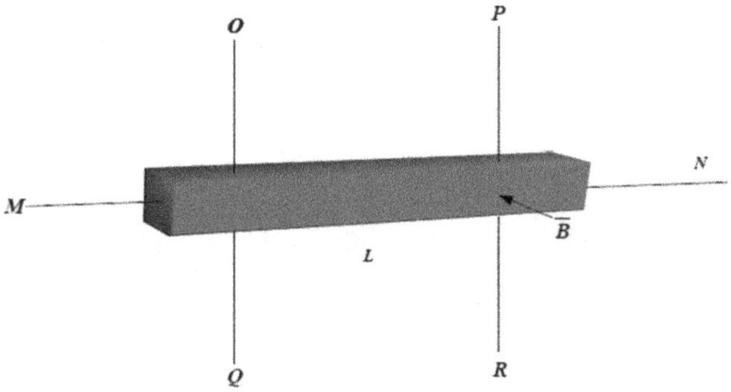

Figure 5.5. Schematic of a van der Pauw's method configuration for measuring the Hall coefficient.

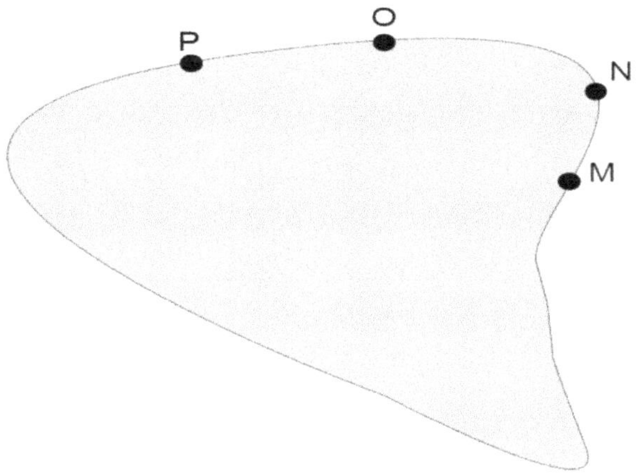

Figure 5.6. Schematic of the van der Pauw configuration.

and O will be positioned along the symmetry axis, while N and P will be arranged to be symmetrical with respect to that axis. Then

$$R_{NO,\,PM} = R_{MN,\,OP}. \tag{5.3}$$

From the reciprocity theorem for the four passive poles, we have in general $R_{NO,\,PM} = R_{PM,\,NO}$, which derives from the particular symmetry considered. Then we obtain equation (5.3) and, finally, the resistivity can be calculated easily from equation (5.2):

$$\rho = \frac{\pi d}{\ln 2} R_{MN,\,OP}.$$

In the case considered, even a single measurement is enough.

In general, however, it is not always possible to express ρ through known functions. A general solution, in any case, can be written in the form

$$\rho = \frac{\pi d}{\ln 2} \frac{R_{MN,\,OP} + R_{MN,\,OP}}{2} f, \tag{5.4}$$

where f is considered as a function of the ratio $R_{MN,\,OP}/R_{NO,\,PM}$, as shown in figure 5.7.

It is evident, therefore, that to determine ρ it is necessary to first determine $R_{MN,\,OP}/R_{NO,\,PM}$, whose value can be found in the graph of figure 5.7, with the corresponding value of f and then we can derive ρ from equation (5.4).

Let us now consider a simply connected sample of arbitrary form. Consider an existing point M, approximately in the centre of the sample and suppose that there is an output current at this point, $2i$. In this case the current density j as a function of the distance r of the centre M will be

$$j = \frac{2i}{S} = \frac{2i}{2\pi r d},$$

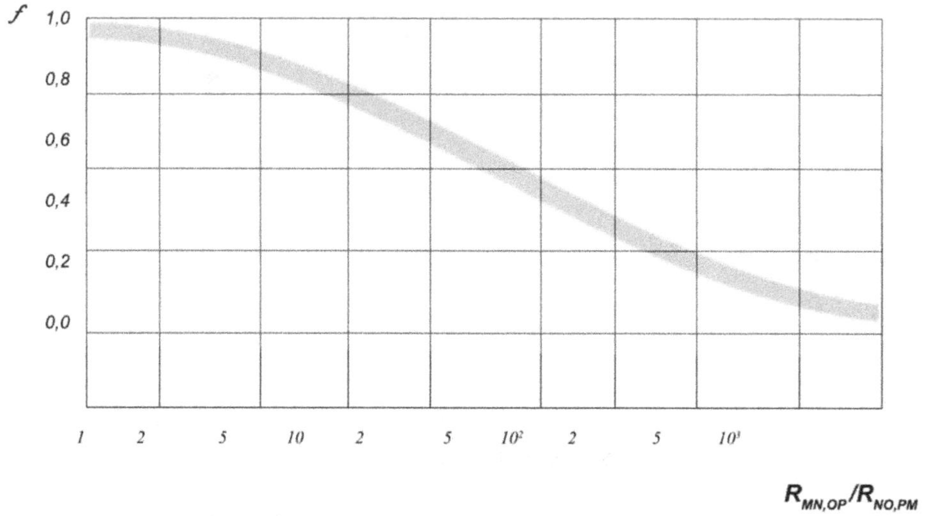

Figure 5.7. The function f versus the ratio $R_{\mathrm{MN,\,OP}}/R_{\mathrm{NO,\,PM}}$.

where d is the thickness of the sample and S is the section considered. From this equation we obtain

$$E = \rho j = \frac{2i}{S} = \frac{2i\rho}{2\pi rd} = \frac{i\rho}{\pi rd}.$$

The voltage drop between points O and P will be (see also figure 5.8)

$$V_{\mathrm{P}} - V_{\mathrm{O}} = \int_{\mathrm{P}}^{\mathrm{N}} E\mathrm{d}r = \int_{\mathrm{P}}^{\mathrm{N}} i\rho\frac{1}{\pi rd}\mathrm{d}x,$$

$$V_{\mathrm{P}} - V_{\mathrm{O}} = \frac{i\rho}{\pi d}\left[\log\left(\frac{a+b}{a+b+c}\right) + \log\left(\frac{b+c}{b}\right)\right]$$
$$V_{\mathrm{P}} - V_{\mathrm{O}} = \frac{i\rho}{\pi d}\log\left[\frac{(a+b)(b+c)}{b(a+b+c)}\right]. \tag{5.5}$$

Then

$$R_{\mathrm{PQ,\,RS}} = \frac{\rho}{\pi d}\log\left[\frac{(a+b)(b+c)}{b(a+b+c)}\right] \tag{5.6}$$

and, therefore, also

$$R_{\mathrm{QR,\,SP}} = \frac{\rho}{\pi d}\log\left[\frac{(a+b)(b+c)}{(ca)}\right]. \tag{5.7}$$

Furthermore,

$$(a+b)(b+c) = b(a+b+c) + ca$$

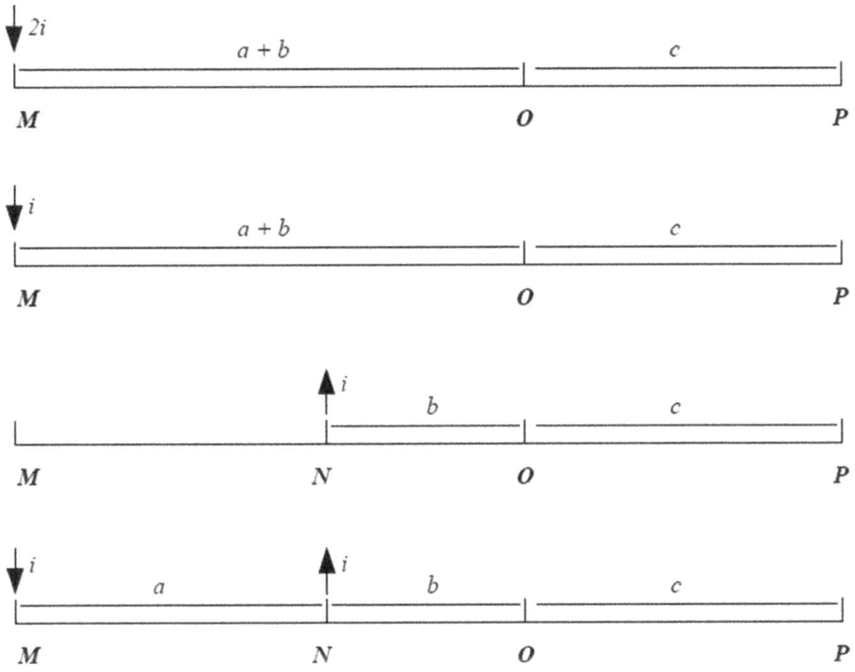

Figure 5.8. A scheme illustrating the derivation of equation (5.2).

and from equation (5.6) we obtain

$$\frac{(a+b)(b+c)}{b(a+b+c)} = \exp\left[\frac{\pi d}{\rho} R_{\mathrm{MN,\ OP}}\right], \tag{5.8}$$

and using equation (5.7) we obtain

$$\frac{(a+b)(b+c)}{ac} = \exp\left[\frac{\pi d}{\rho} R_{\mathrm{MO,\ NP}}\right]. \tag{5.9}$$

Considering the first components of equations (5.8) and (5.9), taking the reciprocal and adding them, we obtain

$$\frac{ac + b(a+b+c)}{(a+b)(b+c)} = \frac{a(b+c) + b(b+c)}{(a+b)(b+c)} = \frac{(a+b)(b+c)}{(a+b)(b+c)} = 1.$$

We now derive equation (5.4) from equation (5.2). Let us consider that

$$1 = \exp\left(-\frac{\pi d}{\rho} R_{\mathrm{MN,\ OP}}\right) + \exp\left(-\frac{\pi d}{\rho} R_{\mathrm{NO,\ PM}}\right).$$

Let us establish, for convenience,

$$\pi d R_{\mathrm{MN,\ OP}} = x_1, \tag{5.10}$$

$$\pi d R_{NO,\ PM} = x_2. \tag{5.11}$$

Then, the formula (5.2) can be written

$$\exp\left(-\frac{x_1}{\rho}\right) + \exp\left(-\frac{x2}{\rho}\right) = 1. \tag{5.12}$$

Finally,

$$\exp\left[-\frac{1}{2\rho}(x_1 + x_2)\right]\cosh\left(\frac{x_1 - x_2}{2\rho}\right) = \frac{1}{2}$$

and

$$\frac{1}{2\rho}(x_1 + x_2) = \frac{\ln 2}{f},$$

from which

$$\rho = \frac{f}{2\ln 2}(x_1 + x_2).$$

Then

$$\exp\left[-\frac{\ln 2}{f}\right]\cosh\left(\frac{x_1 - x_2}{2\rho}\right) = \frac{1}{2},$$

from which

$$\exp\left[-\frac{\ln 2}{f}\right]\cosh\left(\frac{x_1 - x_2}{x_1 + x_2}\frac{\ln 2}{f}\right) = \frac{1}{2},$$

or

$$\exp\left[-\frac{\ln 2}{f}\right]\cosh\left(\frac{(x_1/x_2) - 1}{(x_1/x_2) + 1}\frac{\ln 2}{f}\right) = \frac{1}{2}.$$

The relation between f and x_1/x_2 implies a relation between f and $R_{RM,\ NO}/R_{NO,\ PM}$. By finding ρ from equation (5.12) and by replacing x_1 and x_2 defined by equations (5.10) and (5.11), we have equation (5.4) [2, 3].

5.4.2 A method for measuring the Hall coefficient

The Hall coefficient can be measured with the same method on a thin sheet. We apply a current to one of the contacts, for example to the contact M, and we take the current out of the next contact, for example from contact O. In this way we measure $R_{MO,\ NP}$ after applying a uniform magnetic induction field B in the direction perpendicular to the lamella. The application of B modifies the $R_{MO,\ NP}$ of a quantity $\Delta R_{MO,\ NP}$. It can be shown that the Hall coefficient R_H is given, in this case, by

$$R_{\mathrm{H}} = \frac{d}{B}\Delta R_{\mathrm{MO,\,NP}}. \tag{5.13}$$

It is important to note that the same conditions required for the sample for the measurement of resistivity mentioned previously are applied. The validity of the formula (5.13) depends on the distribution of the current along the flow line which must not change when the field $\vec{B}$ is applied.

For samples with the configuration usually used for currents parallel to the edges of the sample, the currents themselves can be considered uniform. This property is also valid for arbitrarily shaped sheets, provided the above conditions are met. Due to the presence of the magnetic field B, the charge carriers, with charge q, are subjected to a force perpendicular to the flow lines of B. The magnitude of this force can be derived from the Lorentz law and it is equal to

$$F = qvB, \tag{5.14}$$

where v is the velocity of the carriers. The existing relation between the concentration n of the charge carriers, the velocity v and the current density J is

$$v = \frac{J}{nq}. \tag{5.15}$$

From equations (5.14) and (5.15) and dividing by q, we infer that the effect of magnetic fields is equivalent to an apparent electric field E_{H} call the Hall electric field for which we can write, with good approximation (if we neglect the thermal agitation of the electrons),

$$E_{\mathrm{H}} = \frac{1}{nq}JB.$$

Then, E_{H} is proportional to J and B and the proportionality constant $1/nq$ is called the Hall coefficient R_{H}.

Since q is a known quantity, from R_{H} we can calculate the concentration n of charge carriers. After the application of the magnetic field, the electric field inside the sample is no longer in the same direction as the current, but has acquired a transverse component E_{t} which exactly compensates for the apparent electric field due to the Hall effect E_{H}. $\Delta(V_{\mathrm{P}} - V_{\mathrm{N}})$ is the potential difference between P and N which is therefore given by the integral of E_{t} from P to N′, along a path orthogonal to the current through the lamella and then along the periphery, i.e. along a flow line, from N′ to N. The last section does not contribute to the integral. We then have

$$\Delta(V_{\mathrm{P}} - V_{\mathrm{N}}) = \int_{\mathrm{P}}^{\mathrm{N}'} E_{\mathrm{H}}\mathrm{d}s = R_{\mathrm{H}}B\int_{\mathrm{P}}^{\mathrm{N}'} J\mathrm{d}s = R_{\mathrm{H}}B\frac{i_{\mathrm{MO}}}{d},$$

where d is the thickness of the lamella.

Considering that

$$R_{\mathrm{H}}B\frac{i_{\mathrm{MO}}}{d} = \Delta(V_{\mathrm{P}} - V_{\mathrm{N}}),$$

we have

$$R_H = \frac{d}{B}\frac{\Delta(V_P - V_N)}{i_{MO}} = \frac{d}{B}\Delta R_{MO,\,NP},$$

which is exactly equation (5.13) [2].

5.5 Conclusions

Resistivity is one of the most important quantities characterizing the physical electric transport properties of a solid. Since its values can span from 10^{-8} to 10^{15} Ω cm, the experimental procedure to be followed strongly depends on the values of the resistivity to be measured. In this chapter, we have briefly considered only the case of low-resistance measurements with special attention given to the four-probe technique, which is still the method most frequently used in both fundamental and applied research. In particular, major emphasis has been given to the well-known van der Pauw method, a very powerful procedure since it allows us to evaluate the resistivity of a specimen without structuring it in a well-defined geometry, provided that the four contacts are located on the edges of the sample.

References

[1] Keithley 2004 *Low Level Measurements Handbook: Precision DC Current, Voltage, and Resistance Measurements* 6th edn (Cleveland, OH: Keithley Instruments)

[2] van der Pauw L J 1958 A method of measuring specific resistivity and Hall effect of discs of arbitrary shape *Philips Res. Rep.* **13** 1–9

[3] van der Pauw L J 1959 A method of measuring specific resistivity and Hall coefficient on lamellae of arbitray shape *Philips Tech. Rev.* **26** 220–4

IOP Publishing

Modern Physics
A critical approach
Canio Noce

Chapter 6

The electromagnetic theory of thermal radiation

Roberto De Luca and Alessandro Sorgente

At the end of the nineteenth century, the properties of the physical world seemed well defined. The set of gravitational phenomena was perfectly described by Newton's laws, expressed in an elegant form from the mathematical point of view thanks to the Lagrangian and Hamiltonian formalisms.

Thermodynamics allowed us to understand thermal phenomena and to build increasingly efficient thermal machines. Maxwell's electromagnetic theory was able to completely describe electromagnetic phenomena, and the problem of the nature of light that had been debated for centuries was solved. Moreover, continuous technological improvements allowed significant experimental results to be obtained, such as x-rays and the discovery of the electron and the e/m ratio measurement.

The difficulty encountered by classical physics in the interpretation of some experimental effects started to manifest itself in the mechanics of the atom (for example in the explanation of the spectral lines of atoms or in the electron diffraction) and in the attempts to frame the law of radiant heat within the laws of classical physics.

The points of crisis of classical physics are usually indicated using the following three main questions:

1. *The problem of black-body radiation*: the classical electromagnetic theory is not able to explain the colour of light emitted by a warm body.
2. *The problem of specific heats*: according to classical statistical mechanics, each degree of freedom contributes in equal measure to the specific heat of a body. This is in blatant contradiction to experimental evidence, as we will comment upon shortly.
3. *The problem of atomic spectra*: experimental observations show that light emitted, and absorbed, by various chemical elements is restricted to well-defined frequencies which are characteristic of each element. The spectrum, that is the set of frequencies, and its regularities are classically incomprehensible.

doi:10.1088/978-0-7503-2678-0ch6

In this chapter we focus particularly on point 1. The overcoming of the difficulty in explaining the experimental evidence of the radiant heat of a 'black-body' occurred over about twenty years and a complete resolution was reached thanks to the brilliant and revolutionary intuition of Max Planck in 1900 [1].

The first clue for the quantum nature of radiation came from the study of thermal radiation emitted by opaque bodies. When radiation falls on an opaque body, part of it is reflected and the rest is absorbed. However, the atoms contain charges (the electrons) and they are accelerated when oscillating. Consequently, as required by electromagnetic theory, the atoms emit electromagnetic radiation, which reduces the kinetic energy of the oscillations, thus lowering the temperature. When the rate of absorption equals the rate of emission, the temperature is constant and we say that the body is in thermal equilibrium with its surroundings. A good absorber of radiation is therefore also a good emitter. The electromagnetic radiation emitted under these circumstances is called *thermal radiation*. At ordinary temperatures (below about 600 °C) the thermal radiation emitted by a body is not visible and most of the energy is concentrated in wavelengths much longer than those of visible light. As a body is heated, the quantity of thermal radiation emitted increases and the energy radiated extends to shorter wavelengths.

6.1 Thermal radiation

The energy emitted from the surface of a body is in the form of electromagnetic waves in a certain range of frequencies. The radiation emitted or absorbed by a body in the form of heat is called thermal or infrared radiation because it corresponds to a range of wavelengths from 1 μm to 100 μm. Spectroscopic instruments produce detailed analysis of radiation absorption and emission processes, whose spectrum is shown in figure 6.1. It is precisely from the attempt to explain these curves that the limits of classical theory emerge. We thus wish to calculate the intensity of

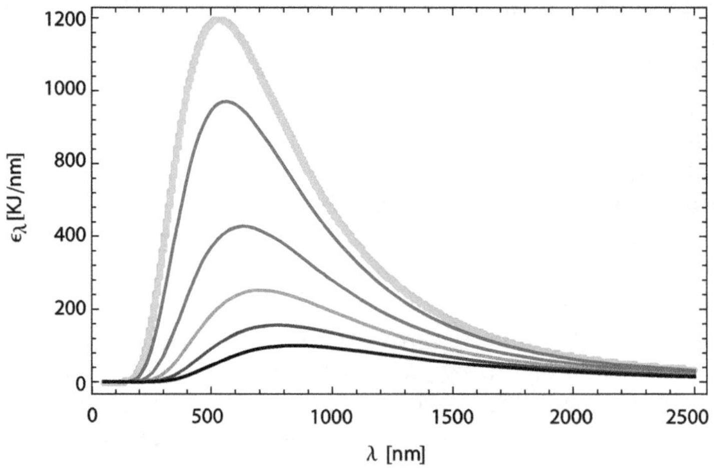

Figure 6.1. Radiation spectrum of a black-body.

electromagnetic radiation in a volume V. The amplitude of the Poynting vector $\vec{I}$ gives the intensity I of the electromagnetic radiation [2]. We recall that the Poynting vector is defined as follows:

$$\vec{I} = \frac{\vec{E} \times \vec{B}}{\mu_0}. \tag{6.1}$$

Suppose the fields in the cavity are plane waves $\vec{E}(\vec{r}, t) = \vec{E}_0 e^{i(\vec{k}\cdot\vec{r}-\omega t)}$ and $\vec{B}(\vec{r}, t) = \vec{B}_0 e^{i(\vec{k}\cdot\vec{r}-\omega t)}$. Substituting these expression in the Maxwell equation,

$$\nabla \times \vec{E} = -\frac{\partial \vec{B}}{\partial t},$$

we have

$$\vec{k} \times \vec{E}_0 = \omega \vec{B}_0, \Rightarrow \vec{B}_0 = \frac{\hat{k}}{c} \times \vec{E}_0, \tag{6.2}$$

being $\frac{\vec{k}}{\omega} = \hat{k}T/\lambda = \hat{k}/c$. Placing equation (6.2) in equation (6.1) we obtain

$$\begin{aligned}
\vec{I} &= \frac{1}{\mu_0 c}\vec{E}_0 \times (\hat{k} \times \vec{E}_0) \\
&= \frac{1}{\mu_0 c}[(\vec{E}_0 \cdot \vec{E}_0)\hat{k} - (\vec{E}_0 \cdot \hat{k})\vec{E}_0] \\
&= \frac{1}{\mu_0 c}E_0^2\hat{k} = \varepsilon_0 c E_0^2 \hat{k}.
\end{aligned} \tag{6.3}$$

The last equation was obtained by exploiting the vector relation $\vec{a} \times (\vec{b} \times \vec{c}) = (\vec{a} \cdot \vec{c})\vec{b} - (\vec{a} \cdot \vec{b})\vec{c}$ and by recalling that the electromagnetic waves are transversal: $(\vec{E}_0 \cdot \hat{k}) = 0$ and $\mu_0 \varepsilon_0 = \frac{1}{c^2}$.

We also know that the density of electromagnetic energy in the vacuum is the sum of a contribution due to the electric and magnetic field such that

$$u = u_E + u_M = \frac{1}{2}\varepsilon_0 E_0^2 + \frac{1}{2\mu_0}B_0^2.$$

We may now write

$$B_0^2 = \vec{B}_0 \cdot \vec{B}_0 = \left(\frac{\hat{k}}{c} \times \vec{E}_0\right) \cdot \left(\frac{\hat{k}}{c} \times \vec{E}_0\right) = \frac{E_0^2}{c^2}.$$

Therefore the energy density is given by the following:

$$u = \frac{1}{2}\varepsilon_0 E_0^2 + \frac{1}{2\mu_0}\frac{E_0^2}{c^2} = \varepsilon_0 E_0^2. \tag{6.4}$$

By considering equation (6.4), equation (6.3) becomes

$$\vec{I} = cu\hat{k}. \tag{6.5}$$

If we wish to obtain the total energy in the cavity we need to integrate the intensity of the emitted radiation over the whole surface. Let us then consider, for simplicity, a spherical surface of radius R as shown in figure 6.2; the radiating power coming out of a small hole having area s_0 is

$$\mathrm{d}P = \vec{s}_0 \cdot \hat{n}\mathrm{d}I',$$

where $\mathrm{d}I' = u'c$ is the fraction of energy contained in the solid angle $\mathrm{d}\Omega' = \sin\theta \mathrm{d}\theta \mathrm{d}\phi$, and $u' = u\frac{\mathrm{d}\Omega'}{4\pi}$. In this way, we have

$$P = \int_S \mathrm{d}P = \int_S s_0 \mathrm{d}I' \cos\theta = \frac{us_0c}{4\pi}\int_0^{2\pi} \mathrm{d}\phi \int_0^{\pi/2} \cos\theta \sin\theta \mathrm{d}\theta = \frac{us_0c}{4}.$$

The light intensity emitted by a black-body through a small hole s_0 will be the power P divided by s_0, such that

$$\varepsilon_{\mathrm{em}} = \frac{cu}{4}. \tag{6.6}$$

6.2 Kirchhoff theorem: definition of a black-body

An important result of classical physics, due to Kirchhoff, states that under conditions of thermal equilibrium the relationship between the emitted power of a black-body and the absorbed power is universal and is proportional to the spectral density of energy in a cavity. Consider a cavity with perfectly reflective walls of volume V (for simplicity we will assume a cube) at temperature T. The walls of this cavity are in thermal equilibrium with the electromagnetic radiation emitted and absorbed by the walls. Let U be the total electromagnetic energy, u the energy density and $u_\nu(\nu, T)$ its spectral density, i.e. $u_\nu(\nu, T)\mathrm{d}\nu$ is the amount of

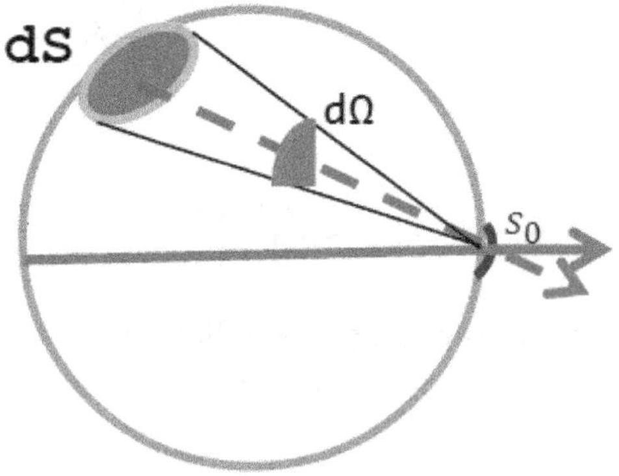

Figure 6.2. Radiation intensity through a hole of section s_0.

electromagnetic energy per unit of volume with frequencies included in the range $[\nu, \nu + d\nu]$. Therefore we have

$$u = \int_0^\infty u_\nu(\nu, T) d\nu.$$

In other words, Kirchhoff's theorem states that the function $u_\nu(\nu, T)$ does not depend on the type of cavity and other parameters, but only on temperature T.

6.2.1 Absorption and emission coefficients

Now consider any body at temperature T and suppose that it emits radiation in a way that is dependent only on temperature. We call $J(\nu, \theta)$ the energy emitted per second per frequency and surface interval in the solid angle, such that

$$\frac{d\varepsilon_{em}}{d\nu d\Omega} = J(\nu, \theta) dS.$$

The quantity J is called the *emission coefficient.*

This same body, subject to radiation, can absorb a certain percentage of the incident radiation. If $I_\nu(\theta)$ is the spectral density of the incident radiation per unit solid angle we have that the incident power density on the unit of surface is given by

$$\frac{d\varepsilon_{in}}{d\nu dS} = cI_\nu \cos\theta \left(\frac{d\Omega}{4\pi}\right).$$

Let $A(\nu, \theta)$ be the fraction of the incident absorbed energy. Let us assume for simplicity that the body is not fluorescent, and all incident radiation is absorbed or reflected. In this situation, at thermal equilibrium, for each solid angle and for each interval frequency a balance between absorbed energy and re-emitted energy is obtained. At thermal equilibrium, the ambient radiation is u_ν, isotropic, therefore we may write

$$\frac{J(\nu, \theta)}{A(\nu, \theta)} = \frac{c}{4\pi} u_\nu \cos\theta.$$

In this way, the ratio between the emission coefficient and the absorption coefficient is universal and does not depend on the body, but only on the spectral density of the radiation at thermal equilibrium. A *black-body* is defined as a body that absorbs all frequencies, for which $A = 1$. $J(\nu, \theta) = \frac{c}{4\pi} u_\nu \cos\theta$ is therefore directly connected to the emission coefficient of a black-body. To make a black-body we use a container (oven) at a uniform temperature in the walls of which a small opening is made. All the radiation that falls from the outside onto the opening enters through it and, after repeated reflections, is completely absorbed. The radiation inside, equal to that which comes out again from the opening, must possess exactly the spectral intensity distribution characteristic of the black-body.

Using only classical thermodynamics and electromagnetism we can establish two laws concerning the dependence of black-body radiation on temperature. The first is the Stefan–Boltzmann law and the second is Wien's law. The first states that the energy density emitted by a black-body is proportional to the fourth power of absolute temperature:

$$u = \sigma T^4, \tag{6.7}$$

where $\sigma = 5.670 \times 10^{-8}$ W m^{-2} T^{-4}. This law was discovered experimentally by Stefan (1879) and was later demonstrated by Boltzmann (1884). The second law, Wien's law, states that the spectral distribution of energy density is given by an equation of the form

$$u_\nu = \nu^3 F\left(\frac{\nu}{T}\right). \tag{6.8}$$

The F function is not known and depends only on the ν/T ratio. The form of F must be derived through comparison with experimental results. We note that the Stefan–Boltzmann law is consistent with Wien's law. In fact, we may write

$$u = \int_0^\infty u_\nu(\nu, T)\mathrm{d}\nu = \int_0^\infty \nu^3 F\left(\frac{\nu}{T}\right)\mathrm{d}\nu.$$

By setting $\nu = xT$, with x not depending on T, we have

$$u = \int_0^\infty T^3 x^3 F(x) T dx = T^4 \int_0^\infty x^3 F(x)\mathrm{d}x = aT^4,$$

where the integral does not depend on T.

6.3 Proof for the Stefan–Boltzmann equation (6.7)

Suppose we have electromagnetic radiation enclosed in a container, closed by a mobile piston that has a reflective surface. The radiation field exerts a pressure p on the piston (radiation pressure). From Maxwell's electromagnetic theory we know that $p = u/3$ [3]. This pressure is due to the momentum of the radiation transferred to the surface in the unit of time due to the reflection of the radiation. Let us then consider radiation as a thermodynamic system. From the principles of thermodynamics, if W is the total energy, S the entropy, V the volume and T the absolute temperature, we have

$$\begin{aligned} T\mathrm{d}S &= \mathrm{d}W + p\mathrm{d}V = \mathrm{d}(uV) + \frac{1}{3}u\mathrm{d}V = V\mathrm{d}u + u\mathrm{d}V + \frac{1}{3}u\mathrm{d}V \\ &= V\frac{\partial u}{\partial T}\mathrm{d}T + \frac{4}{3}u\mathrm{d}V, \end{aligned} \tag{6.9}$$

where we have considered the fact that the internal energy of the radiation depends only on the temperature. On the other hand, $S = S(T, V)$, so that

$$\mathrm{d}S = \frac{\partial S}{\partial T}\mathrm{d}T + \frac{\partial S}{\partial V}\mathrm{d}V. \tag{6.10}$$

Comparing equations (6.9) and (6.10) we obtain

$$\begin{cases} \dfrac{\partial S}{\partial T} = \dfrac{V}{T}\dfrac{\partial u}{\partial T} \\ \dfrac{\partial S}{\partial V} = \dfrac{4}{3}\dfrac{u}{T} \end{cases}. \tag{6.11}$$

We also know that entropy S is a state function so that $\mathrm{d}S$ is an exact differential. Therefore, Schwartz's theorem holds:

$$\frac{\partial^2 S}{\partial V \partial T} = \frac{\partial^2 S}{\partial T \partial V},$$

such that

$$\frac{\partial}{\partial T}\left(\frac{4}{3}\frac{u}{T}\right) = \frac{\partial}{\partial V}\left(\frac{V}{T}\frac{\mathrm{d}u}{\mathrm{d}T}\right),$$

from which, by taking the derivative of both members, we obtain

$$\frac{1}{T}\frac{\partial u}{\partial T} = -\frac{4}{3}\frac{u}{T^2} + \frac{4}{3T}\frac{\partial u}{\partial T}.$$

Simplifying and separating the variables, we have

$$\frac{\mathrm{d}u}{u} = 4\frac{\mathrm{d}T}{T}.$$

Finally, by integrating both members we obtain

$$\ln u = 4 \ln T + c \Rightarrow u = aT^4. \tag{6.12}$$

Thus, from this equation we deduce the Stefan–Boltzmann equation (6.7).

6.4 Proof of Wien's law (6.8)

Wien's law derives mainly from the existence of the Doppler effect for electromagnetic waves [4]. A stationary observer with respect to a wave train sees a frequency shift $\frac{\Delta\nu}{\nu} = (\frac{v}{c})\cos\theta$, where $\frac{\Delta\nu}{\nu}$ is the relative variation of frequency, v is the velocity of the source, and θ is the angle between the latter velocity and the propagation direction of the electromagenetic wave.

Imagine a piston moving slowly with speed v on which a reflecting mirror is mounted. We can think of the incident wave as coming from a stationary source and reflected light as coming from a virtual source placed beyond the mirror. If the real source is at a distance δ from the wall the distance between the virtual source and the real source is 2δ, so when the mirror moves the speed is $2\mathrm{d}\delta/\mathrm{d}t = 2v$ and the Doppler effect of the first order provides for the following frequency shift of the reflected light:

$$\Delta\nu = +\frac{2v}{c}\nu.$$

More generally, due to oblique incidence, we have

$$\nu' - \nu = \frac{2v}{c}\nu \cos\theta \Rightarrow \nu' = \nu\left(1 + \frac{2v}{c}\cos\theta\right). \tag{6.13}$$

The intensity of radiation is also altered by same ratio. In fact, the energy that falls on the mirror of area A in the time $\mathrm{d}t$ is $E = IA\mathrm{d}t$, while the reflected energy is $E' = I'A\mathrm{d}t$. For energy conservation, the reflected energy will be given by that incident energy plus the work of external force $W = pAv\mathrm{d}t$, where $p = (2I/c)\cos\theta$. In this way, we have

$$I' = I\left(1 - \frac{2v}{c}\cos\theta\right).$$

In an adiabatic transformation there is therefore an exchange of energy between the various spectral components of the energy. We wish to write the energy balance of the process, i.e. we would like to calculate the energy change $\mathrm{d}(Vu_\nu)\mathrm{d}\nu$ of the spectral component ν. All energy that falls on the moving mirror in the unit frequency $\mathrm{d}\nu$, in unit time $\mathrm{d}t$ and coming from a solid angle $\mathrm{d}\Omega$ is therefore

$$E = u_\nu \frac{c}{4\pi} A\cos\theta\, \mathrm{d}t\mathrm{d}\nu\mathrm{d}\Omega.$$

The energy E_- removed, due to the Doppler effect, from the component in question, is

$$E_- = u_\nu \frac{c}{4\pi} A\cos\theta\left(1 - \frac{2v}{c}\cos\theta\right)\mathrm{d}t\mathrm{d}\nu\mathrm{d}\Omega.$$

At the same time the energy gain, E_+, can be obtained from equation (6.13), such that

$$\begin{aligned} E_+ &= u'_\nu \frac{c}{4\pi} A\cos\theta\left(1 - \frac{2v}{c}\cos\theta\right)\mathrm{d}t\mathrm{d}\nu'\mathrm{d}\Omega \\ &= \nu u_{\left(1+\frac{2v}{c}\cos\theta\right)} \frac{c}{4\pi} A\mathrm{d}t\mathrm{d}\nu\mathrm{d}\Omega\left(1 - \frac{4v^2}{c^2}\cos^2\theta\right). \end{aligned} \tag{6.14}$$

Neglecting the quadratic terms $(v/c)^2$ and expanding the function u'_ν in Taylor series around $v = 0$, we have

$$\nu u_{\left(1+\frac{2v}{c}\cos\theta\right)} = u_\nu + \nu\frac{\partial u_\nu}{\partial \nu}\frac{2v}{c}\cos\theta + \cdots.$$

Therefore

$$E_+ = \left(u_\nu + \nu\frac{\partial u_\nu}{\partial \nu}\frac{2v}{c}\cos\theta\right)\frac{c}{4\pi} A\cos\theta\, \mathrm{d}t\mathrm{d}\nu\mathrm{d}\Omega. \tag{6.15}$$

Considering that $Av\mathrm{d}t = \mathrm{d}V$, the net energy variation of the radiation in the frequency interval $(\nu, \nu + \mathrm{d}\nu)$ is

$$\Delta E = E_+ - E_- = \frac{\cos^2\theta}{2\pi} \mathrm{d}\Omega \mathrm{d}\nu \mathrm{d}V \left(\nu \frac{\partial u_\nu}{\partial \nu} \right).$$

Furthermore, to obtain the total reflected energy it is necessary to integrate on the whole hemisphere, that is, to add on all the solid angles subtended by the mirror:

$$\begin{aligned} \int_\Omega \Delta E \mathrm{d}\Omega &= \left(\nu \frac{\partial u_\nu}{\partial \nu} \right) \mathrm{d}\nu \mathrm{d}V \int_0^{2\pi} \mathrm{d}\varphi \int_0^{\pi/2} \sin\theta \frac{\cos^2\theta}{2\pi} \mathrm{d}\theta \\ &= \frac{1}{3} \left(\nu \frac{\partial u_\nu}{\partial \nu} \right) \mathrm{d}\nu \mathrm{d}V. \end{aligned} \tag{6.16}$$

On the other hand, the increase in total energy of the radiation due to reflection can also be expressed as follows:

$$\mathrm{d}(Vu_\nu) = u_\nu \mathrm{d}V + V \mathrm{d}u_\nu = u_\nu \mathrm{d}V + V \left(\frac{\partial u_\nu}{\partial \nu} \mathrm{d}\nu + \frac{\partial u_\nu}{\partial V} \mathrm{d}V \right). \tag{6.17}$$

Equating coefficients of $\mathrm{d}V$ in equations (6.16) and (6.17) we obtain

$$\frac{\partial u_\nu}{\partial V} = \frac{1}{3} \nu \frac{\partial u_\nu}{\partial \nu} - u_\nu. \tag{6.18}$$

Equation (6.18) is a differential equation in the unknown u_ν that depends on V and ν. It can be proven that a solution of equation (6.18) has the following form:

$$u_\nu(\nu, V) = \nu^3 \psi(\nu^3 V). \tag{6.19}$$

If the volume change is very slow, we can imagine that the process is adiabatic so that $\mathrm{d}S = 0$. Furthermore, from equation (6.11) and from the Stefan–Boltzmann law, we can write

$$\frac{\partial S}{\partial V} = \frac{4}{3} \frac{aT^4}{T} \Rightarrow S = aVT^3 = C.$$

In this way we have $V = K/T^3$. Replacing this result in equation (6.19) we obtain

$$u_\nu = \nu^3 F(\nu/T). \tag{6.20}$$

This is just the *Wien's formula* of equation (6.8).

6.4.1 Wien's displacement law

The general Wien's formula contains the so-called *displacement law*. This definition derives from the observation of behaviour of spectral radiance as a function of wavelength, as plotted in figure 6.3.

An incandescent body, at a certain temperature T, shows a peak in its radiance $R(\lambda)$ at a particular wavelength $\lambda_{\max}$. For very large wavelengths the radiation intensity is small. For lower temperatures the maximum of $R(\lambda)$ shifts to the right so that

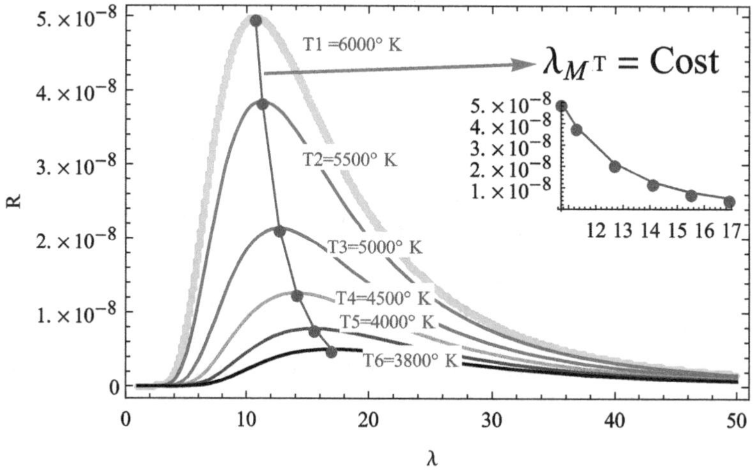

Figure 6.3. Spectral distribution function $R(\lambda)$ measured at different temperatures. The inset represents the displacement law.

$$\lambda_{\max} T = \text{constant}, \tag{6.21}$$

the value of the constant in equation (6.21) has been experimentally determined and its numerical value is 2.98×10^{-3} m K. Equation (6.21) is deduced directly from Wien's law (6.8). Indeed, the maximum of u_ν is obtained by imposing that the derivative of the energy density u with respect to ν is equal to zero, being $\nu = c/\lambda$, we obtain

$$\mathrm{d}u = u_\nu \mathrm{d}\nu = \frac{c^3}{\lambda^3} F\left(\frac{c}{\lambda T}\right)\left(-\frac{c}{T\lambda^2}\right)\mathrm{d}\lambda = -\frac{c^4}{T\lambda^5} F\left(\frac{c}{\lambda T}\right)\mathrm{d}\lambda,$$

therefore

$$u_\lambda \equiv \frac{\partial u}{\partial \lambda} = \frac{c^4}{\lambda^6}\left[5F\left(\frac{c}{\lambda T}\right) + \frac{c}{T\lambda} F'\left(\frac{c}{\lambda T}\right)\right] = 0. \tag{6.22}$$

Setting $c/\lambda T = x$, equation (6.22) is satisfied for $x = x_M$, or $c/\lambda_{\max} T = \text{const}$, so that equation (6.21) is obtained.

6.5 Planck oscillators and the Rayleigh–Jeans law

The thermodynamic and electromagnetic analysis of black-body radiation, studied by Wien, provided a general formula for the spectral density of radiation, as described by equation (6.8), without, however, specifying the explicit form of the function $F(\nu/T)$. This function must be such as to reproduce the experimental results.

To determine the functional form of Wien's function F, it is necessary to construct a physical model that is able to fit the experimental data.

As a simple model of a radiant body, Planck proposed an oscillating dipole that emits electromagnetic waves according to the following mechanism: the atoms of the

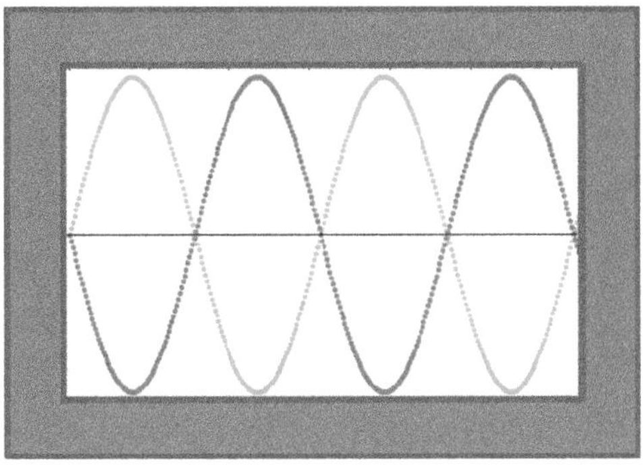

Figure 6.4. Representation of stationary electromagnetic waves in a resonant cavity.

furnace walls, at temperature T, emit thermal radiation in the form of electromagnetic waves; these in turn excite atoms that at the resonance frequency emit electromagnetic waves. The energy contained in the cavity in a frequency interval $(\nu, \nu + d\nu)$ represents the spectral density u_ν.

Consider a cubic cavity of side L (figure 6.4), whose walls are at temperature T and therefore absorb and emit electromagnetic waves [5]. In this way, in a cavity there will be a field of stationary waves. For simplicity we consider, at the moment, a one-dimensional cavity. The stationary waves in it are given by the superposition of incident waves and reflected waves on the wall, such that

$$A\sin(K_x x - \omega t) + A\sin(K_x x + \omega t) = 2A\sin(K_x)\cos(\omega t).$$

The boundary conditions at the wall will be zero for $x = 0$ and $x = L$. Therefore $K_x L = n_x\pi$ or $K_x = n_x\pi/L$. Since the normal modes are perpendicular, we can write the same relation for the other components, $K_y L = n_y\pi$, or $K_y = n_y\pi/L$, $K_z L = n_z\pi$, or $K_z = n_z\pi/L$, where (n_x, n_y, n_z) are integer numbers. Then the module of the wave vector, K, will be

$$K = \sqrt{K_x^2 + K_y^2 + K_z^2} = \frac{\pi}{L}\sqrt{n_x^2 + n_y^2 + n_z^2}. \tag{6.23}$$

On the other hand, $K = \frac{2\pi}{\lambda}$, so that we may rewrite equation (6.23) as follows:

$$\left(\frac{2L}{\lambda}\right)^2 = \left(n_x^2 + n_y^2 + n_z^2\right). \tag{6.24}$$

Equation (6.24) represents the first octant of a sphere, in the space of wave numbers, whose radius is $(\frac{2L}{\lambda})$.

The number of states contained in the octant is equal to its volume (figure 6.5). From equation (6.24), the ground state corresponds to the frequency

$$\nu = \frac{c}{2L}\sqrt{3}.$$

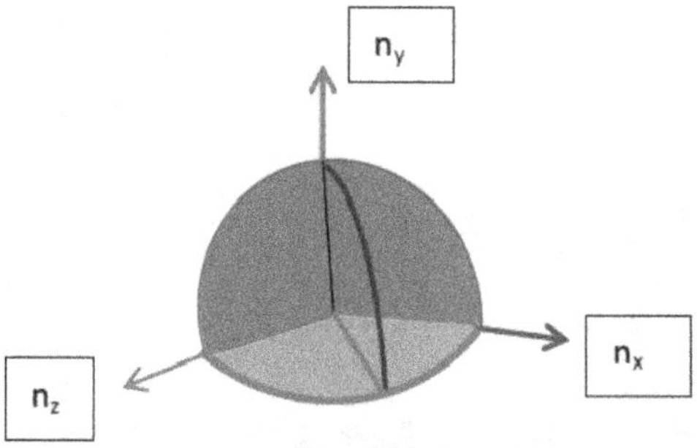

Figure 6.5. The number of states contained in the sphere $(n_x^2 + n_y^2 + n_z^2) = (\frac{2L}{\lambda})^2$ of radius $r = (\frac{2L}{\lambda})$ is equal to the volume of the octant.

Moreover, since $\lambda \ll L$, the number of states falling in the first octant of the sphere of radius L is enormously large, so we can treat n_x, n_y and n_z as continuous variables. Under this hypothesis, the number of states falling in the shell r, $r + \mathrm{d}r$, with $r \leqslant R$ is the volume of the spherical shell. In spherical coordinates the volume element $\mathrm{d}V$ is $dV = r^2 \sin\theta d\theta d\varphi dr$, and the total volume of the spherical octant is

$$V_{\text{oct.}} = \int_{V_{\text{oct.}}} \mathrm{d}V = \int_0^{\pi/2} \mathrm{d}\varphi \int_0^{\pi/2} \sin\theta \mathrm{d}\theta \int_0^R r^2 \mathrm{d}r = \frac{\pi}{6} R^3.$$

In conclusion, the number of modes with frequency less than ν is

$$N_\nu = \frac{1}{8}\left(\frac{4}{3}\pi R^3\right) = \left(\frac{4}{3}\pi\frac{\nu^3}{c^3}\right)L^3.$$

The density of states is therefore

$$n_\nu = \frac{N_\nu}{L^3} = \frac{4}{3}\pi\frac{\nu^3}{c^3}.$$

Taking into account the two states of polarization of the electric field, we have

$$n_\nu = \frac{N_\nu}{L^3} = \frac{8}{3}\pi\frac{\nu^3}{c^3}. \tag{6.25}$$

The number of states n_λ contained in the interval $(\lambda, \lambda + \mathrm{d}\lambda)$ is therefore

$$n_\lambda = \frac{\mathrm{d}n}{\mathrm{d}\lambda} = -\frac{8\pi}{\lambda^4}. \tag{6.26}$$

Since $n_\lambda \mathrm{d}\lambda = -\frac{8\pi}{\lambda^4}\mathrm{d}\lambda$ and $\lambda = c/\nu$, we will have

$$n_\nu \mathrm{d}\nu = -\frac{8\pi}{c^4}\nu^4\left(-\frac{c}{\nu^2}\right)\mathrm{d}\nu = \frac{8\pi}{c^3}\nu^2 \mathrm{d}\nu. \tag{6.27}$$

If w_{m} is the average energy of each oscillator, the energy u_ν in the frequency range $(\nu, \nu + \mathrm{d}\nu)$ is

$$u_\nu \mathrm{d}\nu = w_\mathrm{m}\frac{8\pi}{c^3}\nu^2\mathrm{d}\nu. \tag{6.28}$$

According to Kirchhoff's theorem, in equation (6.28) the parameters e and m of the oscillators do not appear. If one can calculate the average thermal energy of the oscillator, w_m, the total energy of the black-body can be obtained. It is now a well-known fact that a harmonic oscillator in thermal equilibrium has a mean energy $w_\mathrm{m} = kT$. This is just what is stated by the equipartition theorem.

If a system, classically described, is in thermal equilibrium with a heat source at temperature T, the probability P_E of finding the system in a particular energy E is given by

$$P_E = C\exp(-\beta E),$$

where $\beta = \frac{1}{k_\mathrm{B}T}$ and C is a constant of normalization such that

$$\int P_E\mathrm{d}E = 1 \Longrightarrow C = \frac{1}{\int_0^\infty \exp(-\beta E)\mathrm{d}E}.$$

Therefore, the average energy of the black-body is

$$\begin{aligned} w_\mathrm{m} &= \frac{\int_0^\infty E\exp(-\beta E)\mathrm{d}E}{\int_0^\infty \exp(-\beta E)\mathrm{d}E} = -\frac{\partial}{\partial\beta}\ln\int_0^\infty \exp(-\beta E)\mathrm{d}E \\ &= -\frac{\partial}{\partial\beta}\left[-\ln\left(\frac{1}{\beta}\right)\right][\exp(-\beta E)]_0^\infty = \frac{\partial}{\partial\beta}\left[-\ln\left(\frac{1}{\beta}\right)\right] = \frac{1}{\beta} = k_\mathrm{B}T. \end{aligned} \tag{6.29}$$

Replacing the average energy equation (6.29) in (6.28), finally we obtain the *Rayleigh–Jeans law*,

$$u_\nu\mathrm{d}\nu = k_\mathrm{B}T\frac{8\pi}{c^3}\nu^2\mathrm{d}\nu. \tag{6.30}$$

We can now rewrite equation (6.30) as follows:

$$u_\nu\mathrm{d}\nu = k_\mathrm{B}\frac{8\pi}{c^3}\nu^3\frac{T}{\nu}\mathrm{d}\nu. \tag{6.31}$$

Therefore the Rayleigh–Jeans formula determines the Wien function. In the classical limit this function is $F(x) = 1/x$, where $x = \frac{\nu}{T}$. However, this form of F does not exactly represent the experimental data. In contrast, it dramatically deviates for high frequencies, as shown in figure 6.6.

In figure 6.6 we can see that at low frequencies Rayleigh's law is very close to the experimental curves, while it diverges at high frequencies. This means that the model used is consistent with the results of the measurement of cavity radiation for only

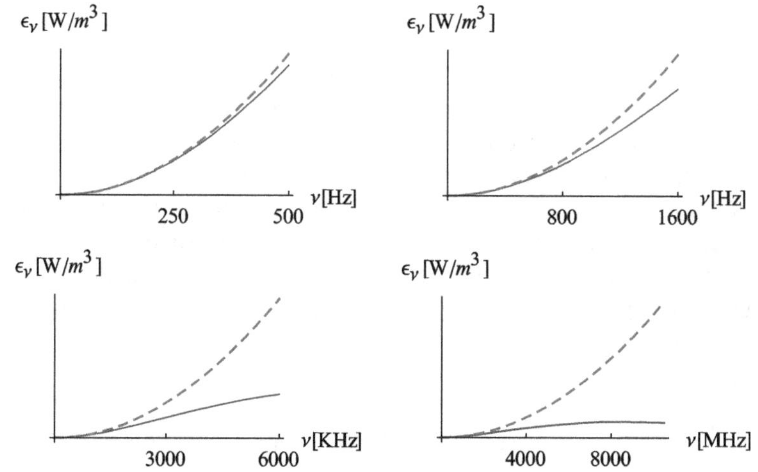

Figure 6.6. The Rayleigh–Jeans law (red dashed lines) and the experimental data (blue lines) as a function of frequency at $T = 3000$ K.

low radiation frequencies. To calculate the total energy we integrate the spectral density on all frequencies:

$$U = \int_0^\infty u_\nu \mathrm{d}\nu = \int_0^\infty k_\mathrm{B} T \frac{8\pi}{c^3} \nu^2 \mathrm{d}\nu = \infty. \tag{6.32}$$

The result of equation (6.32) contains a divergence of the radiation energy. This result contrasts with experience and is often referred to as an *ultraviolet catastrophe*.

In the attempt to find a formula that fit the experimental data (figure 6.7) Wien started from a semi-empirical expression for the average energy of a vibration of frequency ν of the type

$$b(\nu)\mathrm{e}^{-a(\nu)/T}, \tag{6.33}$$

where $a(\nu)$ and $b(\nu)$ are proportional to ν. This formula allowed Wien to obtain a trend of the energy emitted by the black-body at high frequencies, but for not very high frequencies, as shown in figures 6.6 and 6.7. This model also shows remarkable differences compared to the experimental results. The proposed models therefore failed to predict the presence of a peak in the experimental curve.

6.6 Planck's law

In 1900 Max Planck announced that by making somewhat strange assumptions, he could derive a function u_ν that agreed with the experimental data. He first found an empirical function that fit the data and then searched for a way to modify the usual calculation so as to predict his empirical formula. Planck started from the consideration of the existence of atoms, i.e. the fact that there existed a unity of matter [5]. Planck's idea was to also generalize this discrete structure for electromagnetic energy. Planck found that he could derive his empirical formula by calculating the average energy w_m making the following hypothesis: the energy of

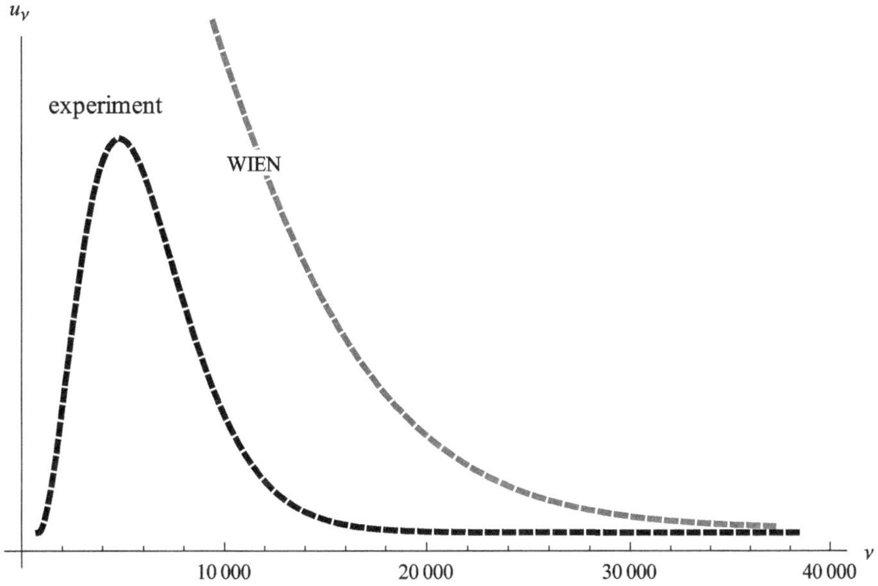

Figure 6.7. Comparison between the experimental curve (black dashed line) of spectral radiance and the curve proposed by Wien (red dashed line).

the oscillating charges, and hence the radiation that they emit, are discrete variables; that is, it can take on only the values 0, ε, 2ε, $\ldots n\varepsilon$, where n is an integer, and further that ε is proportional to the frequency of the oscillators and, hence, to that of the radiation [6]. Planck therefore wrote the energy as

$$U_n = nh\nu \qquad n = 0, 1, 2\ldots n, \tag{6.34}$$

where the proportionality constant h is now called *Planck's constant.* In this way

$$\varepsilon = h\nu \tag{6.35}$$

is the *elementary quantum of energy*. The constant h has the dimension of an *action*, or *energy times time* and its numerical value is estimated by

$$h = 6.626 \times 10^{-34}\ \mathrm{J\ s}.$$

The crucial point of Planck's reasoning is to discretize the energy, therefore calculating the average energy of the black-body is achieved by replacing the integrals with the sums:

$$\int \mathrm{d}E \longrightarrow \varepsilon \sum_n .$$

For the calculation the average energy proceeds in a similar way to that we used for obtaining equation (6.29):

$$w_m = \frac{\sum_n \varepsilon_n e^{-\beta\varepsilon_n}}{\sum_n e^{-\beta\varepsilon_n}} = \frac{\sum_n n\varepsilon_0 e^{-\beta n\varepsilon_0}}{\sum_n e^{-\beta n\varepsilon_0}} = -\frac{\partial}{\partial\beta}\ln\sum_{n=0}^{\infty} e^{-\beta n\varepsilon_0}. \tag{6.36}$$

We note that the last component of equation (6.36) represents the sum of a geometric progression whose rate is $e^{-\beta\varepsilon_0} < 1$. Therefore, we have

$$\sum_{n=0}^{\infty} e^{-\beta n\varepsilon_0} = \frac{1}{1 - e^{-\beta\varepsilon_0}}. \tag{6.37}$$

By replacing equation (6.37) into equation (6.36) and calculating the derivative we obtain

$$-\frac{\partial}{\partial\beta}\ln\frac{1}{1 - e^{-\beta\varepsilon_0}} = \frac{\varepsilon_0}{e^{\beta\varepsilon_0} - 1}. \tag{6.38}$$

By entering this result in equation (6.28) we obtain the famous Planck formula for black-body radiation:

$$u_\nu d\nu = \frac{\varepsilon_0}{e^{\beta\varepsilon_0} - 1}\frac{8\pi}{c^3}\nu^2 d\nu. \tag{6.39}$$

We note that equation (6.39) takes the form of the general Wien law's equation (6.8) if $\varepsilon = h\nu$ and therefore we obtain *Planck's law*:

$$u_\nu = \frac{8\pi}{c^3}\frac{h\nu^3}{e^{\frac{h\nu}{k_B T}} - 1}. \tag{6.40}$$

Equations (6.34), (6.35) and (6.40) can be considered the *birth of quantum mechanics*. The meaning of Planck's formula is this: the energy of the electromagnetic field is quantized. Monochromatic light, with frequency ν (i.e. wavelength $\lambda = c/\nu$) is made by a set of quanta (which Einstein called *photons*), each with energy $h\nu$.

The equation (6.40) accurately interpolates the experimental points at each temperature as shown in figure 6.8. Furthermore, Planck's law contains, as particular cases, Wien's laws, the Rayleigh–Jeans and Stefan–Boltzmann laws, as we shall see in the next paragraph. As an example we compare the average energies of a classical harmonic oscillator and of a quantum harmonic oscillator with a frequency $\nu = k_B T/h$:

$$u_{\text{class}} = k_B T,$$

from Planck's formula

$$u^\nu_{\text{quant}} = \frac{h\nu}{e^{\frac{h\nu}{k_B T}} - 1} = \frac{k_B T}{e^1 - 1} = 0.58 k_B T,$$

$$\frac{u_{\text{quant}}}{u_{\text{class}}} = 0.58.$$

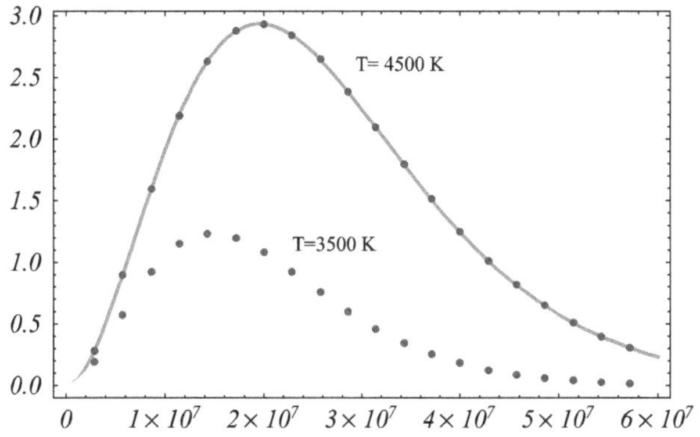

Figure 6.8. Spectral power density versus frequency as predicted by Planck's formula (continuous line). The graph shows perfect agreement between experimental points (purple) and the theoretical curves presented for two different temperatures. The abscissa shows frequency in GHz while the ordinate shows the spectral density power e_ν W $\mathrm{m}^{-3} \times (8\pi h)/c^3 10^{37}$.

6.6.1 Obtaining the Stefan–Boltzmann law from Planck's formula

We wish to deduce the Stefan–Boltzmann law from Planck's law. Here we show that the total energy density in a black-body cavity is proportional to T^4 in accordance with equation (6.7). The advantage of this calculation is to obtain Stefan's constant trough fundamental quantities in Nature. Let us thus set

$$U = \int_0^\infty u_\nu \mathrm{d}\nu = \int_0^\infty \frac{8\pi}{c^3} \frac{h\nu^3}{\mathrm{e}^{\frac{h\nu}{k_\mathrm{B}T}} - 1} \mathrm{d}\nu. \tag{6.41}$$

By replacing $\frac{h\nu}{k_\mathrm{B}T} = y$, from which $\mathrm{d}y = \frac{k_\mathrm{B}T}{h}\mathrm{d}\nu$ and by substituting in equation (6.41), we have

$$U = \frac{8\pi}{c^3 h^3}(k_\mathrm{B}T)^4 \int_0^\infty \frac{y^3}{\mathrm{e}^y - 1}\mathrm{d}y. \tag{6.42}$$

Since the integral is now dimensionless, this shows that U is proportional to T^4. The value of the integral is $\pi^4/15$. Then

$$U = \sigma T^4, \qquad \sigma = \frac{8}{15}\frac{\pi^5 k_\mathrm{B}^4}{c^3 h^3} = 5.67 \times 10^{-8}\ \mathrm{W\ m^{-2}\ K^{-4}}.$$

6.6.2 Special cases of Planck's law

We have seen that at small frequencies (see figure 6.7) the Rayleigh–Jeans law very well approximates the experimental curve for the spectral power density. We can show that Planck's law contains the Rayleigh–Jeans law as a special case. In fact, being $\frac{h\nu}{k_B T} \ll 1$ we can expand the exponential at the denominator of the right-hand side of equation (6.40) as follows

$$e^x \approx 1 + x + O(x^2) \qquad \text{for} \qquad x \ll 1.$$

In this way, equation (6.40) can be written as follows:

$$u_\nu = \frac{8\pi}{c^3}\frac{h\nu^3}{e^{\frac{h\nu}{k_B T}} - 1} \approx \frac{8\pi}{c^3}\frac{h\nu^3}{1 + h\nu/k_B T - 1} = \frac{8\pi}{c^3}k_B T\nu^2. \tag{6.43}$$

Equation (6.43) is exactly the Rayleigh–Jeans law. We observe that the latter does not contain the constant h; this means that this law has validity only in the classical limit.

On the other hand, for very high frequencies we can neglect 1 in the denominator of Planck's law (6.40), so that

$$u_\nu = \frac{8\pi}{c^3}\frac{h\nu^3}{e^{\frac{h\nu}{k_B T}} - 1} \approx \frac{8\pi}{c^3}h\nu^3 e^{-\frac{h\nu}{k_B T}}, \tag{6.44}$$

which corresponds to the semi-empirical formula (6.33) proposed by Wien. The agreement with experimental data is excellent for high frequencies, as one can argue from figure 6.6.

6.6.3 Wien's displacement law from Planck's formula

In section 6.4.1 we have deduced the Wien displacement law by taking a formal derivative of the unknown function of Wien $F(\nu/T)$. In this way, however, it is not possible to determine the Wien constant. Let us now deduce Wien's displacement law by repeating the same procedure, with the function $F(\nu) = \Gamma\frac{\nu^3}{e^{h\nu/k_B T} - 1}$, where $\Gamma = \frac{8\pi h}{c^3}$ is independent from ν. In this way we obtain the numerical value of the Wien constant, which is in excellent agreement with the measured values.

Taking the derivative with respect to ν and setting $\frac{d}{d\nu}F(\nu) = 0$, we have

$$\frac{d}{d\nu}F(\nu) = \Gamma\frac{(3\nu^2(e^{h\nu/k_B T} - 1) - \nu^3 e^{h\nu/k_B T}h/k_B T}{(e^{h\nu/k_B T} - 1)^2} = 0, \tag{6.45}$$

by now setting $x = \frac{h\nu}{k_B T}$, equation (6.45) becomes

$$e^x = \frac{3}{3 - x},$$

the solution for which can be obtained as shown in figure 6.9, where the curves e^x and $\frac{3}{3-x}$ are shown. The intersection point of these curves represents the solution that is being sought. From figure 6.9 we can infer the existence of a solution, but we cannot establish the exact value of the abscissa at the point of intersection. Therefore,

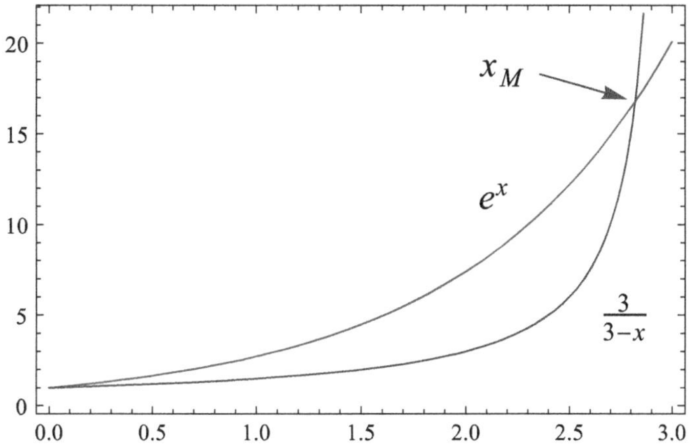

Figure 6.9. Graphic solution of the equation $e^x = \frac{3}{3-x}$.

we use Mathematica software to obtain the numerical solution. The following string gives us the solution

$$\text{FindRoot}\left[\exp(x) - \frac{3}{3-x} = 0, \{x, 2.85\}\right], \Longrightarrow \{x_M = 2.82\}.$$

We are now able to obtain Wien's displacement law:

$$\begin{aligned} \frac{h\nu_{\max}}{k_B T} &= 2.82, \\ \frac{\nu_{\max}}{T} &= \frac{2.82 k_B}{h} = 5.87 \times 10^{10}\ \text{Hz K}^{-1}. \end{aligned} \tag{6.46}$$

Equation (6.46) allows us to define the Wien constant through the fundamental constants k_B and h.

However, the most usual form of the Wien shift is written in terms of the wavelength, as equation (6.21). Recalling equation (6.39), we transform (6.40) in terms of the wavelength λ:

$$u_\nu \mathrm{d}\nu = -u_\lambda \mathrm{d}\lambda,$$

or

$$u_\lambda = -u_\nu \frac{\mathrm{d}\nu}{\mathrm{d}\lambda} = u_\nu \frac{c}{\lambda^2},$$

the minus sign indicates that, although u_λ and u_ν are both positive, $\mathrm{d}\lambda$ and $\mathrm{d}\nu$ have opposite signs. Then, we obtain

$$u_\lambda = \frac{8\pi hc}{\lambda^5} \frac{1}{(e^{hc/\lambda k_B T} - 1)}. \tag{6.47}$$

Equating the first derivative with respect to λ of u_λ to zero, we obtain a transcendental equation which, as in equation (6.45), we solve by means of graphical analysis. Therefore, by writing

$$\frac{d}{d\lambda}\frac{8\pi hc}{\lambda^5}\frac{1}{(e^{hc/\lambda k_B T}-1)} \rightarrow (e^{hc/\lambda k_B T} hc/\lambda k_B T - 5) + 5 = 0, \tag{6.48}$$

and by setting $x = hc/\lambda k_B T$, we have

$$e^x(5 - x) = 5. \tag{6.49}$$

The numerical solution of equation (6.49) is $x_{max} = 4.9651$, as can be observed by looking at figure 6.10; hence

$$\frac{hc}{k_B T \lambda_{max}} = 4.9651. \tag{6.50}$$

Finally, we write

$$\lambda_{max} T = \frac{hc}{4.9651 k_B} = 2.90 \times 10^{-3}\ \text{m K}. \tag{6.51}$$

We explicitly observe that Wien's constant cannot be obtained simply by substitution, in equation (6.46), of $\nu_{max} \longrightarrow \frac{c}{\lambda_{max}}$, because from equations (6.46) and (6.51) we have: $\lambda_{max} \cdot \nu_{max} = 0.568c$. Of course this result is not consistent with the relation $\lambda\nu = c$. Thus the result in equation (6.46) cannot be obtained by the mere substitution $\nu_{max} = \frac{c}{\lambda_{max}}$

6.7 Some applications

In this section we will apply the results presented in the previous sections to solving some simple exercises. We are confident that in this way a general reader can consolidate the ideas and concepts he/she has learned. We indeed think that

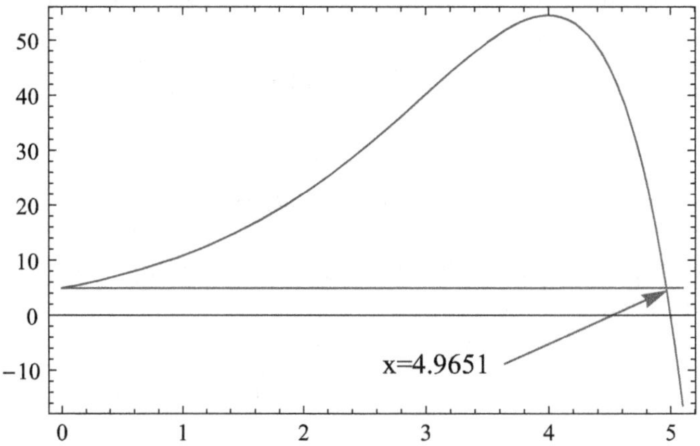

Figure 6.10. Graphic solution of equation $e^x(5 - x) = 5$.

problem-solving is really important for obtaining a better intuition of what is being taught.

6.7.1 The Sun as a black-body

We know that the radius of the Sun $R_\odot$ is 6.96×10^6 m and that the power emitted from its surface is $P_s = 3.77 \times 10^{26}$ W. Assuming the Sun as a black-body we calculate:

1. T_S,
2. λ_{max},
3. $u_{\lambda_{max}}$.

Solution

1.

$$\varepsilon_{em} = \frac{P_s}{4\pi R_\odot^2} = \sigma T_S^4 \Rightarrow T_S = \sqrt[4]{\frac{P_s}{4\pi R_\odot^2 \sigma}}.$$

Replacing the numerical values $T_S = \sqrt[4]{\frac{3.77 \times 10^{26}\ \text{W}}{4\pi 6.96 \times 10^6 \times 5.67 \times 10^{-8}\ \text{W K}^{-4}}} = 5750$ K.

2. $T_S \lambda_{max} = 2.90 \times 10^{-3}$ m K therefore $\lambda_{max} = 2.90 \times 10^{-3}$ m K$/T_S = \frac{2.90 \times 10^{-3}\ \text{m K}}{5750\ \text{K}} =$ 504 nm. This wavelength falls within the green band.

3. $u_{\lambda_{max}} = \frac{8\pi hc}{(\lambda_{max})^5} \frac{1}{(e^{hc/(\lambda_{max})k_B T} - 1)}$, replacing λ_{max} with $2.90 \times 10^{-3}/T_S$ we have $u_{\lambda_{max}} = bT_S^5 = 8.09 \times 10^{13}$ W m^{-3}.

6.7.2 Luminous intensity on Earth

We would like to calculate the luminous intensity on the Earth's surface when exposed to solar radiation.

Solution

The power emitted by the Sun's surface is

$$U_S = \varepsilon_{em} 4\pi R_\odot^2 = 6.193 \times 10^7\ \text{W m}^{-2},$$

where ε_{em} is the power intensity. By the energy conservation we have

$$\varepsilon_{em} 4\pi R_\odot^2 = I_E(4\pi r_{ES}^2).$$

Solving for I_E, we have

$$I_E = \frac{R_\odot^2}{r_{ES}^2} U_S = \frac{6.96 \times 10^6}{1.50 \times 10^{11}} 6.193 \times 10^7 = 1333\ \text{W m}^{-2}.$$

If we consider the absorption of the atmosphere, equal to about 80% of the radiation emitted by the Sun, we have

$$I_{Ground} \approx 0.80 I_E = 1.066\ \text{kW m}^{-2}.$$

6.7.3 TRAPPIST-1

TRAPPIST-1 is a small red dwarf that has 8% of the Sun's mass[1], just above the limit that allows it to trigger the hydrogen fusion to be converted into helium within it. Its actual temperature is just 2550 K, and its radius is 12% of the solar radius. TRAPPIST-1 possesses its own planetary system, with terrestrial-like extra-solar planets.

Solution

By knowing the temperature T_{Tr} of TRAPPIST-1, we may calculate the power density I_{Tr} emitted by this star as follows:

$$I_{Tr} = \sigma T_{Tr}^4 = 5.67 \times 10^{-8}\ \mathrm{W\ m^{-2}\ K^{-4}} \times (2.55 \times 10^3)^4\ \mathrm{K^4}$$
$$= 2.40 \times 10^6\ \mathrm{W\ m^{-2}}$$

$$\lambda_{max}\, T_{Tr} = 2.9 \times 10^{-3} \rightarrow \lambda_{max} = 1140\ \mathrm{nm}.$$

At what average distance is a planet of the TRAPPIST-1 system for which $I = I_{Earth}$?

$$I_{Tr} = \left(\frac{R_0}{r}\right)^2 U_{Tr} \rightarrow r = \sqrt{\frac{U_{Tr}}{I_{Tr}}}\, R_0$$

$$= \sqrt{\frac{2.40 \times 10^6}{1.333 \times 10^3}}(0.117)(6.96 \times 10^8\ \mathrm{m}) = 3.45 \times 10^9\ \mathrm{m}$$

$$\frac{r}{r_{ES}} = \frac{3.45 \times 10^9\ \mathrm{m}}{1.50 \times 10^{11}} \approx 0.023.$$

It could therefore be the third planet of the system TRAPPIST-1.

6.7.4 Comparison of stars

Measurement of the wavelength at which the spectral distribution u_λ from a certain star is maximum indicates that the star's surface temperature is 3000 K. If the star is also found to radiate 100 times the power P_S radiated by the Sun, how big is the star? (T_S = 5750 K).

Solution

Assuming the Sun and the star both radiate as black bodies, their surface temperatures have been determined to be 5750 K and 3000 K, respectively. Measurements also indicate that $P_{star} = 100P_S$. Thus, for the star we have that

$$U_{star} = P_{star}/4\pi R_{star}^2 = 100P_S/4\pi R_{star}^2 = \sigma T_{star}^4.$$

[1] TRAPPIST-1 is an ultra-cool red dwarf star, in the constellation Aquarius, that is slightly larger, but much more massive, than the planet Jupiter. On 22 February 2017, astronomers announced that the planetary system of this star is composed of seven temperate terrestrial planets. Three of the planets orbit within the habitable zone.

On the other hand, for the Sun we have

$$P_S/4\pi R_S^2 = \sigma T_S^4.$$

Thus, we have

$$\frac{\sigma T_{\text{star}}^4}{\sigma T_S^4} = \frac{100P_S}{4\pi R_{\text{star}}^2}\frac{4\pi R_S^2}{P_S} = 100\frac{R_S^2}{R_{\text{star}}^2},$$

then we have

$$\left(\frac{T_S}{T_{\text{star}}}\right)^4 = \frac{R_{\text{star}}^2}{100R_S^2} \to R_{\text{star}} = 10R_S\left(\frac{T_S}{T_{\text{star}}}\right)^2 = 37.4R_S.$$

This star has a radius of about half the radius of the orbit of Mercury—it is a red giant.

References

[1] Kuhn T S 1987 *Black-body Theory and Quantum Discontinuity* (Chicago, IL: University of Chicago Press)

[2] Feynman R, Leighton R and Sands M 2011 *The Feynman Lectures on Physics* vol II (New York: Basic Books)

[3] Novozhilov Y V and Jappa Y A 1986 *Electro-dynamics* (Moscow: MIR)

[4] Feynman R, Leighton R and Sands M 2011 *The Feynman Lectures on Physics* vol I, part 2 (New York: Basic Books)

[5] Born M 1969 *Atomic Physics* (London: Blackie)

[6] Noce C 2018 *Introduzione alla Fisica Moderna* (Roma: Aracne)

IOP Publishing

Modern Physics
A critical approach
Canio Noce

Chapter 7

The dawn of quantum mechanics

Delia Guerra and Maria Teresa Mercaldo

7.1 Introduction

At the end of the nineteenth century, our knowledge of physical phenomena seemed to be complete. For example, A A Michelson stated: 'While it is never safe to affirm that the future of Physical Science has no marvels in store even more astonishing than those of the past, it seems probable that most of the grand underlying principles have been firmly established, and that further advances are to be sought chiefly in the rigorous applications of these principles to all the phenomena which come under our notice' [1]. Indeed, all the discoveries made up to the turn of the twentieth century allowed scientists to divide physical phenomena into two categories, shown schematically in figure 7.1. The properties of matter were explained by 'corpuscular' and deterministic models, while radiation, which was described well by Maxwell's equations, was associated with 'undulatory' phenomena. In particular, Maxwell was the first, using his equations, to propose that light was an electromagnetic phenomenon.

However, in the early twentieth century several categories of experimental evidence were found that could not be arranged into one of these two frameworks. The first of them was black-body radiation, for which wave theory failed to explain the observed spectrum of thermal radiation.

In 1900 Max Planck was able to obtain perfect agreement with experimental data by postulating that the energy exchanges between matter and radiation cannot vary continuously, but only through *quanta*, i.e. discrete quantities, of energies. In particular, he postulated that radiation comes in 'packets' with energy $E = h\nu$, where ν is the frequency and h is Planck's constant, obtained through the analysis of experimental data. We reproduce part of the Nobel Lecture by Planck (1918 Nobel laureate) [2], given on 2 June 1920, on his constant h:

'The explanation of the second universal constant of the radiation law was not so easy. Because it represents the product of energy and time (according to the first calculation it was 6.55×10^{-27} erg s), I described it as the elementary quantum of

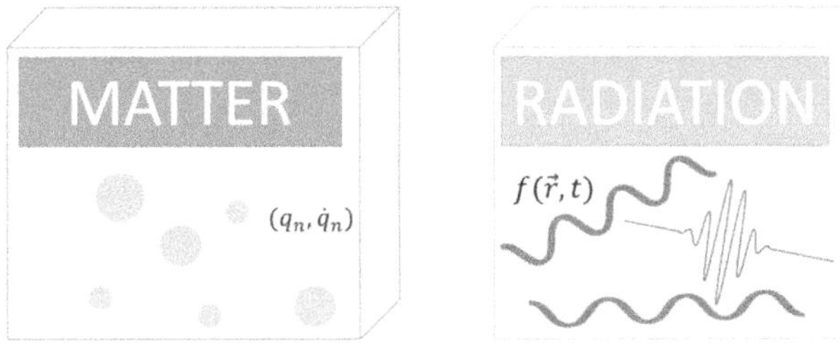

Figure 7.1. The two frameworks in which, at the end of the nineteenth century, all physical phenomena were thought to fall: *matter* (constituted by particles) and *radiation* (described by wave theory).

action. Whilst it was completely indispensable for obtaining the correct expression for entropy, (omissis…) it proved elusive and resistant to all efforts to fit it into the framework of classical theory. As long as it was looked upon as infinitely small, that is, for large energies or long periods of time, everything went well; but in the general case, however, a gap yawned open in some place or other, which was the more striking, the weaker and faster the vibrations that were considered. The foundering of all efforts to bridge the chasm soon left little doubt. Either the quantum of action was a fictional quantity, then the whole deduction of the radiation law was in the main illusory and represented nothing more than an empty non-significant play on formulae, or the derivation of the radiation law was based on a sound physical conception. In this case the quantum of action must play a fundamental role in physics, and here was something entirely new, never before heard of, which seemed called upon to basically revise all our physical thinking, built as this was, since the establishment of the infinitesimal calculus by Leibniz and Newton, upon the acceptance of the continuity of all causative connections. Experiment has decided for the second alternative...'

The explanation of black-body radiation was crucial and we may say it triggered the dawn of quantum mechanics. Indeed, to theoretically explain other experiments, such as the photoelectric effect and the Compton effect, we need to make use of Planck's hypothesis. These experiments are characterized by a 'corpuscular' behaviour of radiation. At about the same time, the observation of atomic spectra encouraged the formulation of several atomic models. Among them, Bohr's model plays a central role. In this model one obtains *quantized* energy levels, which account for the line spectra of atomic elements. After the accumulation of all the evidence which attributes a particle-like nature to radiation, de Broglie formulated another bold hypothesis, where matter also has a wave-like behaviour.

In this chapter, we analyse the phenomenology which contributed to the birth of quantum mechanics. The first sections deal with the photoelectric and Compton effects. In the following sections the atomic spectra and some atomic models (the Thomson, Rutherford and Bohr models) are recalled, with another section on the Frank–Hertz experiment. We do not analyse black-body radiation, which is described in the previous chapter. Finally, de Broglie's hypothesis about 'matter waves' is illustrated, and finally we review the double-slit experiment which, according to Richard Feynman [3], has in it the heart of quantum mechanics.

7.2 The photoelectric effect

In 1887 Heinrich R Hertz observed that when the surface of a metal is illuminated by light electrons are emitted [4]. This phenomenon was later extensively studied by Hallwachs, Augusto Righi, Lenard [5–7] and others and was called the 'photo-electric' effect by the Italian physicist Righi in 1888. The phenomenon and the experimental apparatus needed to observe the photoelectric effect are illustrated schematically in figure 7.2.

When light falls on the metal surface (the emitter in figure 7.2(b)) electrons can be released. The emitted electrons are also called photoelectrons. A variable voltage source establishes a potential difference V_{ext} between the emitter and the collector. When the latter is positive with respect to the emitter, the photoelectrons are accelerated towards the collector and the rate of electron emission can be evaluated in terms of an electric current i that flows in the circuit, which is measured by an ammeter. By increasing V_{ext}, the ammeter measures increasing values of the current, until all the emitted electrons are collected. In this case, the increase of V_{ext} has no effect on the current, which has reached its maximum value, called the saturation current. On the other hand, when the polarity of the battery is reversed, a negative potential is applied to the collector, hence the emitted electrons are repelled by it and only the most energetic ones can reach the collector. By increasing the magnitude of the voltage until $i = 0$, only the electrons with the maximum kinetic energy $K_{\max}$ reach the collector. The value of the potential relative to this case is called the stopping potential V_s. At this point, $K_{\max}$ is equal to the energy eV_s lost by the electrons in order to reach the collector:

$$K_{\max} = eV_s.$$

In the classical picture, the surface of the metal is irradiated by an electromagnetic wave and electrons absorb energy from the wave until the binding energy of the electron to the metal is exceeded. Then, the electron is removed from the surface.

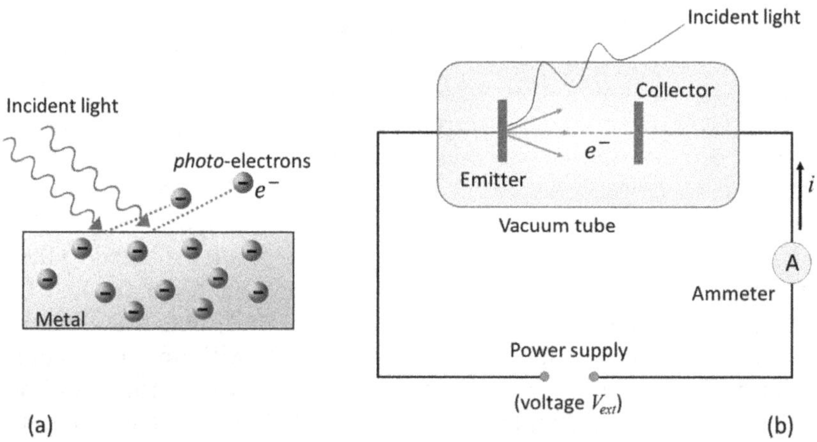

Figure 7.2. Sketch of the photoelectric effect (a) and of the experimental set-up (b) for observing the phenomenon.

The minimum value of the energy required to remove an electron is called the work function Φ. Typical values of Φ are of the order of a few electronvolts.

Experimental results for the photoelectric effect were known in 1902 and there were many aspects which could not be explained in the classical framework:

1. Considering the classical wave theory, K_{max} should be proportional to the intensity of the incident radiation, while the experimental results showed that it is totally independent of the intensity of the light source.
2. Classically, the photoelectric effect should occur for light of any frequency, but it was found that it occurs only if the frequency of the incident light is above a particular value, which is characteristic of the material and is called the cut-off frequency ν_c.
3. It was expected that the time lag between turning on the light and observing the first emitted electron should be of the order of seconds, but experimentally it was proved that the electrons are emitted within 10^{-9} s.

The correct interpretation of the photoelectric effect was given by Einstein in 1905 [8], who was awarded the 1921 Nobel Prize in Physics for this breakthrough. Based on Planck's ideas of black-body radiation, he argued that the energy of a light wave is not continuously distributed over the wave front but is concentrated in localized bundles, which can be considered as 'quanta of light', subsequently called *photons*. The energy of a photon associated with a wave with frequency ν is given by

$$E = h\nu. \tag{7.1}$$

In equation (7.1), h is Planck's constant, and its current *exact*[1] value is $h = 6.626\,070\,15 \times 10^{-34}$ J s. The frequency of a wave ν is linked to its wavelength λ through the relation $\nu = \frac{c}{\lambda}$. As a consequence, we can write the energy equation (7.1) in terms of wavelength in the following way:

$$E = \frac{hc}{\lambda}.$$

The photoelectric effect can be interpreted as a collision between a photon and an electron on the surface of the metal. The energy E of the incident photon is absorbed by the electron and, if $E < \Phi$, the photoelectric effect does not occur. If $E \geqslant \Phi$, an electron is released. In particular, when $E > \Phi$, the exceeded energy appears as kinetic energy of the emitted electron. Its maximum value is given by

$$K_{max} = h\nu - \Phi. \tag{7.2}$$

From equation (7.2), it is clear that K_{max} does not depend on the intensity of the incident light—if this intensity is doubled, the number of photons is doubled and consequently the number of electrons, but not the value of K_{max}, which is the same for all of them. This is the correct explanation of point 1 above.

[1] Since May 2019, the numerical value of the Planck constant has been fixed, hence it is defined to have the exact value given above. (See [9])

When $E = \Phi$, the electron absorbs the minimum energy required to be released. This is related to the cut-off frequency ν_c of point 2 above, whose value can be then determined by substituting for E its expression given by equation (7.1):

$$h\nu_c = \Phi.$$

Then

$$\nu_c = \frac{\Phi}{h}$$

and the corresponding wavelength value is

$$\lambda_c = \frac{hc}{\Phi}.$$

Finally, according to the Einstein interpretation, since the photoelectron is released as a result of an encounter with a *single* photon, its entire energy is delivered *instantaneously* to the *single* electron, and consequently there is no measurable time delay. This explanation accounts for point 3 above.

By using the relation equation (7.2), we have

$$V_s = \frac{h}{e}(\nu - \nu_c),$$

and hence, measuring the stopping potential V_s for different frequencies of the incident light, it is possible to obtain a value for Planck's constant. Such an experiment was made by Robert A Millikan in 1915 [10]. The plot of the experimental data is indeed a straight line, whose slope gives an estimation of h, being the value of the electric charge e already known at that time, thanks to the 'oil drop' experiment by Millikan himself (see chapter 14). In figure 7.3 we present the original graph obtained by Millikan. The obtained value of h was 6.569×10^{-34} J s, which is very close to the currently given value. Millikan was the 1923 Nobel laureate in Physics 'for his work on the elementary charge of electricity and on the photoelectric effect'.

7.3 The Compton effect

The Compton effect, like the photoelectric one, also deals with interaction between electromagnetic radiation and matter. The physicist Arthur Compton was the first to perform an experiment in order to analyse this kind of scattering [11, 12]. In his experiment, a beam of x-rays hits a target made of a light chemical element. In the actual target there are no 'free electrons', but in some materials the outer or valence electrons are very weakly bound to the atoms and their kinetic energies are small compared to the energies of the incident x-rays, so they can be treated as free electrons. The experimental set-up was also equipped with a movable detector, in order to measure the energy of the scattered x-rays at various angles. According to classical wave theory, part of the energy of the incident radiation is given to the electrons of the target as kinetic energy, and the remaining amount is re-irradiated as

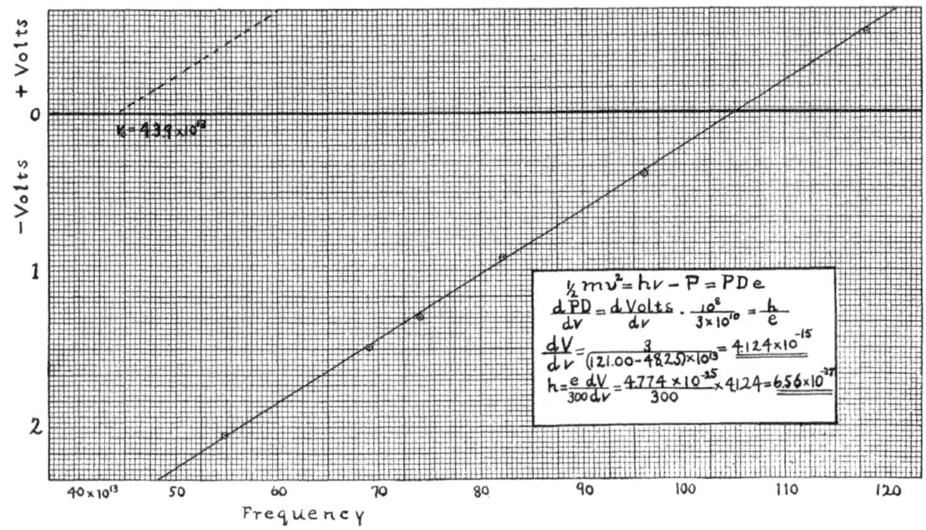

Figure 7.3. Millikan's original plot on the study of the photoelectric effect. Reproduced with permission from [10].

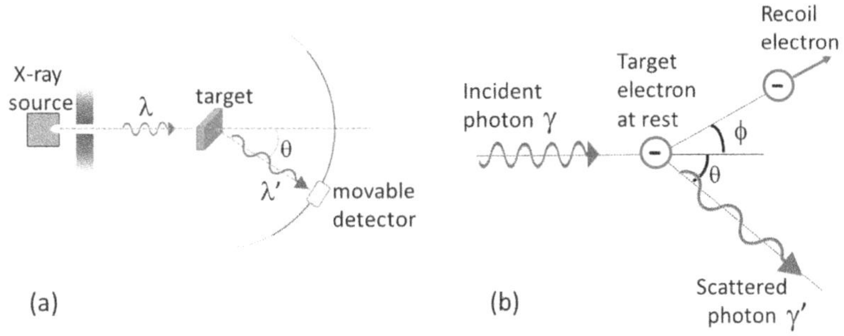

Figure 7.4. (a) Schematic illustration of the Compton scattering experimental apparatus. (b) Compton effect as a *collision* process, where a single photon γ of wavelength λ hits an electron at rest.

an electromagnetic wave which is then less energetic (hence with smaller *intensity*) with respect to the incident one, but with the same wavelength. The results obtained by Compton showed that the scattered waves had a longer wavelength (i.e. a lower frequency) and the increase of the scattered wavelength was independent of the incident wavelength.

Compton interpreted this scattering process as an elastic collision between a photon and an electron, assumed to be at rest. Figure 7.4 shows an illustration of the scattering and of the apparatus used in the experiment. Using energy and the momentum conservation laws, we have

$$E_{\text{in}} = E_{\text{fin}} \quad \Rightarrow \quad E_\gamma + E_{\text{e}} = E'_\gamma + E'_{\text{e}}, \tag{7.3}$$

$$\vec{p}_{\text{in}} = \vec{p}_{\text{fin}} \quad \Rightarrow \quad \vec{p}_\gamma + \vec{p}_{\text{e}} = \vec{p}'_\gamma + \vec{p}'_{\text{e}}. \tag{7.4}$$

In equation (7.3), E_γ and E_e are the energies of the photon and the electron before collision, respectively, while E'_γ and E'_e are the energies after collision. Similarly, $\vec{p}_\gamma$ and $\vec{p}_e$ in equation (7.4) are the momenta of the photon and the electron before collision, and $\vec{p}'_\gamma$ and $\vec{p}'_e$ the momenta after collision. The energy of a photon is given by equation (7.1), while its momentum is given by

$$\vec{p} = \frac{E}{c}\hat{n}, \tag{7.5}$$

where $\hat{n}$ is the propagation direction of the photon. The energy and the momentum of the electron are given by the relativistic relations

$$E = \frac{m_e c^2}{\sqrt{1 - \frac{u^2}{c^2}}} = \gamma m_e c^2, \tag{7.6}$$

$$\vec{p} = \frac{m_e \vec{u}}{\sqrt{1 - \frac{u^2}{c^2}}} = \gamma m_e \vec{u}, \tag{7.7}$$

where m_e is the rest mass of the electron, u is its speed, c is the speed of light and $\gamma = (1 - u^2/c^2)^{-1/2}$ is the Lorentz factor. We can rewrite the conservation laws (7.3) and (7.4) by projecting equation (7.4) along the propagation direction of the photon and its perpendicular direction, and by substituting the explicit expression of energies and momenta, given by equations (7.1), (7.5)–(7.7):

$$h\nu + m_e c^2 = h\nu' + \gamma m_e c^2, \tag{7.8}$$

$$\frac{h\nu}{c} = \frac{h\nu'}{c}\cos\theta + \gamma m_e u \cos\phi, \tag{7.9}$$

$$0 = -\frac{h\nu'}{c}\sin\theta + \gamma m_e u \sin\phi, \tag{7.10}$$

where θ and ϕ are the angles that the scattered photon and electron form with the initial propagation direction of the incident photon (see figure 7.4(b)). The system of equations (7.8)–(7.10) cannot be solved uniquely, since it is under-determined, but we may choose to measure the energy and direction of the scattered photon, and eliminate the quantities related to the recoil electron from the problem. So, after some algebra, the last two equations above can be rewritten as

$$(h\nu)^2 + (h\nu')^2 - 2h^2\nu\nu'\cos\theta = \gamma^2 c^2 (m_e u)^2. \tag{7.11}$$

We conveniently reformulate equation (7.11) (again with algebraic manipulation and squaring both sides of the equation) as

$$(h\nu)^2 + (h\nu')^2 + m_e^2 c^4 - 2h^2\nu\nu' + 2h\nu m_e c^2 - 2h\nu' m_e c^2 = \gamma^2 m_e^2 c^4.$$

Using the relativistic relation between energy and momentum,

$$E^2 = c^2p^2 + m^2c^4,$$

we obtain

$$(\nu - \nu')m_e c^2 = h\nu\nu'(1 - \cos\theta). \tag{7.12}$$

By using the relation between frequency and wavelength $\nu = \frac{c}{\lambda}$, equation (7.12) can also be written as

$$\lambda' - \lambda = \Delta\lambda = \frac{h}{m_e c}(1 - \cos\theta), \tag{7.13}$$

where λ is the wavelength of the incident photon and λ' is the wavelength of the scattered photon. Equation (7.13) gives the shift of the wavelength of a photon due to the interaction with an electron. As the cosine function is bounded in $[-1, 1]$, it is always $\lambda' > \lambda$. This means that the scattered photon has less energy than the incident photon. The difference between the two wavelengths is a function of the scattering angle θ. The coefficient $h/m_e c = 0.002\,426$ nm is known as the Compton wavelength of the electron. In general, the quantity $\frac{h}{m^* c}$ is considered as the Compton wavelength of a particle with mass m^*, however, we stress that it is a change in wavelength ($\Delta\lambda$) and not a *true wavelength*.

The experimental results for the wavelength of the scattered photons at different observation angles θ are shown in figure 7.5. There are two different peaks, one centred at λ', which is in excellent agreement with Compton scattering equation (7.13), while the second peak is centred around the value of the incident wavelength λ. The latter corresponds to scattering processes with electrons tightly bound to the atom, so the incident photon does not lose energy while going through the target. Similar results were also obtained later using gamma-rays as the incident beam. Compton was awarded the Nobel Prize in Physics in 1927 'for his discovery of the effect named after him'. With this experiment Compton was able to convince even the most sceptical physicist of that time of the particle-like nature of radiation.

7.4 Atomic spectra

At the end of nineteenth century there were several studies on atomic spectra. Atoms can emit radiation through continuous and discrete spectra. In the first case, all

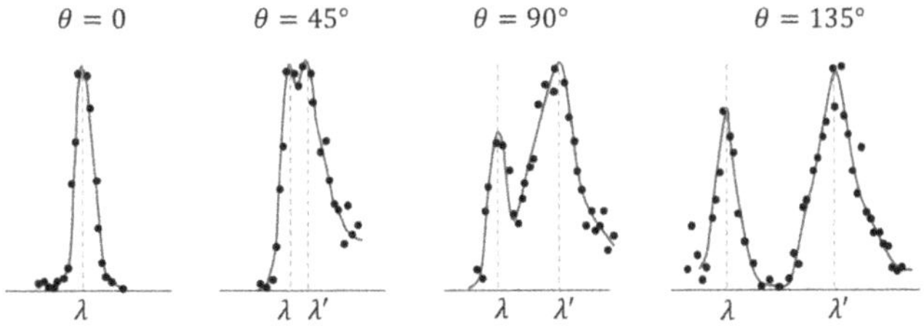

Figure 7.5. Reproduction of Compton's original results for x-ray scattering on a graphite target [12].

wavelengths are emitted—this is the case for the radiation emitted by a hot glowing object, for example. In the second case, only a few discrete wavelengths are emitted. The apparatus which we can use in order to see a discrete spectrum is shown in figure 7.6.

An electric discharge is created in a tube containing the vapour of an element. The produced light passes through a prism, which displays the various wavelengths at different positions. Through this procedure, we obtain the emission spectrum of an element. We can also obtain the absorption spectrum of an element. In this case, a beam of light passes through a sample of gas and only certain wavelengths are absorbed. The resulting spectrum is constituted by several dark lines in correspondence to the absorbed wavelengths. The emission and absorption spectra of the hydrogen are shown in figure 7.7. Johannes J Balmer noticed that the wavelengths relative to the emission spectrum of the hydrogen can be individuated using the following empirical formula, which he proposed in 1885 [13]:

$$\lambda = 364.5\frac{n^2}{n^2 - 4} \qquad n \geqslant 3, \tag{7.14}$$

where n assumes an integer value greater than 3 and λ is measured in nanometres. Equation (7.14) is called the Balmer formula and the associated series of lines is the Balmer series. The wavelength $\lambda = 364.5$ nm corresponds to $n \to \infty$ and is called the series limit. The Balmer formula (7.14) was generalized by Rydberg in 1888 with a similar formula which fits all of the groupings of the lines in the hydrogen spectrum:

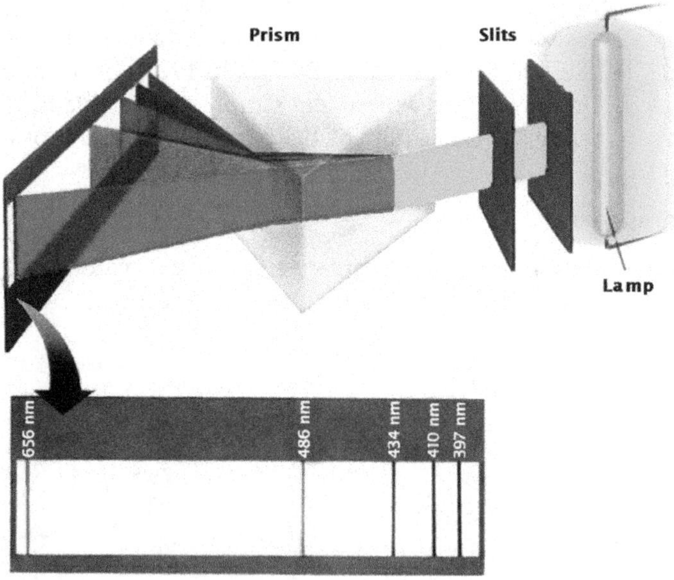

Figure 7.6. Apparatus for observing emission spectra.

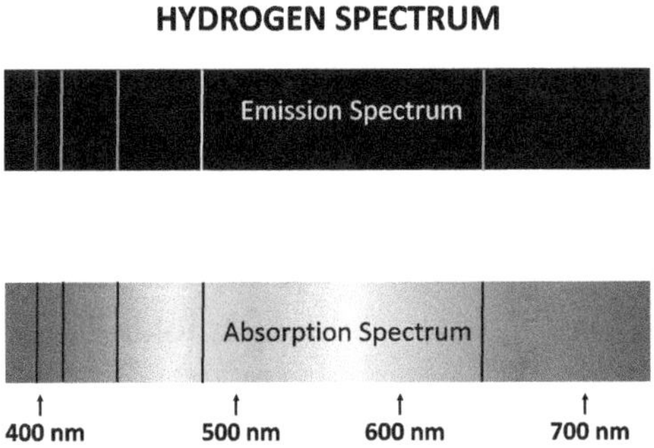

Figure 7.7. Emission and absorption lines of the hydrogen atom in the visible spectrum.

$$\lambda = \lambda_{\text{limit}} \frac{n^2}{n^2 - n_0^2}. \tag{7.15}$$

Here, λ_{limit} is the wavelength of the appropriate series limit, while n can assume integer values starting from $n_0 + 1$. n_0 defines the series—for example, for the Balmer series, related to the emission spectrum of visible light, it is $n_0 = 2$. The spectrum of the hydrogen atom also contains an ultraviolet series, called the Lyman series, for which $n_0 = 1$ and three others in the infrared, known as Paschen ($n_0 = 3$), Brackett ($n_0 = 4$) and Pfund ($n_0 = 5$).

Another property of the hydrogen wavelengths in the spectra is the Ritz combination principle: if we convert the wavelengths of the emission spectrum to frequencies, certain pairs of frequencies added together give other frequencies of the spectrum. For example, the longest wavelength of the Balmer series can be calculated by substituting $n = 3$ in equation (7.14) and its value is 656.1 nm. The corresponding frequency is $\nu_{\text{B},\, n=3} = \frac{c}{\lambda} = 4.57 \times 10^{14}$ Hz. The longest two wavelengths of the Lyman series can be evaluated using equation (7.15) with $n_0 = 1$, $n = 2$ and $n = 3$ obtaining, respectively, $\nu_{\text{L},\, n=2} = 24.67 \times 10^{14}$ Hz and $\nu_{\text{L},\, n=3} = 29.24 \times 10^{14}$ Hz for the frequencies. Note that

$$\nu_{\text{L},\, n=2} + \nu_{\text{B},\, n=3} = \nu_{\text{L},\, n=3}. \tag{7.16}$$

Equation (7.16) is an example of the Ritz principle.

7.5 Atomic models

According to the pre-Socratic philosopher Democritus, an atom is the smallest constituent unit of ordinary matter. John Dalton was the first scientist who tried to describe the atom, relying for his model on the previous works by Lavoisier and Proust. According to the basic concepts of his description, atoms are stable, electrically neutral, and very small. For this reason, the study of their properties was difficult to realize. Dalton imagined an atom as an indivisible and completely full sphere. In 1897, Thomson's experiment demonstrated the existence of electrons, which are the same for all kinds of materials, and determined also the ratio *e*/*m* between the charge and the mass. In addition, Millikan's experiment allowed us to know the value of the charge of an electron, equal to 1.6×10^{-19} C. At that time, the line spectra of atomic elements were already known, thus it was well-known that atoms can emit and absorb electromagnetic radiation, but there was no satisfactory explanation for the phenomenon. Therefore, several models were proposed in order to understand atomic structure; these are illustrated in the following subsections.

7.5.1 The Thomson model

The first atomic model was developed by Joseph John Thomson and it is known as the 'plum-pudding' model. According to this model, Z electrons are embedded in an uniform sphere of radius R and positive charge Ze, where e is the charge of the electron. A schematic diagram is shown in figure 7.8.

We can derive the force on an electron at distance r to the centre of a sphere of charge Ze by Gauss's law:

$$F = \frac{Ze^2}{4\pi\varepsilon_0 R^3} r = kr. \tag{7.17}$$

The obtained force is a restoring one, therefore electrons oscillate about their equilibrium position with a frequency ν given by

$$\nu = \frac{1}{2\pi}\sqrt{\frac{k}{m}},$$

where k is the constant defined in equation (7.17) and m is the mass of the electron. Thus, we might expect that the radiation emitted by an atom would be constituted by electromagnetic waves of frequency equal to ν, but in the observed atomic spectra there are several frequencies, not only one. In addition to this, there is another discrepancy which marks the failure of the Thomson model and it concerns the

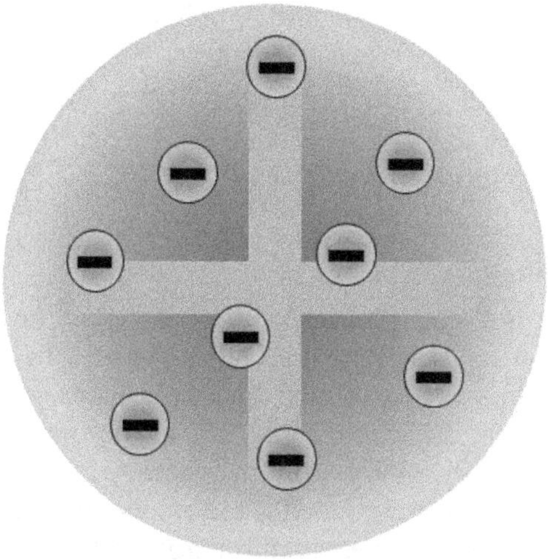

Figure 7.8. The Thomson atomic model: the *plum-pudding* model has electrons surrounded by a volume of positive charge, like negatively charged 'plums' embedded in a positively charged 'pudding' (the blue sphere in the schematic diagram).

scattering of charged particles by atoms. Suppose we send a positively charged particle toward an atom—it experiences a force due to the positive charge of the atom, which is repulsive, and another one due to the negatively charged electrons, which is then attractive. Since the mass of electrons is generally very small compared to the mass of the incident particle, the consequent deflection of the incident particle is caused mainly by the positive charges of the atom. Assuming that the contribution of the electrons is negligible, we can calculate an average value of the scattering angle θ relative to the incident particle and we indicate it as θ_{avg}. A typical scattering experiment consists of a beam of incident particles on a thin foil. Alpha particles, which are particles emitted by radioactive elements such as polonium, uranium and others, are generally used as bullets in this kind of experiment. Assuming that the atomic radius is $R = 0.1$ nm, we imagine having 5 meV alpha particles hitting a foil of gold ($Z = 79$). Under these conditions, the average value of the Thomson scattering angle is equal to $\theta_{avg} \sim 0.01°$. Scattered particles are observed at all possible values of θ in the laboratory. In the above-mentioned experimental set, many bullets are scattered by many atoms and a single bullet is scattered many times; every atom deflects the bullet by an angle equal to θ_{avg}. By statistical calculations, we know that for N scatterings the average net scattering angle θ is given by

$$\theta \sim \sqrt{N}\theta_{avg}. \tag{7.18}$$

For $N = 10^4$ and $\theta_{avg} = 0.01°$, equation (7.18) is consistent with observations. The model fails if we consider the probability that a bullet is scattered with a large angle. In fact, in order to observe scattered bullets with a total angle larger than 90°, we should have 10^4 successive scatterings, each of which pushes the bullet toward larger angles. The probability to obtain that evidence in an experiment is about $\left(\frac{1}{2}\right)^{10\,000} \sim 10^{-3000}$. Hans Geiger and Ernest Marsden [14], under the supervision of Ernest Rutherford in 1910, found experimentally that the probability of an alpha particle being scattered at angles greater than 90° was about 10^{-4}, which is totally inconsistent with the Thomson atomic model.

7.5.2 The Rutherford model

The experimental apparatus used by Geiger and Marsden in the laboratory of Rutherford is illustrated in figure 7.9

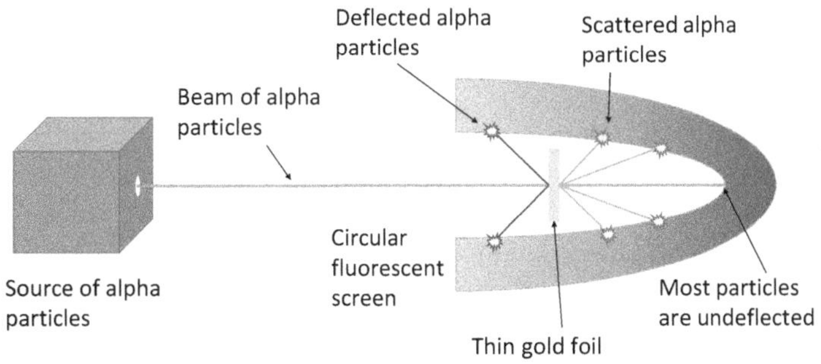

Figure 7.9. The Rutherford experiment.

Alpha particles are emitted by a radioactive material. The obtained collimated beam collides on a gold foil and, passing through it, the alpha particles are scattered in different directions. They hit a screen coated with zinc sulphide, producing fluorescence. The results of the experiment are the following:

1. Most of the alpha particles are not diverted.
2. Some of them are deviated by small angles.
3. A few of them are deviated by large angles.
4. A few of them are scattered directly in the backward direction.

Based on this evidence, Rutherford hypothesized a 'planetary' model of the atom, where most of its mass, with a positive charge, is concentrated in a small region, called the nucleus. The dimension of the nucleus is about 10^{-15} m, hence it is very small compared to the dimension of an atom itself ($\sim 10^{-10}$ m). The number of electrons balances the positive charge, so the atom is electrically neutral, and electrons move around the nucleus on different orbits, like planets around the Sun. The Coulomb force provides the centripetal force required by electrons to move around the nucleus. The previously listed experimental results are then well explained by this model[2]: when an alpha particle hits a nucleus of the atoms of the gold foil, the repulsion is strong and the scattering angle is large—this explains points 3 and 4. Experimentally this event has a low probability of occurring, hence we can infer that the nucleus occupies a small part of the atom. When an alpha particle is not deviated, it passes through the remaining part of the atom, which is basically empty, hence the trajectory of the particle is not substantially modified—this explains points 1 and 2.

However, the Rutherford model also presents two big failures. One issue is regarding the stability of the atom. Moving around the nucleus on a closed elliptical orbit, the electron is accelerated and then it loses energy, emitting radiation. As a consequence, the radius of the orbit will be smaller and smaller until the atom collapses. Hence, the description of the atom through the planetary model of Rutherford, which makes use of the Newtonian laws and Maxwell's equations, leads to an unstable atom. The second problem of the model is that it fails to explain the atomic spectra. Indeed, according to this model, electrons may emit radiation of all wavelengths but, experimentally, as discussed in a previous section, it was found that atoms can absorb or emit only certain specific frequencies, which are characteristic of the element considered.

7.5.3 The Bohr model

Based on the Rutherford model, Niels Bohr proposed another atomic model [15], in which the planetary structure was preserved but several assumptions were made in order to overcome its failures. We consider the Bohr model of the hydrogen atom, which has one electron and a nucleus with a single positive charge equal to e. According to Bohr, the electron moves around the nucleus along a circular orbit of

---

[2] We are giving here only a 'naive' account of this model, without treating the scattering process analytically.

radius r. The force on the electron due to the positive charge of the nucleus is the Coulomb force. This is a centripetal force, then if m is the mass of the electron and v its tangential speed, we have that

$$F = \frac{1}{4\pi\varepsilon_0}\frac{e^2}{r^2} = m\frac{v^2}{r}. \tag{7.19}$$

The total energy of the system is given by

$$E = K + U = \frac{1}{2}mv^2 - \frac{1}{4\pi\varepsilon_0}\frac{e^2}{r}. \tag{7.20}$$

To avoid atom instability, Bohr postulated that the electron may be in certain specific states in which it does not radiate energy. These states are called *stationary states* and are characterized by a quantized angular momentum. Specifically, the angular momentum L assumes values which are integer multiples of $\hbar$. The quantization of L leads to the quantization of the energy levels. If $\vec{r}$ is the vector position of the electron and $\vec{p}$ its linear momentum, in a circular orbit $\vec{r}$ is perpendicular to $\vec{p}$. Then, the angular momentum, defined as $\vec{L} = \vec{r} \times \vec{p}$, has magnitude $L = rp = rmv$. Bohr's postulate of the *quantization of angular momentum* is expressed by

$$mvr = n\hbar \qquad n = 1, 2, 3.... \tag{7.21}$$

Using equations (7.19)–(7.21), we have that

$$r_n = \frac{4\pi\varepsilon_0\hbar^2}{me^2}n^2 = a_0 n^2. \tag{7.22}$$

Equation (7.22) means that the radius r can assume only specific *quantized* values. Here, a_0 is the Bohr radius and it is equal to $a_0 = \frac{4\pi\varepsilon_0\hbar^2}{me^2} = 0.052\,9$ nm. By substituting equation (7.22) in the expression of the total energy equation (7.20), we have that the energy can also assume only quantized values, given by

$$E_n = -\frac{me^4}{32\pi^2\varepsilon_0^2\hbar^2}\frac{1}{n^2}. \tag{7.23}$$

The subscript n is an index which labels the energy levels. For $n = 1$, equation (7.23) gives the lowest energy level, called the ground state, while for $n > 1$ we have the so-called excited states.

In order to explain the atomic spectra in this framework, Bohr postulated that an electron can emit radiation when it moves to a lower energy level or, absorbing radiation, can jump to a higher energy level. In the first case, the electron has less energy than in the initial state and the loss of energy appears as a quantum of radiation $h\nu$. So, if the electron jumps from the state with $n = n_1$ to the state with $n = n_2$, a photon appears and its energy is

$$h\nu = E_{n_1} - E_{n_2}. \tag{7.24}$$

Then, substituting equation (7.23) in equation (7.24), the frequency of the emitted photon is

$$\nu = \frac{me^4}{64\pi^3\varepsilon_0^2\hbar^3}\left(\frac{1}{n_2^2} - \frac{1}{n_1^2}\right). \tag{7.25}$$

The wavelength of the emitted radiation is

$$\lambda = \frac{c}{\nu} = \frac{64\pi^2\varepsilon_0^2\hbar^3 c}{me^4}\left(\frac{n_1^2 n_2^2}{n_1^2 - n_2^2}\right) \equiv \frac{1}{R_\infty}\left(\frac{n_1^2 n_2^2}{n_1^2 - n_2^2}\right). \tag{7.26}$$

In the case of absorption of radiation, according to Bohr's model, the electron moves from a lower energy level to an higher one. The two processes are shown in figure 7.10. R_∞ in equation (7.26) is called the Rydberg constant and its currently accepted value is

$$R_\infty = 1.097 \times 10^7 \text{ m}^{-1}.$$

The Bohr model was the first which was able to give an explanation of atomic spectra, because it actually gives the correct values for the energy levels.

Indeed, if we substitute $n_2 = 2$ in equation (7.26), we obtain

$$\lambda = \frac{4}{R_\infty}\frac{n_1^2}{n_1^2 - 4}. \tag{7.27}$$

Since $\frac{4}{R_\infty} = 364.5$ nm, equation (7.27) corresponds to the Balmer formula equation (7.14). The wavelength in equation (7.27) is relative to the emitted radiation for transitions from a state n_1 to a state $n_2 = 2$. This means that the radiation identified as the Balmer series corresponds to transitions from higher levels to the $n = 2$ level. Similar arguments can be found for the other series of radiation. Bohr's theory is also in agreement with the Ritz principle. If we consider a transition from level n_3 to

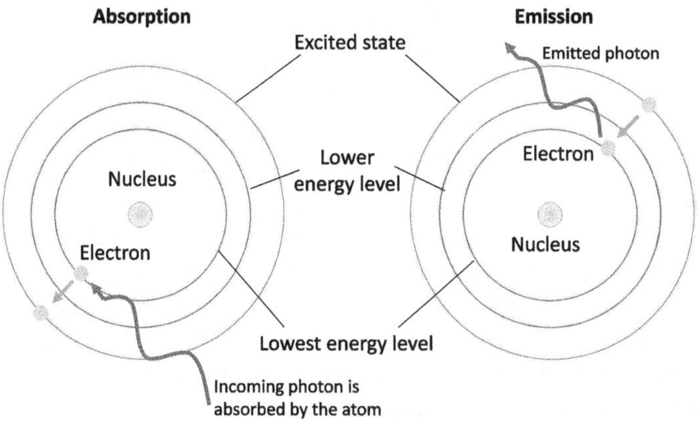

Figure 7.10. The emission (left) and absorption (right) of a photon and the jump of the electron within Bohr's model of the hydrogen atom.

level n_2 and then from n_2 to n_1, we can write explicitly the frequencies of the emitted radiation according to equation (7.25):

$$\nu_{n_3 \to n_2} = cR_\infty\left(\frac{1}{n_3^2} - \frac{1}{n_2^2}\right),$$
$$\nu_{n_2 \to n_1} = cR_\infty\left(\frac{1}{n_2^2} - \frac{1}{n_1^2}\right). \tag{7.28}$$

It is evident that $\nu_{n_3 \to n_2} + \nu_{n_2 \to n_1} = \nu_{n_3 \to n_1}$. This relation is a formal check of the Ritz principle. Another important test for Bohr's theory was the comparison between the Rydberg constant R_∞ and the experimental results. The perfect agreement between theory and experiment ensured the triumph of the theory. In consideration of his important work, Bohr received the Nobel Prize in Physics in 1922.

However, this model is not completely satisfactory, since the condition for stationary states is given as an *ad hoc* assumption, without a theoretical ground. The model is also incomplete: if on one hand it is useful for atoms that contain one electron, it fails for atoms with more electrons, since it does not consider electron–electron interaction. Another serious defect relates to the predictions of the angular momentum, for example, the ground state of hydrogen, for which $n = 1$, has $L = \hbar$ in the model, while the correct value, now known from quantum mechanics, is $L = 0$. Furthermore, this model is still based on the classical concept of trajectory and hence it violates the uncertainty principle[3]. Despite all these difficulties, the Bohr model is very important and marked a definitive turning point with the quantization of dynamic quantities such as speed, angular momentum and energy.

7.6 The Franck–Hertz experiment

In 1914, James Franck and Gustav Hertz carried out an experiment using the apparatus shown in figure 7.11(a) [16], which provided further experimental evidence of the quantization of energy levels in atoms. In fact, they were not aware of Bohr's atomic model at the time they performed the experiment, and their first explanation of the results was not correct.

In the experiment, a tube contains diluted mercury vapour. The cathode emits electrons, which are accelerated towards a positively charged grid by a potential difference V. Behind the grid there is a plate which is kept at a lower potential than the grid, through a small retarding potential V_0, but still positive with respect to the cathode. When electrons reach the plate an ammeter measures the generated current i. The experimental result is presented in figure 7.11(b). As the voltage V is increased, the current i also increases, but when $V \simeq 4.9$ V there is a sudden drop in the current, then a similar drop in current is again observed for $V \simeq 9.8$ V and so on. The explanation for this is as follows. Due to the presence of mercury vapour, there are collisions among electrons and Hg atoms. For small V these collisions are elastic, but when $eV = E_{n_2} - E_{n_1}$, where E_{n_i} are the atomic energy levels, the majority of collisions

[3] This principle is treated in detail in chapter 8.

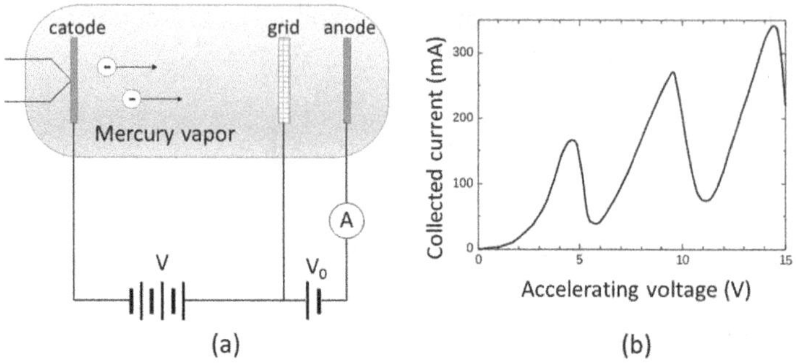

Figure 7.11. (a) Illustration of the Franck–Hertz experimental set-up. Electrons leave the cathode and are accelerated towards the grid; there is a retarding potential V_0 between the grid and the anode. Electrons that reach the anode are recorded measuring the current through an ammeter. (b) Typical plot of collected current as a function of the accelerating voltage in an experiment with mercury.

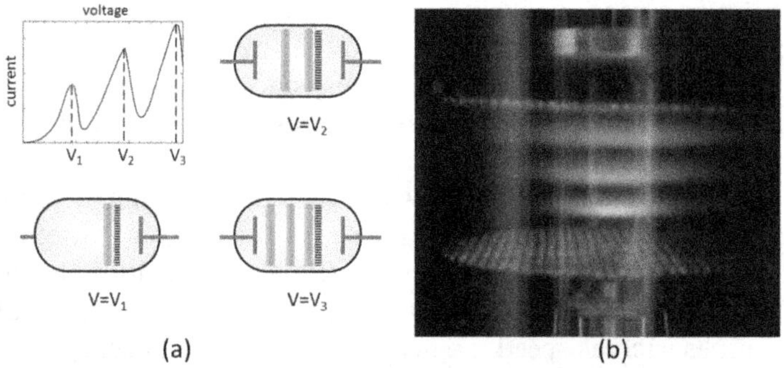

Figure 7.12. (a) Sketch for the appearance of the coloured bands. (b) Photograph of a Franck–Hertz experiment with neon gas showing three luminous orange bands. Image reproduced from [32] under a Creative Commons Attribution-Share Alike 4.0 International license. Author: Infoczo.

will be inelastic, because the electron has enough kinetic energy to excite the atom. In this case, the electron loses its energy, and once it has reached the grid it will not be able to overcome the retarding potential V_0, hence we see a drop in current. Meanwhile, the mercury atoms, after being in the excited state E_{n_2}, will emit radiation (i.e. photons of energy $h\nu = E_{n_2} - E_{n_1}$) to go back to their ground state. In the case of mercury the radiation is in the ultraviolet range. Specifically, the wavelength of the emitted photon is $\lambda = 253.7$ nm.

Indeed, the experiment is more spectacular when one uses neon instead of mercury to fill the tube, because in this case the frequency of the emitted photon is in the range of visible light, close to the orange frequency. Hence, it is possible to *see* where collisions occurred. Indeed, increasing the voltage V, the first visible luminous layer appears very close to the grid, when V has reached the value $V_1 = \Delta E/e$, and accordingly one observes a current drop (see figure 7.12(a)). Increasing V further, the orange band is shifted toward the cathode, since electrons will now reach the required energy earlier to excite the neon atoms, until a second band appears close to the grid. This happens for $V = 2V_1$ and in correspondence to the second drop in the current. Figure 7.12(b) shows a photo of an experiment with neon gas, with three visible orange bands.

This experiment is another cornerstone of modern physics and Franck and Hertz were awarded the 1925 Nobel Prize in Physics 'for their discovery of the laws governing the impact of an electron upon an atom'.

7.7 The wave–particle duality

The nature of light had puzzled many scientists and philosophers throughout the centuries. Democritus had an 'atomistic' view of the Universe, including 'light'. Similarly and at about the same time, in the Vaiśeṣika school of Hindu philosophy it was postulated that all objects in the physical universe were reducible to *paramāṇu* (atoms). Initial studies of optics were 'geometrical' (by Euclid, for example, and also later in the eleventh century by Arab scientist Ibn al-Haytham, considered the father of 'modern optics'). These works were based on a 'corpuscular' view of light. Only in the seventeenth century do we find the first wave-like description of light, by Francesco Maria Grimaldi (1618–1663), René Descartes (1596–1650) and primarily Christiaan Huygens (1629–1695). In the same century Isaac Newton (1642–1727), extensively studied light-related phenomena using the corpuscular theory, and due to its fame and reputation, many scientists of the seventeenth and eighteenth centuries subscribed to the view that light was a bunch of particles. The wave theory of light received a major boost at the beginning of the nineteenth century from the English scientist Thomas Young, thanks to his famous double-slit experiment. The wave theory was then consolidated after the prediction and discovery of electromagnetic waves (see also chapter 1).

However, as we learned in sections 7.2 and 7.3 of this chapter, to explain photoelectric and Compton effects we need to again attribute a corpuscular nature to light. We have to deal, indeed, with a dual nature of light, *the wave–particle duality*, according to which light is a wave or a set of particles, depending on the specific experiment we are carrying out. In 1924 Louis de Broglie expressed an inventive hypothesis in his PhD thesis [17], namely he extended the concept of duality, which characterizes light, to matter. He was awarded the 1929 Nobel Prize in Physics 'for his discovery of the wave nature of electrons'. According to de Broglie, matter can have a wave behaviour and then we can associate a plane wave to each particle, with momentum equal to p, and with wavelength equal to

$$\lambda = \frac{h}{p}. \tag{7.29}$$

Equation (7.29) defines the de Broglie wavelength, where h is Planck's constant. We can use equation (7.29) in order to determine the wavelength associated with different objects. For example, if we consider an Alfa Romeo Giulia car with mass equal to $m = 1500$ kg and moving at speed $v = 100$ m s^{-1}, the corresponding wavelength is equal to

$$\lambda = \frac{h}{mv} = \frac{6.6 \times 10^{-34}\ \text{J s}}{(1.5 \times 10^{3}\ \text{kg})(10^{2}\ \text{m s}^{-1})} = 4.4 \times 10^{-39}\ \text{m}. \tag{7.30}$$

The obtained wavelength for a macroscopic object such as a car is a very small value, this implies that it is impossible, with the technologies available at the present time, to observe it in an experiment. Next, we calculate the wavelength of an electron with kinetic energy equal to $K = 1$ eV. Since its rest energy $E = mc^2 = 5.1 \times 10^5$ eV is much larger than the kinetic energy, we can use non-relativistic kinematics. Then, the quantity of motion p of the electron is equal to $p = \sqrt{2mK} = 5.4 \times 10^{-25}$ kg m s^{-1} and the corresponding wavelength is equal to

$$\lambda = \frac{h}{p} = \frac{6.6 \times 10^{-34}\ \text{J s}}{5.4 \times 10^{-25}\ \text{kg m s}^{-1}} = 1.2 \times 10^{-9}\ \text{m}.$$

The wavelength of an electron is more significant with respect to equation (7.30), it is actually of the same order of magnitude of that of x-rays. This means that, due to the smallness of h, the wave behaviour is observable only for particles with atomic or nuclear size. In the macroscopic world, connected to our experience, particles are well described by Newtonian dynamics while their wave-like behaviour is impossible to detect—they show corpuscular behaviour. Therefore, in light of the previous considerations and the calculation of equation (7.30), which gives an idea about the size of the wavelength associated with macroscopic objects, there is no experiment that can be done to reveal their wave nature. However, in the microcosm, the opposite happens—here, particles also show undulatory behaviour.

In order to reveal the wave nature of matter, we have to use the particles of the atomic world, and 'smoking-gun' evidence would be to see an interference or diffraction pattern. Indeed, soon after the de Broglie's hypothesis, experiments of electron diffraction through crystals were performed. In such experiments, a beam of electrons is accelerated through a potential V. Then, it acquires a non-relativistic kinetic energy K, and the associated momentum p is $p = \sqrt{2mK}$. The obtained results strongly recall those of x-ray diffraction experiments—the electron beam hits the crystal in the same way as a beam of x-rays and the diffraction patterns in the two cases are very much alike, as shown in figure 7.13. The first experimental confirmation concerning the wave nature of electrons was provided by Davisson and Germer in 1926 [18]. Their results give quantitative information about the wavelength associated with an electron, which is compatible with equation (7.29). The experiment is described in detail in chapter 15 of this book. Similar experiments were also performed by G P Thomson [19] (the son of J J Thomson), and in 1937 Davisson and Thomson shared the Nobel Prize in Physics 'for their experimental discovery of the diffraction of electrons by crystals'. Curiously, Joseph Thomson (the father) was awarded a Nobel Prize (1906) for the discovery of electrons as particles (he indeed showed that cathode rays consist of particles), while his son George received his Nobel Prize for the observation of the wave-like behaviour of electrons. Of course, the wave nature can be revealed not only for electrons but for any particle with momentum p, according to equation (7.29).

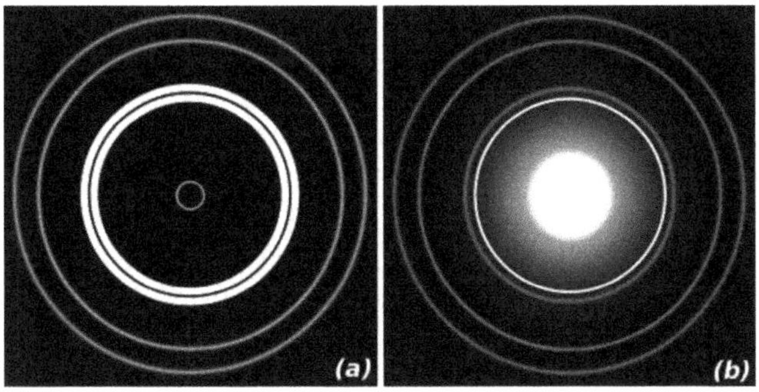

Figure 7.13. (a) Electron and (b) x-ray diffraction patterns.

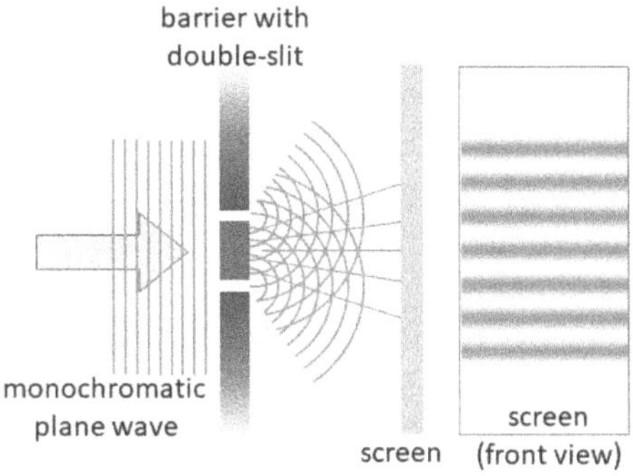

Figure 7.14. A double-slit experiment with waves and the resulting interference pattern.

7.8 The double-slit experiment

As has been already mentioned in the first chapter of this book, the principle of superposition is a distinctive aspect of waves and it leads to interesting phenomena, such as interference and diffraction. Young's double-slit experiment is one of the most famous for seeing the effects of interference between two different waves. As schematically shown in figure 7.14, a monochromatic plane wave hits a barrier with two slits and, at each slit, diffraction of the incident light occurs. Then, the two diffracted waves overlap and as a result one observes an interference pattern on the screen, i.e. a figure with a sequence of bright and dark zones. The bright zones are due to constructive interference. In particular, if x_1 and x_2 are the distances from the point on the screen to the two slits, the condition in order to obtain a constructive interference is

$$|x_1 - x_2| = n\lambda \qquad n \in Z. \tag{7.31}$$

Equation (7.31) is the formal way to indicate that a crest of a wave from one slit arrives on the screen together to another one from the other slit. On the other hand, the dark zones are given by destructive interference. In this case, the crest of a wave from one slit combines with the valley of a wave coming from the other slit. Formally,

$$|x_1 - x_2| = \left(n + \frac{1}{2}\right)\lambda \qquad n \in Z.$$

This experiment was realized by Young in 1801 and was definitive proof of the wave nature of light. At the early stages of quantum mechanics, it was not possible to build a double slit of the right size compared to the wavelength of electrons to be able to see the interference pattern. Only in 1961 was the first double-slit experiment with electrons carried out [20]. In this case electrons were accelerated through a potential equal to 50 kV and hit a double slit of separation equal to 2 μm and width 0.5 μm. If we consider electrons as particles, we expect that the figure on the screen will be a sort of 'shadow' in correspondence to the two slits, as shown schematically in figure 7.15(a). Instead, a pattern is observed like the one in figure 7.15(b), which is an interference image, like the one obtained using light.

Similar experiments have been since improved [21] and also performed with other particles [22], including helium atoms [23].

Differently to waves in a pond, we can send electrons one by one. At the beginning we will see on the screen only isolated spots, which will in the end build up the interference pattern. It is quite natural then to ask through which slit the particle has passed. In order to do this, each slit could be surrounded with an electromagnetic loop, which causes a deflection of the metre when a charged particle or a particle with a magnetic moment passes near the slit. If we do so, the experimental result is no longer an interference pattern! Indeed, whatever device we use to determine through which slits the electron has gone, we will no longer be able to observe the interference fringes. The reason for this disappointing result is that we are investigating a *particle*-like aspect of the electron, hence we do not observe its wave nature. Actually, it is impossible to observe the corpuscular and wave behaviours simultaneously. This is known as *Bohr's complementary principle* [24], which is one of the milestones of quantum mechanics and states that particle-like and wave-like behaviours are mutually exclusive and complementary aspects of quantum systems.

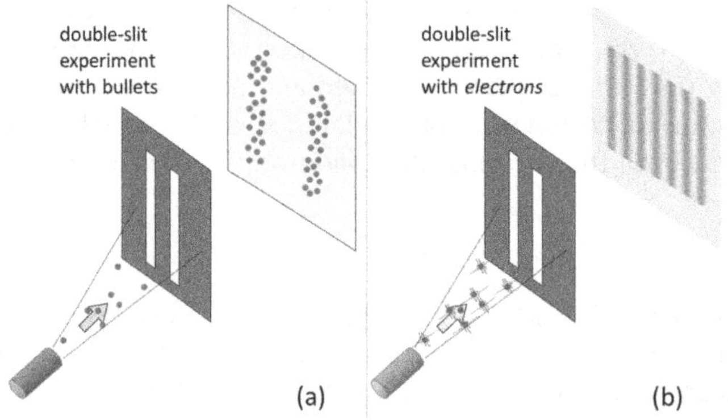

Figure 7.15. Sketch of the double-slit experiment with macroscopic bullets and with microscopic particles (such as electrons, protons, neutrons and so on).

Recently, many experiments have been carried out to investigate the complementarity principle further, mainly using photons [25–27]. A celebrated example is the illustration of wave–particle duality by considering single particles in a two-path interferometer, where one *chooses* either to observe interference fringes, associated with wave-like behaviour, or to know which path of the interferometer has been followed, according to a particle-like behaviour. The paradoxical nature of duality is best captured in the 'delayed-choice' experiment, where the choice between the two complementary measurements is made long after the particle enters the interferometer. In this way, one rules out a too naive view of complementarity, which would assume that the particle could 'know' when entering the apparatus which experimental configuration has been set (a record of interference or determination of which-path information (WPI)). Actually, the complementary quantities—WPI and interference visibility—could be partially determined simultaneously, i.e. one has partial WPI while keeping interference with limited visibility [28, 29]. There is indeed an upper bound to the maximum values of simultaneously determined interference visibility V and path distinguishability D, a parameter that quantifies the available WPI on the quantum system, given by the inequality $V^2 + D^2 \leqslant 1$ [30, 31]. This inequality comprises the extreme 'all-or-nothing' situations: $V = 1$ implies $D = 0$, hence in such a case only the interference pattern is observed, and vice-versa when $D = 1$ one has $V = 0$, giving full knowledge of the path followed.

Finally, to conclude this chapter, we return to the quote by Michelson reported at the very beginning of this chapter: learning the lesson from the years spanned in this chapter (1900–1927), we could definitely affirm that 'the future of Physical Science has a lot of marvels in store, even more astonishing than those of the past'. Indeed, we should bear in mind that the 'story of physics' will continue to evolve: each time we look deeper or refine experimental techniques we will find and learn something new.

References

[1] Michelson A A 1894 *University of Chicago, Annual Register 1894—1895* p 150; available online at https://babel.hathitrust.org/cgi/pt?id=njp.32101065108746&view=1up&seq=166&q1=Michelson

[2] Planck M 1920 *The Genesis and Present State of Development of the Quantum Theory* available online at nobelprize.org/prizes/physics/1918/planck/lecture/

[3] Feynman R P, Leighton R B and Sands M 2011 *The Feynman Lectures on Physics* vol 3 (New York: Basic Books) (also available online at http://www.feynmanlectures.caltech.edu/III_toc.html)

[4] Hertz H 1887 Über den Einfluss des ultravioletten Lichtes auf die electrische Entladung *Ann. Phys.* **267** 983

[5] Hallwachs W 1907 Über die lichtelektrische Ermüdung *Ann. Phys.* **328** 459

[6] Lenard P 1902 Über die lichtelektrische Wirkung *Ann. Phys.* **313** 149

[7] Righi A 1909 *La Materia Radiante e i Raggi Magnetici* (Bologna: Zanichelli)

[8] Einstein A 1905 Über einen die Erzeugung und Verwandlung des Lichtes betreffenden heuristischen Gesichtspunkt *Ann. Phys.* **332** 132

[9] BIPM 2018 Resolutions of the 26th CGPM (General Conference on Weights and Measures), Bureau International des Poids and Mesures (available at: https://www.bipm.org/en/cgpm-2018/)

[10] Millikan R A 1916 A direct photoelectric determination of Planck's '*h*' *Phys. Rev.* **7** 355
[11] Compton A H 1923 A quantum theory of the scattering of x-rays by light elements *Phys. Rev.* **21** 483
[12] Compton A H 1923 The spectrum of scattered x-rays *Phys. Rev.* **22** 409
[13] Balmer J J 1885 Notiz über die Spectrallinien des Wasserstoffs *Ann. Phys. Chem.* **25** 80
[14] Geiger H and Marsden E 1913 The laws of deflexion of a particles through large angles *Philos. Mag.* **25** 604
[15] Bohr N 1913 On the constitution of atoms and molecules *Philos. Mag.* **26** 1
[16] Franck J and Hertz G 1914 Über Zusammenstöße zwischen Elektronen und Molekülen des Quecksilberdampfes und die Ionisierungsspannung Desselben *Verh. Dtsch. Phys. Ges.* **16** 457
[17] De Broglie L 1927 La mécanique ondulatoire et la structure atomique de la matière et du rayonnement (Thèse de Doctorat, 1924) *J. Phys. Radium* **8** 225
[18] Davisson C and Germer L H 1927 The scattering of electrons by a single crystal of nickel *Nature* **119** 558
[19] Thomson G P and Reid A 1927 Diffraction of cathode rays by a thin film *Nature* **119** 890
[20] Jönsson C 1961 Elektroneninterferenzen an mehreren künstlich hergestellten Feinspalten *Z. Phys.* **161** 454
[21] Bach R, Pope D, Liou S-H and Batelaan H 2013 Controlled double-slit electron diffraction *New J. Phys.* **15** 033018
[22] Gähler R and Zeilinger A 1991 Wave optical experiments with very cold neutrons *Am. J. Phys.* **59** 316
[23] Carnal O and Mlynek J 1991 Young's double-slit experiment with atoms: a simple atom interferometer *Phys. Rev. Lett.* **66** 2689
[24] Bohr N 1928 Das Quantenpostulat und die neuere Entwicklung der Atomistik *Naturwissenschaften* **16** 245
[25] Jacques V, Wu E, Grosshans F, Treussart F, Grangier P, Aspect A and Roc J-F 2008 Delayed-choice test of quantum complementarity with interfering single photons *Phys. Rev. Lett.* **100** 220402
[26] Peruzzo A, Shadbolt P, Brunner N, Popescu S and O'Brien J L 2012 A quantum delayed-choice experiment *Science* **338** 634
[27] Dou L-Y, Silberberg Y and Song X-B 2019 Demonstration of complementarity between path information and interference with thermal light *Phys. Rev.* A **99** 013825
[28] Wootters W K and Zurek W H 1979 Complementarity in the double-slit experiment: quantum nonseparability and a quantitative statement of Bohr's principle *Phys. Rev.* D **19** 473
[29] Greenberger D M and Yasin A 1988 Simultaneous wave and particle knowledge in a neutron interferometer *Phys. Lett.* A **128** 391
[30] Jaeger G, Shimony A and Vaidman L 1995 Two interferometric complementarities *Phys. Rev.* A **51** 54
[31] Englert B-G 1996 Fringe visibility and which-way information: an inequality *Phys. Rev. Lett.* **77** 2154
[32] Infoczo 2014 *Wikimedia Commons* https://commons.wikimedia.org/wiki/File:Franck-Hertz-Neon-3.png

Chapter 8

Key concepts in quantum mechanics

Marco Figliolia, Martina Moccaldi and Canio Noce

The birth of a novel physical theory is always induced by inexplicable experimental results, which cause a deep crisis in old models and force physicists to develop different and more general paradigms. The resulting revolution is slow and strenuous because it requires introducing a new point of view and, consequently, formulating all key concepts, methods and problems in a different way. Moreover, it produces the proliferation of alternative interpretations and mathematical formulations, which leads to long-lasting scientific debates. On the other hand, changing the point of view is an exciting intellectual challenge, which allows the scientific community to approach a deeper knowledge of nature and a crisis environment is a fruitful terrain to develop a well-structured and wide-ranging theory.

The revolution leading to quantum mechanics started in the last decades of nineteenth century, when several experiments showed astonishing results. Before these experiments, physicists believed that there was nothing fundamental left to discover. There was indeed a deep knowledge about mechanics, thermodynamics and phenomenological optics, and a common thought was that remaining progress would only pertain to the accuracy of measurements and technological applications, as remarked by Lord Kelvin in 1900. However, his 'famous last words' came only five years before the so-called *annus mirabilis*, when Albert Einstein introduced special relativity and explained the photoelectric effect, giving rise to an extreme transformation of physics.

In this chapter, we will first present a brief overview of the fundamental experiments that caused a deep crisis in the classical paradigms (section 8.1), and then we will describe the basic concepts that characterize quantum mechanics, devoting particular attention to the wave function, a quantity that includes all the main aspects of the novel theory (section 8.2). The essential features of the novel principles and the wave function will be discussed in section 8.3, and the different interpretations of quantum mechanics are investigated in section 8.4. Finally,

doi:10.1088/978-0-7503-2678-0ch8

appendices A and B are devoted to the continuity of the wave function and the derivation of the uncertainty relation, respectively.

8.1 The history of quantum theory

Experiments concerning electromagnetic radiation carried out at the end of the nineteenth century, in particular those in spectroscopy, exhibited results that could not be explained by the classical paradigms (see chapter 7). First, physicists investigated experimentally the atomic structure and nature of matter. Joseph J Thomson computed the ratio between the charge and mass of electrons in 1897 (see chapter 13), Robert A Millikan evaluated the charge of electrons in 1909 (see chapter 14), Ernest Rutherford proved that the atomic model proposed by Thomson was wrong in 1909, and James Franck and Gustav Hertz confirmed Niels Bohr's atomic model in 1914. Simultaneously, experiments in spectroscopy, Max Planck's solution of the black-body problem in 1900 (see chapter 6), Einstein's explanation of the photoelectric effect, the Compton effect and experiments involving x-rays played a fundamental role in understanding how radiation interacts with matter, as well as the so-called 'wave–particle' duality (see chapter 10). This issue was further investigated by Arthur Compton in 1923 and Clinton Davisson and Lester Germer in 1927 (see chapter 15).

All the experimental results highlighted the crucial limits of classical physics in a microscopic context. For this reason, many different novel interpretations have been proposed involving alternative mathematical formulations, but a unique theory is still far from reach. The recent awarding of the Nobel Prize to physicists investigating quantum mechanics (Serge Haroche and David J Wineland, 2012) and the astonishing results in the applications of quantum mechanics (for example, quantum cryptography and quantum computing) show that the revolution that began at the end of the nineteenth century is still ongoing.

We can essentially identify three phases in the evolution of quantum physics [1]: quantum theory, quantum mechanics and quantum field theory. The first consists in the explanation of experiments carried out between the end of the nineteenth century and the beginning of the twentieth century. It is based on four fundamental intuitions: the introduction of Planck's constant, the discretization of physical quantities (in particular, the angular moment and energy), the corpuscular nature of radiation in certain situations and the wave nature of matter in particular circumstances. Quantum mechanics is instead a systematic theory, provided by general principles and a rigorous mathematical formalism, and was mainly developed between 1926 and 1930 thanks to Erwin Schrödinger, Werner Heisenberg, Pascual Jordan, Max Born and Paul Dirac. Finally, the third theory was formulated by Dirac to connect the quantum properties of microscopic particles within the electromagnetic field, which had first been investigated in depth by Michael Faraday and James Clerk Maxwell in the nineteenth century. It was completed by Richard Feynman and was used to understand the connection between the quantum properties of particles and other fields. For this reason, it is called quantum field theory. In the following section, we will investigate in detail the main aspects of the above-mentioned experiments.

8.1.1 Experiments with unexpected results

In Cambridge, in 1896–1897, Thomson performed a significant experiment by employing a cathode ray tube, with an internal condenser and a fluorescent screen used as a detector. He observed some particles on the detector which were radiated by the cathode and deflected with respect to the initial horizontal direction of motion due to the condenser. Despite the ancient belief (introduced by the Greek philosopher Democritus) of indivisible atoms, he suggested that atoms are composed of certain particles exhibiting a negative charge, since they are deflected towards the positive shell of the condenser. Thomson showed that these particles, called *electrons*, are independent of the material used for the cathode and computed the ratio between their charge and mass,

$$\frac{q}{m} = 1.7 \times 10^{11}\ \mathrm{C\ kg^{-1}},$$

by investigating the deflection observed on the detector. He also proved that this ratio is independent of the gas in the tube and the chemical nature of the electrode. However, Thomson's experiment did not allow the charge of electrons or their mass to be computed. These quantities were measured in 1909 by Millikan, who analysed the motion of small oil drops inserted in a camera containing a gas and a condenser connected to a power generator. Applying the classical dynamical laws and repeating the experiment, he discovered that the charge of the oil drops is an integer multiple of an elementary charge, which he assumed to be the charge of electrons ($q = 1.6 \times 10^{-19}$ C).

Thomson's experiment hinted that an atom is composed of negative particles. However, it also has to contain positive particles due to the neutrality of matter. For this reason, Thomson supposed that electrons are submerged in a fluid exhibiting a positive charge and made up most of the atomic mass. However, this model was not coherent with several experimental results, such as those obtained by Hans W Geiger and Ernest Marsden, led by Rutherford. Indeed, they observed that α particles (i.e. the nuclei of helium atoms), after hitting the atoms of a layer of gold, continued their linear motion in most cases. For this reason, Rutherford proposed a planetary model, where electrons describe closed orbits around a small positively charged region, called the nucleus, like the planets moving around the Sun. However, electrons accelerated along closed orbits would lose energy and would collapse into the nucleus, which is not coherent with the stability of matter. To overcome this limitation, in 1913–1914 Bohr formulated a novel model, by introducing certain ad hoc assumptions and employing only classical dynamical laws (and not relativistic ones). He first observed that spectroscopy experiments contradicted the planetary model. Indeed, at the end of the nineteenth century, it was already known that only certain frequencies of electromagnetic radiation are absorbed (or emitted) by a material. These frequencies can be shown on a screen as parallel lines, *spectral lines*, which characterize the emitter. Broadening this idea, Bohr assumed that electrons can move only along certain orbits, called *stationary states*, because there is no emission of radiation. Consequently, the angular moment, characterizing the motion of electrons, can assume only discrete values,

$$L = \frac{nh}{2\pi}, \quad n \in \mathbb{N},$$

depending on Planck's constant h. Hence, the other physical quantities describing the motion of electrons (the radius of the orbit, the speed and, mainly, the energy) are also *quantized*, i.e. they appear as integer multiples of certain quantities. In particular, the energy can be expressed by the following relation:

$$E_n = -\frac{2\pi^2 \, me^4Z^2}{n^2h^2}, \quad n \in \mathbb{N},$$

where m and e are the mass and charge of electrons, respectively, and Z characterizes the material (for example, $Z = 1$ for the hydrogen atom). Energy increases with the parameter n and, in the case of $n = 1$ (the fundamental state energy), it is equal to the ionization energy (which is experimentally known), i.e. the energy required to extract the electron from the hydrogen atom. Moreover, each emission or absorption of radiation causes a transition from a stationary state to another state and the frequency of the radiation (emitted or absorbed) is given by

$$h\nu = E_1 - E_2,$$

where E_i is the energy of the i stationary level. It is significant to observe that Bohr's assumptions led to expressing the radiation frequency of the hydrogen atom as

$$\nu = c \, R_\infty \, Z^2\left(\frac{1}{n_1^2} - \frac{1}{n_2^2}\right),$$

which corresponds to Johann Balmer's series, introduced 30 years ealier. Despite already presenting open problems (elliptic orbits instead of circular ones, behaviour in the presence of electric or magnetic fields), Bohr's model was essential because it introduced the idea of a discrete variability set for the physical quantities involved, which was a crucial difference to classical descriptions. However, it is worth noting that the energy is quantized when electrons can assume only certain positions (constrained states), whereas it can change continuously if electrons can move freely. Bohr's hypotheses about the stationary states and the discretization of energetic atomic levels were confirmed in 1914 by Franck and Hertz, who investigated the energetic exchanges in the case of inelastic collisions of electrons against mercury atoms.

A parallel investigation concerned the interaction between electromagnetic radiation and matter. In this context, a fundamental role was covered by the so-called *black-body* problem. A black-body can be viewed as a box with a small hole and it is generally defined as a body absorbing all the incident radiation. From Kirchhoff's law, it is also a perfect emitter. Applying Stefan–Boltzmann's law, the intensity of the radiation can be derived as follows (Rayleigh–Jeans's law):

$$u(\nu, T)\,\mathrm{d}\nu = kT\frac{8\pi\nu^2}{c^3}\,\mathrm{d}\nu, \tag{8.1}$$

where T is the constant temperature of the cavity and ν is the frequency of the radiation. As a consequence, the intensity of the radiation should diverge as the frequency increases (ultraviolet catastrophe), which differs from the experimental evidence. In 1900, Planck assumed that the exchanges between radiation and matter take place in a discrete manner, so the integration used to obtain equation (8.1) is not allowed. He introduced the constant $h = 6.6 \times 10^{-34}$ J s, exhibiting the dimensions of an action, and computed the intensity of radiation as follows:

$$u(\nu, T) = \frac{8\pi\nu^3 h}{c^2\left(\mathrm{e}^{\beta h\nu} - 1\right)},$$

where $\beta = \frac{1}{kT}$, thus avoiding the problem of divergence. However, Planck did not underline the necessity of modifying the whole theory of electromagnetic radiation.

Five years later Einstein, in his fundamental work [2], extended Planck's crucial intuition to explain the well-known *photoelectric effect*, discovered by Hertz in 1887. It essentially consists in the emission of electrons from a metal hit by a beam of light having a sufficiently short wavelength. It can be observed in gases and solids and with x-rays, γ-rays and ultraviolet radiation. Classically, we can observe that when electromagnetic radiation strikes a substance, electrons are accelerated by the electric field associated with the electromagnetic radiation and receive sufficient energy to leave the surface of the metal. However, no classical theory can explain why the kinetic energy does not depend on the intensity of radiation, but it depends linearly on frequency. If the frequency is lower than a limit value ν_0, which changes according to the metal, there is no emission of electrons.

Moreover, the phenomenon is instantaneous, despite classical expectations. Broadening Planck's ideas, Einstein stated that the energy of a beam of monochromatic light is not uniformly spread out, as suggested by classical wave theory, but it is quantized in *quanta*, now referred to as *photons* following the terminology introduced by Gilbert Lewis in 1922. The energy of a photon is

$$E = h\nu,$$

where h is Planck's constant and ν is the wave frequency, so the kinetic energy of the emitted electron is given by

$$K = h\nu - W_0,$$

where W_0 is the energy that joins the electron to the nucleus. It is clear that electrons are expelled from the metal only if $h\nu > W_0$, otherwise their speed would assume imaginary values, which is not physically acceptable. The strength of Einstein's insight (and the reason for his Nobel Prize in 1921) lies in the proposed dual behaviour of electromagnetic radiation—indeed, it can behave both as a wave and as a particle.

The corpuscular nature of radiation (already hypothesized by Isaac Newton) was investigated also by Compton around 1923. If x-rays (discovered by Wilhelm C Röntgen in 1895) hit electrons, we can experimentally observe two different results: the wavelength of the radiation does not change after the impact (Thomson scattering) or it changes after the impact (Compton scattering). The second case, generally called the *Compton effect*, occurs in particular when monochromatic x-ray radiation strikes lightweight chemical elements, such as carbon. In this circumstance, the resulting radiation is split into two types of radiation (one having the same wavelength and the second exhibiting a wavelength lower than the first). Following the ideas presented by Einstein, Compton conferred dynamical properties to photons, such as the energy and the quantity of motion, in order to study their elastic collisions against electrons.

The surprises were not over yet. In 1927, Davisson and Germer bombarded a crystal with electrons. It was known that electromagnetic radiation, when hitting a crystal, is reflected and produces diffraction patterns on a detector. This phenomenon arises when the distances characterizing the crystal lattice are comparable to the wavelength of the radiation. Unexpectedly, the same effect arises when employing a beam of electrons instead of radiation. Davisson and Germer's experiment confirmed Louis de Broglie's hypothesis dating back to 1924: an electron moving with a quantity of motion p can be associated with a wavelength λ equal to

$$\lambda = \frac{h}{p},$$

called the *de Broglie wavelength*.

The wave nature of electrons was also revealed by the renowned double-slit experiment, first realized by Thomas Young in 1801 to prove the wave nature of light and then later extended to electrons. Let us first consider a wall with two holes and an absorbent board behind it. Shooting bullets against the wall, the probability P of finding them at a certain position x on the board is equal to the sum of the single probabilities of discovering them at x after they have gone through the first or the second hole: P_1 and P_2, respectively (see figure 8.1 for more details). Young performed the same experiment using a light source (nowadays we could employ a laser). A beam of pure-wavelength light hits an opaque screen with two vertical slits, which are small and very close to each other. A detector is located at a second screen, which is perfectly absorbent, and can compute the intensity of the considered wave by measuring the squared height of the wave. In the case of a single slit, light goes through the hole and radiates up to the second screen. Most of its intensity is detected at the centre of the screen and the slit behaves like a virtual wave source (the *Huygens–Fresnel's principle*). However, when both slits are open, the intensity of the wave is not equal to the sum of intensities computed when only one hole is open. Young suggested that light is a wave, thus it obeys the superposition principle. Two slits indeed generate two coherent waves, i.e. waves with a definite wave relationship. The overlapping waves give rise to plenty of different circumstances depending on the distance between the wall and the position x on the screen. The resulting

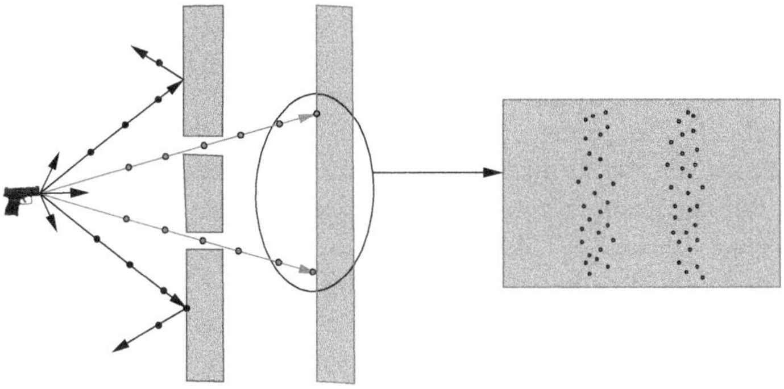

Figure 8.1. Double-slit experiment with bullets. The probability of finding a bullet at a certain position x on the screen is equal to the sum of the probabilities P_1 and P_2 of discovering it after it has crossed the first or the second slit, respectively.

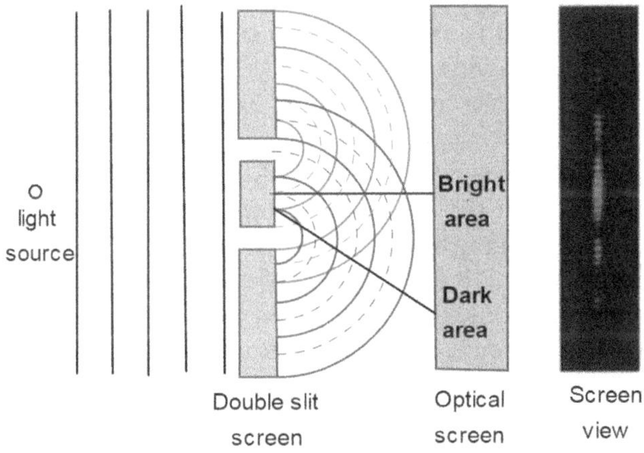

Figure 8.2. Double-slit experiment with waves. The slits behave as wave sources, so they generate two waves, interfering each other. This explains the interference fringes detected on the screen. Credit for the image of the screen view: PASCO.

intensity is maximum when this distance is an integer multiple of the wavelength, thus the crests of the two waves approach the screen at the same time (constructive interference). In contrast, the resulting intensity is null when the path differences are integer multiples of the half-wavelength (destructive interference). This behaviour is detected on the screen through equally spaced bright bands, called *interference fringes*, corresponding to constructive interference and dark bands corresponding to destructive interference (see figure 8.2). Young employed sunlight, but currently the experiment is carried out using monochromatic (single wavelength) light to obtain more obvious effects.

Broadening Young's experiments, in 1965 Feynman presented a thought experiment to prove the wave behaviour of matter. In particular, he proposed repeating Young's experiment employing electrons. Further, Feynman's experiment has actually been carried out, confirming his intuition (see [3, 4] and references therein).

Let us consider a wall with two slits and an absorbent screen, where an acoustic detector reveals the arrival of electrons. The probability distributions for the single- and double-slit cases are computed. Surprisingly, electrons behave like waves and not like bullets. Indeed, interference fringes can be observed on the screen (see figure 8.3). However, by adding a light source between the wall and the board, electrons exhibit corpuscular behaviour. The light source is a path detector since the light is deflected upwards or downward according to the transition of electrons through the upper or the lower slit, respectively.

Therefore, a particle can display both corpuscular and wave features and these behaviours are mutually exclusive. The wave nature (shown, for example, by phenomena such as interference) is exhibited when two events are both possible (in the case of the double-slit experiment, the events are the transition through one hole or the other). When it is possible to distinguish (also only theoretically) the two events, the particle nature arises. This property is generally called the *complementary principle*. Moreover, it is impossible to introduce, in an experimental apparatus, a device that is able to identify the hole the electrons cross without perturbing the experiment itself and destroying the interference.

In summary, both matter and radiation exhibit wave or particle properties according to the circumstances. Macroscopic particles are well described as corpuscles, as the related wavelength is extremely small, thus they satisfy the laws of classical mechanics. On the other hand, microscopic particles can show both features. However, establishing if radiation (or matter) is a wave or a particle is not a well-posed problem. Radiation and matter, indeed, simply behave as a wave or as a particle according to the context.

Interpreting matter as a wave makes it more difficult to identify the exact position of a particle at a microscopic scale. This is an intrinsic feature of quantum mechanics and it is expressed in Heisenberg's principle (8.2), as we will discuss in the following. However, if we introduce a detector in the experimental apparatus, it becomes possible to establish the position of the particle, which behaves in this case as a corpuscle and not as a wave. Hence, the quantum description of a physical system is *a priori* probabilistic, since it has to include all the possible configurations, but *a posteriori* deterministic, since, after the measurement, the time evolution of the system is uniquely determined once the initial conditions are known. Therefore, it appears evident that measurement plays a fundamental role in quantum mechanics, differently to classical mechanics where we do not need to realize a measurement, because all classical physical quantities are measurable. Investigating this role is a highly demanding challenge, involving a wide range of degrees of freedom, the phenomenon of decoherence, interaction with a very complex environment and the emergence of classical behaviour in certain particular limits of quantum mechanics.

All the described experimental evidence is not explicable using classical paradigms, so it appears necessary to formulate a novel, more general theory: the quantum mechanics. The word *quantum* derives from Latin and means 'how much?' It refers to the crucial idea of discretized energy, introduced by Planck and developed by Einstein. Classical physics is still a good approximation to describe the evolution of bodies at a macroscopic scale, but in a microscopic context the quantum theory becomes essential.

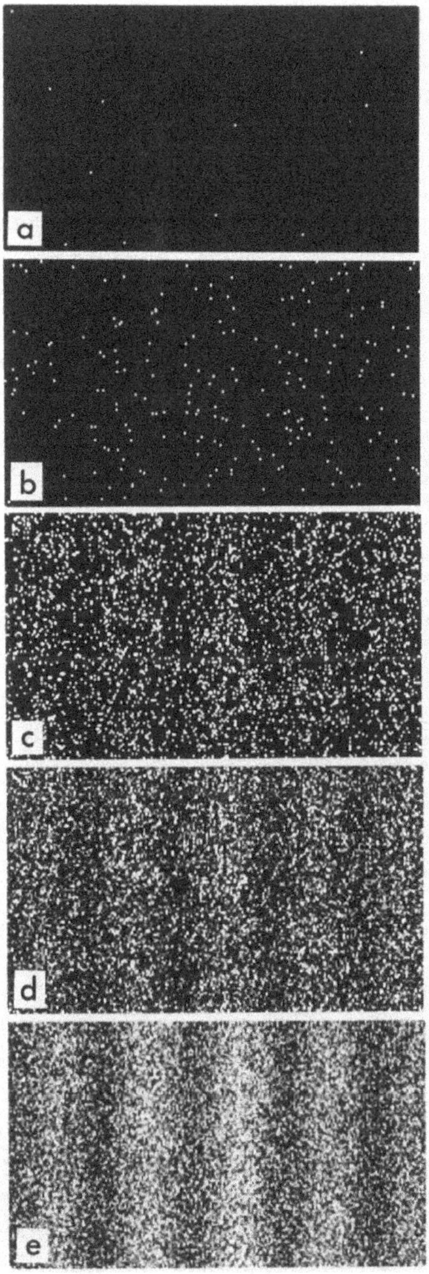

Figure 8.3. Double-slit experiment with electrons. Interference fringes captured on a screen during the double-slit experiment carried out by Akira Tonomura [3]. The numbers of electrons are (a) 10, (b) 200, (c) 6000, (d) 40 000 and (e) 140 000. Provided with kind permission of Dr Tonomura.

8.2 Novel mechanics and novel principles

The revolution of the classical paradigms begins with mechanics [5]. The investigation of the motions of bodies indeed involves all branches of physics, from the movement of electrons up to the movement of planets. Moreover, mechanics has always been the first field of application of the scientific method.

Mechanics studies the movements of all bodies—microscopic or macroscopic, slow or fast. These four categories lead to different fields of investigation, which are defined according to the comparison of the involved physical quantities with two fundamental constants: the speed of light (c) and Planck's constant (h). Therefore, the motions of bodies with a speed significantly less than c and an action greater than h are studied *using classical mechanics*, which investigates all day-to-day bodies. *Relativistic mechanics* is instead the study of bodies with a speed similar to c and an action greater than h. This branch was founded by Einstein, who in his landmark article [6] laid the foundations for restricted relativity. The movements of bodies with a speed significantly less than c and an action comparable to h are studied using *quantum mechanics*, whereas *relativistic quantum mechanics* analyses the motions of bodies with a speed and an action similar to c and h, respectively.

8.2.1 Classical principles

As in the classical case, it is necessary to define some fundamental principles, derived from observations and generally verified, in order to lay the foundations for the novel mechanics. For this reason, let us focus on the principles of classic mechanics and let us understand how they can be modified in order to include the main features of the new theory (see chapter 1).

In a classical context, the main principle is the *principle of relativity*, first described by Galileo Galilei [7]. It establishes that experiments give the same results if they are carried out in different laboratories moving according to a uniform straight translation motion with respect to each other. Therefore, physical laws do not change in such reference frames (the covariance of physical laws).

The second principle requires the following definition: a reference frame is called *inertial* if a material point at rest remains in this condition when it is not subjected to forces. Therefore, if a reference frame is inertial, all reference frames moving according to a uniform straight translation motion with respect to the first are inertial. Indeed, if it was not inertial, a body in the second system would start moving and this would violate the principle of relativity because the experimental *object at rest* would be not the same in the two reference frames. The *first principle of dynamics*, or the *inertial principle*, establishes that a free body in an inertial reference frame is at rest or moves according to a uniform straight translation motion. The principle of relativity implies the first principle but the first principle does not imply the principle of relativity.

The second principle of dynamics states that the force $\vec{f}$ applied to a body determines a proportional acceleration $\vec{a}$, as shown in the following:

$$\vec{f} = m\vec{a}.$$

The proportionality constant m is called the inertial mass, in order to distinguish it from the proportionality constant compared in the definition of the force of weight, i.e. the gravitational mass. This mass can be effectively defined as the feature of a body that unbalances the plates of a scale. It is a scalar quantity and its unit of measurement is the kilogram. It is important to highlight that if $\vec{f} = 0$, the acceleration is null and the speed of the body is constant. Hence, the second principle is coherent with the first and the first principle implies the second one.

The principle of relativity does not focus on particular dynamical experiments, so it has to be valid in all fields of physics, and also in quantum mechanics. The first principle is referred to dynamics but it does not include physical laws, so it has to be valid in all mechanics, also in quantum mechanics. On the other hand, the second principle employs a particular law, so it should be modified in a different context. This dynamical principle is replaced in quantum mechanics by Schrödinger's equation (8.3), discussed in the following. However, quantum mechanics requires a further principle, which implies Schrödinger's equation—Heisenberg's principle (8.2).

8.2.2 The definition of a state

Before providing details about the key principles of quantum mechanics, we have to focus on the concept of a *state* and its meaning in this new context. In general, the term 'state' identifies all the variables that are necessary to describe the evolution of a physical system or to unequivocally identify a physical system. In classical and relativistic mechanics, the state of a certain physical system is determined by its position and momentum. Other crucial physical features, such as energy, can be evaluated by knowing the state. Both theories aim to identify the trajectory of a physical system by knowing the initial conditions. They differ only in the formulation of the second principle of dynamics used to establish this trajectory. For example, if the initial position and velocity are known, we can precisely determine the trajectory of a bullet. Moreover, in both types of mechanics, two identical systems evolve in the same manner if they are characterized by the same initial conditions (the *causality principle*). In problems of electromagnetism, the state of a generic physical system is completely defined through the knowledge of the electric and magnetic fields, for all the positions at any time. Indeed, electromagnetism is devoted to analysing these fields when the distributions of charges and electrical current are known. In thermodynamics, the state of a certain physical system is identified through scalar physical quantities such as temperature, pressure, volume, energy, the number of particles and so on. It is worth noting that all physical quantities characterizing the state of a physical system are measurable. Since the equations characterizing classical and relativistic mechanics, electromagnetism and thermodynamics are linear, the following *superposition principle* is satisfied: if $a(x)$ and $b(x)$ are solutions of a certain evolution equation, any linear combination $\alpha\, a(x) + \beta\, b(x)$ is still a solution, where α and β are constants, x is the position, the time or a certain thermodynamic variable, and $a(x)$ and $b(x)$ are state variables.

We stress that this principle is a strict consequence of the relativity principle, the basic principle of all physical theories.

8.2.3 Quantum principles

Following the ideas developed in the above-mentioned branches of physics, a demanding challenge consists in defining a quantum state. In quantum mechanics we can formulate predictions about the probability of certain results for a measurement, but we cannot determine the evolution of a system exactly. Hence, the strict determinism characterizing classical mechanics cannot be extended easily to quantum mechanics, due to the nature of the studied phenomena. This assumption is rigorously expressed by *Heisenberg's uncertainty principle*, which establishes that the uncertainties on the position of a body and its linear momentum are related to Planck's constant as follows:

$$\Delta x \Delta p_x \geqslant \frac{h}{4\pi}, \tag{8.2}$$

where Δx is the variation of position, Δp_x is the variation of linear momentum and h is Planck's constant. This means that it is not possible to determine both the position and the linear momentum of a body with an arbitrary accuracy, implying that it is impossible to exactly define the trajectory of a system in a classical sense. It is worth noting that this uncertainty does not rely on the methods for carrying out the experiments and the instruments used. Thus, this limit cannot be overcome by improving the experimental procedure, the technologies employed, or the devices and tools used. It appears necessary to introduce a quantity that includes this intrinsic feature of quantum mechanics, as well as the so-called wave–particle dualism discussed in section 8.1: the wave function $\Psi(\vec{r}, t)$, a complex function in three spatial coordinates and the temporal function. The wave function or quantum state is no longer measurable, so it is essential to identify the physical quantities defining the physical state. We assume that $|\Psi(\vec{r}, t)|^2 d\vec{r}$ is the probability of finding a particle in a certain elementary volume centred at the point $\vec{r}$ and having radius $|d\vec{r}|$ at time t (the *Born hypothesis*, see figure 8.4). The nature of the wave function is not a generally accepted concept and it has been a source of deep debate since the very beginning of quantum theory. Schrödinger and de Broglie initially considered the state as a real physical wave, exhibiting some particular quantum features. Hence, phenomena such as interference are easily described. This ontic nature was not included in the so-called Copenhagen interpretation, promoted by Bohr, Heisenberg and Pauli, who believe that a quantum state is a state of knowledge, representing uncertainty about the real physical state of the system. This *epistemic* point of view can clarify the collapse on measurement, because the measurement adds new information. However, if the quantum state corresponds only to the information about the real state of a physical system, the resulting experimental predictions deny those of quantum theory [8]. Hence, they proved using a no-go theorem that the quantum state must be physical in a wide range of realistic approaches of quantum theory.

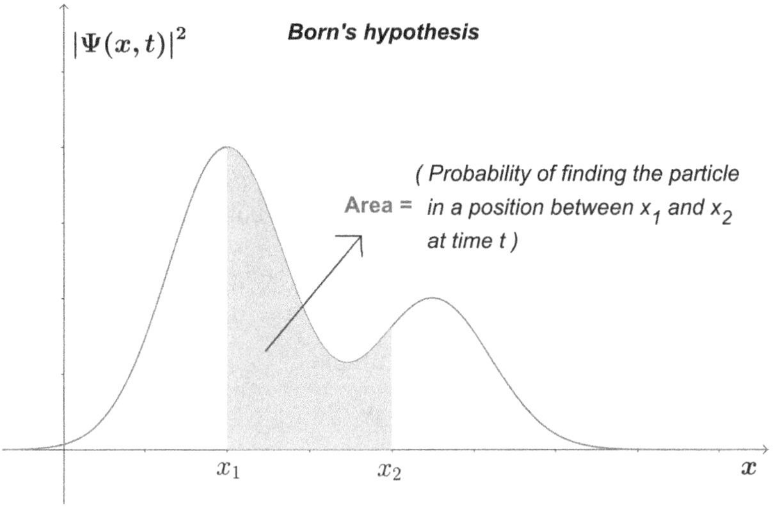

Figure 8.4. Born's hypothesis: the square of the modulus of the wave function gives the probability of obtaining a specific value in the measurement of the position of a particle, at a certain time t.

Once the problem of uncertainty, which is intrinsic in quantum mechanics, is understood, let us focus on the novel dynamical principle replacing the second principle of dynamics: Schrödinger's equation. It has the following form:

$$i\hbar\frac{\partial}{\partial t}\Psi(\vec{r},\, t) = -\frac{\hbar^2}{2m}\,\nabla^2\Psi(\vec{r},\, t) + U(\vec{r},\, t)\Psi(\vec{r},\, t), \tag{8.3}$$

where $\hbar = \frac{h}{2\pi}$ is reduced Planck's constant and $U(\vec{r},\, t)$ is the classical potential energy. It is a parabolic partial differential equation (PDE), for which there is not an existence and uniqueness theorem. For this reason, it has to be solved according to the problem. If $\Psi(\vec{r},\, t)$ is known at a certain time and at a certain position, we can investigate the evolution of the physical system. As already mentioned, the determinism is still verified in quantum mechanics, but only after measurement. Before measurement, the physical system can evolve in all possible ways.

Schrödinger's equation made the quantum theory more complete and it has allowed a plethora of phenomena to be described and understood which could not be explained previously using classical physics. In the following section, we will present some key features of the wave function and we will solve Schrödinger's equation for some significant cases.

8.3 Applications and developments

Here we show some elementary but meaningful applications of this 'novel mechanics' and its principles, which, as already mentioned, imply a definite and in a sense traumatic departure from the classical paradigm. In the previous section we have seen that in non-relativistic quantum mechanics the state of a physical system is completely captured (as far as we know) by a complex-valued function, the wave function $\Psi(\vec{r}, t)$, evolving in time according to Schrödinger's equation (8.3), which we may state as the counterpart of the second principle of dynamics (8.11) in quantum mechanics.

For a single particle having mass m and moving in the presence of a potential $U = U(\vec{r}, t)$, accounting for the different kinds of interactions with the environment, the fundamental equation is expressed by equation (8.3). This equation could be formally obtained starting from the parabolic differential operator

$$L[t, \nabla] \equiv \left[\mathrm{i}\hbar\frac{\partial}{\partial t} + \frac{\hbar^2}{2m}\nabla^2 \right],$$

defined by means of the following substitutions,

$$E \longrightarrow \mathrm{i}\hbar\frac{\partial}{\partial t},$$
$$\vec{p} \longrightarrow -\mathrm{i}\hbar\nabla,$$

into the classical expression of the energy of a free particle $E = \frac{p^2}{2m}$ and then generalizing the equation $L[t, \nabla]\Psi = 0$ to the interacting case ($U \neq 0$) for which $E = \frac{p^2}{2m} + U$, that is

$$(L[t, \nabla] - U)\Psi = 0.$$

Since all the information about a quantum system in this theoretical framework is encapsulated inside the wave function, the analysis only requires writing down Schrödinger's equation for that system (with the appropriate potential and boundary conditions) and solving it. Unfortunately, this is much easier said than done. Indeed, the problem of finding a solution to the PDE equation (8.3) in general, namely for any possible choice of boundary conditions given an arbitrary real-valued potential function $U = U(\vec{r}, t)$, is highly non-trivial. However, we will confine our introductory study to one-dimensional problems, which are much simpler to approach and from which we can nevertheless draw significant indications for the behaviour of particles in the quantum world.

Before analysing a selection of special cases, we want to discuss some mathematical properties of the wave function so that it can represent a genuine, physical state.

8.3.1 Properties of the wave function

Since $\Psi(\vec{r}, t)$ solves Schrödinger's equation (8.3), which is a linear PDE, it obeys the superposition principle: if $\Psi_1(\vec{r}, t)$ and $\Psi_2(\vec{r}, t)$ are solutions, $\Psi = a\Psi_1 + b\Psi_2$, with $a, b \in \mathbb{C}$, is also a solution of the Schrödinger equation and describes a possible physical state of the system. This feature forms the basis of quantum mechanics and its relevance cannot be overrated. Moreover, as reported above, the linearity of the PDE equation (8.3) guarantees that the Schrödinger equation verifies the principle of relativity.

Due to the imaginary factor i in front of the time derivative in equation (8.3), we infer that Ψ may be a complex-valued function. Since $\Psi(\vec{r}, t)$ represents, according to Born's rule (see section 8.2.1), a probability amplitude and its squared modulus $|\Psi(\vec{r}, t)|^2$ is the probability density function (PDF) associated with the random variable $\vec{R}(t)$ expressing the position of the particle detected at time t after a position measurement, we have to assume that $\Psi(\vec{r}, t)$ is a continuous function on its domain $D \subseteq \mathbb{R}^3$. This requirement is essential to assure that such a probabilistic interpretation is well-defined (the probability must be independent of the particular path we follow to arrive at $\vec{r}(t)$). Using Schrödinger's equation, it can also be shown that for the potential functions $U(\vec{r}, t)$, which at a given point are continuous or present a finite jump, the first derivatives of the wave function are also continuous at this point (see appendix B).

Moreover, the normalization condition must be fulfilled as well,

$$\int_D |\Psi(\vec{r}, t)|^2 \mathrm{d}^3 r = 1,$$

implying that the probability of finding the particle on the whole domain D must be equal to one. From a mathematical point of view, any proper wave function must be a 'vector' in the Hilbert space of $L^2(A)$ functions, where $A \subseteq \mathbb{R}^3$, the space of the so-called 'square-integrable functions', for which $\int_A |\Psi(\vec{r}, t)|^2 < +\infty$.

From this perspective, the meaning of what we have said above about the complex nature of the wave function should be more understandable: since each complex number c can be expressed as $c = |c|\mathrm{e}^{\mathrm{i}\theta_c}$, where $|c|$ is the modulus and θ_c is the argument or phase of the complex number, a wave function written as $\Psi(\vec{r}, t) = |\Psi(\vec{r})|\mathrm{e}^{i\theta(\vec{r}, t)}$ would generate a time-independent PDF even if $\Psi(\vec{r}, t)$ remains time-dependent. Complex numbers therefore allow us to confine the time evolution of the state in a wide range of situations (or the change of other parameters connected with physical quantities) in the phase factor of the wave function, in such a way that we can account for the same object both quantum interference and conservation of some physical quantities and probabilities (the sum of probabilities for any possible outcome is always 1).

A remarkable case in which we can separate the time dependence from the spatial dependence is when the potential function is time-independent, i.e. $U(\vec{r}, t) = U(\vec{r})$, which is a very common situation. Under this assumption, we can try to solve equation (8.3) by following the method of separating variables. As a first step, let us write $\Psi(\vec{r}, t) = \phi(\vec{r})f(t)$; then, let us substitute the last expression in the fundamental equation and divide both members of the equation by Ψ (assuming that it is not identically vanishing), obtaining

$$\frac{i\hbar}{f(t)}\frac{\partial f(t)}{\partial t} = -\frac{\hbar^2}{2m\phi(\vec{r})}\nabla^2\phi(\vec{r}) + U(\vec{r}).$$

Now, observing that the left side of the equation depends only on the time variable, while the right side is a function of only the space coordinates, the only possibility for which the equation above can hold for any time and position inside the range fixed by the problem is that both members equal the same constant, E.

Therefore, we have succeeded in separating the time dependence from the spatial dependence, achieving a great mathematical simplification of the initial problem:

$$\frac{\mathrm{d}f(t)}{\mathrm{d}t} = -\mathrm{i}\frac{E}{\hbar}f(t), \tag{8.4}$$

$$-\frac{\hbar^2}{2m\phi(\vec{r})}\nabla^2\phi(\vec{r}) + U(\vec{r})\phi(\vec{r}) = E\phi(\vec{r}). \tag{8.5}$$

Equation (8.4) can be easily integrated, leading to

$$f(t) = f(0)\mathrm{e}^{\frac{-\mathrm{i}E}{\hbar}t},$$

while equation (8.5) is the so-called time-independent Schrödinger equation and it is nothing other than the eigenvalue equation for the energy observable, the Hamiltonian operator $\hat{H} = -\frac{\hbar^2}{2m}\nabla^2 + U(\hat{\vec{r}})$ of the system:

$$\hat{H}\phi(\vec{r}) = E\phi(\vec{r}).$$

A complete state Ψ is called a *stationary state* if it is an eigenstate of the Hamiltonian operator $\hat{H}$, for which the energy has a definite value (E) and the wave function can be expressed as

$$\Psi(\vec{r}, t) = \phi(\vec{r})\mathrm{e}^{\frac{-\mathrm{i}E}{\hbar}t}.$$

Now, we can define any *physical observables* in quantum mechanics as follows: a physical observable G (e.g. energy, linear momentum, angular momentum, etc) is represented by a self-adjoint operator $\hat{G}$ defined on a given Hilbert space and the possible outcomes of a measurement of this observable are all the eigenvalues g of this operator:

$$\hat{G}\phi_g(\vec{r}) = g\phi_g(\vec{r}).$$

The self-adjointness of $\hat{G}$, i.e. $\hat{G} = \hat{G}^\dagger$, assures that the spectrum of eigenvalues and any *expectation value* $\langle G \rangle$ for the observable in the state Ψ, defined as

$$\langle G(t) \rangle \equiv (\Psi, \hat{G}\Psi) = \int_D \Psi(\vec{r}, t)^* \hat{G}\Psi(\vec{r}, t)\mathrm{d}^3r, \tag{8.6}$$

is real, a necessary condition for any directly measurable quantity. The symbol (u, v) indicates the inner product (a c-number or, more generally, a c-function) between the vectors u and v of the Hilbert space, which in the case of equation (8.6) are $u = \Psi$ and $v = \hat{G}\Psi$.

It is also easy to check that when the system is in a stationary state the expected value of any physical observable is time-independent:

$$\frac{\mathrm{d}}{\mathrm{d}t}\langle G(t)\rangle = \frac{\mathrm{d}(\Psi, \hat{G}\Psi)}{\mathrm{d}t} = \frac{\mathrm{d}}{\mathrm{d}t}\left[\int_D \phi(\vec{r})^* \mathrm{e}^{\frac{+\mathrm{i}E}{\hbar}t}\hat{G}\phi(\vec{r})\mathrm{e}^{\frac{-\mathrm{i}E}{\hbar}t}\mathrm{d}^3r\right]$$
$$= \frac{\mathrm{d}}{\mathrm{d}t}\left[\int_D \phi(\vec{r})^*\hat{G}\phi(\vec{r})\mathrm{d}^3r\right] = \frac{\mathrm{d}}{\mathrm{d}t}\langle G(0)\rangle = 0.$$

Another peculiarity of the mathematical formalism of quantum mechanics is that it predicts on general grounds some uncertainty relations among incompatible quantities. The most famous of them is the Heisenberg uncertainty principle

$$\Delta x \Delta p_x \geqslant \frac{\hbar}{2},$$

where, for a given observable G in a state Ψ its uncertainty, defined as the standard deviation of its measured values, is given by

$$\Delta G \equiv \sqrt{\langle G^2\rangle - \langle G\rangle^2}.$$

Assuming this reasonable prescription for uncertainty, it is not difficult to show (see appendix B) that for any two observables, G_1 and G_2, with the system in state Ψ, the following inequality holds:

$$\Delta G_1 \Delta G_2 \geqslant \frac{1}{2}|(\Psi, [\hat{G}_1, \hat{G}_2]\Psi)|, \tag{8.7}$$

where $[\hat{G}_1, \hat{G}_2] \equiv \hat{G}_1\hat{G}_2 - \hat{G}_2, \hat{G}_1$ is the *commutator* of the operators $\hat{G}_1$ and $\hat{G}_2$, which, in general, is not zero, as it is the case of a commutator between two matrices. Only when this commutator (which is an operator) vanishes and the dynamical variables G_1 and G_2 are said to be 'compatible' does the uncertainty relation follow the classical expectations and there is no *a priori* restriction on the accuracy with which we can know the simultaneous values of both observables. Otherwise, if the two operators do not commute, it is not possible to measure both observables at the same time and with arbitrary precision; in this case, relation (8.7) places a theoretical limit on the minimum uncertainty with which these simultaneous measurements can be carried out.

In the special case of the two observables, the position, x, and the linear momentum, p_x, whose operators are defined in the coordinate representation as follows

$$\hat{x}\phi(x_0) = x_0\phi(x_0), \qquad \hat{p}_x\phi(x) = -\mathrm{i}\hbar\frac{\mathrm{d}}{\mathrm{d}x}\phi(x),$$

it is straightforward to evaluate their commutator

$$[\hat{x}, \hat{p}_x]\phi(x) = -\mathrm{i}\hbar\hat{x}\frac{\mathrm{d}}{\mathrm{d}x}\phi(x) + \mathrm{i}\hbar\frac{\mathrm{d}}{\mathrm{d}x}(x\phi(x)) = \mathrm{i}\hbar\phi(x) \Longrightarrow [\hat{x}, \hat{p}_x] = \mathrm{i}\hbar\mathbb{1},$$

where $\mathbb{1}$ indicates the identity operator. Hence, these observables are incompatible and applying the general uncertainty relation equation (8.7) we obtain

$$\Delta x \Delta p_x \geqslant \frac{1}{2}|\langle[\hat{x}, \hat{p}_x]\rangle| = \frac{1}{2}|\langle i\hbar\mathbb{I}\rangle| = \frac{\hbar}{2},$$

which is just the mathematical statement of the celebrated Heisenberg's uncertainty principle (8.2), which is therefore incorporated into the formalism of quantum mechanics.

In the following sections, we will explore a selection of one-dimensional problems to obtain a better grasp on this subject.

8.3.2 Free particles in classical and quantum mechanics

In classical mechanics, a free particle having mass m exhibits a well-defined linear momentum $\vec{p} = m\vec{v}$ and an energy $E = \frac{1}{2}mv^2 = \frac{p^2}{2m}$. Assuming that our frame of reference is inertial (the lab reference), such a particle moves along a straight line with constant speed (i.e. with constant velocity) since, being free, according to Newton's second law, it is verified that

$$\vec{a} = \frac{\vec{F}}{m} = 0 \Longrightarrow \vec{v} = \text{const.}$$

and therefore we can write an equation for the trajectory:

$$\vec{r}(t) = \vec{r}_0 + \vec{v}t = \vec{r}_0 + \frac{\vec{p}}{m}t.$$

Applying a Galilean transformation,

$$\begin{cases} \vec{r}' = \vec{r} - \vec{V}t, \\ t' = t, \end{cases}$$

the velocity of the particle would become $\vec{v}' = \vec{v} - \vec{V}$, but the dynamics would be the same (zero acceleration and uniform motion along a straight line), in accordance with the principle of Galilean relativity.

What is the situation for a quantum free particle? In the same lab frame of reference, the wave equation for a free particle obeys Schrödinger's equation with $U(\vec{r}) = 0$:

$$i\hbar\frac{\partial}{\partial t}\Psi(\vec{r}, t) = -\frac{\hbar^2}{2m}\nabla^2\Psi(\vec{r}, t).$$

Since the problem is separable (U vanishes, so it is also t-independent), it is verified that $\Psi(\vec{r}, t) = \phi(\vec{r})e^{\frac{-iEt}{\hbar}}$, where $\phi(\vec{r})$ solves the time-independent Schrödinger equation:

$$-\frac{\hbar^2}{2m}\nabla^2\phi(\vec{r}) = E\phi(\vec{r}). \tag{8.8}$$

The general solution of equation (8.8) can be expressed as follows (with A and B being complex constants):

$$\phi(\vec{r}) = Ae^{-i\vec{k}\cdot\vec{r}} + Be^{i\vec{k}\cdot\vec{r}},$$

with $k^2 = \vec{k}\cdot\vec{k} = \frac{2mE}{\hbar^2}$. The last constraint brings out an infinite degeneracy (that is, for the given value of energy E an infinite multiplicity of states of the form $e^{\pm i\vec{k}\cdot\vec{r}}$) which is associated with the given value of energy, the same that we would have in the classical case fixing the energy E but not the momentum.

So, if we require that $\phi(\vec{r})$ is also an eigenfunction connected to the eigenvalue $\vec{p}$ for the linear momentum operator $\hat{\vec{p}} = -i\hbar\nabla$,

$$\hat{\vec{p}}\,\phi(\vec{r}) = \vec{p}\,\phi(\vec{r}) \Longrightarrow \nabla\phi(\vec{r}) = \frac{i}{\hbar}\vec{p}\,\phi(\vec{r}) \Longrightarrow \phi(\vec{r}) = Ae^{i\vec{k}\cdot\vec{r}},$$

with $\vec{p} = \hbar\vec{k}$. On the basis of this analysis we infer that the general wave function for a state of energy $E = \frac{p^2}{2m} = \frac{\hbar^2 k^2}{2m}$ and a linear momentum $\vec{p} = \hbar\vec{k}$ has the form

$$\Psi(\vec{r}, t) = Ae^{-i\frac{(Et-\vec{p}\cdot\vec{r})}{\hbar}},$$

while for a state of energy $E = \frac{p^2}{2m} = \frac{\hbar^2 k^2}{2m}$ and a linear momentum $\vec{p} = -\hbar\vec{k}$ the solution is

$$\Psi(\vec{r}, t) = Be^{-i\frac{(Et+\vec{p}\cdot\vec{r})}{\hbar}}.$$

In any case, the wave function is a plane wave directed along the direction $\vec{k}$, the first one moving forward (with amplitude A) and the second one moving backwards (with amplitude B). Since the considered states are eigenstates of the linear momentum operator, it follows that $\Delta p_x = \Delta p_y = \Delta p_z = 0$ (no uncertainty on the value of $\vec{p}$) and to reconcile this result with the Heisenberg's principle the only possibility is that Δx, Δy and Δz are divergent (i.e. the standard deviation of the coordinates is infinite), which means that there is no information about the possible position of a particle with definite momentum. Indeed, the states represented by plane waves are not normalizable, so they are not L^2-functions and virtually describe a particle that can be found with the same probability at any point of space along the direction of the motion. But these states are unphysical (it is experimentally impossible to prepare a particle in a pure state of momentum, i.e. without the slightest uncertainty about its value) and a better model for a real, free quantum particle is provided by a certain superposition of plane waves with momenta inside a more or less narrow interval corresponding to a region $D_p = [p_x; p_x + \delta p_x] \times [p_y; p_y + \delta p_y] \times [p_z; p_z + \delta p_z]$ in the space of momenta:

$$\Psi(\vec{r}, t) = \int_{D_p} A(\vec{p})e^{\frac{-i}{\hbar}(Et-\vec{p}\cdot\vec{r})}d^3p = \int_{\mathbb{R}^3} A(\vec{p})e^{\frac{-i}{\hbar}(Et-\vec{p}\cdot\vec{r})}d^3p, \tag{8.9}$$

with $A(\vec{p}) = 0\ \forall\ \vec{p} \notin D_p$. Considering a wave packet like the one in equation (8.9), the linear momentum spreads over a finite region in the momentum space, resulting in a non-zero dispersion and, correspondingly, the uncertainty about the position

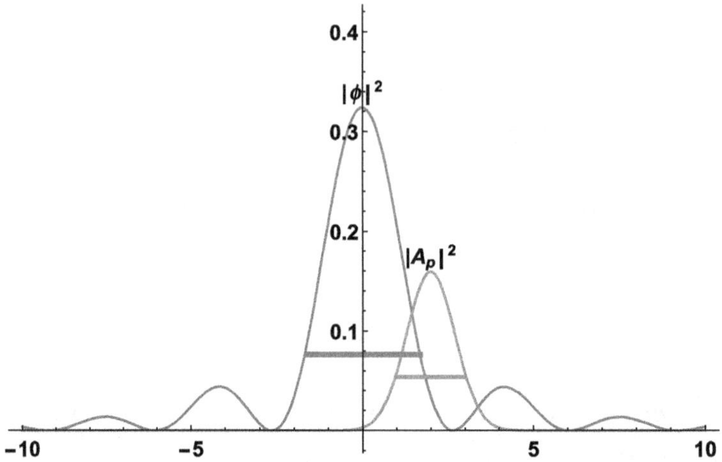

Figure 8.5. Plot of the squared modulus of $\phi(x)$ and $A_p(x)$ describing a free particle moving in one dimension. The thick horizontal lines provide a visual representation of position and momentum uncertainties of the particles (with $\hbar$ set equal to 1).

also reduces to a finite value. In this setting, it can be shown, using the machinery of the Fourier transform, that the product of these spreads satisfies the uncertainty principle, vividly displaying that uncertainty relations originate from the wave aspects of matter. Instead of performing such calculations (the interested reader can find the details in any good textbook on quantum mechanics, e.g. [9, 10]), we have provided a visual example of this fact by plotting the squared modulus of a wave packet and the squared modulus of the wave function A_p in the momentum space, which collects the amplitudes of each plane wave present in the packet (see equation (8.9)). The uncertainties Δx and Δp_x on x and p_x can be estimated as half the length of the horizontal segments drawn in figure 8.5 and their product is greater than 1 (the unit $\hbar$ has been omitted in Δp).

8.3.3 An infinitely deep potential well

Now we are ready to approach one of the simplest but also most significant problems in quantum mechanics, the problem of the infinitely deep potential well, in which a particle is trapped in one dimension between two impenetrable rigid walls (see figure 8.6).

We can model the potential function for this situation as follows:

$$U(x) = \begin{cases} 0, & \text{for } 0 \leqslant x \leqslant L \\ +\infty, & \text{for } \left| x - \dfrac{L}{2} \right| > \dfrac{L}{2} \end{cases}.$$

Classically, a particle placed inside the well cannot escape and travels along straight lines with uniform motion between successive collisions with the walls. In any case, a classical particle trapped in this manner could have any energy value.

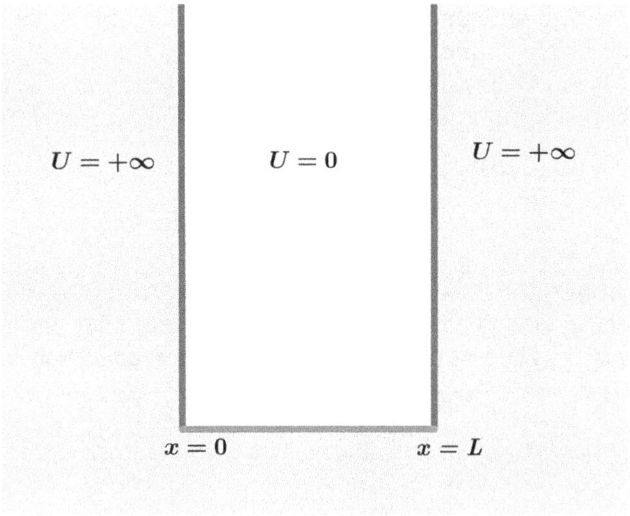

Figure 8.6. Infinite square-well potential energy.

What happens to a quantum particle of mass m, obeying the principles of quantum mechanics? We have to find its wave function by solving equation (8.3). When the potential U is a piecewise continuous function a common and elementary strategy consists in solving the problem in each region where U is continuous (or even constant) and trying to match all the pieces by imposing appropriate boundary conditions. In this special case, we can divide the space into three regions. The first ($x < 0$) and third ($x > L$) are characterized by an infinite potential energy, so we can expect that these regions are absolutely inaccessible to the particle that otherwise would acquire an infinite energy, while in the second region ($x < 0$) the particle is free, and the time-independent Schrödinger equation is written as follows:

$$-\frac{\hbar^2}{2m}\frac{d^2}{dx^2}\phi(x) = E\phi(x) \Longrightarrow \phi''(x) = -k^2\phi(x), \tag{8.10}$$

where $k^2 = \frac{2mE}{\hbar^2}$ and $\phi''(x) = \frac{d^2}{dx^2}\phi(x)$. Note that here it has been assumed that the particle has a non-negative energy E (hence the position for k^2), excluding the case of negative energy on physical grounds. However, we could verify mathematically, following the same path that we are illustrating here, that the only solution compatible with $E < 0$ is the trivial one, that is $\phi(x) = 0$ for any value of x (this means that it is impossible to allocate a particle inside an infinite potential well with negative energy).

Observing that the ordinary differential equation (ODE) (8.10) is the same as that you encounter in classical mechanics when you want to solve the problem of the free harmonic oscillator with an angular frequency k, whose general solution is well-known, we can write directly

$$\phi(x) = A\sin(kx + \theta_0).$$

Now it is time to impose boundary conditions for this problem. Since the regions I and III are inaccessible, there $\phi(x) = 0$ identically ($\phi(x) = \phi(x) = 0 \ \forall \ x$ s. t. $x < 0 \vee x > L$), but the wave function must be globally continuous, hence the following conditions must be imposed at connection points $x = 0$ and $x = L$:

$$\begin{aligned} \phi(0^+) &\equiv \lim_{x \to 0^+} \phi(x) = \phi(0^-) = 0, \\ \phi(L^-) &\equiv \lim_{x \to L^-} \phi(x) = \phi(L^+) = 0. \end{aligned} \tag{8.11}$$

The first gives $A\sin(\theta_0) = 0$ (with $A \neq 0$), so $\theta_0 = m\pi$, with $m \in \mathbb{Z}$; without loss of generality we set $\theta_0 = 0$. The second condition implies that $\sin(kL) = 0$, that is $kL = n\pi$ with $n \in \mathbb{Z}$. The last equality is a quantization condition for the energy of the particle, in fact,

$$k^2L^2 = n^2\pi^2 \Longrightarrow \frac{2\,mEL^2}{\hbar^2} = n^2\pi^2 \Longrightarrow E = \frac{n^2\hbar^2\pi^2}{2\,mL^2} = \frac{n^2h^2}{8\,mL^2}.$$

This is a clear sign of a departure from classicality: a particle confined inside an infinite potential well cannot have any possible value of energy, but is *quantized*, i.e. it can assume only those values that are particular integer multiples of a specific quantity, the *quantum* of energy $E_1 = \frac{h^2}{8\,mL^2}$. To sum it up, the spectrum of energy is discrete and consists only of those energies E_n that can be expressed as the quantum of energy E_1 multiplied by a perfect square (1, 4, 9, 16, 25, etc):

$$E_n = n^2 E_1.$$

Why do we not also consider the integer value $n = 0$? Well, in this case we would have $E_0 = 0$, but this value on closer inspection is impossible to obtain because the corresponding spatial wave function would be $\phi_0(x) = A\sin(0x) = 0$ (for $E = 0$ we have $k = 0$ as well), therefore a particle in such a state of zero energy could not be detected at all. However, we could have predicted this fact without using the equations, simply by appealing to Heisenberg's principle: since the particle is confined inside a region with a finite extension L, the uncertainty of its position is at most L; if the particle had zero energy in the absence of potential energy, the only possibility would be that the speed is zero, and thus the momentum would also be exactly zero, with $\Delta p = 0$ (sharp prediction), thus it is verified that $\Delta x \Delta p \geqslant L \cdot 0 = 0$, which would constitute an evident infringement of the uncertainty principle. From this perspective, we can consider that the minimum, non-vanishing energy E_1 that the particle has in the lowest energy state, the so-called ground state, and that constitutes the so-called zero-point energy (ZPE), is necessary to 'save' the uncertainty principle, without which the particle could lie still anywhere in the finite hole, making itself identifiable in space and time as a Newtonian material point. Before ending this section, we want to provide an example of the procedure of normalization, which serves to fix the amplitude parameter A, left undetermined even after setting the boundary conditions.

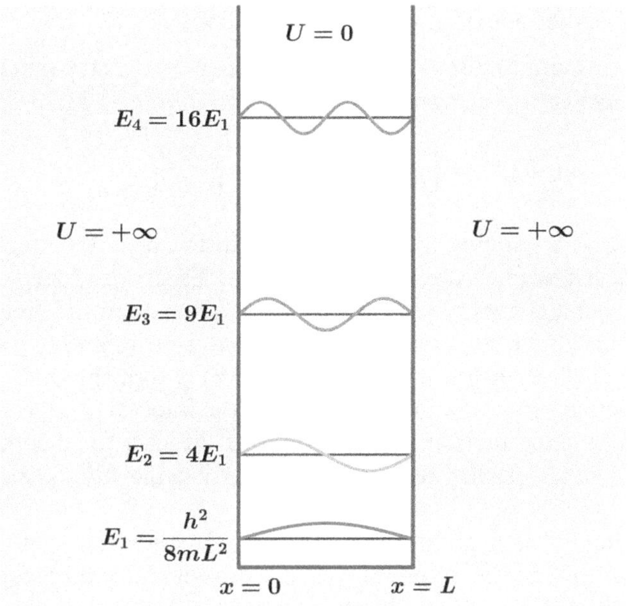

Figure 8.7. The first four energy levels and spatial wave functions for a particle inside an infinite potential well.

Since we are dealing with stationary states, it is verified that

$$\begin{aligned}\int_{\mathbb{R}} |\Psi(x, t)|^2 \mathrm{d}x &= \int_{\mathbb{R}} |\phi(x)|^2 \mathrm{d}x \\ &= \int_0^L |A \sin(kx)|^2 \mathrm{d}x = 1 \implies |A| = \frac{1}{\sqrt{\int_0^L \sin^2(kx)\mathrm{d}x}},\end{aligned}$$

and since $kL = n\pi$,

$$\int_0^L \sin^2(kx)\mathrm{d}x = \int_0^L \frac{1 - \cos(2kx)}{2}\mathrm{d}x = \frac{L}{2} \implies |A| = \sqrt{\frac{2}{L}}.$$

We can also observe that the constant phase α in $A = |A|e^{i\alpha}$ is arbitrary and inessential, this means that we can put $\alpha = 0$ without loss of generality. Moreover, if we consider negative integers n the energy remains the same, but also the wave function does not change in an essential way because the only variation occurring by taking $-n$ instead of n is a global sign in $\phi_n(x)$, sign that can be reabsorbed by choosing the phase angle $\alpha = \pi$ (since $e^{i\pi} = -1$).

Therefore, we can write the full expression of the wave function for the nth energy level E_n as

$$\Psi(x, t) = \sqrt{\frac{2}{L}} \sin\left(\frac{n\pi x}{L}\right) e^{\frac{-i\hbar\pi^2 n^2}{2mL^2}t}.$$

The first four energy levels are depicted in figure 8.7.

8.3.4 The surprises do not stop: quantum tunnelling

In this section, we want to discuss the behaviour of a quantum particle facing a finite potential barrier, like the one presented in figure 8.8:

$$U(x) = \begin{cases} U_0, & \text{for } 0 \leqslant x \leqslant a \\ 0, & \text{for } x < 0 \ \vee \ x > a \end{cases}.$$

Classically, a particle moving along the x-axis in the positive direction (i.e. to the right) would exceed or not exceed the barrier placed between 0 and a depending on whether its mechanical energy (E_{tot}) is higher or lower than the 'height' (U_0) of the barrier itself, since kinetic energy (E_{k}) must be non-negative and so for every allowed position x it must be verified that $E_{\text{k}} = E_{\text{tot}} - U(x) \geqslant 0$ (it should be clear that here it is not necessary to think of a material barrier: the interval $[0; a]$ is instead a region in which a repulsive interaction force is recorded, at least for incoming particles).

As you could easily predict, for a quantum particle the story is not so simple and neat.

Since the potential is a piecewise continuous function, we can adopt the same strategy used for the infinite potential well. First we identify three regions where the function is constant or in any case continuous:

$$\text{I: } x < 0; \quad \text{II: } 0 \leqslant x \leqslant a; \quad \text{III: } x > a.$$

Then we try to solve equation (8.5) in each of the three regions and at the end we will connect the three solutions by imposing the appropriate boundary conditions at the limits of any region, in this case at $x = 0$ and $x = a$. In regions I and III we have $U(x) = 0$, so the time-independent Schrödinger's equation becomes (to simplify the notation we omit arguments when this does not create ambiguity):

$$\phi'' = -\frac{2\,mE}{\hbar^2}\phi = -k^2\phi,$$

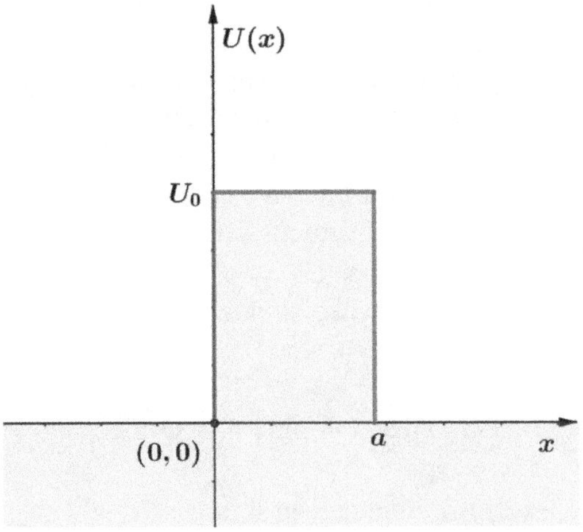

Figure 8.8. Finite rectangular potential barrier.

with $k = \sqrt{\frac{2mE}{\hbar^2}}$.

In region II $U(x) = U_0$ and equation (8.5) can be written as follows:

$$\phi'' = \frac{2m}{\hbar^2}(U_0 - E)\phi = \beta^2\phi. \tag{8.12}$$

In the last position $\beta = \sqrt{\frac{2m}{\hbar^2}(U_0 - E)}$ is a real number since here we are interested only in the case $E < U_0$, which represents the situation where the classical particle would have no chance of moving from region I to region III and would be inexorably reflected by the potential barrier. The case $E \geqslant U_0$ appears a little less interesting and can be dealt with in an analogous way, even if the quantum particle could be reflected by the barrier also not exhibiting a sufficient energy to overcome it, at least according to what is foreseen by classical mechanics.

The most general solution of the ODE (8.12) has the form

$$\phi_{\text{II}}(x) = C\text{e}^{-\beta x} + D\text{e}^{\beta x},$$

so the spatial wave function for this problem exhibits the following aspect:

$$\phi(x) = \begin{cases} \phi_{\text{I}}(x) = A\text{e}^{-\text{i}kx} + B\text{e}^{-\text{i}kx}, & \text{for } x < 0 \\ \phi_{\text{II}}(x) = C\text{e}^{-\beta x} + D\text{e}^{\beta x}, & \text{for } 0 \leqslant x \leqslant a, \\ \phi_{\text{III}}(x) = F\text{e}^{-\text{i}kx} + G\text{e}^{-\text{i}kx}, & \text{for } x > a \end{cases}$$

where A, B, C, D, F and G are complex constants to be determined.

The first simplification can be derived from the following observation: the plane waves that overlap in regions I and III represent a progressive wave with amplitude A and a regressive wave (moving backwards) with amplitude B, and this is fine for region I, but such a superposition is not necessary for region III, where, if the particle could arrive, it would only go forward. Therefore, using this physical argument, we are entitled to set $G = 0$. Now we are ready to set the boundary conditions and, in addition to the continuity requirement on ϕ, as the discontinuities in $U(x)$ at $x = 0$ and $x = a$ are jump discontinuities, we can also impose the continuity of the first derivative of the wave function ϕ' at these points (see appendix A for a simple proof).

Using the same notation as in the previous section, we have for the continuity of ϕ,

$$\begin{aligned} \phi(0^-) = \phi_{\text{I}}(0^-) = \phi_{\text{II}}(0) = \phi(0^+) &\Longrightarrow A + B = C + D, \\ \phi(a^-) = \phi_{\text{II}}(a) = \phi_{\text{III}}(a^+) = \phi(a +) &\Longrightarrow C\text{e}^{\beta a} + D\text{e}^{-\beta a} = F\text{e}^{\text{i}ka}, \end{aligned} \tag{8.13}$$

and for the continuity of ϕ',

$$\begin{aligned} \phi'(0^-) = \phi'_{\text{I}}(0^-) = \phi'_{\text{II}}(0) = \phi'(0^+) &\Longrightarrow \text{i}k(A - B) = \beta(C - D), \\ \phi'(a^-) = \phi'_{\text{II}}(a) = \phi'_{\text{III}}(a^+) &\Longrightarrow \beta(C\text{e}^{\beta a} - D^{-\beta a}) = \text{i}kF\text{e}^{\text{i}ka}. \\ = \phi'(a^+) & \end{aligned} \tag{8.14}$$

We can note that the amplitude A must necessarily be non-zero if we want the particle to interact directly with the barrier, so we can divide each of the above equations by A to obtain the following system of linear equations in the new unknowns $b = B/A$, $c = C/A$, $d = D/A$ and $f = F/A$:[1]

$$\begin{cases} 1 + b = c + d, \\ 1 - b = \dfrac{-i\beta}{k}(c - d), \\ ce^{\beta a} + de^{-\beta a} = fe^{ika}, \\ ce^{\beta a} - de^{-\beta a} = \dfrac{ik}{\beta}fe^{ika}. \end{cases} \tag{8.15}$$

Before solving the system (8.15), it is appropriate to note that since, after a measurement of position, the particle either is found in region I or in region III (it cannot be detected in region II if we want $E < U_0$, since the kinetic energy in any measurement must be positive[2]), the probability $R \equiv |b|^2$ that it is reflected by the barrier plus the probability $T \equiv |f|^2$ that it is transmitted beyond the barrier must be equal to 100%.

In light of this undulatory interpretation the two real coefficients R and T are called, respectively, the *reflection coefficient* and *transmission coefficient*, such that

$$R + T = 1.$$

Now, let us try to find an explicit expression for R and T by solving the system above.

Summing the first two equations of (8.15) we obtain

$$c = \frac{2 - \left(1 + \frac{i\beta}{k}\right)d}{1 - \frac{i\beta}{k}},$$

and we can substitute this expression in the last two equations of (8.15), thus forming a 2×2 linear system in the unknowns d and f:

[1] This choice is also useful to circumvent the normalization problem that we would have encountered acting differently, after all we have considered (for the sake of simplicity) plane waves solutions in the unlimited regions instead of wave packets, as a more rigorous treatment would be required.

[2] This is a subtle point—if we want to measure the energy of a particle with an initial energy $E < U_0$ when it is inside a narrow region, for example the classically forbidden region of our problem, we must also perform a measurement of position with a certain accuracy. Thus, the value of energy that we measure is $E' \geqslant U_0$, which differs from E as it is affected by the disturbance due to the previous position measurement; this perturbation is present and unavoidable because energy and position are not compatible observables.

$$\begin{cases} 2d\mathrm{e}^{-\beta a} = f\mathrm{e}^{\mathrm{i}ka}\left(1 - \mathrm{i}\dfrac{k}{\beta}\right) \\ \dfrac{4 - 2\left(1 + \frac{\mathrm{i}\beta}{k}\right)d}{1 - \frac{\mathrm{i}\beta}{k}}\mathrm{e}^{\beta a} = f\mathrm{e}^{\mathrm{i}ka}\left(1 + \mathrm{i}\dfrac{k}{\beta}\right) \end{cases}. \tag{8.16}$$

At this point we can solve for f comparing the two equations in (8.16):

$$f = \frac{4\mathrm{e}^{-\mathrm{i}ka}}{\mathrm{e}^{\beta a}\left(2 + \frac{\mathrm{i}\beta}{k} - \frac{\mathrm{i}k}{\beta}\right) + \mathrm{e}^{-\beta a}\left(2 - \frac{\mathrm{i}\beta}{l} + \frac{\mathrm{i}k}{\beta}\right)} = \frac{\mathrm{e}^{-\mathrm{i}ka}}{\cosh(\beta a) + \frac{\mathrm{i}}{2}\left(\frac{\beta}{k} - \frac{k}{\beta}\right)\sinh(\beta a)}.$$

Now, recalling the definitions of k and β, we can finally calculate the transmission coefficient $T = |f|^2$:

$$T = \left[\cosh^2(\beta a) - \frac{1}{4}\left(\frac{\beta^2 - k^2}{\beta k}\right)^2 \sinh^2(\beta a)\right]^{-1} = \left[1 + \frac{U_0^2 \sinh^2(\beta a)}{4E(U_0 - E)}\right]^{-1}. \tag{8.17}$$

The reflection coefficient $R = |b|^2$ can be obtained by completing the system resolution or, more simply, using the relation $R = 1 - T$. In any case, it can be expressed as follows:

$$R = \left[1 + \frac{4E(U_0 - E)}{U_0^2 \sinh^2(\beta a)}\right]^{-1}. \tag{8.18}$$

After some long and tedious calculations we have discovered quite a strange effect: a quantum particle moving towards a finite potential barrier higher than the energy of the particle itself has a finite probability of crossing such a barrier, which is a classically inaccessible region and in which not even a quantum particle can be detected. It is just as if there were a short-cut, a hole in the barrier, hence the name 'quantum tunnelling' for this bizarre phenomenon.

For the sake of completeness, it is worth noting that the analysis carried out so far can be suitably generalized. In fact, if we deal with more realistic potential barriers and we are able to find a continuous and positive function $U = U(x)$ on a certain interval $[a; b]$ (whose graph describes the barrier under examination quite well), then we can still evaluate the total transmission probability $T(E)$ across the barrier using the following formula (its derivation relies on more advanced methods which are beyond the scope of this chapter):

$$T(E) = \mathrm{e}^{-2\int_a^b \frac{\sqrt{2m(E-U(x))}}{\hbar}\mathrm{d}x}.$$

Now, let us make some remark about the importance and magnitude of this effect. If we have $U_0 \gg E$ the transmission coefficient (i.e. the probability of overcoming the potential barrier) tends to be very small and vanishes for an infinitely high barrier; the same happens for a very thick barrier ($a \gg 1$). However, this effect is definitely real, in many situations involving atoms, nuclei and subatomic particles (e.g. electrons) the tunnelling probability T results are significantly different from zero and can also get very close to 1.

Indeed, several physical phenomena and technological applications exploit this effect strongly: for example, it plays a crucial role in radioactive decay, in the cold emission of electrons from metals and in the stellar fusion process, where it facilitates the penetration of the Coulombian barrier. Furthermore it is thanks to this effect that today we can 'see' atoms using a scanning tunnelling microscope (STM, see figure 8.9). It is (also) because of the tunnelling effect that technologists and engineers can design and operate Josephson's junctions and the great variety of superconductor and semiconductor electronic devices available today.

To be more concrete, let us consider an electron moving within an atom ($a \approx 10^{-10}$ m) and suppose that it is facing a potential barrier of 10 eV. If it has an energy of 8 eV (2 eV less than the barrier), then the probability T of overtaking the barrier is about 50%. Let us also examine the case of an atomic nucleus of mass $m = 1 \times 10^{-27}$ kg moving in a region of about 1 fm $=10^{-15}$ m. If the potential barrier has an U_0 of 10 MeV and the energy of the nucleus is equal to 9 MeV, then the transmission coefficient is slightly less than 93%, which is like saying that the barrier is almost transparent for the considered nucleus.

What happens if we try to repeat these numerical experiments with macroscopic quantities, for example to calculate the probability that a car crosses a potential barrier consisting of a fairly high mountain? What you obtain by applying the formula (8.17) are ridiculously small numbers which probably have no physical meaning, but that make us understand how size matters, in particular in relation to the magnitude of the quantum of action h, which is an excellent benchmark to realize whether quantum effects are relevant or not.

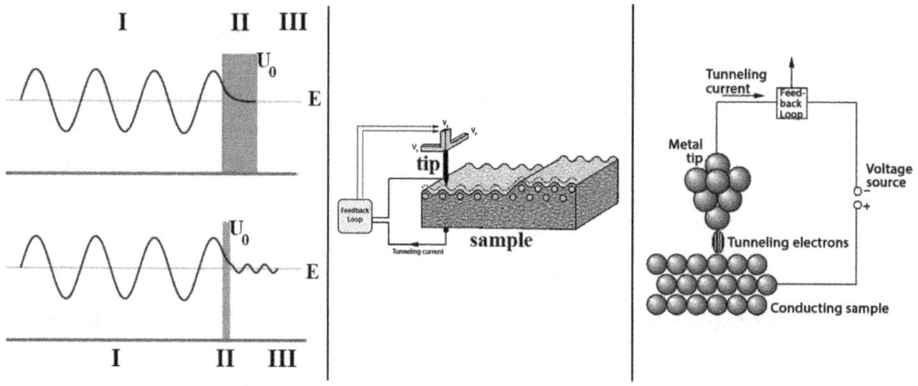

Figure 8.9. Left: a pictorial representation of a wave function describing the phenomenon of barrier penetration (quantum tunnelling). The other two figures illustrate the basic scheme of an STM.

8.3.5 The harmonic oscillator: an overview

After this very partial review of one-dimensional problems, it is worth mentioning the quantum harmonic oscillator, given its extraordinary relevance for the subsequent development of quantum theory and theoretical physics. However, many details will be omitted because the topic is not very simple and there is no room for a detailed discussion here .

The Hamiltonian function of the classical harmonic oscillator with mass m and angular frequency ω in one dimension is

$$H = \frac{p^2}{2m} + \frac{m\omega^2}{2}x^2,$$

so the corresponding time-independent Schrödinger equation in this case is

$$\hat{H}\phi(x) = \left[-\frac{\hbar^2}{2m}\frac{\mathrm{d}^2}{\mathrm{d}x^2} + \frac{m\omega^2}{2}x^2\right]\phi(x) = E\phi(x), \tag{8.19}$$

where $\xi_0 = \sqrt{\frac{\hbar}{2m}}$ is a characteristic length for this problem.

Now, instead of following the same route as the previous sections trying to solve the ODE equation (8.19), we want to mention an algebraic method that has wide applications in quantum field theory, which is the proper relativistic version of quantum mechanics.

This method also has the advantage of allowing you to obtain the energy spectrum of the harmonic oscillator much more simply than the method that requires the full resolution of the differential equation. It is carried out as follows (figure 8.10).

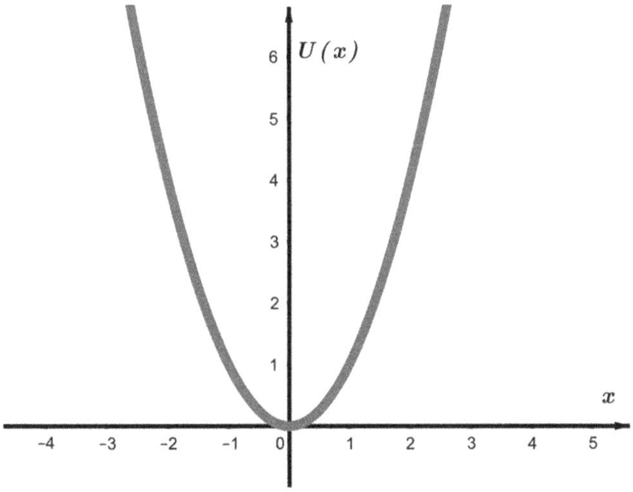

Figure 8.10. Potential energy of the harmonic oscillator.

1. Define the two non-Hermitian (i.e. $\hat{G} \neq \hat{G}^{\dagger}$) operators $\hat{a}$ and $\hat{a}^{\dagger}$:

$$\hat{a} = \frac{m\omega\hat{x} + \mathrm{i}\hat{p}}{\sqrt{2m\omega\hbar}}, \qquad \hat{a}^{\dagger} = \frac{m\omega\hat{x} - \mathrm{i}\hat{p}}{\sqrt{2m\omega\hbar}}. \tag{8.20}$$

Since the commutator $[\hat{x}, \hat{p}]$ between $\hat{x}$ and $\hat{p}$ is equal to $\mathrm{i}\hbar\hat{I}$ (we remember that $\hat{I}$ represents the identity operator), it is easy to check that $[\hat{a}, \hat{a}^{\dagger}] = \hat{I}$.

2. Express the position and momentum operators (which are self-adjoint) in terms of $\hat{a}$ and $\hat{a}^{\dagger}$:

$$\hat{x} = \sqrt{\frac{\hbar}{2\,m\omega}}(\hat{a} + \hat{a}^{\dagger}) = \xi_0(\hat{a} + \hat{a}^{\dagger}), \qquad \hat{p} = \mathrm{i}\sqrt{\frac{\hbar m\omega}{2}}(\hat{a}^{\dagger} - \hat{a}) = \mathrm{i}\xi_0 m\omega(\hat{a}^{\dagger} - \hat{a}), \tag{8.21}$$

and substitute these expressions into the Hamiltonian operator $\hat{H} = \frac{\hat{p}^2}{2m} + \frac{m\omega^2\hat{x}^2}{2}$ to obtain

$$\hat{H} = \frac{\hbar\omega}{2}(\hat{a}\hat{a}^{\dagger} + \hat{a}^{\dagger}\hat{a}) = \frac{\hbar\omega}{2}(\hat{a}^{\dagger}\hat{a} + \hat{a}^{\dagger}\hat{a} + \hat{I}) = \hbar\omega\left(\hat{a}^{\dagger}\hat{a} + \frac{1}{2}\hat{I}\right),$$

in the last step we have used the commutation relation $[\hat{a}, \hat{a}^{\dagger}] = \hat{a}\hat{a}^{\dagger} - \hat{a}^{\dagger}\hat{a} = \hat{I}$.

Thus we have expressed the energy operator in terms of the so-called *number operator* $\hat{N} \equiv \hat{a}^{\dagger}\hat{a}$, that is

$$\hat{H} = \hbar\omega\left(\hat{N} + \frac{1}{2}\hat{I}\right).$$

3. Find the eigenvalues and eigenvectors (that are the eigenfunctions in the coordinate representation) of the operator $\hat{N}$ (a longer task).

The eigenvalue equation for $\hat{N}$ is

$$\hat{N}u_{\nu} = \nu u_{\nu},$$

and since the squared norm of any vector is non-negative

$$0 \leqslant (\hat{a}u_{\nu}, \hat{a}u_{\nu}) = (u_{\nu}, \hat{a}^{\dagger}\hat{a}u_{\nu}) = (u_{\nu}, \hat{N}u_{\nu}) = (u_{\nu}, \nu u_{\nu}) = \nu(u_{\nu}, u_{\nu}),$$

we conclude that the eigenvalues ν of $\hat{N}$ cannot be negative. It is possible that $\nu = 0$, in fact for the previous chain of relations $(\hat{a}u_0, \hat{a}u_0) = 0$, which occurs only when the vector $\hat{a}u_0 = 0$, so

$$\hat{a}u_0 = 0 \Longleftrightarrow \frac{m\omega\hat{x} + \mathrm{i}\hat{p}}{\sqrt{2m\omega\hbar}}u_0 = 0 \Longleftrightarrow \left(\frac{\mathrm{d}}{\mathrm{d}x} + \frac{x}{{\xi_0}^2}\right)u_0 = 0.$$

Solving the last ODE, we obtain (after normalization) the eigenfunction associated with $\nu = 0$:

$$u_0(x) = \sqrt{\frac{1}{\sqrt{\pi}\,\xi_0}}\ \exp-\frac{x^2}{2\xi_0^2}.$$

Now we observe that the commutators of $\hat{N}$ and $\hat{a}$ and $\hat{a}^\dagger$ have the following pattern (the property of commutators $[AB, C] = [A, C]B + A[B, C]$ turns out to be useful):

$$\begin{aligned}[\hat{N}, \hat{a}] &= [\hat{a}^\dagger, \hat{a}]\hat{a} + \hat{a}^\dagger[\hat{a}, \hat{a}] = -\hat{a},\\ [\hat{N}, \hat{a}^\dagger] &= [\hat{a}^\dagger, \hat{a}^\dagger]\hat{a} + \hat{a}^\dagger[\hat{a}, \hat{a}^\dagger] = \hat{a}^\dagger.\end{aligned} \tag{8.22}$$

From these last properties we can deduce that if u_ν is an eigenfunction of $\hat{N}$ with eigenvalue ν, then $\hat{a}^\dagger u_\nu$ is an eigenfunction of $\nu + 1$. Indeed,

$$\hat{N}\hat{a}^\dagger u_\nu = (\hat{a}^\dagger + \hat{a}^\dagger\hat{N})u_\nu = \hat{a}^\dagger u_\nu + \nu\hat{a}^\dagger u_\nu = (\nu + 1)\hat{a}^\dagger u_\nu.$$

To preserve normalization (an exercise: prove it starting from $(\hat{a}^\dagger u_\nu, \hat{a}^\dagger u_\nu) = (u_\nu, \hat{a}\hat{a}^\dagger u_\nu)$) we have to write

$$\hat{a}^\dagger u_\nu = \sqrt{\nu + 1}\, u_{\nu+1}.$$

Likewise, we can prove that $\hat{a}u_\nu$ is an eigenfunction of $\nu - 1$ and that to preserve normalization we have to write

$$\hat{a}u_\nu = \sqrt{\nu}\, u_{\nu-1}.$$

Now, we can build any discrete eigenstate starting from u_0, in fact $u_1 = \frac{1}{\sqrt{1}}\hat{a}^\dagger u_0$, $u_2 = \frac{1}{\sqrt{2\cdot 1}}(\hat{a}^\dagger)^2 u_0$, $u_3 = \frac{1}{\sqrt{3!}}(\hat{a}^\dagger)^3 u_0$ and so on; hence, for a generic n,

$$u_n = \frac{1}{\sqrt{n!}}(\hat{a}^\dagger)^n u_0. \tag{8.23}$$

Are those reported in equation (8.23) all the possible eigenstates for $\hat{N}$? Can there be eigenstates associated with non-integer eigenvalues for the number operator?

The answer is 'no'. Indeed, if such eigenstates existed then, applying to one of them the operator $\hat{a}$ in succession a suitable number of times, we would obtain a proper eigenstate whose eigenvalue would lie between zero and one, and applying $\hat{a}$ another time we would obtain an eigenstate associated with a negative eigenvalue, which we have proved to be impossible. To avoid this contradiction, we must conclude that the indices ν and all the possible eigenvalues of $\hat{N}$ can only take natural values.

Thus this discussion solves our problem. We have deduced that the spectrum of the (self-adjoint) number operator $\hat{N}$ is composed of all natural numbers $(0, 1, 2, 3, \ldots)$ and only those, therefore the energy levels of the quantum harmonic oscillator are given by

$$E_n = \hbar\omega\left(n + \frac{1}{2}\right), \qquad \text{with} \quad n = 0, 1, 2, 3, \ldots.$$

The first five energy levels are shown in figure 8.11. Again, we have found that for bound states the energy levels are discrete, in other words, the energy is quantized.

The energy eigenvectors (that we can also call eigenfunctions in this representation) are exactly the same as the number operator and can be obtained by repeatedly applying the operator $\hat{a}^\dagger$ to the ground state u_0:

$$u_n = \frac{1}{\sqrt{n!}}(\hat{a}^\dagger)^n u_0 = (n!\sqrt{\pi}\xi_0)^{-1/2}(\hat{a}^\dagger)^n \exp\left(-\frac{x^2}{2\xi_0^2}\right).$$

Carrying out the calculations in greater detail, it could be shown that the generic eigenfunction for the nth energy level, $u_n(x)$, is given by the product of a particular n-degree polynomial (called the *Hermite polynomial*) and an exponential function of the Gaussian type. In the figure below the first five eigenfunctions and the corresponding energy levels are shown. It should not have escaped the attentive reader's attention that the eigenfunctions of the harmonic oscillator have well-defined symmetries with respect to the spatial coordinates, in fact they are always either even or odd functions. This fact obviously is not a coincidence, as will be explained later.

Since the operators $\hat{a}^\dagger$ and $\hat{a}$ when acting on a state of energy E_n create or remove a quantum of energy equal to $\hbar\omega = |E_{n\pm1} - E_n|$, these operators are called, respectively, *creation* and *annihilation* operators, or even, less frequently, *ladder operators*.

As mentioned before, the harmonic oscillator with the creation and annihilation operators constitutes a paradigmatic system, in fact in quantum field theory, the branch of theoretical physics that reconciles quantum mechanics and special

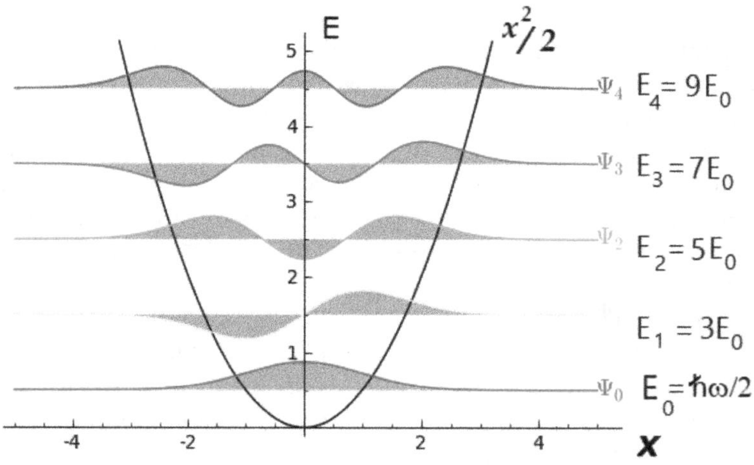

Figure 8.11. Visualization of the first five energy levels of the harmonic oscillator and the corresponding eigenfunctions.

relativity, the quantum field associated with a given elementary particle can be seen as an infinite superposition of $\hat{a}$ and $\hat{a}^\dagger$ operators, one for each dynamical degree of freedom of the field, so these ladder operators can create or destroy a certain number of quanta of the field depending on the specific state considered. To quote Sidney Coleman, a prominent theoretical physicist: 'The career of a young theoretical physicist consists of treating the harmonic oscillator in ever-increasing levels of abstraction'.

Before closing this discussion, it is worth dwelling once again on the concept of the ZPE. Also in this case, the problem foresees a minimum value of energy $E_0 = \frac{\hbar\omega}{2}$ that is greater than zero, in contrast to the classic case that admits the still particle at the minimum point of the potential. As has already been said for the infinite potential well, the ZPE is necessary to make sure that the dynamics of the system is compatible with the uncertainty principle, so there will always be quantum fluctuations that prevent the particle from lying in the minimum of the potential. However, from this point of view, the harmonic oscillator is special since, when the system is in the ground state ($n = 0$), the product of the uncertainties on position and momentum is exactly $\frac{\hbar}{2}$, which is the minimum possible value according to Heisenberg's principle.

8.3.6 General discussion of 1D problems in quantum mechanics

In this section, we want to summarize some general properties of the one-dimensional Schrödinger equation for an arbitrary time-independent potential $U = U(x)$.

For the sake of clarity, we rewrite the time-independent Schrödinger equation as follows:

$$\frac{d^2\phi(x)}{dx^2} = \frac{2m}{\hbar^2}(U(x) - E)\phi(x) \equiv \alpha(x)\phi(x). \tag{8.24}$$

Note that, unless explicitly stated, the following properties also apply to the 3D case:

1. We have already said that the solutions of equation (8.24) must be monodrome and continuous, even when the potential energy is not continuous or is singular at some point.
2. It has also been said that even the first derivatives of the wave function must be continuous when the potential energy is continuous or presents discontinuities of the first kind; the exceptions are the points where the potential energy is divergent.
3. Another property of equation (8.24) is that its solution ϕ is finite everywhere if the potential energy results are finite too. Instead, on every region A where the potential energy diverges the spatial wave function $\phi(x)$ must vanish.
4. Any energy eigenvalue E must satisfy the condition $E \geqslant U_{\min}$, where $U_{\min}$ is, if it exists, the minimum value of the potential energy function.

   Indeed, the expected value $E = \langle H \rangle \equiv (\Psi, \hat{H}\Psi)$ of the energy in a state Ψ is $E = \langle H \rangle = \langle \hat{T} + U \rangle = \langle T \rangle + \langle U \rangle \geqslant U_{\min}$, since $\langle U \rangle \geqslant U_{\min}$ and $\langle T \rangle \geqslant 0$ because $\hat{T}$ is a positive operator. If Ψ is a stationary state, then

$\langle H \rangle$ is one of the energy eigenvalues. On the other hand, if it was $E < U_{min}$, then the condition $a(x) > 0$ would be valid everywhere, so the wave function would diverge at infinity, which is not physically acceptable.

5. Eigenvalues of the discrete spectrum correspond to a limited motion (*bound states*), whereas those belonging to a continuous spectrum correspond to an unlimited motion.

   Indeed, wave functions of the continuous spectrum are not normalizable, so if A and B are two infinite regions of space, the ratio

$$\frac{\int_A |\phi(u)|^2 \mathrm{d}u}{\int_B |\phi(u)|^2 \mathrm{d}u}$$

   can be non-zero and could represent the ratio between the probability that the particle is detected in A and the probability that it is found in B. This ratio becomes zero if A is a finite region, unless ϕ is a normalizable wave function relative to a discrete spectrum. In this case, the probability of detecting the particle very far from the origin becomes negligible, although in general it is not vanishing.

6. If the potential energy function is vanishing at infinity (i.e. $\lim_{|\vec{r}| \to \infty} U(\vec{r}) = 0$), then the bound states are found for negative energy eigenvalues ($E < 0$), whereas the states of the continuous spectrum are selected in correspondence with positive energy eigenvalues ($E < 0$).

   To see this, we can argue as follows: if for negative energies there were states of the continuous spectrum, then for these energies the motion would not be bounded and our quantum particle could also be found very far from the origin with high probability. In this condition the particle would behave as if it were a free particle, equipped essentially with only kinetic energy ($U \approx 0$), so the energy constant E should be positive, contrary to what was assumed. In the same way, we could argue that we cannot have bound states with $E > 0$.

   From this property we can also infer that if $U(\vec{r})$ vanishes at infinity and in addition is positive everywhere, then any energy eigenvalue is positive and the discrete spectrum is absent.

7. Since the energy E and the kinetic energy T are not compatibly observable as the corresponding operators do not commute,

$$[\hat{H}, \hat{T}] = [\hat{T} + \hat{U}, \hat{T}] = [\hat{U}, \hat{T}] = \frac{i\hbar}{m} \nabla \hat{U} \cdot \hat{\vec{p}} \neq 0,$$

   then it is possible for a particle in a stationary state with energy E to be detected by means of a measurement of position in the classically prohibited region, that is the set of points $\vec{r} \in \mathbb{R}^3$ for which $U(\vec{r}) > E$.

   Indeed there is no contradiction with the fact that kinetic energy has a positive spectrum, because when the energy has a definite value, E, the kinetic energy does not, and after a measurement of position both linear momentum and kinetic energy are affected by a disturbance, in accordance with the uncertainty relations.

Observe that if a particle is actually measured within a classically forbidden region, the experimental values will always be compatible with the classical expectation of positive kinetic energy, that is, with reference to the potential barrier, with a total energy $E' \geqslant U_0$, even if before the measurement we had $E < U_0$; this happens because the energy and position observables are also not compatible.

8. When the potential energy in three spatial dimensions (3D) has the form

$$U(\vec{r}) = U_1(x) + U_2(y) + U_3(z),$$

then we can write the Hamiltonian operator as a sum of three separate one-dimensional Hamiltonian operators,

$$\hat{H}(\hat{\vec{r}}, \hat{\vec{p}}) = \hat{H}_1(\hat{x}, \hat{p}_x) + \hat{H}_2(\hat{y}, \hat{p}_y) + \hat{H}_3(\hat{z}, \hat{p}_z),$$

so our problem of solving the time-independent Schrödinger equation in 3D can be reduced to three independent one-dimensional problems, that is the equation (8.24), using again the *method of separation of variables*. We assume that the complete spatial wave function of the original problem can be factorized as follows:

$$\phi(\vec{r}) = \phi_1(x)\phi_2(y)\phi_3(z).$$

Thereby we derive that the spatial one-dimensional wave functions for each coordinate, $\phi_1(x)$, $\phi_2(y)$ and $\phi_3(z)$, must satisfy the equations

$$\frac{d^2\phi_1(x)}{dx^2} = \frac{2m}{\hbar^2}(U_1(x) - E_1)\phi_1(x),$$
$$\frac{d^2\phi_2(y)}{dy^2} = \frac{2m}{\hbar^2}(U_2(y) - E_2)\phi_2(y),$$
$$\frac{d^2\phi_3(z)}{dz^2} = \frac{2m}{\hbar^2}(U_3(z) - E_3)\phi_3(z),$$

so any possible value E of the energy spectrum can be decomposed as the sum of the three terms coming from the one-dimensional eigenvalue problems written above:

$$E = E_1 + E_2 + E_3.$$

For example, dealing with an infinite potential well in 3D is immediate, a model that proves to be very useful to describe a particle trapped in a box of dimensions $a \times b \times c$ (U is zero inside the region of 3D space $[0; a] \times [0; b] \times [0; c]$ and is infinite elsewhere) and, as a first approximation, a particle confined within a finite region of space. In this case, a state can be identified by the triplet of quantum numbers $\vec{n} = (n_x, n_y, n_z) \in (\mathbb{Z}^+)^3$ and the total energy is

$$E_{\vec{n}} = \frac{h^2}{8m}\left(\frac{n_x^2}{a^2} + \frac{n_y^2}{b^2} + \frac{n_z^2}{c^2}\right),$$

whereas the spatial wave function for this state can be written as

$$\phi_{\vec{n}}(\vec{x}) = \sqrt{\frac{8}{abc}}\,\sin\left(\frac{n_x\pi x}{a}\right)\sin\left(\frac{n_y\pi y}{b}\right)\sin\left(\frac{n_z\pi z}{c}\right).$$

It should be noted that in this case it could happen that the same energy level corresponds to different states (this is certain if there are two or more equal dimensions, e.g. $a = b$), which means that such energy levels are *degenerate*. Another noteworthy example of a separable 3D problem is the 3D harmonic oscillator, for which we have $U(x, y, z) = \frac{1}{2}k_1x^2 + \frac{1}{2}k_2y^2 + \frac{1}{2}k_3z^2$, thus the results of the above approach are very useful.

9. For one-dimensional problems, all energy levels of the discrete spectrum are non-degenerate.

   To prove this fact we suppose that ϕ_1 and ϕ_2 are two eigenfunctions of the energy eigenvalue E, so from equation (8.24) the following relation holds:

$$\frac{\phi_1''}{\phi_1} = \alpha(x) = \frac{\phi_2''}{\phi_2} \Longrightarrow \phi_1''\phi_2 - \phi_2''\phi_1 = \frac{\mathrm{d}}{\mathrm{d}x}(\phi_1'\phi_2 - \phi_2'\phi_1) = 0.$$

   The last condition is equivalent to stating

$$\phi_1'\phi_2 - \phi_2'\phi_1 = c,$$

   where c is a complex constant. Now we can exploit the fact that the two eigenstates belong to a discrete spectrum, which we can consider to be bound states, for which the probability to escape towards infinity tends to zero. So the following boundary conditions at infinity must hold:

$$\lim_{|x|\to\infty} \phi_1(x) = \lim_{|x|\to\infty} \phi_2(x) = 0.$$

   Taking the limit for $|x| \to \infty$ of $\phi_1'\phi_2 - \phi_2'\phi_1$ we obtain that $c = 0$, so, by integrating the ODE

$$\phi_1'\phi_2 - \phi_2'\phi_1 = 0,$$

   we obtain

$$\phi_2 = A\phi_1, \quad \text{with } A \in \mathbb{C}.$$

   Therefore, the two functions, which differ only by a constant factor that can be eliminated by imposing the normalization constraint, represent the same quantum state. Note that the same conclusion holds even if the motion results are limited only for $x < 0$ (or only for $x > 0$).

10. The following *oscillation theorem* holds for the wave function of a discrete spectrum: if $\phi_n(x)$ is the eigenfunction corresponding to the $(n + 1)$th energy eigenvalue E_n (the eigenvalues being arranged in order of increasing magnitude), then $\phi_n(x)$ has n nodes, that is it vanishes n times for a finite value of x (for a detailed proof of this and other related results see [11]).

    From this theorem it follows that the ground state eigenfunction has zero nodes, so if it has a well-defined parity it must be an even function.
11. Suppose that the potential function has the following limits at infinity:

$$\lim_{x\to-\infty} U(x) = U_- \qquad \text{and} \qquad \lim_{x\to+\infty} U(x) = U_+,$$

with $|U_+| < +\infty$, $|U_-| < +\infty$. Now, we can set one of these finite limits as equal to zero (since the zero of potential energy is arbitrary), for example the lowest one, say U_+, and the other limit becomes $U_0 = U_- - U+$. Under these conditions it is not difficult to show by appealing to the asymptotic forms of the wave function that the energy spectrum $\{E\}$ is:
    - discrete and non-degenerate when $E < 0$;
    - continuous and non-degenerate when $0 < E < U_0$;
    - continuous and doubly degenerate when $E > U_0$.

Figure 8.12 summarizes the possible cases for a one-dimensional potential with finite limits at infinity.

From a more practical point of view, one could obtain the results cited above by observing that in those spatial regions where the function $\alpha(x)$ in equation (8.24) is

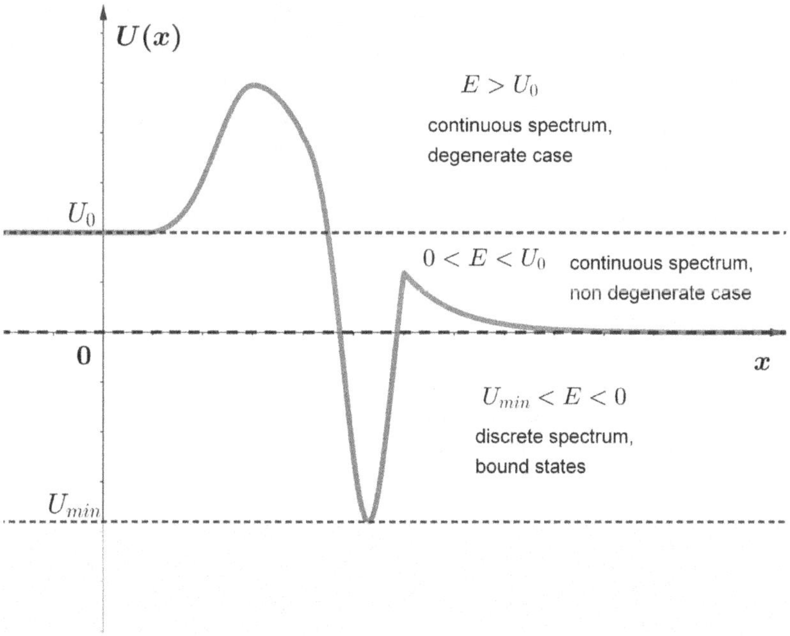

Figure 8.12. Possible cases for one-dimensional problems depending on the value of energy.

negative the second derivative of ϕ has a sign opposite to that of ϕ, that is the graph of the (real) wave function must turn its concavity towards the x-axis, as a sinusoidal function would. Conversely, in those regions where $\alpha(x) > 0$ the solution ϕ would be convex to the x-axis, as an exponential function would. To join the solutions found in these two types of regions, namely where $\alpha(x)$ is positive or negative, we impose the continuity conditions on ϕ and ϕ' at the turning points (if the potential is not singular). Therefore, to single out bound states we should find the range of energy eigenvalues E for which the corresponding wave functions fall off exponentially at infinity. If the potential is also even, we can restrict our search to even or odd functions.

It is also interesting to note that for singular potentials in the origin $x = 0$, if $U(x)$ diverges as $1/x^a$ for $x \to 0$ with $0 < a < 2$, then the spectrum is bounded from below. Otherwise ($a > 2$) the spectrum is unbounded from below, in other words there is no minimum value for energy eigenvalues (a physically undesirable situation in most cases).

8.4 Interpretational issues

Quantum mechanics reached maturity as a physical theory by 1927 and since then it has been refined in its mathematical structure, and extended and applied in a great variety of contexts and circumstances, but no experiment has ever questioned the fundamental principles discussed in section 8.2.1, that is the probabilistic interpretation of the wave function and the uncertainty relations. Indeed, despite the fact that in general quantum mechanics allows only probabilistic predictions about the possible outcomes of a certain measurement, it can be considered the most successful theory so far developed in terms of predictive effectiveness. However, this huge amount of success has not prevented the birth of a growing debate on the interpretation of the theory, starting from the discussion of some technical points (such as the definition of measurement or locality) up to the ultimate meaning of the theory itself.

This debate, which had already begun before quantum mechanics was considered a well-established theory, is continuing today with renewed vigour, involving physicists, mathematicians and philosophers, but also psychologists, writers and mystics, undoubtedly attracted by the existential meanings that quantum theory heralds, and it seems very unlikely that a general consensus will be reached in the near future.

In this section we want to give a quick overview of these interpretation issues by retracing some of the most significant experimental and conceptual stages concerning the foundations of quantum theory.

8.4.1 The measurement problem and the Copenhagen interpretation

The first source of controversy in quantum mechanics is that the theory only attributes probabilities to possible outcomes of an experiment, and this peculiarity was attacked by many critics of the theory, the most eminent of whom was undoubtedly Einstein [12].

Einstein, during his long and legendary exchanges with Niels Bohr[3], father and strenuous defender of quantum theory, stressed that this probabilistic aspect indicated that the theory, although useful and successful in many respects, was not complete, claiming also that it was not a fundamental theory of matter and interactions in the same way that statistical thermodynamics was not a fundamental theory with respect to classical mechanics. It was on one of these occasions that Einstein uttered the well-known phrase 'God does not play dice', with which he wanted to express his unease about a theoretical framework that suggested the impossibility of predicting the values of certain quantities exactly or that the outcomes of certain measures were intrinsically random.

The reply given by Bohr and the other physicists who had contributed to the development of the novel mechanics, in particular Heisenberg, Pauli and Born (Schrödinger instead soon took positions closer to those of Einstein), was a radical rethinking of the role of the measuring apparatus and the experimenter, which in some way became part of the principles of the theory and, more generally, a reformulation of the aims of physics itself.

According to this paradigm, which became known as 'the Copenhagen interpretation' of quantum mechanics because of the enormous influence that Bohr and his closest collaborators had on its diffusion and establishment, the purpose of science 'is not to disclose the real essence of the phenomena but only to track down, so far as it is possible, relations between the manifold aspects of our experience' [14]. More specifically, a physical theory is required to reproduce the results of experiments, that is, what we can know about nature when we interrogate it with specific measuring devices, while any question about the real state of the system when it is not measured would be illegitimate and meaningless. In this setting, it is the very act of measuring that causes the system under examination to assume a well-defined value for the physical quantity under examination, so in general it is not possible to attribute any degree of physical reality to such a quantity before the measurement. For example, the question of where the electron really is before actually making a position measurement would be meaningless and beyond the domain of science, according to the proponents of the Copenhagen interpretation, and the same could be said for every physical observable when the state of the system has not already been prepared with a well-defined value of that quantity. The postulate that translates operationally this view is the one that takes into account the so-called *collapse of the wave function* (or of the state vector, in a more general setting). According to this hypothesis, when a quantum system is measured it interacts with a *classical instrument*, which forces the measured quantity of the quantum system to assume only one of the possible eigenvalues for that observable, so that the wave function instantly changes, i.e. collapses, in the eigenfunction associated with the selected eigenvalue. Hence, the Copenhagen interpretation is completely agnostic about the ultimate reality of natural phenomena and it does not pretend that the mathematical objects of its formal structure have a 'real' physical meaning beyond

[3] For a historically very accurate, but also engaging and accessible exposition of the developments of quantum mechanics and of the Einstein–Bohr debate, we refer the reader to the text *Quantum* by Manjit Kumar [13].

the scope for which they were introduced. From this point of view, the whole formalism can be considered as a calculation device to obtain expected values, correlations and probabilities of events related to possible experiments.

It should be noted that while the temporal evolution of the state vector is linear, unitary (that is, it preserves normalization over time) and deterministic,

$$\Psi(t) = \hat{U}\Psi(0),$$

according to what Schrödinger's equation prescribes (the linear and unitary[4] operator $\hat{U}$ sends $\Psi(0)$ in $\Psi(t)$), the evolution of the state during the measurement process (namely the collapse) is non-linear, non-unitary, irreversible and random (in the equation below we have introduced an operator $\hat{R}$ that associates a state vector to an eigenstate):

$$\phi_k = \hat{R}\Psi(t).$$

The only thing that can be said, according to standard quantum mechanics, is that in the process of state reduction (or state collapse) the eigenstate ϕ_k that is selected when the measurement takes place had a probability $\mathcal{P}_k$ of being selected given by the squared modulus of the scalar product (ϕ_k, Ψ) between the initial state Ψ and the eigenstate ϕ_k considered:

$$\hat{G}\phi_n(x) = g_n\phi_n(x),$$
$$\Psi(x, 0) = \sum_n c_n(0)\phi_n(x) \overset{U}{\Longrightarrow} \Psi(x, t) = \sum_n c_n(t)\phi_n(x) \overset{R}{\Longrightarrow} \phi_k(x),$$
$$\mathcal{P}_k = \mathcal{P}(G = g_k) = |(\phi_k, \Psi)|^2 = |c_k|^2.$$

Therefore, the theoretical scheme proposed by the followers of the Copenhagen school assumed the existence of two parallel processes of time evolution: a '*U*-evolution' clearly defined by Schrdinger's which is equation, linear, unitary and deterministic, and an '*R*-evolution' which is non-linear, non-unitary, irreversible and not explicitly defined, and which manifests itself with the collapse of the state vector at the time of measurement and from which the probabilistic aspects arise. Another very peculiar aspect of this *R*-evolution is that it requires an interaction with a macroscopic measuring device which obeys the laws of classical mechanics. Faced with this picture, the scepticism and disappointment of Einstein, and also of other physicists such as Schrödinger and De Broglie, were not so much due to the appearance of probability as a substitute for a total determinism—after all, many years earlier it was Einstein himself who used a probabilistic model to describe the emission and absorption of light by atoms and molecules. The real 'red line' for Einstein resided in the total renunciation of realism on the atomic scale that the new paradigm more or less evidently required, to the extent that, as already mentioned, a quantum system that finds itself in a superposition of (orthogonal) eigenstates of a

[4] An operator $\hat{U}$ on a Hilbert state A is *unitary* if $\hat{U}^{\dagger} = \hat{U}^{-1}$ s.t $(\hat{U}v, \hat{U}w) = (v, \hat{U}^{\dagger}\hat{U}w) = (v, w) \quad \forall v, w \in A$.

certain observable has no real, definite property concerning this observable until a measurement is performed. This was philosophically unacceptable for Einstein. It can be also observed that the relationship between quantum mechanics and classical mechanics, which in principle should be a special case of the first, is rather twisted in the so-called 'orthodox' interpretation of the theory. In fact, as has already been said, to have well-defined results from an experiment the measuring apparatus must be considered as a *classical object*, otherwise the superposition of states would continue forever and there would never be a unique result from a measurement. But even accepting the fact that a more fundamental theory needs its own limiting case in order to be usable, one might wonder exactly where and unambiguously the boundary between the classical world and the quantum world lies.

This subject, which has been emphasized for the first time through one of the most famous 'paradoxes' of quantum mechanics, that of Schrödinger's cat (which will be the subject of a short review later), is probably the main open problem of the theory, which in fact, citing an effective expression by John Stuart Bell [15], is a theory that works fine FAPP (i.e. 'for all practical purposes').

To summarize what has been said so far, the general nature of the controversy between the realists like Einstein and the followers of the Copenhagen interpretation lies essentially in considering quantum probabilities as *epistemic* probabilities, i.e. being due to insufficient knowledge of the objective reality (which would imply that the quantum theory is 'incomplete' since it would not capture all the elements of physical reality), or non-epistemic (*ontic*), i.e. as an intrinsic, unavoidable feature of nature.

Einstein, after trying unsuccessfully in the early years of quantum mechanics to disprove directly, through a sequence of thought experiments, its load-bearing column, which is the Heisenberg uncertainty relation, tried to give a lethal blow to the now established quantum theory by means of another, very subtle and well designed thought experiment, the famous *Einstein–Podolsky–Rosen (EPR) paradox*, which, along with the already mentioned paradox of the Schrödinger's cat, is analysed in the next section.

8.4.2 Quantum paradoxes

The EPR paradox. A paradox is an apparently self-contradictory statement, the underlying meaning of which is revealed only by careful scrutiny. On the other hand, if the logical contradiction remains even after a thorough analysis of the terms of the question, then this probably indicates a problem in the starting assumptions or in the consistency of the interpretative model adopted to examine such a question.

In 1935 Einstein, Boris Podolsky and Nathan Rosen published a paper titled 'Can quantum-mechanical description of physical reality be considered complete?' [16], whose aim was to show, through a thought experiment with paradoxical results, that quantum mechanics, although correct, was not a complete theory, namely that the wave function of the system does not contain all the information about the underlying physical reality (see figure 8.13).

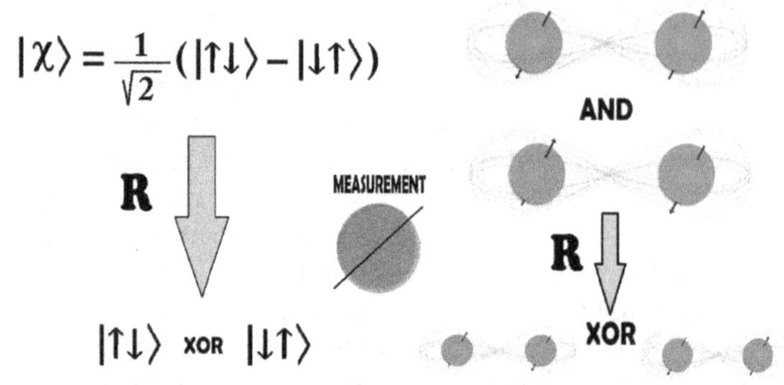

Figure 8.13. Schematic representation of the EPR thought experiment according to standard quantum mechanics.

Here we give a brief account of the EPR argument using a simpler and easier to test version provided by David Bohm (1917–1992) in his textbook on quantum mechanics [17]. This argument consists in almost simultaneously measuring a given component of the spin of the electron, say S_z, for a two-electron system. The physical observable called 'spin' $\vec{S}$ is an intrinsic angular momentum without any classical analogue and the corresponding self-adjoint operator $\hat{\vec{S}}$ has Cartesian components obeying the typical commutation relations of angular momenta in quantum mechanics:

$$\begin{aligned} [\hat{S}_x, \hat{S}_y] &= i\hbar\hat{S}_z, \\ [\hat{S}_y, \hat{S}_z] &= i\hbar\hat{S}_x, \\ [\hat{S}_z, \hat{S}_x] &= i\hbar\hat{S}_y. \end{aligned} \tag{8.25}$$

Without going into the details of the treatment of angular momentum in quantum mechanics, here we limit ourselves to mentioning only a few consequences of the algebraic relations (8.25) between spin operators:

a. The spin components are incompatible observables (i.e. if we measure S_x the information previously acquired about S_y or S_z is destroyed).
b. In quantum mechanics only integer or half-integer multiples of $\hbar$ are possible as possible values of $S = |\vec{s}|\hbar$.
c. The spin component results are quantized as well, indeed they can only assume the $2s + 1$ values differing by $\hbar$ between $S_{max} = s\hbar$ and $S_{min} = -s\hbar$, i.e. $S = \{-s, -s + 1, -s + 2, \ldots, s - 1, s\}\hbar$. In the case of fermions of spin $s = 1/2$ (we stress that this s value is an intrinsic property of these particles and cannot be modified), the components S_x, S_y and S_z can only assume $+1/2$ and $-1/2$ as possible values (in units of $\hbar$).

To indicate the two eigenstates of S_z (and the same applies also to S_x and S_y) we use the following up–down notation:

$$\hat{S}_z|z; \uparrow\rangle = \frac{\hbar}{2}|z; \uparrow\rangle, \qquad \hat{S}_z|z; \downarrow\rangle = -\frac{\hbar}{2}|z; \downarrow\rangle,$$

which is conventional for two-state systems concerning spin. If there is no risk of ambiguity, we will simplify the notation as follows: $|z; \downarrow\rangle = |\downarrow\rangle$, $|z; \uparrow\rangle = |\uparrow\rangle$.

Moreover, if the system is composed of two electrons and we refer to a state of the system as a whole, such a state $|\chi\rangle$ [5] for the z-component $S_z = S_{1,z} + S_{2,z}$ of the total spin can be written as a linear combination of the tensor-products of all the possible single-electron spin eigenstates (note that the order here matters: $|\downarrow\rangle|\uparrow\rangle \neq |\uparrow\rangle|\downarrow\rangle$), that is,

$$|\chi\rangle = a|\downarrow\rangle|\downarrow\rangle + b|\downarrow\rangle|\uparrow\rangle + c|\uparrow\rangle|\downarrow\rangle + d|\uparrow\rangle|\uparrow\rangle,$$

where a, b, c and d are proper complex coefficients s.t. $|a|^2 + |b|^2 + |c|^2 + |d|^2 = 1$.

Having said this, let us suppose we prepare our system of two electrons in the so-called 'singlet state' $|\chi\rangle \equiv |0; 0\rangle$, compatible with a total spin of the system equal to zero. Such a state has only mixed components as non-zero, moreover they are equally likely:

$$|0; 0\rangle = \frac{1}{\sqrt{2}}(|\downarrow\rangle|\uparrow\rangle - |\uparrow\rangle|\downarrow\rangle).$$

Now, according to the standard interpretation of quantum mechanics, when we measure the z-component of the first electron spin and obtain, say, $s_{1,z} = +1/2$, then we are sure that we have selected the state of the system $|\uparrow\rangle|\downarrow\rangle$ since the singlet state collapses *instantaneously* to such a state. Therefore, from now on the other electron *must* have a well-defined value for $s_{2,z}$, that is $s_{2,z} = -1/2$, and when such an electron is subjected to a measurement of $S_{2,z}$ this value will be obtained with a probability of 100%.

It is precisely at this point that the impressive EPR attack takes place. If we make the following very reasonable requests for a theory that claims to be correct and complete, that is the assumptions of *locality*, i.e. nothing can travel more rapidly than light in vacuum, and *realism*, i.e. all the elements of physical reality are properly contained in the theory, then we are compelled to assume that the spin values $(+1/2, -1/2)$ of the two electrons were already held by the two electrons of this system *before* the first spin measurement. Why? Because if we allow the two electrons to get far away enough without letting them interact with anything, which in principle is absolutely possible, then the system remains in the singlet state and the standard quantum mechanics predicts that one measurement on part of the system (an electron) constrains the value of the other electronic z-spin, even if the two electrons are so far apart that no physical signal has the time to reach the other electron before the second measurement takes place. Hence, if we accept the initial, general assumptions of EPR, we are compelled to conclude either that for very close

[5] From now on we use the Dirac notation $|\cdot\rangle$ to indicate a generic vector of the Hilbert space of the system's states. The wave function $\phi(x)$ can be seen as a component (in the infinite-dimensional base of the coordinates $\{|x\rangle\}$) of a state vector ϕ in its abstract form: $|\phi\rangle = \int_{\mathbb{R}} \phi(x)|x\rangle \mathrm{d}x$.

measurements in time by means of detectors very far away in space (i.e. for measurements defining a space-like interval, in the jargon of special relativity) the up–down correlation is no longer verified, or that the electrons had from the beginning the observed opposite values of spin along the z-direction, even if quantum theory does not account for it. Since it is experimentally verified that the correlations occur without exception (and at the time of the EPR paper it was strongly expected, also because otherwise the total angular momentum would not be conserved), then the only possible conclusion seems to be that the quantum theory is incomplete, since it does not take into account further dynamical variables that would make the values of the spin predetermined. Every other option would foresee those that Einstein called 'spooky action at a distance', which are in clear contrast to the locality hypothesis[6]. Indeed, in the article cited above on EPR stated the following: 'We are thus forced to conclude that the quantum-mechanical description of physical reality given by wave functions is not complete.'

The argument put forward by EPR was undoubtedly very convincing and, unlike those initially devised by Einstein to directly refute the uncertainty principle, it did not seem to have any flaws linked to some specific details of the example taken into consideration, because the principles on which it is based are quite general and widely accepted. Bohr, after long reflection, replied about five months later to the EPR paper with an article in the same journal and, meaningfully, with the same title. However, although almost all supporters of the orthodox interpretation of the theory felt that Bohr had won once again, the most attentive observers of the time and most scholars today agree that Bohr's response was ineffective, rather obscure and unable to dismantle the EPR attack. So, although not explicitly stated, the EPR argument invoked a deterministic completion of ordinary quantum mechanics through (local) 'hidden variables'. However, the issue of finding such a completion was shelved for many years because of the great number of achievements obtained by quantum theory when applied to solids, atoms, molecules and their interactions with radiation, and also because a mathematical result by the leading Hungarian physicist and mathematician John von Neumann. It first appeared in his textbook on the mathematical foundations of quantum mechanics [18], and was interpreted by most physicists for over 20 years as an no-go theorem for any possible hidden variable theory that was able to experimentally reproduce the same statistical predictions of ordinary quantum mechanics. However, David Bohm ignored von Neumann's result and in 1952 presented in two papers [19, 20] his formulation of a theory of hidden variables, the so-called *De Broglie–Bohm pilot wave theory*, which was capable of explaining the correlations of the EPR thought experiment without 'spooky actions at a distance'. Indeed, Bohm himself and then Bell in more detail noticed that in von Neumann's theorem about the impossibility of hidden variables there were unnecessary hypotheses, which thus made the result much less general.

[6] Note that this argument about locality is completely independent from the dynamical law regulating the time evolution of states between two successive measurements, that is equation (8.3), which as we know is not relativistic. Indeed, as we said before, the collapse of the state vector is a separate evolutionary process and is also present in the relativistic extension of quantum theory.

In any case, the mere possibility of such an interpretation showed that it was not necessary to abandon a precise, rational and objective description of individual systems at the quantum scale. However, at this point, how could one discriminate between ordinary quantum mechanics and a theory based on (local) hidden variables, given that the latter was designed to reproduce the same results as the former? Many physicists thought that it was simply a matter of taste and that no experiment would ever highlight an observable difference between the two interpretations.

In fact, to answer this no longer idle question it was necessary to have all the ingenuity of the brilliant Irish theoretical physicist Bell, who recognized the limitations of von Neumann's no-go theorem on hidden variables and therefore sought a way to find predictions that were able to discriminate experimentally ordinary quantum mechanics from its alternative based on hidden variables. Bell's efforts were rewarded by his most famous achievement, that of the inequality that bears his name, published in 1964 in an paper titled 'On the Einstein–Podolsky–Rosen paradox' [21].

To derive its inequality, Bell formulated the following prescription for locality: the probability of obtaining a pair of results in a joint measurement of the observables $\mathcal{C}$ and $\mathcal{D}$ at the two extremes A and B of an EPR-like apparatus is simply the product of the probabilities of obtaining either one of the results γ and δ, independently of whether or not the other measurement is performed. If we refer to the z-component (or also the x-component) of the spin observable and we choose to measure it when the detectors at A and B are oriented along the directions of $\vec{c}$ and $\vec{d}$, we can write the formula

$$p_\lambda^{\mathrm{AB}}(\vec{c}, \vec{d}; \gamma, \delta) = p_\lambda^{\mathrm{A}}(\vec{c}; \gamma) p_\lambda^{\mathrm{B}}(\vec{d}; \delta), \tag{8.26}$$

where, following the notation adopted in the enlightening book *Sneaking a Look at God's Cards* on the foundations of quantum mechanics written by Giancarlo Ghirardi [22]:

- $p_\lambda^{\mathrm{AB}}(\vec{c}, \vec{d}; \gamma, \delta)$ is the joint probability of obtaining the outcomes γ and δ in an EPR-like experiment in which, at the two extrema A and B (where the detectors are located), the observable S_z is measured with the detector's axis oriented along the directions $\vec{c}$ and $\vec{d}$ in the case of a physical system characterized by the parameter λ, that here accounts for all the variables, explicit or hidden, that specify the physical state of the system as exhaustively as the formalism allows.
- $p_\lambda^{\mathrm{A}}(\vec{c}; \gamma)$ is the probability of obtaining the result γ when a measurement of S_z is performed on the system in A when the detector's axis is directed along the direction of $\vec{c}$.
- $p_\lambda^{\mathrm{B}}(\vec{d}; \delta)$ is the probability of obtaining the result δ when a measurement of S_z is performed on the system in B when the detector's axis is directed along the direction of $\vec{d}$.

Now, we are ready to present Bell's main achievement more precisely: any (deterministic or probabilistic) theory that fulfils the condition equation (8.26) cannot reproduce all the probabilities that quantum mechanics predicts for experimental outcomes, not even when assuming that the preparation of the system does not completely determine the variables. Actually, starting from this assumption Bell's inequality can be derived by appropriately combining the probabilities of joint events. Without giving all the details, for which we refer to Ghirardi's text cited above, for fixed orientations of the measurement devices given by $\vec{c}$ and $\vec{d}$ one defines the following function $E_\lambda(\vec{c}, \vec{d})$:

$$\begin{aligned} E_\lambda(\vec{c}, \vec{d}) \equiv\, & p_\lambda^{AB}(\vec{c}, \vec{d}; \uparrow, \uparrow) - p_\lambda^{AB}(\vec{c}, \vec{d}; \downarrow, \uparrow) \\ & - p_\lambda^{AB}(\vec{c}, \vec{d}; \uparrow, \downarrow) + p_\lambda^{AB}(\vec{c}, \vec{d}; \downarrow, \downarrow). \end{aligned}$$

Now, using the locality requirement (8.26) it is possible to show after a few steps of elementary algebra that

$$|E_\lambda(\vec{a}, \vec{b}) - E_\lambda(\vec{a}, \vec{d})| + |E_\lambda(\vec{c}, \vec{b}) + E_\lambda(\vec{c}, \vec{d})| < 2. \tag{8.27}$$

Moreover, if we consider an ensemble of systems with a distribution of hidden variables[7] $\{(\lambda_1, f_1); (\lambda_2, f_2); \ldots; (\lambda_n, f_n)\}$, where f_i is the relative frequency of the hidden variable value λ_i in such an ensemble, and consider the ensemble average of the functions $E_\lambda(\vec{a}, \vec{b})$,

$$E(\vec{a}, \vec{b}) = \sum_i f_i E_{\lambda_i}(\vec{a}, \vec{b}),$$

it is not difficult to prove that equation (8.27) holds also for the averaged quantities $E(\vec{a}, \vec{b})$:

$$|E(\vec{a}, \vec{b}) - E(\vec{a}, \vec{d})| + |E(\vec{c}, \vec{b}) + E(\vec{c}, \vec{d})| < 2. \tag{8.28}$$

Relation equation (8.28) is *Bell's inequality*, valid for any possible choice of vectors $\vec{a}$, $\vec{b}$, $\vec{c}$ and $\vec{d}$. It can be shown that in the case of a couple of electronic spins quantum mechanics predicts $E(\vec{a}, \vec{b}) = \cos[2(\widehat{a, b})]$, so Bell's inequality becomes

$$|\cos[2(\widehat{a, b})] - \cos[2(\widehat{a, d})]| + |\cos[2(\widehat{c, b})] + \cos[2(\widehat{c, d})]| < 2,$$

where $\widehat{a, b}$ is the angle between vectors $\vec{a}$ and $\vec{b}$. It is therefore possible to choose the angles between the z-axis and the directions of detector analysers in such a way as to violate the inequality just written, for example with $\alpha = (\widehat{a, z}) = 0°$, $\beta = (\widehat{b, z}) = 30°$, $\gamma = (\widehat{c, z}) = 45°$ and $\delta = (\widehat{d, z}) = 60°$, we obtain

$$\begin{aligned} & |\cos[2(\alpha - \beta)] - \cos[2(\alpha - \delta)]| + |\cos[2(\gamma - \beta)] + \cos[2(\gamma - \delta)]| \\ & = |\cos(-60°) - \cos(-120°)| + |\cos(30°) + \cos(-30°)| = 1 + \sqrt{3} > 2. \end{aligned}$$

[7] For simplicity we assume a discrete distribution of λ_i, but it is completely unnecessary for our aims.

This violation of Bell's inequality has actually been verified in a large number of experiments since the pioneering ones conducted by Alain Aspect and his team in 1981–1982 [23, 24]. One of the most recent and significant goes back to 2015 [25] and claims to be loophole-free. This implies that the alternative theories based on local hidden variables do not reproduce the statistical experimental results that ordinary quantum mechanics can explain comprehensively. However, we must underline that this claim is (a.s.) valid only for theories based on the EPR hypothesis of *local realism*, while other non-local versions of hidden variables (such as the Bohmian mechanics) are still compatible with the observations and results are as far as we know experimentally indistinguishable from standard quantum mechanics.

It should be also emphasized that what has been highlighted so far shows that quantum mechanics is a theory that does not respect a principle of strong locality. Indeed, the quantum correlations between measurements observed in systems described by *entangled states*, i.e. quantum states that cannot be factorized into states describing two or more apparently independent parts of the system itself, occur without any delay, instantaneously. However, it can be seen easily that even though a sort of super-luminal effect exists, this effect cannot be exploited to transfer information at speeds higher than that of light in vacuum.

In conclusion, although ordinary quantum mechanics in the end was not really weakened by the EPR challenge, this attack certainly had the merit of shedding light on some properties of quantum theory that had not been well understood, also disclosing the incredibly fruitful and amazing phenomenon of *quantum entanglement*, which has given rise to new areas of quantum information and computation, whose increasing relevance in engineering and technology is also currently manifest to everyone (see chapters 11 and 12). It should also be added that, beyond the experimental successes, there is still no in-depth understanding of the aspects of non-locality and non-separability that characterize entanglement and quantum theory in general—quantum physics is far from unlocking all its mysteries.

Schrödinger's cat. In this paragraph, we want to address another mental experiment that presents paradoxical aspects and that has had great resonance, even beyond the educational and popular contexts, to become an emblem of the oddity of quantum physics in popular culture. The Austrian physicist Schrödinger, like Einstein, was particularly critical towards the Copenhagen interpretation even though he was one of the main architects of the quantum revolution. Also in his case, the main object of criticism was not so much quantum uncertainty, but rather the lack of space–time continuity entailed by the hypothesis of the collapse of the state vector associated with the measurement process, so creating non-trivial problems for locality, as seen in the EPR paradox. In this regard, his phrase addressed to Bohr is often quoted: 'If we are going to stick to this damned quantum-jumping, then I regret that I ever had anything to do with quantum theory.'

To highlight the limits of the 'orthodox' interpretation, in 1935 Schrödinger devised [26] (the same year as the EPR paper) the famous conceptual experiment of the cat in a box, which is outlined briefly here. The experiment consisted in creating a system isolated from the external environment, as follows: a quantum system (for example a sample of radioactive material) which had, within an hour, a 50%

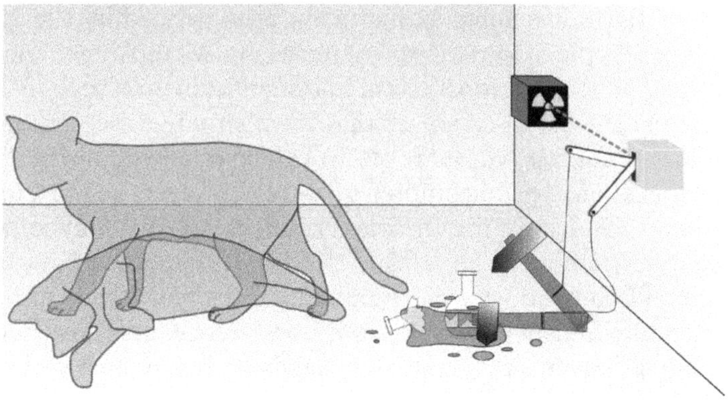

Figure 8.14. A pictorial representation of the Schrödinger's cat experimental set-up (source: Wikipedia, author: Dhatfield https://upload.wikimedia.org/wikipedia/commons/thumb/9/91/Schrodingers_cat.svg/1920px-Schrodingers_cat.svg.png).

probability of decaying into a different state from the initial one; a detector (e.g. a Geiger counter) capable of measuring whether or not the decay has occurred; a vial of sealed poison, connected to the detector just mentioned by a system capable of breaking it after decay; and, of course, the cat, which after an hour would have died with a 50% chance. The situation is illustrated in figure 8.14. Now, the argument goes as follows: since a radioactive atom obeys quantum mechanics, it can decay with a certain probability at any instant of time, so, before a measurement, it is found in a superposition of the states 'atom decayed' $|D\rangle$ and 'atom not decayed' $|ND\rangle$. But when the atom (assuming for now that there is only one radioactive atom capable of activating the deadly mechanism) has not yet decayed, then the poison flask is still whole in the box and the cat is alive, while on the contrary the cat is poisoned and dead.

What is paradoxical in this situation? Well, if we demand that quantum mechanics is fundamental and that therefore it must describe the evolution of all the physical systems, even if they are macroscopic, then the state of the cat before the measurement must correlate to the state of the atom on which the result of the experiment depends, therefore, on the states 'not decayed' and 'decayed', corresponding to the states of the cat 'alive', $|A\rangle$, and 'dead', $|NA\rangle$. In conclusion, the system before the measurement would be represented by an entangled state like

$$|\Psi\rangle = \frac{1}{\sqrt{2}}(|D\rangle|NA\rangle + |ND\rangle|A\rangle),$$

which collapses into a quantum state where the existence of life in the cat is well-defined, that is $|D\rangle|NA\rangle$ or $|D\rangle|NA\rangle$, only after a measurement has been made, for instance after the act of opening of the box by an observer. So, where is the problem? The paradoxical result is that, before the measurement, quantum mechanics asserts that the poor cat in the box lives in some overlap of 'living cat' and 'dead cat', a state which for our classical understanding of physical reality is absolutely meaningless. Other critical observations follow closely: at what exact stage does the collapse of the entangled state describing the 'atom + cat' system occur, when an observer opens the box or becomes aware of the state of the cat or atom? And why cannot the cat itself (or a human substitute) 'make' the measurement, causing the quantum state to immediately collapse? If there is this ambiguity about where and when the collapse actually occurs, is it not a signal that this notion is superfluous and artificial? Certainly, some might argue that naively extending the quantum formalism to macroscopic objects may be an abuse, but it is also true that in the absence of a clear distinction between the nano-world and macro-world we are not allowed to limit the theory to the first context only.

The supporters of the Copenhagen interpretation limited themselves to say that, whatever happened to the cat in the box, quantum mechanics foresees that the cat, when a measurement is carried out, is found to be alive or dead with a certain probability and that, once the box is opened, either an alive cat or a dead cat is found. Therefore, no strange overlap occurs at the moment of measurement and the theory, for its purposes, would reproduce the experimental data without problems. After all, the superposition of orthogonal states at the level of the state vector (i.e. of the wave function, in a less general setting) is not an issue, because state vectors and wave functions are mathematical quantities without any directly measurable property, they are only mathematical tools that are useful to predict probabilities and expected values of physical quantities. In this perspective, what happens to the cat before the measurement is not a legitimate scientific question, thus we can think what we want, because quantum theory, according to Bohr and colleagues, does not aspire to describe the physical world as it really is, but to understand what we can know about the physical world.

In any case, even if we want to accept this very 'reticent' approach, the problem of clearly defining at what stage the measurement[8] actually takes place, remains open: where is the boundary between the part of the world labelled as classical and the one that instead obeys the principles of quantum mechanics? Unfortunately, ordinary quantum theory does not provide an objective criterion to tackle this issue (we recall once again Bell's opinion in which *quantum mechanics is good FAPP*) and this was the reason why relatively different ideas were affirmed even among the supporters of standard quantum theory. For example, Eugene Wigner and von Neumann argued that for the collapse of the wave function to take place the presence of a *conscious observer* was necessary (therefore a cat or another animal could not collapse the state vector). But this way of seeing, if on the one hand it avoided making an explicit and well-defined distinction between the classical and quantum worlds, on the other hand it brought into physics an even more elusive and difficult to define concept, *consciousness*. On the other hand, even if the measuring apparatus itself were treated as a quantum object, then the system 'cat + atom + apparatus' should once again be described by an entangled state of the form

$$|\Psi\rangle = \frac{1}{\sqrt{2}}(|D\rangle|NA\rangle|r = \text{`decayed'}\rangle + |ND\rangle|A\rangle|r = \text{*not `decayed'}\rangle),$$

[8] In Bohr's view, a measurement is an interaction under controlled conditions between a quantum system and a classical system used for measuring purposes.

which would place us in a strange overlap of 'atom-decayed—dead cat—measuring apparatus', that marks 'decayed' and 'non-decayed atom—live cat—measuring apparatus', that marks 'not decayed'.

Since the publication of Schrödinger's paper in 1935, rivers of ink have been poured over the paradox of the cat in search of a resolution that could be more convincing and complete than the answers provided by the followers of the standard theory. Some interesting attempts are based on the so-called *mechanism of decoherence*, that aims to explain the collapse of the vector state (for example, an entangled state) through a 'loss of coherence' of the quantum state after a characteristic time linked to the degree of complexity of the system and its interactions with the environment (in general, for a macroscopic system one expects such a time to be extremely short). Nevertheless, even though these attempts may work fine from a practical point of view, they do not yet provide a really satisfactory answer from a theoretical perspective. Hence it can be said that the problem of measurement, so effectively illustrated by the cat paradox, more than 80 years after its formulation definitely remains an open question.

8.4.3 Alternative interpretations and 'ontology' of the state

Quantum mechanics, as we have said, is an extremely successful theory, for which there are various formalizations at various degrees of abstraction (we have presented here the one that appears simpler to understand for a neophyte) and a considerable number of interpretations. Just to name a few:

- Copenhagen interpretation.
- Consciousness causes collapse.
- Consistent histories.
- de Broglie–Bohm theory.
- Many-minds interpretation.
- Many-worlds interpretation.
- Objective collapse theories.
- Relational quantum mechanics.
- Stochastic interpretation.

This is not the place to describe in detail these approaches to quantum theory, which are all experimentally equivalent, but we want to add just a few words on one of the most popular and discussed among them: the many-worlds interpretation of quantum mechanics.

This interpretation, introduced for the first time by Hugh Everett III in his doctoral thesis in 1957 under the less evocative name of *'Relative State' Formulation of Quantum Mechanics* [27], was aimed at solving the 'measurement problem' using only what was firmly established in ordinary quantum mechanics. Indeed, Everett's solution consisted in rejecting the more controversial hypothesis concerning the collapse of the state vector, maintaining only the superposition principle and the unitary temporal evolution of the states according to Schrödinger's equation. In this framework one can define a 'universal wave function', which contains a superposition of all possible states of any individual particle in the Universe. As time goes by, some of these superpositions vanish, making certain realities distinct and isolated from one another. So, in this context, the process of measurement does not 'create' physical reality, but simply coincides with a branching of reality into all the possible scenarios compatible with the initial state of the Universe. There are no more sudden collapses of the wave function, but we must postulate that the Universe forks at any single-particle measurement in such a way that any possible outcome is realized in some world. The price to be paid to do without the collapse of the wave function is therefore the very strong hypothesis that an infinite and continuously growing number of universes exist somewhere and, moreover, that there are always almost perfect copies of them, which differ only for the instantaneous values of a single physical quantity.

This formulation would make quantum mechanics a realistic, deterministic and (presumably) local theory, but was initially discarded as absurd and ridiculous by a large part of the physics community, which is why Everett abandoned his research career in physics. However, thanks to the work of the theoretical physicist Bryce DeWitt, who in the early 1970s supported and popularized Everett's view, referring to it as the many-worlds interpretation, a certain consensus began to arise around this interpretation. Should we take this interpretation seriously? Well, apart from any aesthetic consideration, a scientific theory should always be experimentally falsifiable and in this case it seems very difficult to conceive of an experiment capable of testing the existence of other universes and also the values assumed in each of them by a given observable. That said, even today the many-worlds interpretation remains the preferred one for several influential physicists.

To complete this section, let us talk about the ontological status of the wave function (or, more generally, of the state vector) in the various interpretations of quantum theory. In the literature on the foundations of quantum mechanics we often encounter the attributes 'epistemic' and 'ontic'. The first essentially refers to quantities that represent only our knowledge of the world and are unable to single out a unique state of the underlying physical reality, while the second refers to quantities that capture the real physical state of the system, existing independently of our study. In particular, these adjectives can be attributed to the wave function ψ, which, depending on the interpretations, can be classified as ψ-epistemic or ψ-ontic. A ψ-epistemic wave function could be compared to the electrodynamic potentials in classical electromagnetism, that is to say to auxiliary quantities to which a correspondent of physical reality is typically not directly attributed, but which are useful to determine other directly measurable quantities, reflecting real properties of the system (in this example the electromagnetic field). For example, the wave function in the context of the Copenhagen interpretation would be ψ-epistemic, but, since the quantum theory here is assumed to be a complete theory of nature, the state of the system should also be defined in this setting as 'complete', because it is supposed that any substantial and testable attempt to complete the theory would not be compatible with experiments. As already mentioned, in the Copenhagen interpretation the uncertainties about physical observables reflect an inherent property of nature, therefore their origin is ontic, not epistemic. On the other hand, in a hidden variable theory the wave function would still be an epistemic object, but in this case the state it represents cannot be regarded as complete, as there would be variables not considered by the theory that, if known, would be able to uniquely identify the physical state of the system. This is therefore the case of a ψ-epistemic incomplete state. What can we say instead of the quantum state in the many-worlds interpretation? The peculiarity of this version of quantum theory lies in the fact that the 'universal wave function' here is considered an ontic quantity of the system, indeed it would reflect the real state of the Universe for every instant of time, even when it shows a superposition of orthogonal states. This would then be the case of a ψ-ontic complete state. In the case of Bohmian mechanics we would instead have to deal with incomplete ontic states, because the real physical state of the system would be captured both by the wave function and by the state of motion concerning the described particles. At this point it is natural to ask the question whether the choice between these interpretations is only a matter of personal taste or whether it can be determined on the basis of experimental facts. In this sense some interesting results have been achieved recently, in particular the work of Pusey, Barrett and Rudolph [8], according to whom, if quantum mechanics is a correct theory, the wave function cannot be a completely epistemic object, but must be ontic, i.e. must reflect some or all of the elements of physical reality. Clearly this result is based on some initial hypotheses about the preparation of states that may even turn out to be overly binding in the future, but it is interesting that issues considered until a few years ago to be exclusively philosophical now are returning to the domain of physical sciences.

Appendix

A On the continuity of the first derivative of the wave function

In this appendix we want to discuss an important property of the wave function, more precisely with regard to the first derivative. For simplicity we will limit our analysis to the one-dimensional case to avoid having to deal with complex functions of more real variables, however, the discourse could also be generalized to this case.

Let us consider the time-independent Schrödinger equation and suppose that the potential energy function $U(x)$ is continuous at all points of an interval $[a, b]$ except for a point x_0 within this interval. If we integrate both members of the equation between $x_0 - \epsilon > a$ and $x_0 + \epsilon < b$, where ϵ is an arbitrary, small positive quantity, we have

$$-\frac{\hbar^2}{2m}\int_{x_0-\epsilon}^{x_0+\epsilon}\frac{\mathrm{d}^2\phi(x)}{\mathrm{d}x^2}\mathrm{d}x = E\int_{x_0-\epsilon}^{x_0+\epsilon}\phi(x)\mathrm{d}x - \int_{x_0-\epsilon}^{x_0+\epsilon}U(x)\phi(x)\mathrm{d}x. \qquad (A.1)$$

If we take the limit of equation (A.1) as $\epsilon \to 0$, we have that the first integral on the right-hand side vanishes since the integrand $\phi(x)$ must be a continuous function (a necessary assumption also on physical grounds). Moreover, $\phi(x)$ is supposed to be twice differentiable in $[a, x_0[\cup]x_0, b]$, therefore the relation can be written as follows:

$$\lim_{\epsilon\to 0}\int_{x_0-\epsilon}^{x_0+\epsilon}\frac{d^2\phi(x)}{dx^2}dx = \lim_{\epsilon\to 0}\left[\frac{d\phi(x)}{dx}\right]_{x_0-\epsilon}^{x_0+\epsilon} = \frac{2m}{\hbar^2}\lim_{\epsilon\to 0}\int_{x_0-\epsilon}^{x_0+\epsilon}U(x)\phi(x)dx.$$

Now, if we deal with a potential energy that is at least bounded in $[a, b]$, we can find a positive constant K such that $|U(x)| < K$ for any $x \in [a, b]$, hence

$$\begin{aligned}\lim_{\epsilon\to 0}\left|\left[\frac{d\phi(x)}{dx}\right]_{x_0-\epsilon}^{x_0+\epsilon}\right| &= \frac{2m}{\hbar^2}\lim_{\epsilon\to 0}\left|\int_{x_0-\epsilon}^{x_0+\epsilon}U(x)\phi(x)dx\right| \\ &\leqslant \frac{2m}{\hbar^2}\lim_{\epsilon\to 0}\int_{x_0-\epsilon}^{x_0+\epsilon}|U(x)||\phi(x)|dx \\ &< \frac{2\,mK}{\hbar^2}\lim_{\epsilon\to 0}\int_{x_0-\epsilon}^{x_0+\epsilon}|\phi(x)|dx = 0;\end{aligned} \tag{A.2}$$

and we conclude that the first derivative of the wave function is continuous even at a point $x = x_0$ where the potential energy is bounded but discontinuous:

$$\lim_{\epsilon\to 0}\frac{d\phi}{dx}(x_0+\epsilon) = \lim_{\epsilon\to 0}\frac{d\phi}{dx}(x_0-\epsilon) = \phi'(x_0).$$

Clearly, this analysis does not apply when the potential energy is divergent at the point x_0. In this case the first derivative of the wave function exhibits a discontinuity, as we have already seen, for example, for the infinite potential well at the boundary of the $U = 0$ region.

B Derivation of the uncertainty relations

In this appendix we want to show how the uncertainty relations, such as that discovered by Heisenberg, follow mathematically from the axioms of the theory. As we have said in section 8.3, physical observables are represented by *self-adjoint* operators defined on an appropriate Hilbert space $\mathcal{H}$, therefore, two observables G_1 and G_2 correspond to the operators $\hat{G}_1$ and $\hat{G}_1$ such that, for any possible state (wave function) $\phi \in \mathcal{H}$

$$(\phi,\hat{G}_1\phi) = (\hat{G}_1^\dagger\phi,\phi) = (\hat{G}_1\phi,\phi); \qquad (\phi,\hat{G}_2\phi) = (\hat{G}_2^\dagger\phi,\phi) = (\hat{G}_2\phi,\phi).$$

The following derivation makes use of the *Schwarz inequality* of the scalar product $(\cdot,\cdot)$ between Hilbert space vectors, i.e.

$$|(\phi,\psi)|^2 \leqslant (\phi,\phi)(\psi,\psi), \qquad \forall\phi,\psi\in\mathcal{H}. \tag{B.1}$$

Now, let us assume that our system is in a state ϕ and define the two states $\chi_1 = \hat{G}_1\phi$ and $\chi_2 = \hat{G}_2\phi$. From equation (B.1) and making use of the Hermiticity property of $\hat{G}_1$ and $\hat{G}_2$, we have

$$(\chi_1,\chi_1)(\chi_2,\chi_2) \geqslant |(\chi_1,\chi_2)|^2 = |(\hat{G}_1\phi,\hat{G}_2\phi)|^2 = |(\phi,\hat{G}_1\hat{G}_2\phi)|^2,$$

so, still using the Hermiticity of the $\hat{G}_1$ and $\hat{G}_2$ operators, it turns out that

$$\left(\phi,\, \hat{G}_1^2\phi\right)\left(\phi,\, \hat{G}_2^2\phi\right) \geqslant |(\phi,\, \hat{G}_1\hat{G}_2\phi)|^2. \tag{B.2}$$

Recalling the statistical interpretation of the wave function, the expected value of the quantity G_1, if we perform a large number of identical measurements with the system prepared each time in the state $\phi = \sum_i c_{1,i}\phi_{1,i}$, is

$$\langle G_1 \rangle = (\phi,\, \hat{G}_1\phi) = \int_D \phi^*(q)\hat{G}_1\phi(q)\mathrm{d}q = \sum_i g_{1,\,i}|c_{1,\,i}|^2,$$

where $\phi_{1,i}$ is the ith eigenfunction associated with the ith eigenvalue $g_{1,i}$.

In the situation described, a sensible definition for the statistical uncertainty ΔG_1 on G_1 is provided by the standard deviation

$$\Delta G_1 \equiv \sqrt{(\phi,\, (\hat{G}_1 - \langle G_1 \rangle)^2\phi)} = \sqrt{\left(\phi,\, \left(\hat{G}_1^2 - \langle G_1 \rangle^2\right)\phi\right)} = \sqrt{\left(\phi,\, \hat{G}_1^2\phi\right) - \langle G_1 \rangle^2}$$
$$= \sqrt{\langle G_1^2 \rangle - \langle G_1 \rangle^2}.$$

Clearly, $\langle G_1^2 \rangle - \langle G_1 \rangle^2 \geqslant 0$, i.e. $\langle G_1^2 \rangle \geqslant \langle G_1 \rangle^2$.

The same relations also hold for the second observable G_2.

If we evaluate the product $(\Delta G_1)^2(\Delta G_2)^2 = (\phi,\, (\hat{G}_1 - \langle G_1 \rangle)^2\phi)(\phi,\, (\hat{G}_2 - \langle G_2 \rangle)^2\phi)$ and apply the inequality equation (B.2) with operators $\hat{X} \equiv \hat{G}_1 - \langle G_1 \rangle$ and $\hat{Y} \equiv \hat{G}_2 - \langle G_2 \rangle$ instead of $\hat{G}_1$ and $\hat{G}_2$ respectively, we obtain

$$(\Delta G_1)^2(\Delta G_2)^2 \geqslant |(\phi,\, (\hat{G}_1 - \langle G_1 \rangle)(\hat{G}_2 - \langle G_2 \rangle)\phi)|^2 = |(\phi,\, \hat{X}\hat{Y}\phi)|^2. \tag{B.3}$$

At this point it turns out to be useful to introduce the following algebraic identity between operators:

$$\hat{G}_1\hat{G}_2 = \frac{1}{2}[\hat{G}_1,\, \hat{G}_2] + \frac{1}{2}\left\{\hat{G}_1,\, \hat{G}_2\right\},$$

that is, the product of any two operators can be expressed as the sum of their commutator $[\hat{G}_1,\, \hat{G}_2] \equiv \hat{G}_1\hat{G}_2 - \hat{G}_2\hat{G}_1$ divided by two and their anticommutators $\{\hat{G}_1,\, \hat{G}_2\} \equiv \hat{G}_1\hat{G}_2 + \hat{G}_2\hat{G}_1$ divided by two.

We can observe that while an anticommutator of two Hermitian operators is also Hermitian[9],

$$\{\hat{G}_1,\, \hat{G}_2\}^\dagger = \{\hat{G}_1,\, \hat{G}_2\},$$

the commutator is *anti-Hermitian*, i.e.

[9] This ensues plainly from the following property of the *Hermitian conjugation*, which in a sense is inherited from matrix transposition:

$$(\hat{A}\hat{B})^\dagger = \hat{B}^\dagger\hat{A}^\dagger.$$

Indeed, $((\hat{A}\hat{B})^\dagger\phi,\, \psi) = (\phi,\, \hat{A}\hat{B}\psi) = (\hat{A}^\dagger\phi,\, \hat{B}\psi) = (\hat{B}^\dagger\hat{A}^\dagger\phi,\, \psi) \Longrightarrow (\hat{A}\hat{B})^\dagger = \hat{B}^\dagger\hat{A}^\dagger$.

$$[\hat{G}_1, \hat{G}_2]^\dagger = -[\hat{G}_1, \hat{G}_2].$$

It can be shown quite easily that while any Hermitian operator has real expectation values, an anti-Hermitian operator $\hat{A}$ has purely imaginary expectation values. Indeed,

$$(\phi, \hat{A}\phi) = (\hat{A}^\dagger\phi, \phi) = -(\hat{A}\phi, \phi) = -(\phi, \hat{A}\phi)^*.$$

So, using the sesquilinearity of the inner product $(\cdot, \cdot)$, the complex number $c = (\phi, \hat{A}\phi)$ is the opposite of its complex conjugate $c^*=(\phi, \hat{A}\phi)^*$, namely it is an imaginary number.

Now, returning to the inequality equation (B.3), we can write the right-hand side using the above decomposition for the product $\hat{X}\hat{Y} = (\hat{G}_1 - \langle G_1\rangle)(\hat{G}_2 - \langle G_2\rangle)$:

$$\begin{aligned}(\Delta G_1)^2(\Delta G_2)^2 &\geqslant \left|\frac{1}{2}(\phi, [\hat{X}, \hat{Y}]\phi) + \frac{1}{2}(\phi, \{\hat{X}, \hat{Y}\}\phi)\right|^2 \\ &= \frac{1}{4}|(\phi, [(\hat{G}_1 - \langle G_1\rangle), (\hat{G}_2 - \langle G_2\rangle)]\phi) \\ &\quad + (\phi, \{(\hat{G}_1 - \langle G_1\rangle), (\hat{G}_2 - \langle G_2\rangle)\}\phi)|^2.\end{aligned} \tag{B.4}$$

Since the modulus of a complex number is always greater or at most equal to the imaginary part of the same number, we can write

$$(\Delta G_1)^2(\Delta G_2)^2 \geqslant \left|\frac{1}{2}(\phi, [\hat{X}, \hat{Y}]\phi) + \frac{1}{2}(\phi, \{\hat{X}, \hat{Y}\}\phi)\right|^2 \geqslant \frac{1}{4}|(\phi, [\hat{X}, \hat{Y}]\phi)|^2. \tag{B.5}$$

Now we are almost done, we just have to evaluate the commutator $[\hat{X}, Y]$, but this is straightforward (here we restore the identity operator $\hat{I}$, previously omitted):

$$\begin{aligned}[\hat{X}, \hat{Y}] &= [\hat{G}_1 - \langle G_1\rangle\hat{I}, \hat{G}_2 - \langle G_2\rangle\hat{I}] \\ &= [\hat{G}_1, \hat{G}_2] - \langle G_1\rangle[\hat{I}, \hat{G}_2] - \langle G_2\rangle[\hat{G}_1, \hat{I}] + \langle G_1\rangle\langle G_2\rangle[\hat{I}, \hat{I}] \\ &= [\hat{G}_1, \hat{G}_2].\end{aligned} \tag{B.6}$$

Thus, substituting the last result in equation (B.5) and taking the square root of both members, we finally obtain the desired uncertainty relations in their general form for any pair of non-commuting self-adjoint operators (i.e. incompatible observables):

$$\Delta G_1\,\Delta G_2 \geqslant \frac{1}{2}|(\phi, [\hat{G}_1, \hat{G}_2]\phi)| = \frac{1}{2}|\langle[\hat{G}_1, \hat{G}_2]\rangle|.$$

References

[1] Besson U and Malgieri M 2018 *Insegnare la Fisica Moderna* (Roma: Carocci)

[2] Einstein A 1905 Über einen die Erzeugung und Verwandlung des Lichtes betreffenden heurischen Gesichtpunkt *Ann. Phys.* **332** 132–48

[3] Tonomura A, Endo J, Matsuda T, Kawasaki T and Ezawa H 1989 Demonstration of single-electron buildup of an interference pattern *Am. J. Phys.* **57** 117–20
[4] Liou S-H, Bach R, Pope D and Batelaan H 2013 Controlled double-slit electron diffraction *New J. Phys.* **15** 033018–24
[5] Noce C 2018 *Introduzione alla Fisica Moderna* (Roma: Aracne)
[6] Einstein A 1905 Zur Elektrodynamik bewegter Körper *Ann. Phys.* **322** 891–921
[7] Galilei G 1632 *Dialogo sopra i due massimi sistemi del mondo*
[8] Rudolph T, Pusey M F and Barrett J 2012 On the reality of the quantum state *Nat. Phys.* **8** 475–8
[9] Sakurai J J 1994 *Modern Quantum Mechanics* revised edn (Reading, MA: Addison-Wesley)
[10] Messiah A 2014 *Quantum Mechanics* (New York: Dover)
[11] Courant R and Hilbert D 1953 *Methods of Mathematical Physics* (New York: Interscience)
[12] Ferraioli A G and Noce C 2019 The measurement problem in quantum mechanics *Sci. Philos.* **7** 41–58
[13] Kumar M 2008 *Quantum: Einstein, Bohr, and the Great Debate About the Nature of Reality* (London: Icon)
[14] Bohr N 1934 *Atomic Theory and the Description of Nature* (Cambridge: Cambridge University Press)
[15] Bell J S 2004 *Speakable and Unspeakable in Quantum Mechanics: Collected Papers on Quantum Philosophy* 2nd edn (Cambridge: Cambridge University Press)
[16] Einstein A, Podolsky B and Rosen N 1935 Can quantum-mechanical description of physical reality be considered complete? *Phys. Rev.* **47** 10
[17] Bohm D 1951 *Quantum Theory* (Englewood Cliffs, NJ: Prentice Hall)
[18] von Neumann J 1955 *Mathematical Foundations of Quantum Mechanics* (Princeton, NJ: Princeton University Press)
[19] Bohm D 1952 A suggested interpretation of the quantum theory in terms of 'hidden' variables. I *Phys. Rev.* **85** 166
[20] Bohm D 1952 A suggested interpretation of the quantum theory in terms of 'hidden' variables. II *Phys. Rev.* **85** 180
[21] Bell J S 1964 On the Einstein Podolsky Rosen paradox *Physics* **1** 195–200
[22] Ghirardi G 2007 *Sneaking a Look at God's Cards: Unraveling the Mysteries of Quantum Mechanics* (Princeton, NJ: Princeton University Press)
[23] Grangier P, Aspect A and Roger G 1981 Experimental tests of realistic local theories via Bell's theorem *Phys. Rev. Lett.* **47** 460
[24] Grangier P, Aspect A and Roger G 1982 Experimental realization of Einstein–Podolsky–Rosen–Bohm Gedankenexperiment: a new violation of Bell's inequalities *Phys. Rev. Lett.* **49** 91–4
[25] Hensen B *et al* 2015 Loophole-free Bell inequality violation using electron spins separated by 1.3 kilometres *Nature* **526** 682–6
[26] Schrödinger E 1935 Die gegenwärtige Situation in der Quantenmechanik (The present situation in quantum mechanics) *Naturwissenschaften* **23** 807–12; 823–8; 844–9
[27] Everett H III 1957 'Relative state' formulation of quantum mechanics *Rev. Mod. Phys.* **29** 454

IOP Publishing

Modern Physics
A critical approach
Canio Noce

Chapter 9

Early attempts to make many-particle physics simple

Marco Figliolia and Alfonso Romano

9.1 Introduction

This chapter aims to provide a concise introduction to the basic concepts and methods of many-body physics and should be suitable for undergraduate science students interested in a first exposure to modern physics.

Our presentation begins with a quick review of some important results of thermodynamics and classical statistical mechanics, with particular reference to the simple but versatile model of the ideal gas and to the classical theory of specific heat for gases and solids. In the following section we analyse the classical treatment of electrical and thermal conduction in metals within the so-called Drude model, focusing on its successes in explaining some experimental data and, in particular, on its limitations.

We then move on to the introduction of some preliminary notions of quantum statistics in order to discuss the first significant improvements of the classical models, obtained by implementing quantum principles. In this context we investigate the Einstein model for the specific heat of solids, and then we move on to the more accurate Debye model, which corrects the results concerning the behaviour of the specific heat in the low-temperature regime.

We want to stress that our goal is not to give an organic presentation of the fundamental aspects of many-particle physics, but to show at a reasonably in-depth level some of the most significant elementary approaches, models and techniques that classical and quantum physics provide to describe such systems.

9.2 Kinetic theory of gases and specific heats: the classical treatment

9.2.1 Statistical mechanics and thermodynamics: from micro to macro

Thermodynamics is a fundamental branch of physics dealing with the relationships between heat, work and other forms of energy and aims to describe the states and physical properties of macroscopic systems by means of a limited set of state variables, such as temperature (T), pressure (P), volume (V) or the number of elementary units of a particular homogeneous component (N).

Thermodynamics is a *macroscopic theory* because it is formulated in terms of parameters that characterize the system as a whole and are consistently determinable only for systems made up of a sufficient number of elementary units (atoms or molecules), typically at least of the order of the *Avogadro number*[1]. A complete set of state variables fixes the macroscopic state of the system univocally and when their values are well-defined at a given time we say that the system is in a state of *thermodynamic equilibrium*. Classical thermodynamics[2] handles those processes (thermodynamic transformations) that make the system go from an initial to a final equilibrium state, regardless of whether the intermediate states are equilibrium states or not (in the first case the processes are called *quasi-static*). The fundamental laws ruling thermodynamic world are the following:

- *Zeroth law of thermodynamics.*
  If two thermodynamic systems are each in thermal equilibrium with a third one, then they are in thermal equilibrium with each other.
- *First law of thermodynamics.*
  The change in the internal energy (U) of a thermodynamic system is equal to the heat (Q) added to the system minus the work (W) done by the system:
  $$\Delta U = Q - W.$$
- *Second law of thermodynamics.*
  In any cyclic process (the initial and final states are the same) it is impossible to extract an amount of heat Q_H from a reservoir and use it all to do work W, i.e. it is impossible to construct a (cyclic) heat engine with efficiency $\eta = \frac{W}{Q_H}$ equal to 1.
- *Third law of thermodynamics (Nernst's theorem).*
  The entropy S of a system approaches a constant value as its temperature T approaches absolute zero.

We assume that the reader already has an adequate knowledge of the basics of thermodynamics, so we will not discuss the laws just reported and their numerous remarkable applications extensively. Nonetheless, we will make a few remarks on them to foster a deeper understanding of the following sections.

The zeroth law was introduced to provide a consistent definition of the temperature as a state variable: the temperature T is the physical (intensive) quantity

[1] The Avogadro number $N_A = 6.022\,140\,76 \times 10^{23}$ is the number of constituent particles that are contained in one *mole* (mol) of a particular substance.

[2] There are other ramifications of thermodynamics focusing on non-equilibrium processes, but the related studies are outside the scope of this chapter.

measured by a thermometer and has the same value for each pair of systems in thermal contact (i.e. able to exchange heat) when they have reached thermal equilibrium.

The first law is a balance equation, an application of the principle of energy conservation to thermodynamic processes for which energy can be transferred not only by doing work, but also through heat flows between systems at different temperatures. This law leads to the introduction of a state function, the internal energy of the thermodynamic system (U), whose variation between two equilibrium states (say A and B) is independent of the particular process (reversible or not) taking the system from A to B. It should be remembered that while internal energy is a state function, heat and work are not since they are quantities associated with transformations and not with equilibrium states.

In differential form the first law of thermodynamics is written as follows:

$$\mathrm{d}U = \delta Q - \delta W.$$

The symbol δ is used here to indicate that, differently to the internal energy U, heat and work are not exact differentials[3] since they are not state variables. While the first law rules out those processes implying the creation of energy from nothing, that is, it excludes *perpetual motion of the first kind*, the second law tells us that it is impossible to build a perfect thermal engine that is able to cyclically convert into mechanical work all the heat absorbed from a thermal reservoir. This means that any thermal engine cannot reach, even in theory, 100% efficiency. In this case we say that *perpetual motion of the second kind* is impossible.

This law, much subtler than the previous one, turns out to be absolutely crucial for understanding the time evolution of thermodynamic systems. It can be stated in various equivalent formulations, among which a very useful one relies on the introduction of a new extensive state function, the *entropy* (S), defined by the following relation:

$$S_{\mathrm{B}} - S_{\mathrm{A}} = \int_{\mathrm{A}}^{\mathrm{B}} \frac{\delta Q}{T}\bigg|_{\mathrm{rev}}.$$

This formula defines the variation of the entropy state function between any two equilibrium states A and B. As the subscript 'rev' suggests, the integral must be calculated along any of the possible reversible transformations connecting the extrema A and B of the actual path. If we fix the entropy value for a system in a reference state A as O, we can associate to any possible equilibrium state B a well-defined value of the entropy S_{B}. A possible choice is to set $S_O = 0$ when the absolute temperature T of the system is equal to 0 K, in conformity with the third law of thermodynamics.

[3] A differential form (more precisely a one-form) is a linear combination of the differentials of the coordinate variables, for instance in a three-dimensional space: $\omega = a(x, y, z)\mathrm{d}x + b(x, y, z)\mathrm{d}y + c(x, y, z)\mathrm{d}z$. A differential one-form is *exact* when it can be expressed as the total differential of a multi-variable function, i.e. in three dimensions: $\omega = \mathrm{d}F = \frac{\partial F}{\partial x}\mathrm{d}x + \frac{\partial F}{\partial y}\mathrm{d}y + \frac{\partial F}{\partial z}\mathrm{d}z$.

Using the notion of entropy, the second law of thermodynamics states more neatly that for any *isolated* thermodynamic system its entropy can only increase, or at most remain constant in the case of reversible processes:

$$S_B - S_A > 0 \quad \text{with } t_B > t_A \text{, for irreversible processes}$$
$$S_B - S_A = 0 \quad \text{with } t_B > t_A \text{, for reversible processes.}$$

Here t_A and t_B indicate the instants of time in which the initial and the final equilibrium states are reached. So, an isolated thermodynamic system evolving in time goes towards an equilibrium state in which the entropy attains its maximum value.

The second law constitutes a severe constraint forbidding a wide spectrum of thermodynamic processes that could have been feasible according to the first law of thermodynamics only. It provides a direction for the time evolution of the macroscopic systems; in fact it is customary to assert, evocatively as well as vaguely, that the second law accounts for the 'arrow of time', namely the observation that almost all macroscopic physical phenomena in everyday life occur in one direction only.

Now, we might ask: even if the laws of thermodynamics are a magnificent synthesis of a multitude of experimental observations and disregard the particular composition and internal structure of a macroscopic system, is it possible to derive or justify them by making some hypotheses about the microscopic components of the system itself? The answer is 'yes', although with some limitations, and this connection between the microscopic and macroscopic worlds is basically the main purpose of statistical thermodynamics and, more broadly, of statistical physics.

Statistical thermodynamics arose in the second half of the nineteenth century with the pioneering works on gas kinematics and dynamics by the Scottish mathematical physicist James Clerk Maxwell (1831–1879) and, in particular, by the Austrian physicist Ludwig Eduard Boltzmann (1844–1906).

In its bare essence, this approach to thermodynamic systems consists in viewing them in terms of their elementary constituents, which interact and move according to a given set of dynamical laws, such as, for instance, Hamilton's equations of motion (just a more mathematically refined version of Newton's second law of classical mechanics) or the Schrödinger equation implementing quantum mechanical dynamics. The microscopic state of an N-particle system, usually called a *microstate*, in classical mechanics is represented by a point $x = (q_1, q_2, \dots q_{3N}; p_1, p_2, \dots, p_{3N})$ in the $6N$-dimensional phase space, where q_i and p_i are the generalized coordinates and the conjugate momenta, respectively. It evolves deterministically according to the laws of classical mechanics and corresponds to a given macroscopic state (*macrostate*) identified by an independent number of state variables directly accessible to the experimenter. In general, for equilibrium states there is a large number of different microstates associated with the same macrostate. In order to develop a statistical analysis, it is necessary to make a fundamental assumption concerning these microstates, known as the *equal a priori probability postulate*:

- for an isolated system with fixed energy and composition, the system can be found with equal probability in any microstate compatible with the considered macroscopic state.

Thanks to this postulate, it is possible to define a *statistical ensemble* of the given system, that is, a probability distribution defined over all the possible microstates associated with a given macrostate. It should be clear that the concrete definition of this distribution relies on the mathematical model describing the space of states (Ω), that in quantum mechanics is different from that of classical mechanics (i.e. a suitable region of the phase space). In any case, knowledge of the probability distribution $\rho(s)$ enables us to calculate macroscopic quantities (G) that can be expressed as ensemble averages of functions (f_G) defined on microscopic states (s):

$$G = \langle f_G \rangle \equiv \int_\Omega f_G(s)\rho(s)\mathrm{d}s.$$

Depending on the macroscopic variables one uses to single out an equilibrium state, different kinds of ensembles can be defined in statistical thermodynamics. Each of them has particular statistical properties and matches different macroscopic constraints imposed on the system. The most commonly used ones are the microcanonical, the canonical and the grand canonical ensembles, pictorially represented in figure 9.1.

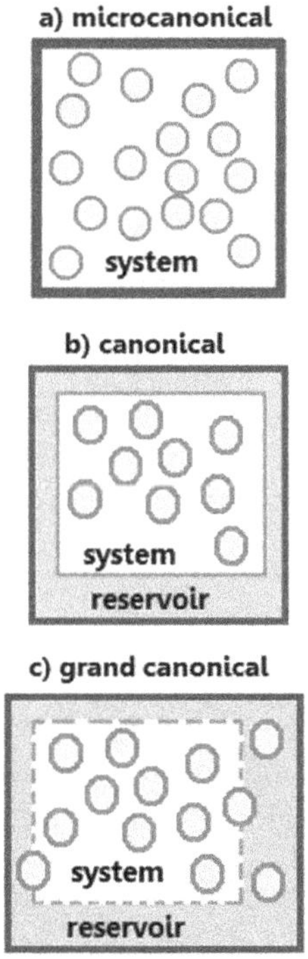

Figure 9.1. Schematic representations of the three fundamental statistical ensembles.

The *microcanonical ensemble* employs as state variables N, V and U, and deals with isolated systems for which the total energy (U) and the number of particles (N) are fixed at specific values, this holding for each member of the ensemble.

The *canonical ensemble* uses as state variables N, V and T, and deals with systems whose energy is not specified but the number of particles (N) and the temperature (T) are fixed. This ensemble proves to be useful when dealing with closed systems (i.e. systems that are unable to exchange particles with the environment) which are in thermal contact with a heat reservoir at the temperature T.

The *grand canonical ensemble* uses as state variables V, T and the so-called 'chemical potential' μ, namely the energy variation that the system experiences for the acquisition of a single particle at fixed volume and temperature. In this case, neither the energy nor the particle number have definite values, whereas the temperature and the chemical potential are kept fixed. This kind of ensemble is suitable for describing open systems in contact with heat and particle reservoirs, so that energy and particle exchange are both allowed.

In order to recover the results of classical thermodynamics one can show, following Boltzmann, that an appropriate statistical definition of the entropy is

$$S = k_B \ln(W). \tag{9.1}$$

Here $k_B = 1.380\,65 \times 10^{23}$ J K^{-1} is the Boltzmann constant and W is the function counting the number of microstates associated with a given macrostate, which are all assumed to be equally likely, for a given energy. Equation (9.1) gives the absolute value of the entropy and relates this quantity to the probability of finding the system in a given macroscopic state. It tells us that the larger the number of microstates associated with a given macrostate is, the higher the corresponding value of its entropy.

In this fairly reductionistic perspective, the second law of thermodynamics claims that isolated systems evolve in time towards macrostates for which the function W is maximized, since they are also the ones with higher entropy. It is just this interpretation of entropy that one appeals to when connecting, somewhat improperly, this quantity to the disorder state of a thermodynamic system.

After this cursory introduction to the fundamental ideas of statistical thermodynamics, some of which will be invoked and developed in later sections, we are now ready to present one of the earliest and most elementary applications of the statistical approach to many-particle systems, namely, the ideal gas model.

9.2.2 Kinetic theory of gases: a first glance

Macroscopic bodies are aggregates of huge numbers of particles and, depending on the strength of the inter-particle interactions, they may exist in solid, liquid, gaseous and plasma[4] states of matter (neglecting uncommon and more exotic phases).

[4] Plasma is a highly ionized gas of free electrons and positive ions; it is the main state of stellar matter and probably the most abundant form of matter in the Universe.

In gases particles interact very weakly; this means that we expect their kinetic energy to be much higher than the potential energy related to the interaction forces. A convenient model to study gases which, despite its simplicity, leads to some reliable results, is the *ideal gas* model, where atoms and molecules are considered punctiform, randomly moving and interacting only via elastic collisions occurring between them and with the walls of the box containing the gas. It reproduces quite well the behaviour of noble gases in not too extreme conditions of temperature and pressure and, more generally, but also less accurately, of other real gases that are sufficiently rarefied and whose temperatures are high enough (what the words 'sufficiently' and 'enough' mean in practical terms depends on the particular gas and, obviously, also on the degree of accuracy one seeks). It is found experimentally that when these conditions are realized, the behaviour of a gas at equilibrium obeys the well-known *state equation of ideal gases*:

$$PV = nRT. \tag{9.2}$$

Here P, V and T are the pressure, volume and temperature of the gas, respectively, $R = k_B N_A = 8.314 \text{ J mol}^{-1} \text{ K}^{-1}$ is the *universal constant of ideal gases* and $n = N/N_A$ is the number of moles.

From a statistical point of view, the equilibrium pressure P results from the time average of the combined action of many randomly intermittent microforces exerted by the atoms or molecules on the rigid walls of the container which contains the gas. Hence let us develop this picture to establish a connection between the microscopic and macroscopic variables.

We assume that each atom or molecule of the gas is a point of mass m following the laws of classical mechanics and that its collisions with the other particles and with the walls of the container are perfectly elastic. Since only the latter contribute to the pressure, we can safely ignore any collision between the particles. To simplify our derivation we also assume that the container is a perfect sphere; we can do this because it can be shown from thermodynamic arguments that its particular shape does not play any role.

When a particle hits the inner surface of the sphere, its linear momentum changes from $\vec{p} = m\vec{v}$ to $\vec{p}\,' = m\vec{v}\,'$ (see figure 9.2), so that the impulse of the average contact force $\vec{f}$ the container exerts on the particle is $\vec{I} = \vec{f}\,\Delta t = \vec{p}\,' - \vec{p} = \Delta\vec{p}$, where Δt is the collision time. Moreover, from the third law of dynamics we have that the average force exerted by the particle on the wall during the collision is the opposite of the force exerted by the wall on the particles, $\vec{f}\,' = -\vec{f} = -\frac{\Delta\vec{p}}{\Delta t}$. For the hypothesis of elasticity of the collisions, the particle's momentum does not change in modulus, $|\vec{p}\,| = |\vec{p}\,'| = p$. It is then easy to see that the momentum variation is in modulus equal to $|\Delta\vec{p}\,| = 2p\cos(\alpha)$, where α is half of the angle between the outgoing momentum $\vec{p'}$ and the ingoing momentum $\vec{p}$. It should also be clear from elementary geometric considerations that the direction of the vector $\Delta\vec{p}$ is radially inward for any possible value of α.

We can now say that the average contribution $f \equiv |\vec{f}\,| = |\vec{f}\,'|$ of this particle to the modulus of the total radial force exerted on the spherical container by all the

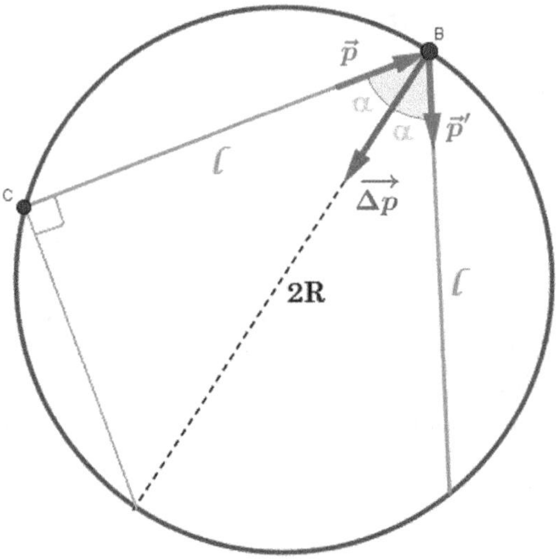

Figure 9.2. Schematic representation of the collision of a particle on the inner surface of a spherical container. The momentum of the particle is denoted by a red arrow.

particles of the gas is obtained by dividing the change in momentum caused by a collision, that is $|\Delta\vec{p}\,|$, by the time $\Delta\tau$ elapsed since the last collision. Considering that the motion between two successive collisions is assumed to be uniform, one has $\Delta\tau = \frac{l}{v}$, where l is the average distance travelled by the particle between two successive collisions and v is its speed. We thus have

$$f = \frac{|\Delta\vec{p}\,|}{\Delta\tau} = 2\,mv\cos(\alpha)\frac{v}{l} = 2\,mv^2\frac{\cos(\alpha)}{l} = \frac{mv^2}{R},$$

where in the last step we have used the trigonometric relation $l = 2R\cos\alpha$ (see again figure 9.2).

We see that the average force due to the particles colliding with the inner surface of the spherical container thus depends only on its speed v. We can thus compute the total pressure P summing up the normal component of the forces associated with all the gas particles, $f_i = mv_i^2/R$, and dividing this sum by the surface area of the container $S = 4\pi R^2$,

$$P = \sum_{i=1}^{N}\frac{f_i}{S} = \sum_{i=1}^{N}\frac{mv_i^2}{R\cdot 4\pi R^2} = \frac{m}{4\pi R^3}\sum_{i=1}^{N}v_i^2 = \frac{Nm}{3V}\frac{1}{N}\sum_{i=1}^{N}v_i^2, \tag{9.3}$$

where $V = \frac{4}{3}\pi R^3$. The quantity $\frac{1}{N}\sum_{i=1}^{N}v_i^2$ appearing in this equation is the square of the average quadratic velocity

$$v_{\text{rms}} = \sqrt{\frac{1}{N}\sum_{i=1}^{N}v_i^2},$$

so that we can rewrite equation (9.3) as follows:

$$PV = \frac{Nmv_{\rm rms}^2}{3}. \tag{9.4}$$

Equation (9.4) is known as the *Joule–Clausius–Krönig* equation and connects the product of two macroscopic parameters, volume and pressure, to microscopic quantities such as the velocity and the mass of the particles. Furthermore, through comparison with the ideal gas law, equation (9.2), we find

$$\frac{Nmv_{\rm rms}^2}{3} = nRT = Nk_{\rm B}T,$$

from which we obtain an important relation between the average molecular kinetic energy $\bar{\mathcal{E}}_{\rm k}$ and the temperature T:

$$\mathcal{E}_{\rm k} = \frac{1}{N}\sum_{i=1}^{N}\frac{1}{2}mv_i^2 = \frac{1}{2}mv_{\rm rms}^2 = \frac{3}{2}k_{\rm B}T.$$

This equation is remarkable since it connects the absolute temperature T of an ideal gas at equilibrium to the average kinetic energy of a particle in the gas. It is also worth noting that the above results imply that the internal energy U of the gas, equal to the total kinetic energy $E_{\rm k} = N\bar{\mathcal{E}}_{\rm k}$, and the average quadratic velocity $v_{\rm rms}$ are both simple functions of the temperature:

$$U = E_{\rm k} = \frac{3}{2}nRT \tag{9.5}$$

$$v_{\rm rms} = \sqrt{\frac{3k_{\rm B}T}{m}}. \tag{9.6}$$

These results, derived here with an elementary statistical approach, are in perfect agreement with the *theorem of equipartition of energy*, which states: molecules in a gas at thermal equilibrium have, for each independent degree of freedom associated with their motion, an average energy equal to $\frac{1}{2}k_{\rm B}T$, where T is the equilibrium temperature.

An elementary proof of this result can be found in standard textbooks on statistical mechanics (see for instance those quoted in [1, 2]); here we refer to it to describe the behaviour of ideal gases whose internal energy is entirely kinetic in nature. For a gas with molecules having f degrees of freedom the total energy is then

$$U = \frac{f}{2}Nk_{\rm B}T = \frac{f}{2}nRT. \tag{9.7}$$

For a monoatomic gas we have $f = 3$ since a point particle possesses only three translational degrees of freedom, so that we recover the previously derived result (equation (9.5)). On the other hand, a diatomic molecule has, in addition to the three translational degrees of freedom, two rotational ones (there are not three since one rotational movement is degenerate). Moreover, if we do not consider it as a perfectly rigid body, one additional vibrational degree of freedom must also be taken into account. Therefore, $f = 3 + 2 = 5$ and $U = \frac{5}{2}nR$ when we refer to rigid molecules, while $f = 3 + 2 + 1 = 6$ and $U = 3nR$ if we adopt for them an elastic model. Likewise polyatomic molecules whose vibrational modes can be regarded as 'frozen', have a number of degrees of freedom ranging from five to six in the absence of particular rotational symmetries. When the rigidity hypothesis is relaxed, the number of degrees of freedom adds up to $f = 3 + 3 + (3s - 6) = 3s$, s being the number of atoms, i.e. we have three translational, three rotational and $3s - 6$ vibrational degrees of freedom, so that the total internal energy becomes $U = \frac{3s}{2}nRT$. However, by studying the specific heat of gases we will see that such predictions are not always fulfilled and we will try to understand why.

Before going on to analyse the equilibrium distribution of the molecular velocities, it is worth noting that the expression of the average quadratic velocity given by equation (9.6) holds for diatomic and polyatomic gases as well. This is because the translational contribution to the average kinetic energy of a single particle is always $\mathcal{E}_{\mathrm{k,\,trans}} = \frac{3}{2}k_{\mathrm{B}}T$.

9.2.3 The Maxwell–Boltzmann distribution

In this section we want to examine the distribution law of the particle velocities in an ideal gas at thermodynamical equilibrium. The knowledge of such a function will enable us to understand how the speeds are peaked and spread over the entire range of allowed values. We also point out that we are still in the realm of classical physics, so the identical particles of the gas are treated as distinguishable entities. In order to avoid all the machinery of statistical mechanics, we will follow a somewhat heuristic path in the derivation of the distribution function.

The vectors representative of the velocities of the particles lie in a 3D space, whose volume element in Cartesian coordinates $Ov_xv_yv_z$ is $\mathrm{d}^3v = \mathrm{d}v_x\mathrm{d}v_y\mathrm{d}v_z$. The number of particles $\mathrm{d}N(\vec{v})$ having velocities in the range $[\vec{v}, \vec{v} + \mathrm{d}\vec{v}]$ is then expressed in the form

$$\mathrm{d}N(\vec{v}) = n(\vec{v})\mathrm{d}^3v,$$

where $n(\vec{v})$ is the *density function of the velocities* which associates to any $\vec{v}$ the number of particles having velocity falling in a unitary volume centred around the 'point' $\vec{v}$.

Dividing this quantity by the number of particles N, we obtain the probability $\mathrm{d}P(\vec{v})$ of finding the velocity of a given particle in the volume element d^3v:

$$\mathrm{d}P(\vec{v}) = \frac{\mathrm{d}N(\vec{v})}{N} = p(\vec{v})\mathrm{d}v_x\mathrm{d}v_y\mathrm{d}v_z. \tag{9.8}$$

Here we have defined the *probability density function* (PDF)

$$p(\vec{v}) = \frac{n(\vec{v})}{N},$$

which is the quantity that we will determine in the following.

We start by making some simplification based on physical assumptions. Since in an ideal gas the particles are non-interacting, their energy is purely kinetic: $\mathcal{E} = \frac{1}{2}mv^2$. Moreover, at equilibrium all the spatial directions are equivalent and thus the particle velocity distribution is expected to exhibit perfect isotropy, i.e. $p(\vec{v})$ depends only on the modulus of v, $p(\vec{v}) \equiv p(v)$. This spherical symmetry allows us to express the fraction of the particles with velocities inside a spherical shell of radii v and $v + \mathrm{d}v$ as the product of $p(v)$ by the volume of the shell $\mathrm{d}V_{\text{shell}} = 4\pi v^2 \mathrm{d}v$:

$$\frac{\mathrm{d}N(v)}{N} = p(v)\mathrm{d}V_{\text{shell}} = 4\pi v^2 p(v)\mathrm{d}v \equiv f(v)\mathrm{d}v. \tag{9.9}$$

Here we have defined the distribution function $f(v)$, which is such that $f(v)\mathrm{d}v$ is the fraction of the particles with a modulus of the velocities in the interval $[v,\ v + \mathrm{d}v]$.

Now, let us assume that two particles, with velocities $\vec{v}_1$ and $\vec{v}_2$ and energies $\mathcal{E}_1$ and $\mathcal{E}_2$, collide and acquire the final velocities $\vec{v}_1'$ and $\vec{v}_2'$ and the final energies $\mathcal{E}_1'$ and $\mathcal{E}_2'$. Since the energy value of a particle is independent of the energy of another particle, the probability $\mathcal{P}$ of this event is simply given by

$$\mathcal{P} = C\ p(\mathcal{E}_1)\ p(\mathcal{E}_2),$$

where the PDFs are here expressed as functions of $\mathcal{E} = \frac{1}{2}mv^2$ and the proportionality constant C does not depend on energies or velocities. Analogous reasoning applies when the same two particles come out of the collision. One then refers to the probability $\mathcal{P}'$ given by

$$\mathcal{P}' = C'\ p(\mathcal{E}_1')\ p(\mathcal{E}_2').$$

In classical mechanics elastic collisions are perfectly reversible, so we can start with the energies $(\mathcal{E}_1', \mathcal{E}_2')$ and the velocities $(-\vec{v}_1', -\vec{v}_2')$ to end up with the energies $(\mathcal{E}_1, \mathcal{E}_2)$ and the velocities $(-\vec{v}_1, -\vec{v}_2)$ (see figure 9.3). We thus have $C = C'$. Then we use the

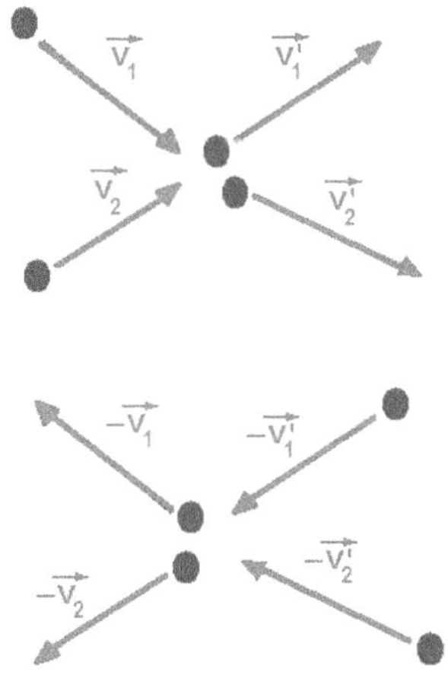

Figure 9.3. Direct and reverse collisions.

assumption that the gas is in thermodynamic equilibrium, implying that the energy probability distribution must be stationary. Hence the process ($\mathcal{E}_1$, $\mathcal{E}_2$; $\vec{v}_1$; $\vec{v}_2$)$\longrightarrow$($\mathcal{E}_1'$, $\mathcal{E}_2'$; $\vec{v}_1'$; $\vec{v}_2'$) cannot be statistically favoured over its reverse ($\mathcal{E}_1'$, $\mathcal{E}_2'$; $-\vec{v}_1'$; $-\vec{v}_2'$)$\longrightarrow$($\mathcal{E}_1$, $\mathcal{E}_2$; $-\vec{v}_1$; $-\vec{v}_2$); this allows us to conclude that $\mathcal{P} = \mathcal{P}'$ and thus

$$p(\mathcal{E}_1)p(\mathcal{E}_2) = p(\mathcal{E}_1')p(\mathcal{E}_2'). \tag{9.10}$$

In elastic collisions, mechanical energy is conserved, so that if we set $\mathcal{E} = \mathcal{E}_1 + \mathcal{E}_2 = \mathcal{E}_1' + \mathcal{E}_2'$ and $\mathcal{E}_1' = \mathcal{E}_1 + u$, equation (9.10) becomes

$$p(\mathcal{E}_1)p(\mathcal{E} - \mathcal{E}_1) = p(\mathcal{E}_1 + u)p(\mathcal{E} - \mathcal{E}_1 - u).$$

We see that our function must obey the rule $p(x)p(y) = p(x + u)(y - u)$ for any admissible value of x, y and $u \leqslant y$. It should not take too much effort to understand that such a function is an exponential one,

$$p(\mathcal{E}) = A\mathrm{e}^{-\beta\mathcal{E}}$$

or, equivalently,

$$p(v) = A\mathrm{e}^{-\frac{\beta mv^2}{2}}. \tag{9.11}$$

Here A and β are two constants related to inherent properties of the particles (the mass) and to some state parameter characterizing the equilibrium. They must strictly be positive numbers since $g(\mathcal{E})$ is a PDF and the probability of finding particles with extremely large energies (that is, in the limit $\mathcal{E} \to +\infty$) must drop to zero. The constant A can be settled by imposing the probability normalization condition

$$\int_0^\infty \mathrm{d}P(v) = \int_0^\infty f(v)\mathrm{d}v = 1,$$

leading to

$$\int_0^\infty A\mathrm{e}^{-\frac{\beta mv^2}{2}} 4\pi v^2 \mathrm{d}v = 1.$$

The calculation in easily performed in terms of the Gaussian integral[5]

$$\int_0^\infty x^2 \mathrm{e}^{-\alpha x^2}\mathrm{d}x = \frac{1}{4}\sqrt{\frac{\pi}{\alpha^3}}$$

and one obtains

[5] To calculate this integral quickly from the fundamental Gaussian integral $I(\alpha) = \int_0^\infty \mathrm{e}^{-\alpha x^2}\mathrm{d}x = \frac{1}{2}\sqrt{\frac{\pi}{\alpha}}$ we can use the well-known 'Feynman's trick' of differentiating under the integral sign with respect to the parameter α:

$$\int_0^\infty x^2 \mathrm{e}^{-\alpha x^2}\mathrm{d}x = -\frac{\partial I(\alpha)}{\partial \alpha} = \frac{1}{4}\sqrt{\frac{\pi}{\alpha^3}}.$$

$$A = \frac{1}{4\pi \int_0^{\infty} e^{-\frac{\beta m v^2}{2}} v^2 dv} = \sqrt{\left(\frac{m\beta}{2\pi}\right)^3}.$$

To give a proper interpretation of the parameter β, we use equation (9.11) to evaluate the average kinetic energy $\bar{\mathcal{E}}_k$:

$$\bar{\mathcal{E}}_k = \int_0^{\infty} \frac{1}{2} m v^2 p(v) 4\pi v^2 dv = 2\pi m \sqrt{\left(\frac{m\beta}{2\pi}\right)^3} \int_0^{\infty} v^4 e^{-\frac{\beta m v^2}{2}} dv$$
$$= 2\pi m \sqrt{\left(\frac{m\beta}{2\pi}\right)^3} \frac{3}{8} \sqrt{\frac{2^5 \pi}{(m\beta)^5}} = \frac{3}{2}\beta^{-1}.$$

Recalling that when the gas is at the temperature T the average kinetic energy per particle is $\bar{\mathcal{E}}_k = \frac{3}{2}k_B T$, we find

$$\frac{3}{2}\beta^{-1} = \frac{3}{2}k_B T \quad \Longrightarrow \quad \beta = \frac{1}{k_B T}.$$

Therefore, the *Maxwell–Boltzmann velocity distribution* has the following expression:

$$f(v) = 4\pi v^2 p(v) = 4\pi \sqrt{\left(\frac{m}{2\pi k_B T}\right)^3} v^2 e^{-\frac{m v^2}{2 k_B T}}. \tag{9.12}$$

It is also useful to introduce the distribution p_c for each of the Cartesian components of the particle velocity. This can be achieved observing that $v^2 = v_x^2 + v_y^2 + v_z^2$ and

$$\begin{aligned} p(v) &= \sqrt{\left(\frac{m}{2\pi k_B T}\right)^3} e^{-\frac{m v^2}{2 k_B T}} \\ &= \sqrt{\left(\frac{m}{2\pi k_B T}\right)^3} e^{-\frac{m v_x^2}{2 k_B T}} e^{-\frac{m v_y^2}{2 k_B T}} e^{-\frac{m v_z^2}{2 k_B T}} = p_c(v_x) p_c(v_y) p_c(v_z), \end{aligned} \tag{9.13}$$

where

$$p_c(u) = \sqrt{\frac{m}{2\pi k_B T}} e^{-\frac{m u^2}{2 k_B T}}.$$

Combining equations (9.8) and (9.9) we have

$$\int_0^{\infty} f(v) dv = \int_{R^3} p(\vec{v}) d^3 v,$$

with $d^3 v = dv_x dv_y dv_z$. Then, recalling that $p(\vec{v}) \equiv p(v)$ and imposing the normalization condition we also have

$$\int_{R^3} p(v) d^3 v = 1.$$

Taking into account equation (9.13), we finally have

$$\int_{-\infty}^{+\infty} p_c(v_x)\, dv_x \int_{-\infty}^{+\infty} p_c(v_y)\, dv_y \int_{-\infty}^{+\infty} p_c(v_z)\, dv_z = \left(\int_{-\infty}^{+\infty} p_c(v_x) dv_x\right)^3 = 1$$

so that the distributions p_c are also correctly normalized

$$\int_{-\infty}^{+\infty} p_c(u) du = 1 \qquad u = v_x,\, v_y,\, v_z.$$

From equation (9.12) we can also derive the *Maxwell–Boltzmann energy distribution* $F_{MB}(\mathcal{E})$. Indeed, if we set $f(v)dv = F_{MB}(\mathcal{E})d\mathcal{E}$ and observe again that $\mathcal{E} = \frac{1}{2}mv^2$, such that

$$dv = \frac{d\mathcal{E}}{\sqrt{2m\mathcal{E}}},$$

then we obtain the expression we are looking for:

$$F_{MB}(\mathcal{E}) = \frac{2}{\sqrt{\pi(k_B T)^3}} \sqrt{\mathcal{E}}\, e^{-\frac{\mathcal{E}}{k_B T}}. \tag{9.14}$$

Now, equipped with these tools, we are able to compute the mean value of isotropic velocity-dependent quantities $G(v)$ or, in the absence of isotropy, of functions $H(v_x, v_y, v_z)$ of the velocity components:

$$\bar{G} = \int_0^{\infty} G(v) f(v) dv$$
$$\bar{H} = \int_{R^3} H(v_x, v_y, v_z) p_c(v_x) p_c(v_y) p_c(v_z) d^3v.$$

Moreover, it is interesting to see how the shape of the Maxwell–Boltzmann distribution changes with temperature. We find that the asymmetric bell-shaped curve representing the distribution tends to flatten as the temperature increases (see figure 9.4), while at low temperatures the distribution is more peaked around the

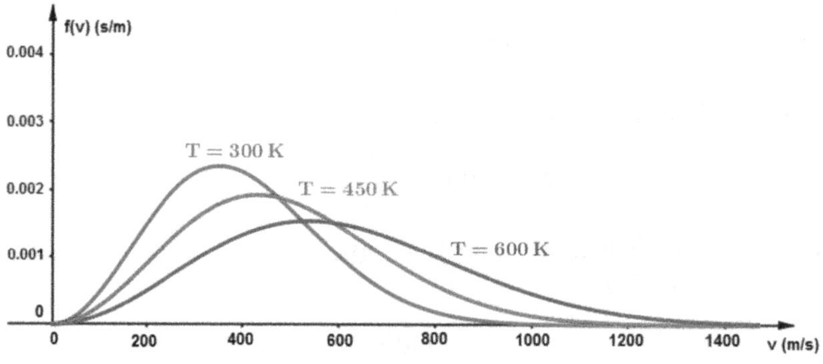

Figure 9.4. Maxwell–Boltzmann distribution for different temperatures.

most probable velocity $v_{\rm mp}$. It is also worth noting that $v_{\rm mp}$, which corresponds to the value of v at which the distribution takes its maximum value, is

$$v_{\rm mp} = \sqrt{\frac{2k_{\rm B}T}{m}} = \sqrt{\frac{2}{3}}\sqrt{\frac{3k_{\rm B}T}{m}} = \sqrt{\frac{2}{3}}\,v_{\rm rms},$$

while the average velocity $\bar{v}$ is

$$\bar{v} = \int_0^\infty v f(v)\mathrm{d}v = \sqrt{\frac{8k_{\rm B}T}{\pi m}} = \sqrt{\frac{8}{3\pi}}\sqrt{\frac{3k_{\rm B}T}{m}} = \sqrt{\frac{8}{3\pi}}\,v_{\rm rms}.$$

Hence we have $v_{\rm mp} < \bar{v} < v_{\rm rms}$, as illustrated in figure 9.5 for a sample of the noble gas Argon-40 at $T = 230$ K.

9.2.4 Specific heats of gases and solids

In this section we present the classical theory of the specific heat for gases and solids as an application and a testing ground for the theoretical arguments illustrated so far. Let us start from ideal gases.

The specific heat in general is the amount of heat per unit mass required to raise the temperature by one degree (kelvin or celsius). For gases, which are highly compressible, it is crucial to also specify the kind of quasi-static transformation carried out by the gaseous system while absorbing heat from a source. Therefore, we have two definitions of the specific heat:

- The *molar specific heat at constant volume*

$$C_V = \frac{1}{n}\frac{\mathrm{d}Q}{\mathrm{d}T}\bigg|_{V=\rm const} = \frac{1}{n}\frac{\mathrm{d}U}{\mathrm{d}T}. \tag{9.15}$$

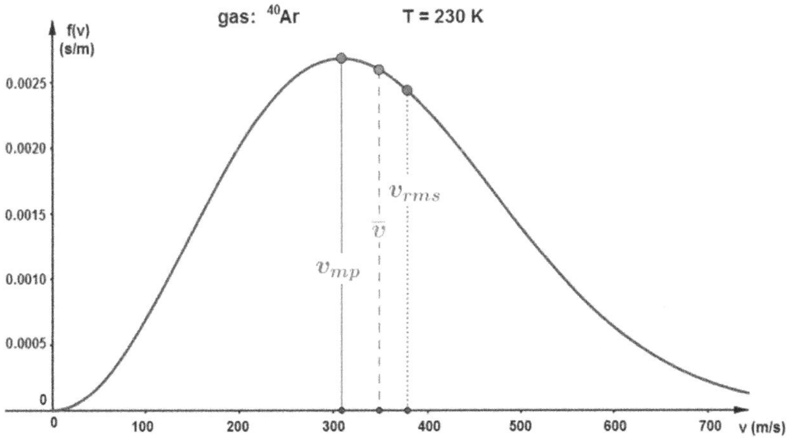

Figure 9.5. Most probable velocity $v_{\rm mp}$, average velocity $\bar{v}$ and average quadratic velocity $v_{\rm rms}$ for a sample of Argon-40 at $T = 230$ K.

- The *molar specific heat at constant pressure*

$$C_P = \frac{1}{n}\frac{\mathrm{d}Q}{\mathrm{d}T}\bigg|_{P=\text{const}} = \frac{1}{n}\frac{\mathrm{d}H}{\mathrm{d}T}\bigg|_{P=\text{const}}.$$

Here n is the number of moles of the gas, T is the absolute temperature, U is the internal energy of the gas and $H = PV + U$ is its enthalpy function.

In general, both C_V and C_P are functions of the temperature and other state variables, but for ideal gases they are independent of state parameters and, furthermore, obey the well-known *Mayer's relation*:

$$C_P = C_V + R.$$

We have seen that for a monoatomic ideal gas the total internal energy is $U = \frac{3}{2}nRT$, and this provides us with two very neat predictions: $C_V = \frac{1}{n}\frac{\mathrm{d}U}{\mathrm{d}T} = \frac{3}{2}R$ and $C_P = C_V + R = \frac{5}{2}R$. For diatomic gases we can rely on the equipartition theorem to obtain for strictly rigid molecules an internal energy $U = \frac{5}{2}nRT$, hence $C_V = \frac{5}{2}R$ and $C_P = \frac{7}{2}R$. If vibrational degrees of freedom were also admitted (the semirigid molecule model), we would be led to $C_V = 3R$ and $C_P = 4R$. Finally, going to ideal gases made up of polyatomic molecules with f degrees of freedom, the total internal energy U is given by equation (9.7), so that the predictions for the specific heat at constant volume and constant pressure are, respectively:

$$C_V = \frac{f}{2}RT$$
$$C_P = \frac{f+2}{2}RT.$$

Are such predictions in agreement with the experimental data?

For noble gases the agreement is excellent over a wide range of temperatures (the table in figure 9.6(a) reports a representative selection of measurements), but for diatomic and polyatomic molecules the classical predictions are not always matched, even when the ideal gas law applies fairly well to the equilibrium state of the system. This disagreement is evident when one examines the behaviour of the specific heat at constant volume or pressure as functions of the temperature and realizes that C_V and C_P are definitely not constant, but have plots resembling a staircase in which at most one of the steps agrees with the classical prediction. So, how could we interpret, for instance, the puzzling behaviour of the specific heat of molecular hydrogen shown in figure 9.6(b)? It seems that high temperatures are able to trigger some of the degrees of freedom that at lower temperatures appear to be inactive. Therefore, experiments seem to indicate that not all the degrees of freedom can be treated on the same ground as the equipartition theorem states, but some of them manifest themselves before others. In the case of hydrogen, only when the temperature is greater than 200 K and lower than 600 K are the degrees of freedom that appear to be effectively active the expected five (three translational and two rotational), thus giving $C_V = 2.5\ R$. On the other hand, for temperatures below 100 K hydrogen appears

		C_v/R	C_p/R	$\gamma = C_p/C_v$
Monoatomic	Ideal gas	1.50000	2.50000	1.666666
	He	1.50	2.50	1.67
	Ne	1.50	2.50	1.67
	Ar	1.50	2.50	1.67
	Xe	1.52	2.51	1.65
Diatomic	Ideal gas	2.50000	3.50000	1.400000
	H_2	2.44	3.42	1.40
	N_2	2.50	3.50	1.40
	O_2	2.54	3.54	1.39
	Cl_2	3.1	4.1	1.32
	CO	2.51	3.51	1.40
	HCl	2.6	3.6	1.38
	NO	2.51	3.51	1.40
Triatomic	Ideal gas	3.50000	4.50000	1.285714
	CO_2	3.48	4.48	1.29
	SO_2	3.97	4.97	1.25
	H_2O	3.05	4.05	1.33
	N_2O	3.7	4.7	1.27

R = 8.3145 J/(K × mol) ; T = 298 K ; P = 100 kPa

(a)

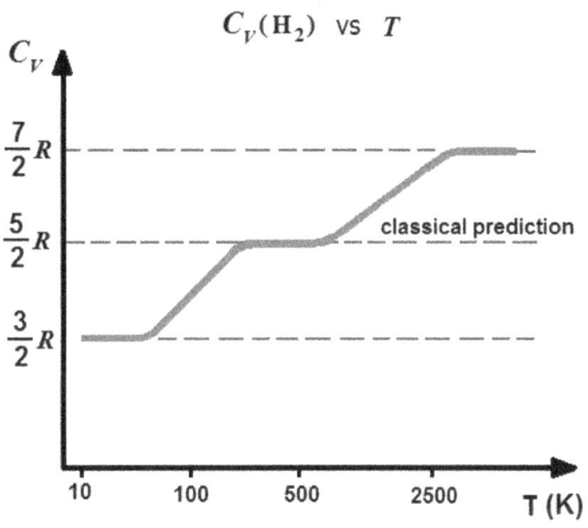

(b)

Figure 9.6. (a) A table showing the molar specific heat and the ratio $\gamma = C_P/C_V$ for different gases at $T = 298$ K and $P = 100$ kPa. (b) Molar specific heat at a constant volume of molecular hydrogen; only in the range of intermediate temperatures does the classical prediction hold.

to behave as a monoatomic gas, that is, as if the rotational degrees of freedom were 'frozen out'. Clearly, this behaviour cannot be explained within the theoretical framework presented above. We also see that at temperatures higher than 3000 K it seems that two additional degrees of freedom become activated. We could imagine that in this range of temperatures H_2 molecules stop being perfectly rigid and also manifest molecular vibrations, thus varying their relative distance. But again, our theory based on classical mechanics cannot provide an explanation of this behaviour.

Now, let us analyse what happens for solids. Unlike in gases, atoms in solids are not free to move in space, but are bound to the sites of the crystalline lattice, as shown for example in figure 9.7 (here we leave out the case of amorphous solids). There they continuously swing around their equilibrium positions, with amplitudes becoming greater the higher the temperature of the system is. This behaviour can be seen as a consequence of the 'neighbouring' action of the periodic potential associated with the electrostatic interactions with the surrounding atoms. However, for our purposes it is enough to study this phenomenon by adopting an effective model not requiring exact knowledge of the real potential function $V(x, y, z)$. Indeed, we know that the equilibrium position of each lattice site coincides with a local minimum of the potential energy associated with its interaction with the other atoms of the lattice (in particular the closest ones). This implies that for small oscillations one can assume that each atom experiences a quadratic potential. Referring for simplicity to the one-dimensional case, this becomes evident if one considers the Taylor expansion of the potential energy $V(x)$ around the stable equilibrium position $x = x_0$, arrested to the second order:

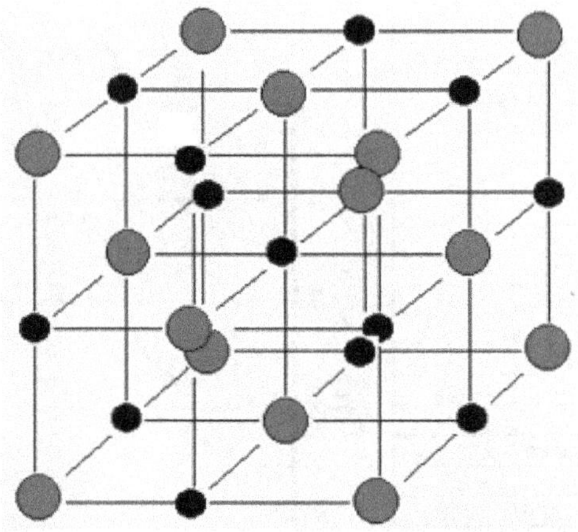

Figure 9.7. Schematic representation of a crystalline lattice composed of two types of atoms.

$$V(x) = V(x_0) + \frac{dV}{dx}\Big|_{x=x_0}(x - x_0) + \frac{1}{2}\frac{d^2V}{dx^2}\Big|_{x=x_0}(x - x_0)^2 + O(|x - x_0|^3)$$
$$\approx V(x_0) + \frac{1}{2}\frac{d^2V}{dx^2}\Big|_{x=x_0}(x - x_0)^2 \equiv V_0 + \frac{1}{2}k(x - x_0)^2.$$

Here $V_0 \equiv V(x_0)$, $k \equiv \frac{d^2V}{dx^2}\Big|_{x=x_0}$ and $\frac{dV}{dx}\Big|_{x=x_0} = 0$ since x_0 is a (regular) local minimum. Rescaling energies and coordinates in such a way as to have $V_0 = 0$ and $x_0 = 0$, respectively, one ends up with the standard harmonic oscillator potential:

$$V(x) = \frac{1}{2}kx^2.$$

Following the same route for a potential function $V(x, y, z)$ defined in R^3, it can be shown that under suitable conditions its behaviour in proximity to an equilibrium position corresponding to $\vec{r}_0 = (0, 0, 0)$ is well described by the expression

$$V(x, y, z) \approx \frac{1}{2}k_1x^2 + \frac{1}{2}k_2y^2 + \frac{1}{2}k_3z^2,$$

where $k_1 = \frac{\partial^2V}{\partial x^2}\Big|_0$, $k_2 = \frac{\partial^2V}{\partial y^2}\Big|_0$ and $k_3 = \frac{\partial^2V}{\partial z^2}\Big|_0$.

The total energy of an atom in the crystal can thus be written as a sum of six quadratic terms:

$$E(\vec{r}, \vec{v}) = \frac{1}{2}mv_x^2 + \frac{1}{2}mv_y^2 + \frac{1}{2}mv_z^2 + \frac{1}{2}k_1x^2 + \frac{1}{2}k_2y^2 + \frac{1}{2}k_3z^2. \tag{9.16}$$

We can then write the total internal energy U of a lattice made of N atoms invoking the equipartition theorem in its more general form, i.e. extended to atoms or molecules that can no longer be regarded as free: for any system at thermodynamic

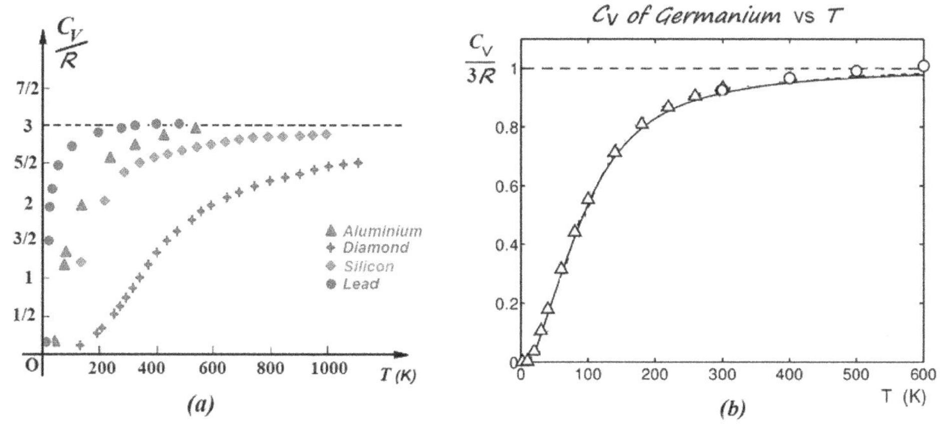

Figure 9.8. (a) Experimental values of the molar specific heat for diamond, aluminium, silicon and lead as functions of temperature; the dashed line indicates the value predicted by the Dulong–Petit law. (b) Temperature dependence of the molar specific heat of germanium.

equilibrium, a contribution $\bar{\mathcal{E}} = \frac{1}{2}k_BT$ to the total energy is associated with each quadratic term in the classical expression of the energy. Hence, when the energy of each atom can be expressed in the form of equation (9.16), one has

$$U = 6N\bar{\mathcal{E}} = 6N\frac{1}{2}k_BT = 3nN_Ak_BT = 3nRT,$$

where $N_A = R/k_B$ is the Avogadro number. Then, according to the definition in equation (9.15), we obtain

$$C_V = 3R = 24.94 \text{ J mol}^{-1}\text{ K}^{-1}.$$

The fact that the molar specific heat[6] for any crystalline solid (e.g. a metal) turns out to be constant and equal to $3R$ is the so-called *Dulong–Petit law*, from the name of the two French physicists who formulated it in 1819. This law is in fairly good agreement with the experiments for almost all solids at sufficiently high temperatures. However, there is a significant departure from this behaviour at low temperatures, where it is evident that C_V decreases monotonically as T goes down, tending to zero as $T \to 0$ with a T^3 law (see figure 9.8). Once again, the comparison with the experimental data makes us understand that the model discussed above works properly only within a certain range of temperatures, below which it proves to be totally inadequate. Since the results that we have deduced so far are based on classical mechanics and on statistical methods consistent with it, we must conclude that we need a new theoretical framework to overcome these difficulties.

[6] Note that here $C_V \approx C_P$ since solids are almost incompressible.

9.3 Transport properties of electrons in metals

In this section we want to deal with the electric conduction in solids, together with some related transport properties, within the so-called *Drude model.* It is a classical model that was proposed in 1900 by Paul Drude to explain the electronic properties of matter (in particular of metals) from a microscopic point of view, using the same statistical methods that we saw at work in the kinetic theory of gases.

According to Drude, electrons can be seen as a gas of non-interacting particles composed of about 10^{23} elementary units per cm^3, but, unlike what happens for the atoms of the ideal gases investigated in kinetic theory, electrons can respond to an external electromagnetic field because of their charge. It is also assumed that they can collide with the positively charged ions of the crystalline lattice (see figure 9.9), in addition to experiencing collisions with the other electrons. The hypotheses underlying the Drude model are the following:

- Between two consecutive collisions, electrostatic electron–electron and electron–ion interactions are assumed to be negligible and electrons move in agreement with the laws of classical mechanics. In particular, in the absence of external fields they move in a straight line.
- The only collisions that are able to change the electron velocities are those with the ions of the lattice.
- The probability that an electron experiences a collision with a ion in a time interval $\mathrm{d}t$ is equal to $\frac{\mathrm{d}t}{\tau}$, where the *relaxation time* τ is the average time between two consecutive collisions.
- The electron velocity after any collision is randomly directed. Its magnitude depends on the temperature at the place where the collision occurred, but does not depend on its value prior to that collision (i.e. collisions are 'memoryless' processes).

The effect of the electronic collisions with the lattice ions can be reinterpreted as an additional drag force acting on a given electron moving throughout the solid.

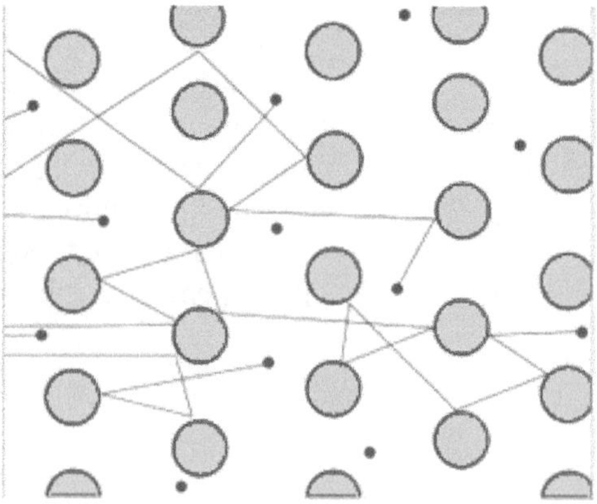

Figure 9.9. Pictorial representation of the motion of the electrons in the Drude model.

Indeed, the probability that an electron experiences no collision within a time $\mathrm{d}t$ is equal to $\left(1 - \frac{\mathrm{d}t}{\tau}\right)$, so that the linear momentum of the electron $\vec{p} = m\vec{v}$, $\vec{v}$ being the average electronic velocity, obeys the relation

$$\vec{p}(t + \mathrm{d}t) = \left(1 - \frac{\mathrm{d}t}{\tau}\right)(\vec{p}(t) + \mathrm{d}\vec{p}(t)) = \left(1 - \frac{\mathrm{d}t}{\tau}\right)(\vec{p}(t) + \vec{F}(t)\mathrm{d}t),$$

where $\mathrm{d}\vec{p}(t) = \vec{F}(t)\mathrm{d}t$ is the variation of the linear momentum due to the external force $\vec{F}(t)$ acting on the electron. Therefore, after some elementary algebra, we obtain the equation

$$\frac{\mathrm{d}\vec{p}(t)}{\mathrm{d}t} = \vec{F}(t) - \frac{\vec{p}(t)}{\tau}. \tag{9.17}$$

A simple case that we can analyse by means of this equation concerns the motion of an electron in the presence of a constant electric field $\vec{E}$. In this case we have $\vec{F}(t) = q\vec{E}$, so that, using $\vec{v} = \vec{p}/m$, equation (9.17) becomes

$$\frac{\mathrm{d}\vec{v}(t)}{\mathrm{d}t} = \frac{q\vec{E}}{m} - \frac{\vec{v}(t)}{\tau}.$$

This first-order linear differential equation can easily be integrated by elementary methods, giving the solution

$$\vec{v}(t) = \frac{q\tau}{m}\vec{E}\left(1 - \mathrm{e}^{-\frac{t}{\tau}}\right) + \vec{v}_0\,\mathrm{e}^{-\frac{t}{\tau}},$$

where $\vec{v}_0$ is an integration constant. For $t \gg \tau$ we obtain the steady state value $\vec{v}_\mathrm{d}$ of $\vec{v}(t)$, usually called the *drift velocity*:

$$\vec{v}_\mathrm{d} = \frac{q\tau}{m}\vec{E}.$$

The current density vector $\vec{J}$ can then be expressed as

$$\vec{J} = nq\vec{v}_\mathrm{d} = \frac{nq^2\tau}{m}\vec{E} \equiv \sigma_0\vec{E}, \tag{9.18}$$

where q is the charge of the carriers ($q = -e$ for electrons), $n = N/V$ is their number density and $\sigma_0 = nq^2\tau/m$ is the (DC) *electrical conductivity* of the conductor. Equation (9.18) can also be rewritten as

$$\vec{E} = \frac{1}{\sigma_0}\vec{J} = \rho_0\vec{J}, \tag{9.19}$$

where

$$\rho_0 \equiv \frac{1}{\sigma_0} = \frac{m}{nq^2\tau} \tag{9.20}$$

is the *electrical resistivity* of the material. Equation (9.19) is nothing more than a microscopic (or local) version of the first Ohm's law, according to which the potential difference ΔV between two points of a conductor is directly proportional to the current I flowing through the conductor between the same two points; the constant of proportionality is the resistance R of the conductor (see figure 9.10):

$$\tau = \frac{m}{\rho_0 n q^2}.$$

For typical values of the resistivity ($\rho_0 \sim 10^{-8}\,\Omega\,\mathrm{m}$ in many metals), we obtain values of τ of the order of 10^{-14} s. This implies that the average distance travelled by electrons between two successive collisions, the so-called *mean free path* $\lambda = \langle v \rangle \tau$, is of the order of a few angströms if one uses for $\langle v \rangle$ the values, of the order of 10^7 cm s^{-1}, that can be deduced from the relation

$$\frac{1}{2} m \langle v \rangle^2 = \frac{3}{2} k_\mathrm{B} T.$$

Considering that such a value of λ is comparable to the typical interatomic distances, the result appears to be consistent with the Drude hypothesis that the dominant collisions are those that electrons undergo from the lattice ions. However, at low temperatures this classical estimate turns out to be incorrect, given that λ in pure samples is experimentally found to be up to thousands of times the interatomic spacing between ions.

Classical Hall effect. The second relevant application of equation (9.17) concerns the Hall effect, which develops when a uniform electric field $\vec{E}$ induces a current through a conducting strip in the presence of a uniform magnetic field $\vec{B}$ directed perpendicularly to $\vec{E}$ and thus to the current density vector $\vec{J}$. Referring to the experimental set-up outlined in figure 9.11, we have that the Lorentz force due to the presence of the magnetic field deflects electrons in the negative y-direction. As they

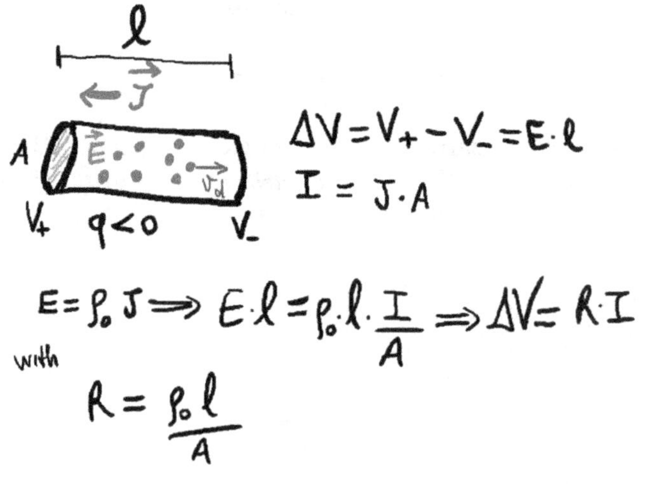

Figure 9.10. Derivation of the first Ohm's law from its local form.

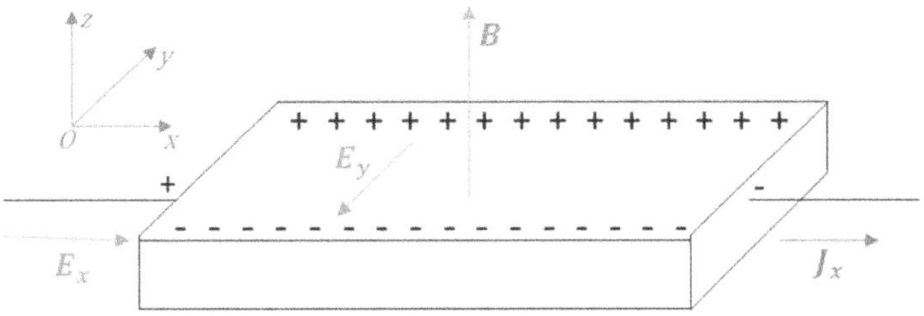

Figure 9.11. Hall effect set-up.

accumulate on the corresponding face, an electric field builds up in the y-direction and hinders their further accumulation. Once the equilibrium is reached, this transverse field E_y, usually called the *Hall field*, balances the Lorentz force and thus current only flows in the x-direction, i.e. $\vec{J} = J_x\hat{x}$. An electron of charge q and momentum $\vec{p}$ will thus experience an electromagnetic force

$$\vec{F}_{\text{e.m.}} = q\vec{E} + \frac{q}{m}\vec{p} \times \vec{B}$$

with the configuration of the fields $\vec{E}$ and $\vec{B}$ given by

$$\vec{E} = E_x\hat{x} + E_y\hat{y} \qquad \vec{B} = B_z\hat{z}.$$

Equation (9.17) then becomes

$$\frac{\mathrm{d}\vec{p}(t)}{\mathrm{d}t} = q\left(\vec{E} + \frac{1}{m}\vec{p}(t) \times \vec{B}\right) - \frac{\vec{p}(t)}{\tau}.$$

Under stationary conditions (current independent of time) $\frac{\mathrm{d}\vec{p}(t)}{\mathrm{d}t} = 0$, so that one has

$$\vec{p}(t) = q\tau\left(\vec{E} + \frac{1}{m}\vec{p}(t) \times \vec{B}\right).$$

Writing this equation in terms of the current density $\vec{J} = nq\vec{p}/m$ and solving with respect to the electric field $\vec{E}$, one finally obtains

$$\vec{E} = \rho_0\vec{J} - \frac{1}{nq}\vec{J} \times \vec{B}, \tag{9.21}$$

where we have used equation (9.20). We note that the total electric field $\vec{E}$ and the current density $\vec{J}$ are still in linear relation, but they are no longer directly proportional, i.e. parallel, because of the presence of the magnetic field $\vec{B}$. This is also evident when we project equation (9.21) in the x- and y-directions:

$$E_x = \rho_0 J_x - \frac{B_z}{nq}J_y$$

$$E_y = \frac{B_z}{nq}J_x + \rho_0 J_y.$$

We notice that these equations can also be written in matrix form $\vec{E} = \underline{\rho}\vec{J}$, where the resistivity matrix $\underline{\rho}$ is defined as

$$\underline{\rho} = \begin{pmatrix} \frac{m}{nq^2\tau} & -\frac{B_z}{nq} \\ \frac{B_z}{nq} & \frac{m}{nq^2\tau} \end{pmatrix} \equiv \begin{pmatrix} \rho_{xx} & \rho_{xy} \\ \rho_{yx} & \rho_{yy} \end{pmatrix}$$

(note that $\rho_{xy} = -\rho_{yx}$). Once the equilibrium is reached, the transverse current J_y vanishes and thus we have

$$E_x = \rho_0 J_x$$
$$E_y \equiv E_{\mathrm{H}} = \frac{B_z}{nq} J_x.$$

Here we have denoted by E_{H} the transverse electric field E_y, which, as previously stated, is usually called the Hall field. One also defines the ratio

$$R_{\mathrm{H}} = \frac{E_y}{J_x B_z} = \frac{1}{nq}$$

known as the *Hall coefficient* R_{H}, whose sign depends on the sign of the charge carriers ($q = -e < 0$ for most metals).

From experiments on the conductivity of different materials in the presence of a magnetic field, it was seen that the measurements of the Hall coefficient R_{H} in general did not match the theoretical predictions, exhibiting a clear dependence on the temperature, the magnetic field and the purity of the sample. As one can see from the table in figure 9.12, the experimental agreement with the Drude model appears good enough only for alkaline metals (entries with a light blue background), worse for 'noble' metals (entries with a light green background), while it is completely absent, with a discordance even on the sign of the charge carriers, in the other cases (orange background).

Finally, the diagonal coefficient ρ_{xx} is called *magnetoresistivity* and, in this model, is independent of the magnetic field B_z. However, the model prediction again does not fit very well with the experimental data. Indeed, in metallic compounds (see for example figure 9.13) the magnitude of the magnetoresistance depends quite strongly on the applied magnetic field, signalling once again some inadequacy of the Drude model.

9.3.1 Thermal conduction in the Drude model

It is well known that when two bodies at different temperatures are placed in thermal contact there is a heat flow from the warmer body to the colder one (see figure 9.14). The rate of this energy transfer depends on the temperature gradient between the two points at different temperatures and on the geometrical and intrinsic properties of the conductor making the thermal contact. In fact, assuming that such a

Hall coefficients of selected elements at $T \sim 10\ K$ and $B \sim 1\ T$		
Metal	**Valence**	$-1/R_H ne$
Li	1	0.8
Na	1	1.2
K	1	1.1
Rb	1	1.0
Cs	1	0.9
Cu	1	1.5
Ag	1	1.3
Au	1	1.5
Be	2	-0.2
Mg	2	-0.4
In	3	-0.3
Al	3	-0.3

Figure 9.12. Ratio between the Drude expression of the Hall coefficient R_H and its measured value for different elements.

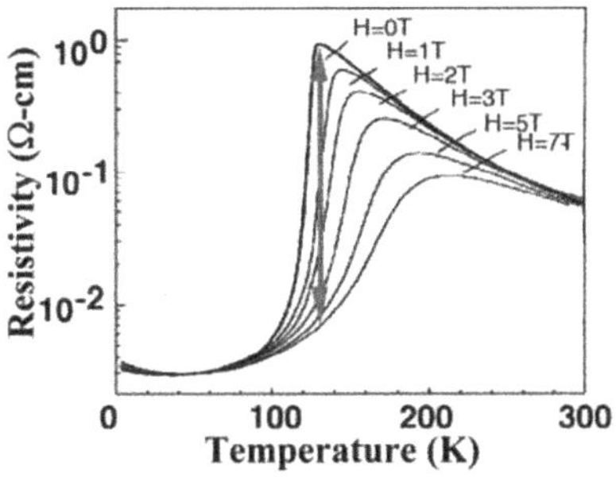

Figure 9.13. Temperature dependence of the resistivity of a sample of $La_{2-2x}Sr_{1+2x}Mn_2O_7$ for different values of the magnetic field H [3].

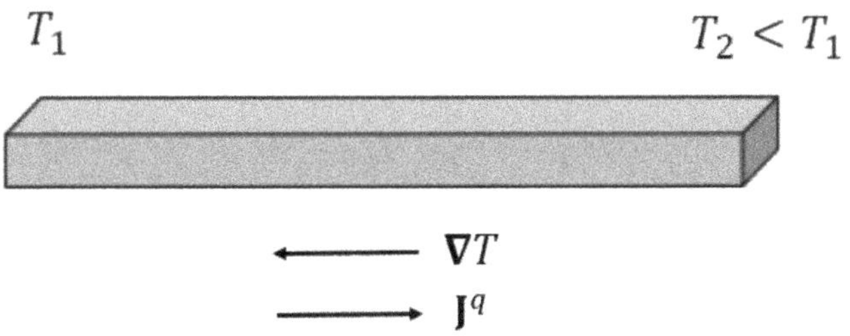

Figure 9.14. Heat flow through a bar whose extremes are kept at different temperatures.

conductor is a bar whose extremes are at temperatures T_1 and T_2 ($T_1 > T_2$), we have that, at least for moderate temperature gradients, *Fourier's law* holds:

$$\vec{J}^q = -\kappa\vec{\nabla}T. \tag{9.22}$$

Here we have defined the following quantities (the corresponding SI units are indicated in brackets):

- $\vec{J}^q$ is the *thermal current density*, i.e. a vector parallel to the direction of the heat flow, whose magnitude gives the amount of heat Q flowing every second across a unit area perpendicular to the flow (W m^{-2}).
- $\vec{\nabla}T = \frac{\partial T}{\partial x}\hat{x} + \frac{\partial T}{\partial y}\hat{y} + \frac{\partial T}{\partial z}\hat{z}$ is the gradient of the temperature scalar field $T = T(x, y, z, t)$, in general depending on space and time (K m^{-1}).
- κ is the *thermal conductivity*, a quantity linking the cause of the heat flow ($\vec{\nabla}T$) and the effect produced ($\vec{J}^q$); for isotropic materials it is a constant within a not too wide range of temperatures (W m^{-1} K^{-1}).

The minus sign in equation (9.22) means that the thermal current flows opposite to the direction of the temperature gradient (with $\kappa > 0$). Within the Drude model, it is possible to show (see for instance [4, 5] and appendix A for a detailed derivation) that the thermal conductivity κ can be expressed in terms of microscopic quantities such as the mean square velocity $\langle v^2\rangle$ of the charge carriers (typically electrons), the relaxation time τ and the electronic specific heat at constant volume c_V:

$$\kappa = \frac{1}{3}\langle v^2\rangle\tau c_V. \tag{9.23}$$

From this equation we can derive a relevant relation between thermal and electrical conductivity, discovered empirically by Gustav Wiedemann and Rudolph Franz in 1853. They found that for a wide variety of materials the ratio between κ and σ_0 at a fixed temperature is a constant independent of the particular material. A few years later (1872) Ludvig Lorenz found that this ratio is directly proportional to temperature,

$$\frac{\kappa}{\sigma_0} = LT,$$

where L is known as the *Lorenz constant*. Using the theoretical predictions of the Drude model and recalling from kinetic theory that for non-interacting electrons $\frac{1}{2}m\langle v^2\rangle = \frac{3}{2}k_B T = \frac{c_V}{n}T$, we have ($q = -e$):

$$\frac{\kappa}{\sigma_0} = \frac{\frac{1}{3}\langle v^2\rangle\tau c_V}{\frac{ne^2\tau}{m}} = \frac{\frac{1}{3}m\langle v^2\rangle n\frac{3}{2}k_B}{ne^2} = \frac{3}{2}\left(\frac{k_B}{e}\right)^2 T.$$

The Drude model thus provides the following estimate of the Lorenz constant:

$$L = \frac{3}{2}\left(\frac{k_B}{e}\right)^2 = 1.11 \times 10^{-8}\ \text{W}\,\Omega\,\text{K}^{-2}.$$

Experimental thermal conductivities and Lorenz constants for some metals				
Element	$T = 273\ K$		$T = 373\ K$	
	$\kappa\ \left(\frac{W}{cm \cdot K}\right)$	$L = \kappa/(\sigma_0 T)\ \left(10^{-8} \frac{W \cdot \Omega}{K^2}\right)$	$\kappa\ \left(\frac{W}{cm \cdot K}\right)$	$L = \kappa/(\sigma_0 T)\ \left(10^{-8} \frac{W \cdot \Omega}{K^2}\right)$
Li	0.71	2.22	0.73	2.43
Cu	3.85	2.20	3.82	2.29
Ag	4.18	2.31	4.17	2.38
Au	3.1	2.32	3.1	2.36
Be	2.3	2.36	1.7	2.42
Mg	1.5	2.14	1.5	2.25
Nb	0.52	2.90	0.54	2.78
Fe	0.80	2.61	0.73	2.88
Zn	1.13	2.28	1.1	2.30
Al	2.38	2.14	2.30	2.19
Ti	0.5	2.75	0.45	2.75
In	0.88	2.58	0.80	2.60
Pb	0.38	2.64	0.35	2.53
Bi	0.09	3.53	0.08	3.35
Sb	0.18	2.57	0.17	2.69

Figure 9.15. Experimental values of the thermal conductivity and the Lorenz number for several elements.

As one can see from the table in figure 9.15, this value of L is almost exactly half of the experimental value, an agreement that in the original calculations performed by Drude was practically perfect for the omission of a factor 2. However, this result should be considered fortuitous and connected to the fact that the relaxation time τ does not enter in the constant L. In fact, it was the consequence of two errors of about one hundred concerning c_V and $\langle v^2 \rangle$ that compensate each other—at room temperature experiments show that c_V is about one hundred times smaller than the classical prediction, while $\langle v^2 \rangle$ is about one hundred times larger.

Before closing this discussion we want to analyse another effect for which the Drude model provides a theoretical estimation: the *Seebeck effect*. It was discovered by Thomas Johann Seebeck in 1821 and consists in the formation of an electric field, the thermoelectric field $\vec{E}_Q$, in the presence of a temperature gradient $\vec{\nabla} T$ inside a conductor [4]. More precisely, $\vec{E}_Q$ and $\vec{\nabla} T$ are directly proportional, and the proportionality constant is known as the *thermopower* Q:

$$\vec{E}_Q = Q\, \vec{\nabla} T.$$

To derive this equation we use again, for the sake of simplicity, a one-dimensional model. The mean electronic velocity v_Q due to the temperature gradient $\mathrm{d}T/\mathrm{d}x$ can be estimated as follows (here we must abandon the simplification that the average velocity v is constant over distances of the order of the mean free path):

$$v_Q = \frac{1}{2}(v(x - v\tau) - v(x + v\tau)) = -v\tau\frac{\mathrm{d}v}{\mathrm{d}x} = -v\tau\frac{\mathrm{d}v}{\mathrm{d}T}\frac{\mathrm{d}T}{\mathrm{d}x} = -\frac{1}{2}\tau\frac{\mathrm{d}v^2}{\mathrm{d}T}\frac{\mathrm{d}T}{\mathrm{d}x}.$$

Taking into account that $\langle v_x^2 \rangle = \langle v_y^2 \rangle = \langle v_z^2 \rangle = \frac{1}{3}\langle \vec{v}^2 \rangle$, we can generalize this relation to the three-dimensional case:

$$\vec{v}_Q = -\frac{\tau}{6}\frac{\mathrm{d}v^2}{\mathrm{d}T}\vec{\nabla}T.$$

The temperature gradient induces a charge imbalance that generates an electrostatic field $\vec{E}$, so that the carriers acquire a drift velocity

$$\vec{v}_\mathrm{d} = \frac{q\tau}{m}\vec{E}$$

and the stationary condition is achieved when

$$\vec{v}_Q + \vec{v}_\mathrm{d} = 0.$$

Using this condition, we can connect $\vec{E}_Q$ and $\vec{\nabla}T$ and find the following expression for Q

$$\vec{E}_Q = \frac{1}{3q}\frac{\mathrm{d}\left(\frac{1}{2}mv^2\right)}{\mathrm{d}T}\vec{\nabla}T = \frac{c_V}{3nq}\vec{\nabla}T,$$

which implies

$$Q = \frac{c_V}{3nq}.$$

Moreover, for $q = -e$ (carriers are electrons) and using the relation $c_V = \frac{3}{2}nk_\mathrm{B}$ coming from the kinetic theory naively, we obtain the following result:

$$Q = -\frac{k_\mathrm{B}}{2e} = -0.43 \times 10^{-4}\ \mathrm{V\ K^{-1}}.$$

This theoretical prediction for Q is bigger by a factor of about 100 than the typical values of thermopower measured at room temperature for ordinary metals. This is a consequence of the fact that, differently to what happened for the Lorenz number, the error on the classical estimate of the specific heat in this case is not compensated. From this result, the substantial inadequacy of the Drude model to deal with the theory of conductivity in solids turns out to be fairly clear.

9.4 A taste of quantum statistics

9.4.1 Classical versus quantum statistics

In the previous sections we have seen that statistical methods allow the state of very complex systems to be described through the evaluation of the average values of some of the most relevant physical quantities. This approach can only be pursued if one introduces suitable PDFs defined on the space of the possible microscopic states (*microstates*) of the system under examination. The purpose of this section is thus to give a brief account of the differences between the statistical behaviours of classical and quantum many-body systems, introducing the related statistical distributions and discussing their main properties.

A distribution function for a given physical quantity Q (e.g. energy) that can assume the values $\{q_i\}_{i\in I}$ tells us, knowing the macroscopic state of equilibrium of the system (*macrostate*), which is the average number of particles taking each of the values q_i that the quantity Q can assume. For a system composed of N identical particles with total energy E at the temperature T, the general procedure to obtain the equilibrium distribution goes as follows. Suppose that the possible energy levels are $\mathcal{E}_1, \mathcal{E}_2, \ldots, \mathcal{E}_r$; let $g_1, g_2, \ldots, g_r$ be the corresponding degeneracies, i.e. g_i is the numbers of single-particle states with energy $\mathcal{E}_i$, and let $n_1, n_2, \ldots, n_r$ be the numbers of particles in each energy level, i.e. n_1 is the number of particles with energy $\mathcal{E}_1$, n_2 is the same quantity referred to the energy $\mathcal{E}_2$, and so on. Denoting by N and E the total number of particles and the total energy of the system, respectively, we have that the integers n_i must satisfy the conditions

$$\sum_{i=1}^{r} \bar{n}_i = N \tag{9.24}$$

$$\sum_{i=1}^{r} \bar{n}_i \mathcal{E}_i = E. \tag{9.25}$$

If $P_1, P_2, \ldots, P_r$ are the numbers of ways in which n_i particles with energy $\mathcal{E}_i$ can be put into the corresponding g_i states, then the total multiplicity (also called *statistical weight*) of the configuration corresponding to the occupation numbers $\underline{n} \equiv (n_1, n_2, \ldots, n_r)$ is

$$W(\underline{n}) = P_1 P_2 \ldots P_r = \prod_{i=1}^{r} P_i.$$

For a very large number of particles, we can assume with a negligible error that the most probable set of values $(\bar{n}_1, \bar{n}_2, \ldots, \bar{n}_r)$ of the occupation numbers coincides with the one that maximizes $W(\underline{n})$ under the constraint equations (9.24) and (9.25). For a classical ideal gas in thermal equilibrium at the temperature T, these values are proportional to the exponential Boltzmann factor, multiplied by the corresponding degeneracy factors,

$$\bar{n}_i = C g_i \mathrm{e}^{-\frac{\mathcal{E}_i}{k_B T}} \qquad i = 1, 2, \ldots, r, \tag{9.26}$$

where the constant of proportionality C is determined by imposing the constraint equation (9.24):

$$\sum_{i=1}^{r} C g_i \mathrm{e}^{-\frac{\mathcal{E}_i}{k_B T}} = N.$$

This leads to $C = N/Z$, where the quantity Z is the so-called *partition function*

$$Z = \sum_{i=1}^{r} g_i \mathrm{e}^{-\frac{\mathcal{E}_i}{k_B T}},$$

so that equation (9.26) becomes

$$\bar{n}_i = \frac{N}{Z} g_i \mathrm{e}^{-\frac{\mathcal{E}_i}{k_B T}}.$$

We can thus say that the probability $\mathcal{P}_i$ that a particle is in a state with energy $\mathcal{E}_i$ is

$$\mathcal{P}_i \equiv \frac{\bar{n}_i}{N} = g_i f_{\mathrm{MB}}(\mathcal{E}_i), \tag{9.27}$$

where g_i counts the number of available states with energy $\mathcal{E}_i$ and the *statistical Boltzmann factor*

$$f_{\mathrm{MB}}(\mathcal{E}_i) = \frac{1}{Z} \mathrm{e}^{-\frac{\mathcal{E}_i}{k_B T}}$$

gives the average occupation number of a single-particle state with the same energy.

A similar expression holds when one considers a continuum energy spectrum, i.e. a very large number of energy levels separated by very small energy gaps. In this case, denoting by *dn* the number of particles with energies falling within the infinitesimal range $[\mathcal{E}, \mathcal{E} + \mathrm{d}\mathcal{E}]$, we can write, in analogy to equation (9.27),

$$F_{\mathrm{MB}}(\mathcal{E})\mathrm{d}\mathcal{E} = \frac{\mathrm{d}n}{N} = g(\mathcal{E}) f_{\mathrm{MB}}(\mathcal{E})\mathrm{d}\mathcal{E},$$

where $F_{\mathrm{MB}}(\mathcal{E})$ is the Maxwell–Boltzmann energy distribution function and $g(\mathcal{E})$ is a function, known as the *density of states*, defined in such a way that $g(\mathcal{E})\mathrm{d}\mathcal{E}$ gives the number of states between the energies $\mathcal{E}$ and $\mathcal{E} + \mathrm{d}\mathcal{E}$. From a comparison with equation (9.14), one can see that

$$g(\mathcal{E}) = \frac{2}{\sqrt{\pi (k_B T)^3}} \sqrt{\mathcal{E}}.$$

Classical statistical mechanics works quite well with gases at relatively low pressures and relatively high temperatures, i.e. when their behaviour approaches the one predicted by the ideal gas theory. However, it soon gets into trouble when dealing with solids and liquids. So, what are the physical assumptions on which classical statistical mechanics relies and that must be abandoned to correctly describe the behaviour of quantum systems[7]?

For liquids and solids the assumption of non-interacting particles is often far from being acceptable—the free space available for each particle is much smaller than in gases (the density is much larger), atoms and molecules are packed closely together and the effects of the quantum statistics, drastically different from the classical

[7] It is worth recalling that when we talk about 'quantum systems' we are actually referring to multi-particle systems that can be properly described only by using the principles of quantum physics.

Maxwell–Boltzmann one, are much more visible. We recall that these effects can ultimately be ascribed to the Heisenberg uncertainty principle which, in turn, makes the notion of trajectory inconsistent. This leads to the classification of the elementary particles in two fundamental categories by virtue of their intrinsic angular momentum s, called *spin*. They are called *bosons* if s is an integer multiple of $\hbar$ and *fermions* if s is a half-integer multiple of $\hbar$ (here $\hbar = h/2\pi$, with h being the Planck constant). In particular, electrons are fermions and, as such, must obey the *Pauli exclusion principle*, according to which two particles cannot share the same quantum state or, to put it equivalently, they cannot have the same set of quantum numbers.

The Pauli principle thus severely limits the allowed energy values and therefore influences notably the distribution of a system of quantum particles over the allowed single-particle states. In this context, it is worth mentioning another crucial difference between classical and quantum behaviour. Classically, there is no restriction on a given energy spectrum, in the sense that the energy of a system of particles varies continuously within a given interval of allowed values. On the other hand, in the quantum case only discrete values of the energy are allowed when the system is confined spatially, this giving rise to discrete energy spectra which never occur in classical systems.

Let us now look at quantum statistics in some detail.

9.4.2 Bose–Einstein statistics

As previously mentioned, the uncertainty principle makes the concept of trajectory meaningless, so that in quantum approaches identical particles must necessarily be considered undistinguishable. For this reason, when we calculate the statistical factors P_i introduced in the previous section to calculate the statistical weight $W(\underline{n}) = \prod_{i=1}^{r} P_i$ for a given set of occupation numbers $\underline{n} \equiv (n_1, n_2,\ldots,n_r)$, we have to count only the different ways in which we can arrange n_i particle in g_i states, without distinguishing, as in the Maxwell–Boltzmann case, which particle goes to occupy each specific (degenerate) state. As shown in the two examples reported in figure 9.16, this leads to a severe reduction of the number of states to be considered distinct. Assuming that the integers $\{n_i\}$ can take any non-negative value, a case leading to the so-called *Bose–Einstein statistics*, it can be shown that the weight factor $W(\underline{n})$ is given by the following expression [5]:

$$W_{\mathrm{BE}}(\underline{n}) = \prod_{i=1}^{r} P_i = \prod_{i=1}^{r} \frac{(n_i + g_i - 1)!}{n_i!(g_i - 1)!}.$$

Recalling elementary combinatorics, we see that the above form of P_i corresponds to the number of combinations with repetition of n_i objects (particles in the ith energy level) taken g_i at a time, where g_i is, as before, the degeneration of the ith level (i.e. the number of distinct states associated with that particular energy value).

Now the problem of determining the equilibrium configuration $\underline{\bar{n}} = (\bar{n}_1, \bar{n}_2, \ldots, \bar{n}_r)$ by maximizing the function $W_{\mathrm{BE}}(\underline{n})$ (or, for computational convenience, $\ln(W_{\mathrm{BE}}(\underline{n}))$) under the constraints expressed by equations (9.24) and (9.25), is of a technical nature and can be addressed in a relatively simple way by using the method of Lagrange

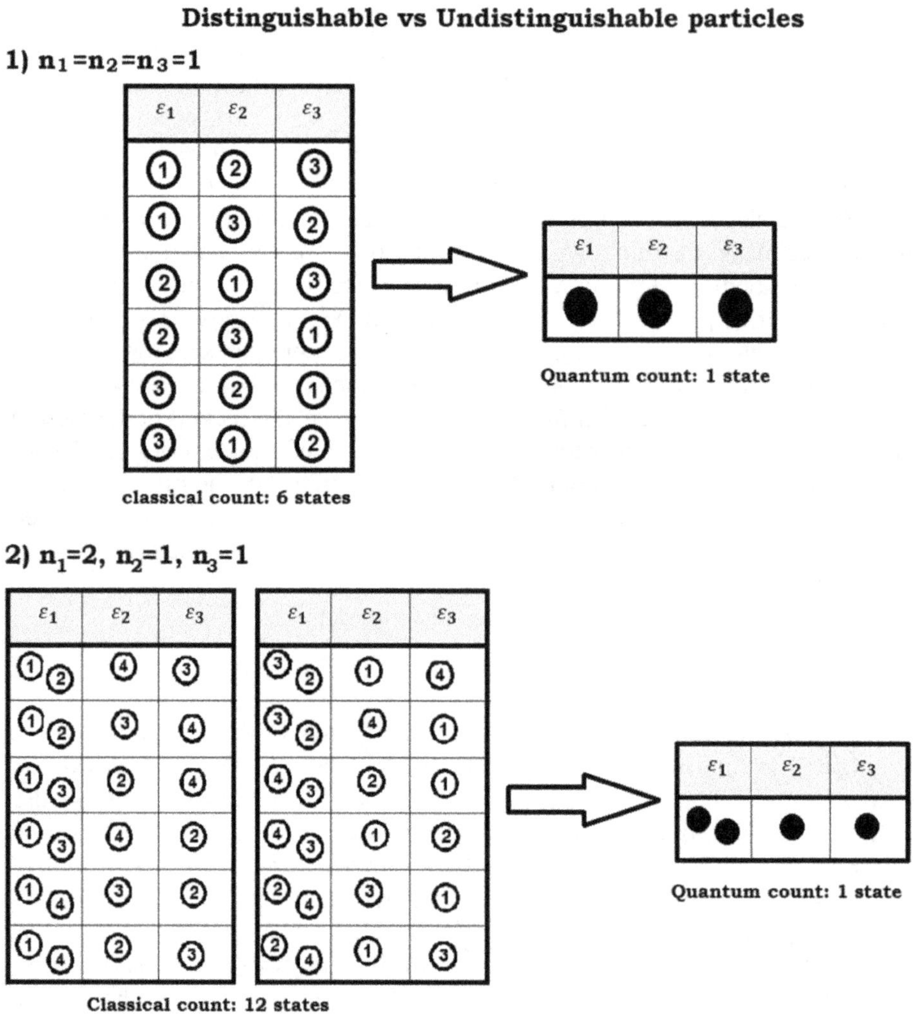

Figure 9.16. An illustration showing two simple cases where the indistinguishably of the particles affects the number of physical states very strongly. In these examples we assume no degeneracy for the three energy levels.

multipliers (see appendix B for the analogous calculation concerning the Fermi–Dirac distribution). The final result is expressed by the Bose–Einstein distribution

$$\bar{n}_i = \frac{g_i}{A_{\mathrm{BE}} e^{\frac{\mathcal{E}_i}{k_\mathrm{B}T}} - 1} \qquad i = 1, 2, \ldots, r \tag{9.28}$$

giving the average number of bosons with energy $\mathcal{E}_i$. The normalization factor A_{BE} is obtained in the form $A_{\mathrm{BE}} = e^{-\frac{\mu}{k_\mathrm{B}T}}$, where the chemical potential μ is a temperature-dependent quantity, already introduced in section 9.2.1, equal to the energy that is absorbed or released by the system when the number of particles varies by one unit. It actually controls the equilibrium of open systems that can exchange matter with the environment, in the sense that no net particle flux between the system and the environment occurs when their chemical potentials coincide. The final expression of the Bose–Einstein distribution is thus

$$f_{\mathrm{BE}}(\mathcal{E}) = \frac{1}{e^{\frac{\mathcal{E}-\mu}{k_{\mathrm{B}}T}} - 1}.$$

In the continuum limit, this quantity is replaced by the number of particles $n_{\mathrm{BE}}(\mathcal{E})$ per unit energy interval centred at the energy $\mathcal{E}$, so that $\mathrm{d}N = n_{\mathrm{BE}}(\mathcal{E})\mathrm{d}\mathcal{E}$ is the number of particles with energies within the range $[\mathcal{E}, \mathcal{E} + \mathrm{d}\mathcal{E}]$, with

$$n_{\mathrm{BE}}(\mathcal{E}) = g(\mathcal{E})f_{\mathrm{BE}}(\mathcal{E}) = g(\mathcal{E})\frac{1}{e^{\frac{\mathcal{E}-\mu}{k_{\mathrm{B}}T}} - 1}.$$

As in the previous section, the degeneracy factor g_i appearing in equation (9.28) has been replaced by the density of states function $g(\mathcal{E})$. This distribution is essentially the one encountered in the statistical treatment of black-body radiation, so we refer the reader to chapter 6 for further details.

9.4.3 Fermi–Dirac statistics

As previously said, the statistics used to deal with particles with half-integer spin, such as for example electrons, is the Fermi–Dirac one. It explicitly takes into account the Pauli exclusion principle, according to which no more than one particle can occupy a given single-particle quantum state. This constraint is implemented assuming that the occupation numbers $\{n_i\}$ can only take the values 0 and 1.

Once again we evaluate the number P_i of ways in which it is possible to arrange n_i particles in the g_i states that correspond to the ith energy level $\mathcal{E}_i$. Since n_i out of g_i states are occupied and $g_i - n_i$ are not, we have that the multiplicity factor P_i is given by

$$P_i = \frac{g_i!}{n_i!(g_i - n_i)!}.$$

The overall weight factor $W(\underline{n}) = \prod_{i=1}^{r} P_i$ corresponding to the set of occupation numbers $\underline{n} = (n_1, n_2, \ldots, n_r)$ is then

$$W_{\mathrm{FD}}(\underline{n}) = \prod_{i=1}^{r} \frac{g_i!}{n_i!(g_i - n_i)!}. \tag{9.29}$$

At this point we can follow the same route outlined above for the Bose–Einstein statistics, i.e. we obtain the Fermi–Dirac distribution function by maximizing $\ln(W_{\mathrm{FD}}(\underline{n}))$ under the constraints expressed by equations (9.24) and (9.25). Such a procedure is summarized in some detail in appendix B and leads to

$$\bar{n}_i = \frac{g_i}{A_{\mathrm{FD}}\, e^{\frac{\mathcal{E}_i}{k_{\mathrm{B}}T}} + 1} \qquad \text{with } i = 1, 2, \ldots, r, \tag{9.30}$$

where $\bar{n}_i$ is the equilibrium value of n_i and the normalization factor $A_{\rm FD}$ is again equal to $e^{-\frac{\mu}{k_B T}}$, with μ being the chemical potential. If we assume that $\mathcal{E}$ is a continuous variable, the above expression is replaced by

$$n_{\rm FD}(\mathcal{E}) = \frac{g(\mathcal{E})}{e^{\frac{\mathcal{E}-\mu}{k_B T}} + 1} \equiv g(\mathcal{E}) f_{\rm FD}(\mathcal{E}),$$

so that $n_{\rm FD}(\mathcal{E})$ is the number of fermions per unit energy interval and $n_{\rm FD}(\mathcal{E})\mathrm{d}\mathcal{E}$ gives the number of fermions with energies in the interval $[\mathcal{E}, \mathcal{E} + \mathrm{d}\mathcal{E}]$. As in the previous cases, $g(\mathcal{E})$ is the density of states function, whereas

$$f_{\rm FD}(\mathcal{E}) = \frac{1}{e^{\frac{\mathcal{E}-\mu}{k_B T}} + 1},$$

is the so-called *Fermi function*, giving the probability that a state of energy $\mathcal{E}$ is occupied by one fermion.

It is worth noting that the chemical potential $\mu \equiv \mu(T)$ appearing in the above expression is connected to a physical quantity, the *Fermi energy* $\mathcal{E}_{\rm F}$, which plays a very relevant role in the context of the fermionic systems. It is the energy of the highest occupied single-particle state at absolute zero temperature and is such that

$$\mathcal{E}_{\rm F} = \mu(T = 0).$$

Since the chemical potential typically varies quite slowly near $T = 0$ K, we can assume that for $T \sim 0$ K the value of μ is constant and equal to $\mathcal{E}_F$, so that we can rewrite the Fermi function as follows:

$$f_{\rm FD}(\mathcal{E}) = \frac{1}{e^{\frac{\mathcal{E}-\mathcal{E}_{\rm F}}{k_B T}} + 1}.$$

Figure 9.17 shows its behaviour as a function of the energy $\mathcal{E}$ at $T = 0$ K and at two finite values of T, having assumed for μ a value of 2 eV (typical values for metals are

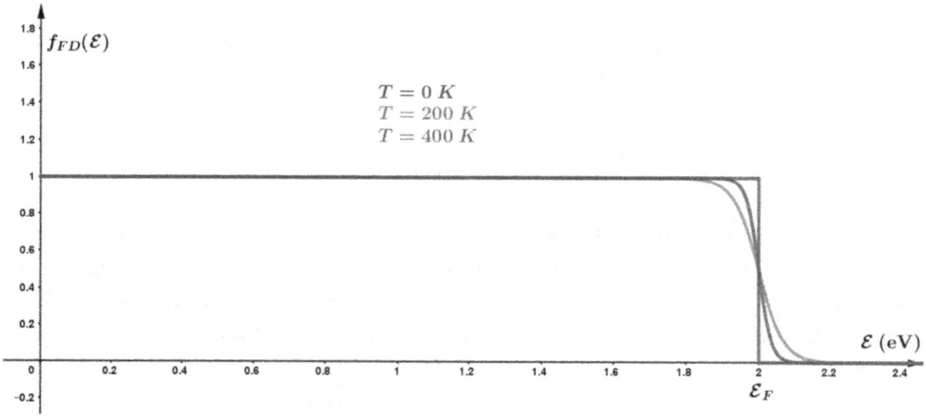

Figure 9.17. Temperature dependence of the Fermi function for $T = 0$, $T = 200$ K and $T = 400$ K.

between 2 eV and 10 eV). We can see that at zero temperature $f_{\rm FD}(\mathcal{E})$ is a step function, this signalling that all the single-particle states below $\mathcal{E}_{\rm F}$ are occupied and all those above are empty. At finite temperatures this situation is modified by the thermal excitation of some particle from below to above. However, we note that as soon as we consider temperatures not much higher than room temperature, the Fermi function deviates from that at $T = 0$ only in a small energy range, of the order of $k_{\rm B}T \sim 10^{-2}$ eV, around $\mathcal{E}_{\rm F}$.

9.4.4 The specific heat of solids

In this section we present an improved version of the theory of the specific heat of solids illustrated in section 9.2.4, taking into account what we have learned about quantum statistics. We first discuss the simpler model due to Einstein (1907), then move on to the model due to Debye (1912), which overcomes some of the limitations of Einstein's approach.

The Einstein model. In the Einstein model, the atoms of a crystalline solid are treated as three-dimensional independent harmonic oscillators, vibrating around their reticular sites with a unique frequency $\nu = \frac{1}{2\pi}\sqrt{\frac{k}{m}}$, where k is the elastic constant of the oscillator and m is its mass. Einstein adopted the quantization hypothesis introduced by Planck to explain the spectrum of the black-body radiation, so that the energy of each oscillator could only exist as an integer multiple of $h\nu$:

$$\mathcal{E}_n = nh\nu.$$

From this assumption, Einstein was able to calculate the average value of the energy at temperature T using the classical Maxwell–Boltzmann distribution:

$$\begin{aligned}\langle\mathcal{E}\rangle &\equiv \frac{\sum_{n=0}^{\infty}\mathcal{E}_n e^{-\beta\mathcal{E}_n}}{Z} = \frac{\sum_{n=0}^{\infty}\mathcal{E}_n e^{-\beta\mathcal{E}_n}}{\sum_{n=0}^{\infty}e^{-\beta\mathcal{E}_n}} = -\frac{d}{d\beta}\ln(Z) = -\frac{d}{d\beta}\ln\left(\sum\nolimits_{n=0}^{\infty}e^{-\beta nh\nu}\right)\\ &= -\frac{d}{d\beta}\ln\left(\sum\nolimits_{n=0}^{\infty}(e^{-\beta h\nu})^n\right) = -\frac{d}{d\beta}\ln\left(\frac{1}{1-e^{-\beta h\nu}}\right) = \frac{h\nu}{e^{\beta h\nu}-1}.\end{aligned}$$

Here $\beta = 1/k_{\rm B}T$ and we have used the fact that $\sum_n e^{-n\beta h\nu}$ is a convergent geometric series with the common ratio $q = e^{-\beta h\nu} < 1$ and $q > 0$. It is also immediate to see that in the limit of zero energy spacing ($h\nu \to 0$) we have $e^{\beta h\nu} \simeq 1 + \beta h\nu$ and thus $\langle\mathcal{E}\rangle \to k_{\rm B}T$, in agreement with the equipartition theorem.

From this result, the total internal energy of the N oscillators is simply calculated as

$$U = 3N\langle\mathcal{E}\rangle = 3N\frac{h\nu}{e^{\beta h\nu}-1},$$

where the factor 3 accounts for the three degrees of freedom of each oscillator. We can now calculate the molar specific heat at constant volume enforcing the usual definition:

$$c_V = \frac{1}{n}\frac{\mathrm{d}U}{\mathrm{d}T}\bigg|_V = 3R\left(\frac{h\nu}{k_B T}\right)^2 \frac{e^{\beta h\nu}}{(e^{\beta h\nu}-1)^2} = 3R\left(\frac{T_E}{T}\right)^2 \frac{e^{\frac{T_E}{T}}}{\left(e^{\frac{T_E}{T}}-1\right)^2}. \quad (9.31)$$

Here $Nk_B/n = N_A k_B = R$ and we have defined the *Einstein temperature* $T_E = h\nu/k_B$.

We can see that in the high temperature limit ($T \gg T_E$), equation (9.31) correctly reduces to the Dulong–Petit law $c_V = 3R$ experimentally verified in solids. On the other hand, for low temperatures ($T \ll T_E$) c_V exhibits an exponentially decreasing behaviour ($c_V \sim e^{-T_E/T}$) not fitting the experimental data, which instead suggest a T^3-dependence of c_V.

The Debye model. A few years after the publication of the Einstein model, Peter Debye developed a more accurate approach for the analysis of the specific heat in solids. He realized that the main drawback of the Einstein theory was that only a single oscillation frequency for the atoms of the crystalline lattice was considered. Instead, in a solid there are collective modes originating from the interactions between the atoms, the so-called normal modes, each being characterized by a specific oscillation frequency. Stated in other terms, since atoms do not move independently of each other, there are exchanges of energy in the form of elastic waves. Such waves can be treated as particles moving through the crystal, the so-called *phonons*, whose energies are quantized as the normal modes of oscillation of a vibrating string between two fixed extrema.

If we suppose that the solid is a cube of edge L, we can write the conditions of stationarity at the boundaries of the cube in the three space directions, quite similar to those valid for a vibrating string (see figure 9.18). They amount to requiring that the edge of the cube is a multiple integer of half of the wavelength λ,

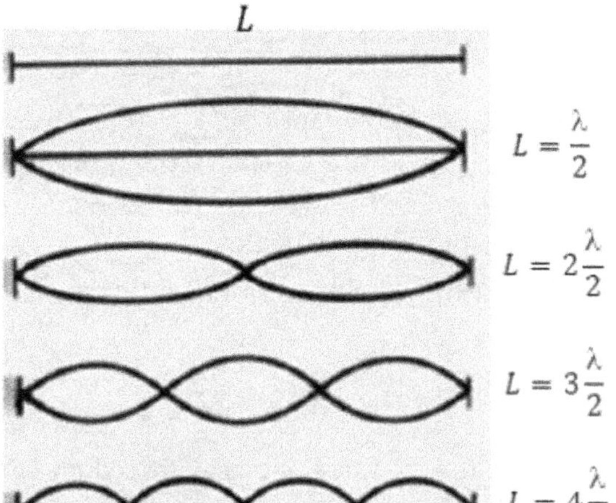

Figure 9.18. Stationary waves on a string.

$$L = n_x\frac{\lambda}{2} \qquad L = n_y\frac{\lambda}{2} \qquad L = n_z\frac{\lambda}{2}, \tag{9.32}$$

where n_x, n_y, n_z are positive integers. Using the wave-like relation $\lambda = 2\pi/k_i$, where k_i $(i = x, y, z)$ are the components of the wavevector $\vec{k}$, equations (9.32) become

$$k_x = n_x\frac{\pi}{L} \qquad k_y = n_y\frac{\pi}{L} \qquad k_z = n_z\frac{\pi}{L}.$$

From these relations we obtain the square of the modulus of the wavevector $\vec{k} = \frac{\pi}{L}(n_x\hat{x} + n_y\hat{y} + n_z\hat{z})$:

$$k^2 = \frac{\pi^2}{L^2}\left(n_x^2 + n_y^2 + n_z^2\right).$$

Going from the wavevectors k to the frequency ν by means of the relation $k = 2\pi\nu/v$, where v is the wave propagation velocity, we obtain

$$n_x^2 + n_y^2 + n_z^2 = \frac{4L^2}{v^2}\nu^2. \tag{9.33}$$

Equation (9.33) represents the equation of a sphere of radius $R = \frac{2L}{v}\nu$ in the space $(O\ n_x\ n_y\ n_z)$, such that each point of coordinates (n_x, n_y, n_z) falling inside this sphere represents a particular allowed mode of oscillation. Given the large values taken by these integers, we can approximate the volume of the octant where they are all positive with the number $N(\nu)$ of different stationary waves with frequencies between 0 and ν (see figure 9.19):

$$N(\nu) = \frac{1}{8}\frac{4\pi}{3}R^3 = \frac{1}{8}\frac{4\pi}{3}\frac{8L^3v^3}{\nu^3} = \frac{4\pi L^3}{3v^3}\nu^3.$$

Correspondingly, the number $g(\nu)\mathrm{d}\nu$ of stationary waves with frequencies between ν and $v + \mathrm{d}v$ will be given by one eighth of the volume of the shell of thickness $\mathrm{d}v$ formed by the two concentric spheres of radii ν and $v + \mathrm{d}v$:

$$g(\nu)\mathrm{d}\nu = \frac{1}{8}4\pi R^2\mathrm{d}R = \frac{\pi}{2}\left(\frac{2L}{\nu}\nu\right)^2\frac{2L}{\nu}\nu = \frac{4\pi L^3}{v^3}\nu^2\mathrm{d}\nu \tag{9.34}$$

(of course, $g(\nu) = \mathrm{d}N(\nu)/\mathrm{d}\nu$).

The density of phononic normal modes expressed by equation (9.34) is formally the same as the density of states used by Rayleigh and Jeans to study black-body radiation. However, a relevant difference is that in this case the number of oscillation modes is limited to $3N$, i.e. the number of oscillators (N) times the number of degrees of freedom of each of them (three). This means that we have to introduce a cut-off in the frequencies, usually defined as the *Debye frequency* ν_D, such that

$$\int_0^{\nu_\mathrm{D}}\mathrm{d}N = \int_0^{\nu_\mathrm{D}}g(\nu)\mathrm{d}\nu = \int_0^{\nu_\mathrm{D}}\frac{4\pi L^3}{v^3}\nu^2\mathrm{d}\nu = 3N.$$

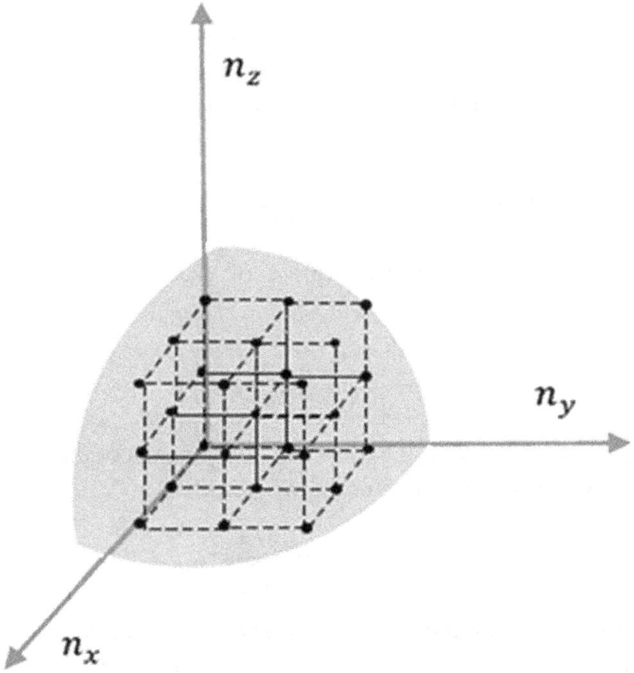

Figure 9.19. Pictorial representation of the allowed values of the integers n_x, n_y and n_z defined in equation (9.32).

Calculating the integral we obtain

$$\frac{4\pi L^3}{3v^3}\nu_D^3 = 3N$$

so that the expression for the Debye frequency is

$$\nu_D = 3\sqrt{\frac{9N}{4\pi L^3}}\,v = 3\sqrt{\frac{9n}{4\pi}}\,v, \tag{9.35}$$

where $n = N/V$ is the density of the atoms in the solid. We can now determine the internal energy of the system[8]:

$$U = \int_0^{\nu_D} g(\nu)\langle\mathcal{E}(\nu)\rangle \mathrm{d}\nu = \frac{4\pi L^3}{3v^3}\int_0^{\nu_D} \nu^2 \frac{h\nu}{\mathrm{e}^{\beta h\nu} - 1}\mathrm{d}\nu,$$

[8] We observe that the average value $\langle\mathcal{E}(\nu)\rangle$ in this as in the previous estimate does not include the zero point energy, but for our scope this is not a source of trouble since this constant energy term does not affect the specific heat.

where we have used equation (9.28) taking into account that phonons behave like particles obeying the Bose–Einstein statistics. By performing the variable substitution

$$\nu = \frac{1}{\beta h}x = \frac{k_B T}{h}x$$

and using equation (9.35) of the Debye frequency we obtain

$$U = \frac{4\pi L^3 h}{v^3}\left(\frac{k_B T}{h}\right)^3 \int_0^{h\nu_d/k_B T} \frac{x^3}{e^x - 1}\,dx.$$

Finally, defining the *Debye temperature*

$$T_D = \frac{h\nu_D}{k_B}$$

the expression of the total internal energy[9] becomes:

$$U = 9Nk_B T\left(\frac{T}{T_D}\right)^3 \int_0^{T_D/T} \frac{x^3}{e^x - 1}dx.$$

The integral on the right-hand side cannot be solved exactly with elementary methods, but we can study the following relevant limiting cases.

• $T \gg T_D$ (the classical limit).

In this case in the integral we have $0 < x < \frac{T_D}{T} \ll 1$, so that we can use for the exponential the approximation $e^x \simeq 1 + x$. Then we have

$$U \approx 9Nk_B T\left(\frac{T}{T_D}\right)^3 \int_0^{T_D/T} \frac{x^3}{1 + x - 1} = 9Nk_B T\left(\frac{T}{T_D}\right)^3 \frac{1}{3}\left(\frac{T_D}{T}\right)^3 = 3Nk_B T,$$

in this way recovering the Dulong–Petit law for the molar specific heat:

$$c_V = \frac{1}{n}\frac{dU}{dT}\bigg|_V = 3k_B N_A = 3R.$$

• $T \ll T_D$ (the quantum limit).

In this case the upper limit in the integral becomes very large, so that we have

$$\int_0^{T_D/T} \frac{x^3}{e^x - 1} \simeq \int_0^{+\infty} \frac{x^3}{e^x - 1} = \frac{\pi^4}{15}.$$

The expression of the internal energy then becomes

$$U \approx 9Nk_B T\left(\frac{T}{T_D}\right)^3 \frac{\pi^4}{15} = \frac{3}{5}\pi^4 Nk_B T\left(\frac{T}{T_D}\right)^3.$$

[9] We recall that we are neglecting the electronic contribution.

Evaluating the molar specific heat, we finally obtain

$$c_V = \frac{1}{n}\frac{\mathrm{d}U}{\mathrm{d}T}\bigg|_V = \frac{12}{5}\pi^4 R\left(\frac{T}{T_{\mathrm{D}}}\right)^3.$$

We thus find that in the low-temperature quantum limit the Debye model correctly predicts a specific heat tending to zero with a cubic temperature dependency. However, it should be noted that for T just a few degrees above absolute zero, the dominant term is experimentally found to be linear in T, and not cubic. Actually, a linear contribution to c_V which at temperatures very close to absolute zero dominates over the phononic one, arises when the dynamics of the conduction electrons, neglected in the Einstein and the Debye models, is explicitly taken into account. This has to be done within models, such as the one due to Sommerfeld, where the electrons are not treated as classic particles any longer, but rather are assumed to obey quantum Fermi–Dirac statistics ensuring the validity of the Pauli principle.

Appendices

A. Derivation of equation (9.23)

Let us assume that conduction electrons can only move along a one-dimensional bar lying in the x-direction. At a given point x of the bar, half of these electrons move in the positive direction (rightward) and half in the opposite direction. Suppose that the thermal energy per electron $\mathcal{E}$ is a function of the temperature T at a given point of the bar, and that electrons going to the right come from a portion of the bar where temperatures are higher than those of the portion toward which they move, i.e. $T(x - a) > T(x + a)$ for any $a > 0$ (rightward moving electrons are more energetic than leftward ones, see also figure 9.14). Since their last collision, half of the electrons arriving at point x from the left have on average an energy $\mathcal{E}(T(x - v\tau))$, whereas the other half arriving from the right have an average energy $\mathcal{E}(T(x + v\tau))$, where $l = v\tau$ is the electron mean free path and v their average speed. Hence the heat current at x can be estimated as

$$J_x^q = \frac{n}{2}[v\mathcal{E}(T(x - v\tau)) - v\mathcal{E}(T(x + v\tau))],$$

where n is the electron density. If temperature changes mildly over a mean free path, we can use a first-order Taylor expansion for $T(x - v\tau)$ and $T(x + v\tau)$, in this way obtaining

$$J_x^q \simeq \frac{nv}{2}\left[\mathcal{E}(T(x)) + \frac{\mathrm{d}\mathcal{E}}{\mathrm{d}T}\frac{\mathrm{d}T}{\mathrm{d}x}(-v\tau) - \left(\mathcal{E}(T(x)) + \frac{\mathrm{d}\mathcal{E}}{\mathrm{d}T}\frac{\mathrm{d}T}{\mathrm{d}x}v\tau\right)\right] = -nv^2\tau\frac{\mathrm{d}\mathcal{E}}{\mathrm{d}T}\frac{\mathrm{d}T}{\mathrm{d}x}.$$

Considering that the internal energy per unit volume is $u = N\mathcal{E}/V = n\mathcal{E}$ and thus, by definition, $c_V = n\frac{\mathrm{d}\mathcal{E}}{\mathrm{d}T}\Big|_V$, we can write

$$J_x^q = -v^2 \tau c_V \frac{\mathrm{d}T}{\mathrm{d}x}.$$

This relation can be extended to a three-dimensional conductor using the isotropy assumption about the velocity distribution, according to which $\langle v_x^2 \rangle = \langle v_y^2 \rangle = \langle v_z^2 \rangle = \frac{1}{3}\langle v^2 \rangle$. We then arrive at the equation

$$\vec{J}^q = -\frac{1}{3}\langle v^2 \rangle \tau c_V \vec{\nabla} T$$

and from the comparison with Fourier's law we obtain

$$\kappa = \frac{1}{3}\langle v^2 \rangle \tau c_V,$$

i.e. equation (9.23).

B. Derivation of equation (9.30)

Having found that the statistical weight $W_{\mathrm{FD}}(\underline{n})$ for a system of N identical fermions is given by equation (9.29), we identify the most probable set $\bar{\underline{n}} = (\bar{n}_1, \bar{n}_2, \ldots, \bar{n}_r)$ of the occupation numbers by maximizing $\ln(W_{\mathrm{FD}}(\underline{n}))$. This has to be done under the constraints represented by equations (9.24) and (9.25) which fix the values of the total number of particles N and the total energy E, respectively.

We proceed by calculating the variation of $\ln(W_{\mathrm{FD}}(\underline{n}))$ when the occupation numbers n_i deviate by δn_i from the most likely values, that is,

$$\delta(\ln(W_{\mathrm{FD}})) = \delta\left(\ln\left(\prod_{i=1}^{r} \frac{g_i!}{n_i!(g_i - n_i)!}\right)\right) = \sum_{i=1}^{r} \delta(\ln(g_i!) - \ln(n_i!) - \ln((g_i - n_i)!)).$$

If the set of occupation numbers is maximal, the variation $\delta(\ln(W_{\mathrm{FD}}))$ should vanish. The variational procedure can be greatly simplified using the Stirling approximation for $\ln(n!)$:

$$\ln(n!) \approx n(\ln(n) - 1) \qquad n \to +\infty.$$

Since the number of particles is assumed to be at least of the order of $N_{\mathrm{A}} \sim 10^{24}$, the numbers n_i and the degeneracies g_i are large enough to safely apply this approximation (for $n = 100$ the error is already less than 1%). We thus have

$$\begin{aligned}\delta(\ln(W_{\mathrm{FD}})) &= \sum_{i=1}^{r} \delta[g_i \ln g_i - n_i \ln n_i - (g_i - n_i)\ln(g_i - n_i)] \\ &= \sum_{i=1}^{r} [-\ln(n_i) + \ln(g_i - n_i)]\delta n_i = 0.\end{aligned}$$

Note that at this point we cannot simply impose that the coefficients of δn_i must vanish in order to have $\delta(\ln(W_{\mathrm{FD}})) = 0$. In fact, because of the constraints expressed by equations (9.24) and (9.25), the δn_i no longer represent linearly independent variations, but rather must satisfy the following conditions:

$$\sum_{i=1}^{r} \delta n_i = \delta N = 0$$

$$\sum_{i=1}^{r} \mathcal{E}_i \, \delta n_i = \delta E = 0.$$

Introducing two unknown quantities α and β, usually called *Lagrange multipliers*, we can combine the three above equations as follows:

$$\delta \ln(W_{\mathrm{FD}}) - \alpha\delta N - \beta\delta E = \sum_{i=1}^{r}(-\ln(n_i) + \ln(g_i - n_i) - \alpha - \beta\mathcal{E}_i)\delta n_i = 0.$$

Now we are free to choose the two multipliers in such a way that two of the terms in the above sum vanish. In this way only $r - 2$ terms remain in the sum and thus the corresponding δn_i can be varied independently from each other. For this reason, the only way the above equation is satisfied for arbitrary variations δn_i is that all the coefficients must be zero. We are then left with the following equations for n_i $(i = 1, 2, \ldots, r)$:

$$\ln\left(\frac{g_i - n_i}{n_i}\right) = \alpha + \beta\mathcal{E}_i \quad \Longrightarrow \quad n_i = \frac{g_i}{\mathrm{e}^{\alpha}\mathrm{e}^{\beta\mathcal{E}_i} + 1}.$$

The last expression coincides with equation (9.30), provided that we make the identification $\beta = \frac{1}{k_{\mathrm{B}}T}$.

References

[1] Pathria R K and Beale P D 2011 *Statistical Mechanics* 3rd edn (New York: Academic)

[2] Reif F 1965 *Fundamentals of Statistical and Thermal Physics* (New York: McGraw-Hill)

[3] Adapted from: Chuang Y-D, Gromko A D, Dessau D S, Kimura T and Tokura Y 2001 *Science* **292** 1509

[4] Ashcroft N W and Mermin N D 1976 *Solid State Physics* (Philadelphia, PA: Saunders)

[5] Sears F W and Salinger G L 1975 *Thermodynamics, Kinetic Theory, and Statistical Thermodynamics* (Reading, MA: Addison-Wesley)

Chapter 10

How to look deep inside matter: scanning electron microscopy

Francesco Avitabile and Antonio Vecchione

10.1 Introduction

Currently, an experimental physicist can take advantage of many analysis techniques that can be used daily in experimental physics laboratories. Moreover, to carry out experiments of a certain relevance and complexity of realization, large facilities such as synchrotron and nuclear power plants can be used. In experimental research in condensed matter physics, solid-state sample analysis is of fundamental importance. Among the analyses which can be performed in the laboratory, scanning electron microscopy plays a fundamental role in several research fields. This chapter is devoted to providing an initial approach for beginners to understand scanning electron microscopy and some related microanalysis techniques, such as energy dispersive and wavelength dispersive x-ray spectroscopy.

10.2 Microscopy

Since ancient times, man has always faced the problem of being able to see objects with ever greater finesse and detail. Over the centuries, first alchemy and chemistry, and then biology and medicine, needed to investigate very small objects and entities such as the structures of cells, bacteria, viruses, etc. One of the first known microscopes, created by the Dutch Anton von Leeuwenhoek, consisted simply of a single convex lens and an adjustable sample holder. A microscope is an optical system that provides a magnified image of an object. In the continuous evolution of the optical microscope, the addition of lenses and various devices made it possible to improve the resolution. However, all the microscopes that use visible light have a theoretical resolution limit strictly dependent on the wavelength of the radiation used and, even with all the possible precautions, it is not possible to greatly exceed the resolution of hundreds of nanometres. With the discovery that electrons accelerated by electric fields travel in a vacuum in a straight line like light and

doi:10.1088/978-0-7503-2678-0ch10

that it is possible to associate to them a characteristic wavelength about 100 000 times smaller than that of visible light led Ernst Rubka to the realization of the first transmission electron microscope (TEM) allowing a resolution of up to 0.1 nm. The natural evolution of this electron microscope was the scanning electron microscope (SEM) which scans the surface of a sample with a finely focused 'probe'.

10.2.1 The optical microscope and its limitations

The optical microscope is a system composed of one or more lenses that works in the visible radiation spectrum. The simplest microscopes consist of a single convex lens, but if one wants to obtain higher magnifications it is preferable to use a two-lens system, so that the magnification obtained by the first lens is increased by the second. For a convex lens, if f is the focal distance, q is the object distance and p is the image distance, the relation between the f, p and q is $1/f = 1/q + 1/p$ leading to a magnification $M = q/p = (p - f)/f$.

A diagram for a two-stage microscope is shown in figure 10.1. On the left, the object O is placed at distance q_1 from the centre of the lens. The first lens, called the objective, provides an inverted image B of the object I at distance p_1 to the centre of the lens. The second lens, called the projector, placed at distance q_2 from B, provides an enlarged image C, at distance p_2 from its centre that can be viewed on a screen or on a photographic plate. Therefore, the overall magnification will be equal to $M = (q_1/p_1)(q_2/p_2) = (p_1 - f_1)(p_2 - f_2)/f_1f_2$. If larger magnifications are required, further stages can be obtained by adding other objective lenses. In the case that you want to look directly at the enlarged image, the projector will be replaced with a lens called ocular. This lens diverges the refracted rays, so it will form an enlarged virtual image visible to the eye. Most of the magnification will be given by the objective, for which M varies between 10 and 100, while the magnification of the eyepiece is about 10. The total magnification is determined by multiplying the magnification of the objective by that of the eyepiece. In general, microscopes have multiple objectives with different focal distances, therefore, the magnification can be

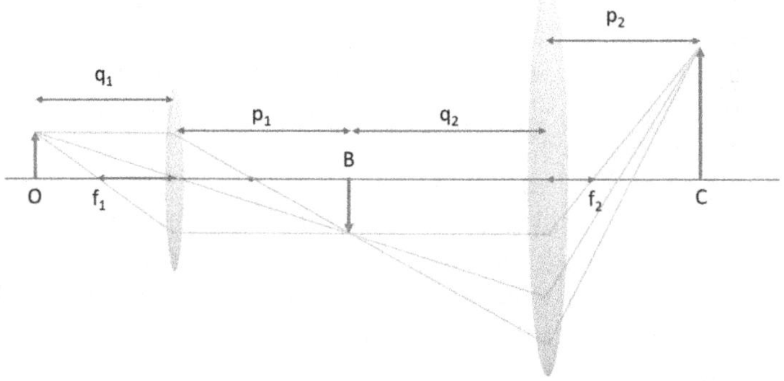

Figure 10.1. Diagram of a two-stage microscope with a projector as the second lens. The object placed at O is magnified and its image is projected at C.

changed by changing the objective. Furthermore, sometimes it is also possible to change the distances p and q by adjusting screw knobs (see figure 10.2).

Lighting system. The optical microscope has a lighting system which is usually located under the object to be observed. It consists of a mirror that reflects sunlight or a lamplight on the object. If the sample is sufficiently transparent, for example thin sections of rocks, minerals or crystals, the light rays are conveyed onto the sample from behind. In the case of opaque samples, such as ceramics or semiconductors, the light comes from the front. The lighting system also includes a condenser, which collects the diverging light from the source and focuses it onto the sample which appears brighter, also allowing control of the incidence angle and, therefore, of the contrast. Moreover, a diaphragm regulates the amount of reflected light and, therefore, the area of sample that is illuminated. Figure 10.3 shows an example of an optical microscope image taken of grains of sand obtained using the dark-field technique.

Polarized light optical microscopy. In some studies, such as in the mineralogical or materials sciences, it may be useful to observe a sample illuminated by polarized light because it allows an in-depth optical characterization of the constituent compounds. In a beam of polarized light, the waves have vibration planes oriented in a single direction and are parallel to each other. The polarization of light is possible thanks to a polarizer placed inside the instrument. In order to be penetrated by light, the rock or other types of samples are cut into very thin layers (usually around tens of microns). A polarized light optical microscope (PLOM) has a second polarizing filter, called analyser, placed between the objective and the eyepiece,

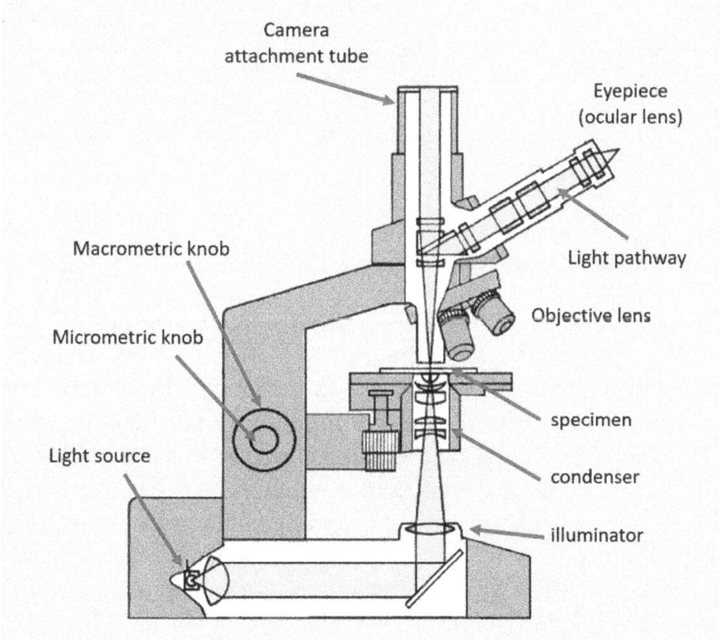

Figure 10.2. The internal operating scheme of an optical microscope.

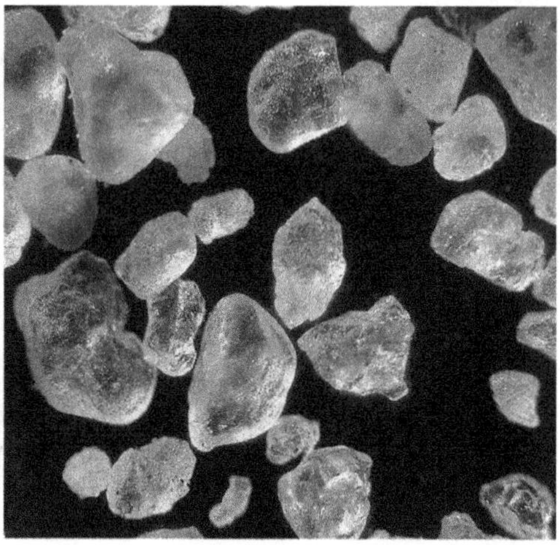

Figure 10.3. An image of grains of sand acquired by the dark-field technique; magnification 200× [1] Wikimedia Common. Author: Jan Homann.

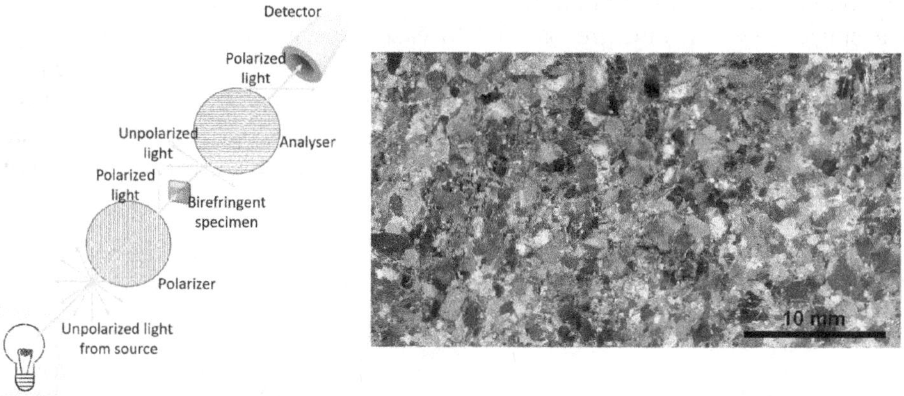

Figure 10.4. Left: illustration of the polarizer housing in a PLOM. Right: scanned image of thin section from Siilinjärvi apatite ore in cross polarized transmitted light [2] Wikimedia Common. Author: Kallerna.

which again polarizes the light that has passed through the sample, leaving only part of the radiation to pass through (see figure 10.4, left panel). As the area of minerals exposed to light changes, there will be a different interaction that contains information regarding the crystals present in the section, different to those obtained using only the first filter. Due to the limited depth of field of this analysis, the roughness of the sample must be reduced as much as possible. For this purpose, the sample can be polished with abrasive fabrics and diamond suspensions of different roughnesses of the order of microns. The limits of this technique are due to the spatial resolution of about 0.5 μm. An example of a polarized light optical image is shown in figure 10.4 (right panel).

Resolution. It has already been said that the magnification of a microscope can be increased by adding one or more lenses to the objective–ocular system. However, the size of the smallest details that can be distinguished through an optical microscope is some hundreds of nanometres. The resolution of a microscope is defined as the closest distance in which two distinct points can be seen as separate entities. In addition to the presence of distortions due to lens imperfections, resolution is limited by the effects of light diffraction. The light inside a microscope is forced to pass through slits of various sizes and, as is well known, it is diffracted. Therefore, the beam of light that would have been seen in the form of a luminous spot in this case appears in the form of a series of concentric circles that are known as Airy rings. For each wavelength $d \sim 1/D$, where d is the diameter of the central spot and D is the diameter of the objective aperture [3]. In order to obtain the best resolution, the central spot must be as small as possible to reduce the spot size, therefore, the aperture must be increased as much as possible. According to Rayleigh, two points are barely distinguishable as two distinct entities if the maximum peak of one occurs in the first minimum of the other (see figure 10.5), and thus the resolution limit is $d_{\min} = d/2$. This length can be derived from diffraction theory and is given by the Rayleigh criterion leading to $d_{\min} = 0.61\lambda/(n \sin\theta) = 0.61\lambda/\mathrm{NA}$ (see table 10.1), where λ is the light wavelength, n is the refractive index of the medium (typically air or a special oil placed between the objective and the object if a greater resolution is required), $\mathrm{NA} = n \sin\theta$ is the numerical aperture and θ is the semi-angle at which the light from the sample diffuses through the objective aperture (see figure 10.6). Once the maximum possible aperture is reached and thus a higher refractive index, another way to improve the resolution of the microscope is to use light of a shorter wavelength, for example green light or light as close as possible to ultraviolet.

10.2.2 Scanning electron microscopy

As seen above, the resolution of a microscope, that is the minimum distance at which we can distinguish two different points, is proportional to the wavelength of the radiation. In an optical microscope, the minimum distance is limited by the fact that

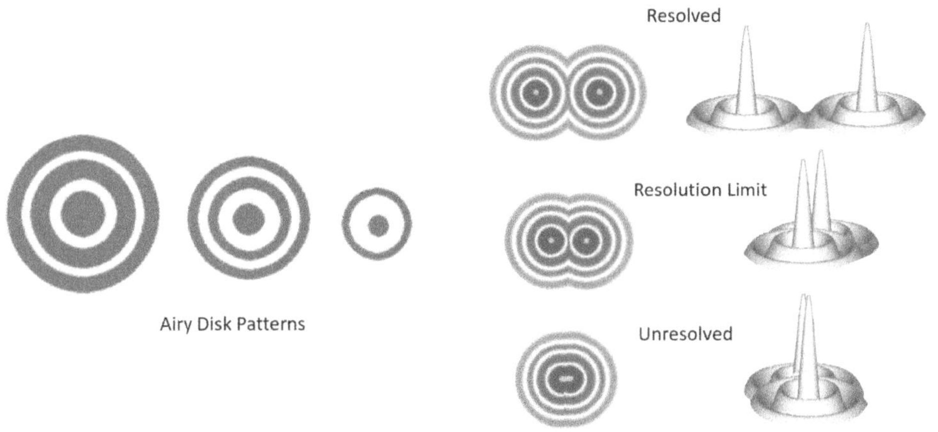

Figure 10.5. Left: Airy rings obtained by different apertures. Right: examples of Airy patterns with resolved points, at the limit of resolution and unresolved points.

Table 10.1. Resolution of an optical microscope at varying wavelengths.

Wavelength (nm)	Resolution (μm)
360	0.19
400	0.21
450	0.24
500	0.26
550	0.29
600	0.32
650	0.34
700	0.37

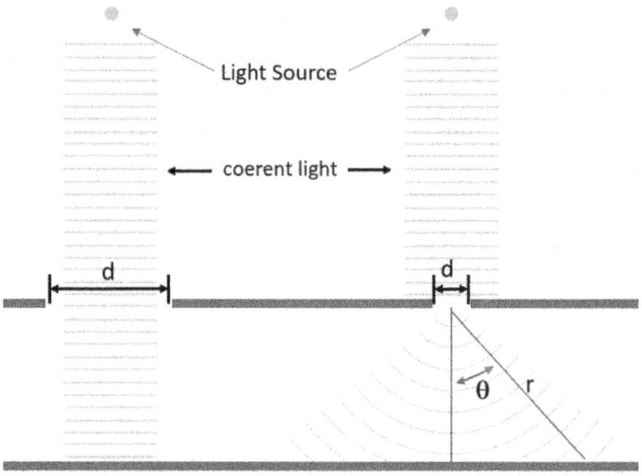

Figure 10.6. A sketch of (left) an undiffracted electromagnetic wave $d \gg \lambda$ and (right) $d < \lambda$ where we can observe the diffracted light intensity distribution.

the minimum light wavelength is about 400 nm. However, we must also consider that matter, due to the de Broglie hypothesis [4], exhibits wave-like behaviour with a wavelength given by the relationship $\lambda = h/p$, were h is the Planck constant and $p = mv$ is the momentum. For an electron accelerated by 20 keV, taking into account also the relativistic effects, the wavelength λ associated with an electron is about 0.0086 nm [3]. The SEM uses electrons instead of light to form an image. The magnification and resolution of an electron microscope are far better than those of an optical microscope. Moreover, another important improved feature is the depth of field, that is, the distance interval in which an object remains in sharp focus (see table 10.2).

The combination of these three characteristics makes the SEM one of the most used tools in the study of solid-state samples, in the semiconductor industry and in other fields of research.

Table 10.2. A comparison between optical and electronic microscopes; the values are purely indicative of those commonly commercialised.

Microscope	Magnification	Depth of field	Resolution
OM	4×–1000×	40 μm–0.2 μm	~0.2 μm
SEM	10×–3 000 000×	4 mm–0.4 μm	1–10 nm

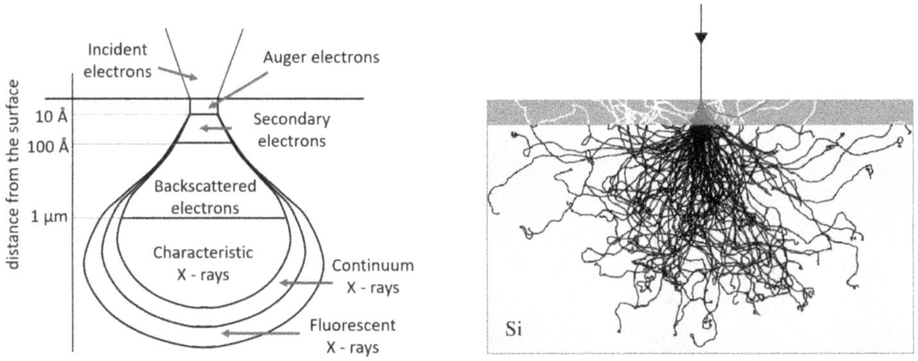

Figure 10.7. Left: the interaction volume; typically, for an electron beam with (E = 20 kV, Z = 28), we have the electron beam penetration $R = 5\ \mu$m, secondary electrons between 10 and 50 nm, backscattered electrons within 1–2 μm, Auger electrons between 1–3 μm and x-rays $<5\ \mu$m. Right: Monte Carlo simulation of 300 keV electron 'zig-zag' trajectories in silicon. The white lines are the backscattered electrons and the red lines are the electrons absorbed inside the sample. Reprinted from [5], copyright 2009, with permission from Elsevier.

Electron interaction with samples. To understand the functioning of an SEM, it is necessary to understand the nature of interactions between an incident electron beam and the surface of a sample. Depending on its energy and on the thickness of the sample, after having travelled a certain free path, the penetrating electrons will undergo a scattering event that deviates them from their trajectory. If the scattering does not change the primary electron energy, the scattering is defined as elastic, otherwise, if it absorbs part of the primary electron energy the scattering is inelastic. When penetrating a bulk sample, the incident (called also primary) electrons undergo many inelastic scattering events before losing all of their energy or leaving the sample by coming out of the surface. The scattering events can be various. For example, an incident electron on a sample can increase the atom vibrations around the lattice sites, losing energy by increasing the temperature of the sample itself. Alternatively, the electron can excite one of the innermost shell electrons giving it the energy it needs to jump to an outer shell. Another scattering phenomenon is due to the interaction with the 'sea' of conduction electrons of a metal or with the bonding electrons of a non-metal. In a scattering sequence, until it loses all its energy, the incident electron follows a typical zig-zag trajectory which can be calculated using a Monte Carlo method that takes into account the probability associated with each scattering angle (see the right panel of figure 10.7). It is estimated that only a small part of the incident electrons is backscattered and leaves the sample while the greater part is braked and absorbed inside the sample. The volume within which the electron is stopped is called the *interaction volume* and involves a region such as that shown in the left panel of figure 10.7. The penetration length depends on the electron energy and the atomic number of the investigated sample (see figure 10.8). When a primary electron beam hits the sample, there are also several secondary effects. These effects, by definition, are those detected outside the sample and consist in the emission of electrons or electromagnetic radiation. Below are listed some of the most important secondary effects.

Secondary electrons (SEs). These are electrons emerging from the surface of the sample with very low energy (less than 50 eV). SEs are therefore made up of incident electrons that have undergone many scatterings or, more likely, they are electrons coming from close to the sample surface to which enough energy has been supplied by the electron incident beam to leave the surface. The *yield* of SEs, that is the number of SEs per incident electron, is very high, approximately equal to or higher than 1 [3]. Thanks to their abundance, SEs are used in SEM imaging (see below).

Backscattered electrons (BSEs). BSEs are primary electrons that have an elastic interaction with the atomic nuclei of the specimen and leave the surface with high energy (thousands of electronvolts) and a large escape depth. They are fewer than secondary electrons and are used for diffraction and SEM analysis.

Cathodoluminescence (CL), characteristic x-rays and Auger electrons (AEs). Due to the impact of primary electrons, an outermost electron can be ejected from an atom. At some later time, it could be replaced by an electron belonging to a more external shell which jumps to the lowest energy level by releasing a photon, typically of the visible spectrum. This effect is called cathodoluminescence. If the ejected electron belongs to an innermost shell, when an electron from an outermost shell jumps into the vacant position, energy is released in the form of an electromagnetic wave in the x-ray spectrum. The energy of the x-ray is the difference between the energies of the two excited states and its value is characteristic of each atomic species. For this reason, x-rays are used for chemical analysis. Furthermore, it may occur that this energy can be collected by an outermost energy level electron which leaves the atom carrying excess energy in the form of kinetic energy. This process is Auger emission (figure 10.9).

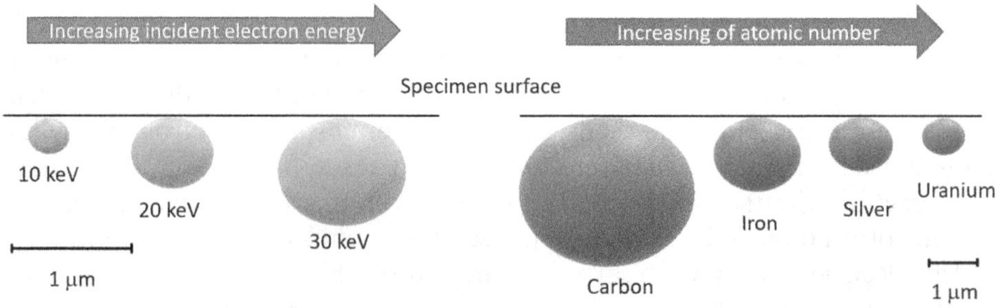

Figure 10.8. The interaction volume depends on the acceleration voltage and the atomic number of the specimen.

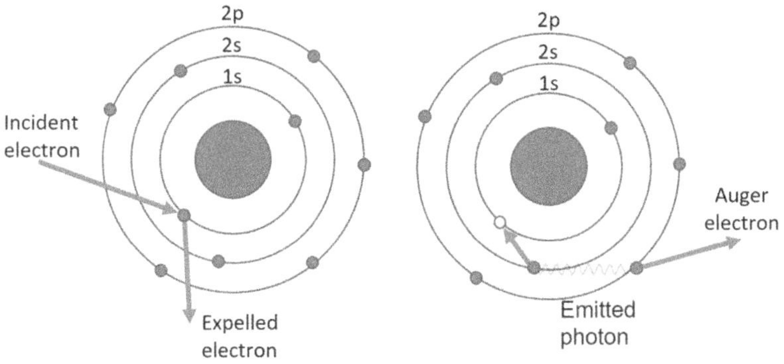

Figure 10.9. Schematic drawings of an electron collision with an atom and the subsequent emission of x-rays or AEs.

The continuous x-ray spectrum (bremsstrahlung). Primary electrons can also lose their energy due to the deviation they could undergo in interacting with a nucleus. In this case they emit a photon whose energy is equal to the loss of kinetic energy of the electron (bremsstrahlung = braking radiation). The energy of the x-ray emitted can vary continuously up to the total kinetic energy of the electron and is therefore no longer characteristic of the target atomic species.

10.2.3 SEM components

The SEM is a widely used instrument to obtain information on bulk specimens. A schematic drawing of an SEM is shown in figure 10.10. It mainly consists of the following components [6].

Electron gun. This is placed at the top of the instrument and provides a stable beam of electrons with adjustable energy. In the simplest version, the electron gun consists of a tungsten thermionic emitter (a hairpin shaped filament about 0.1 mm in diameter) heated electrically to about 2700 K, giving electrons sufficient thermal energy to overcome the potential barrier at the surface. A crossover limits the effective emitting area of the filament and thus focuses the emitted electrons which are accelerated from the negative potential of the filament to the ground potential of the anode (typically in the range 1–30 kV). The anode consists of a plate at ground potential with a small aperture which allows a fraction of the electrons to proceed down a column.

Electromagnetic lenses. Below the electron gun are two (or more) electromagnetic lenses, the condenser and the objective lens. An electromagnetic lens consists of a coil of copper wire, carrying a direct current generating a magnetic field, and its main task is to demagnify the electron beam and to control its diameter. In fact, interaction with this field causes electrons to be deflected towards the axis, similarly to the effect caused by the convex glass lenses used for focusing light. The strength of the lens can be controlled by varying the current in the coil. The beam path in this instrument has been designed so that beam crossover is avoided between the source and the specimen.

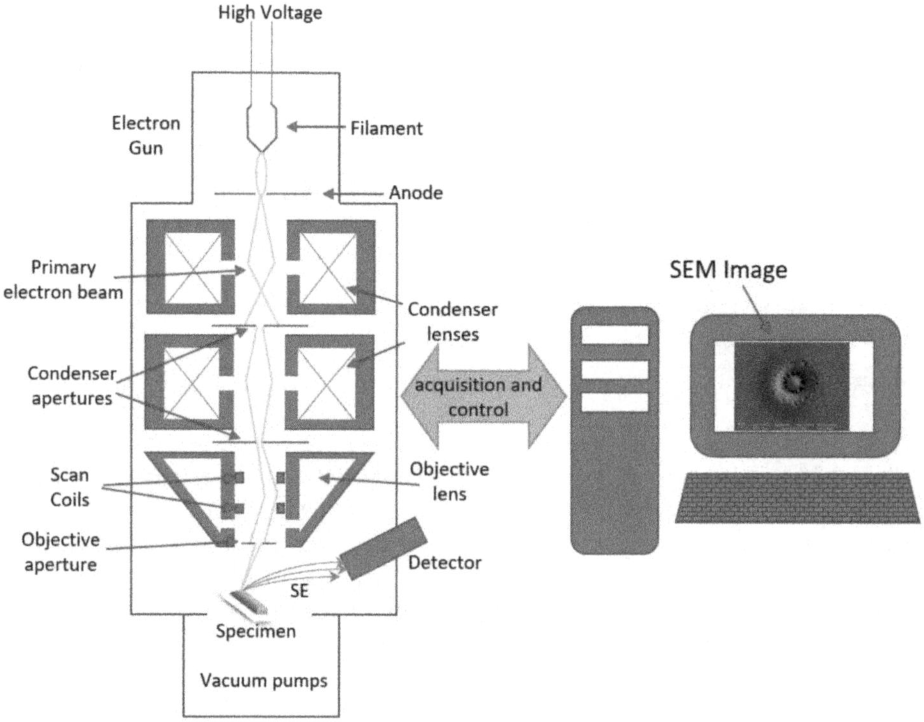

Figure 10.10. Schematic drawing of the internal functioning of an SEM.

Aperture limiters. These limit the beam diameter in the final lens to control spherical aberration, i.e. the lens focuses outer rays more strongly than those close to the axis resulting in an enlargement of the beam at the focus. The minimum diameter of the beam that can be reached with a high-performance SEM is generally around 2 nm [7].

Scanning coils. Scanning images are produced by sweeping the beam across the sample while displaying the output of an electron detector. The electron beam is deflected by means of coils located above the final lens. The electron beam is scanned in a similar way to an old-style television receiver, that is, in a rectangular set of straight lines known as a raster. The line scan frequencies determine the number of lines in the image (typically 500–2000). In modern instruments, the same effect is obtained by digitally controlling the beam position on the sample and the resultant image is displayed on a computer screen.

Sample holder. The samples for SEM are commonly fixed on a metal disk mounted on a mechanism for which linear movements are possible in the x- and y-directions, which allows different areas of the sample to be viewed. Linear movement in the z-direction is used to position the surface of the sample at the required height with respect to the final electronic lens. Furthermore, inclination and rotation movements allow the adjustment of the orientation of the sample with respect to the different detectors in order to optimize the image.

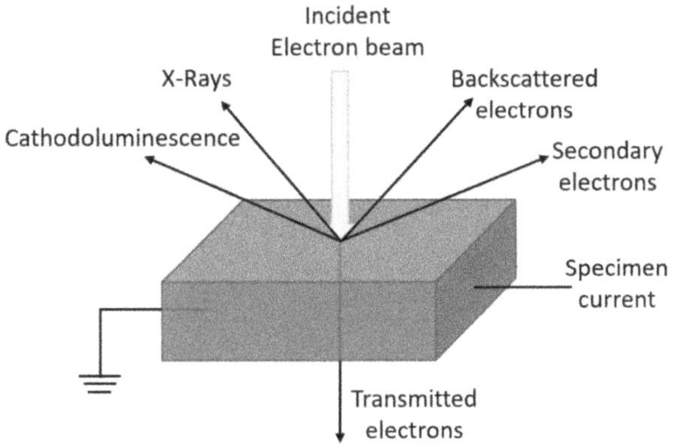

Figure 10.11. Some signals used in SEM analysis.

Vacuum system. Instruments that use an electron beam must be evacuated well enough not only to avoid damage to the electron source and high voltage failures but also to allow the electrons to reach the sample without being scattered. Therefore, the operating pressure must be less than 10^{-5} mbar. This pressure can be reached by a turbo molecular pump backed by a mechanical rotary pump.

Specimen coating. If non-conductive samples are bombarded by an electron beam, they can acquire a negative charge. This charge easily deflects the secondary electrons because of their low energy and the image will be unstable. Insulating specimens can be coated with a conductive material to prevent charging, ensuring that the coating is properly conductive and connected to the holder. To do this, the insulating samples are coated by sputtering with gold, silver or carbon. Environmental SEM (ESEM) using specialized electron detectors and differential pumping systems is also available. They allow for the transfer of the electron incident beam from the high vacuum in the electron gun area to the high pressures attainable in sample chamber. ESEM is designed for imaging specimens in their natural state (as in the case of organic samples) avoiding any coating.

10.2.4 SEM imaging

When the electronic beam hits the specimen, several signals are generated (see figure 10.11). All generated signals are emitted simultaneously and each carries different information. Both SEs and BSEs are used for imaging while x-rays are used for chemical analysis. All SEMs normally have facilities for detecting SEs and BSEs. X-rays are analysed using energy dispersive x-ray spectrometry (EDS) and wavelength x-ray spectrometry (WDS).

Magnification. The magnification is the ratio between the length C of the scan displayed on the screen and the corresponding length x of the scan on the sample. Figure 10.12 shows an example of two different enlargements of the same area of a sample.

SE imaging Everhart–Thornley detector. SEs emitted in the interaction between primary electrons and matter are detected by means of a scintillator, which produces

Figure 10.12. SEM images of TiO_2 nanoparticle assembled film obtained using ultra-fast laser ablation. The image on the left shows a portion of the sample with a magnification of 60k×. The white rectangle is expanded in the image on the right with a magnification of 250k×.

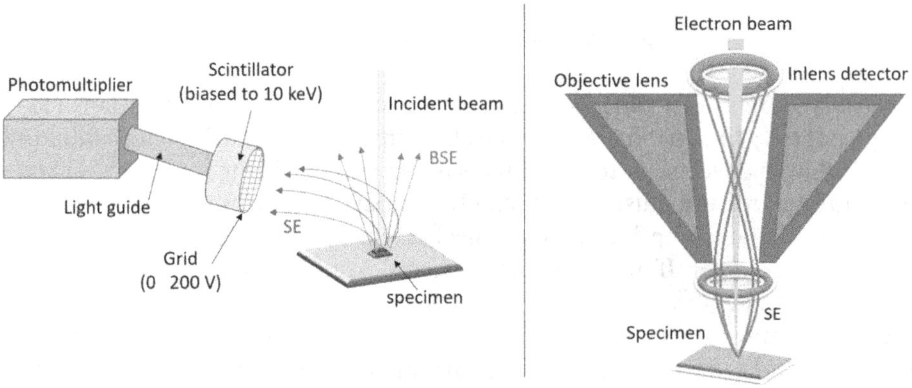

Figure 10.13. Schematic representation of the Everhart–Thornley detector (left) and in-lens detector (right).

light when it is bombarded with electrons. In turn, the light is converted into an electrical signal by a photomultiplier. This system, shown in the left panel of figure 10.13, is known as the Everhart–Thornley detector (or ET detector) [3]. As we saw in section 10.2.2, SEs are emitted from areas very close to the surface of the sample (surface distance $d < 10$ nm) with energies within 50 eV and, therefore, in order to get to the scintillator, they are first accelerated towards the scintillator by applying a bias voltage of ~10 keV to it. Further acceleration is achieved by a grid placed around the scintillator to which a potential of about 200 V is applied. If the angle between the incident beam and the surface of the sample (tilt angle) increases, SE yield increases as well. In fact, near the edges, due to the proximity to the surface of a larger portion of the interaction volume, more electrons are emitted and therefore the area appears brighter. This is called the *edge effect* and is more evident when a high acceleration voltage is used, due to the greater penetration of electrons. For opposite reasons, the valleys appear less bright, therefore, the joint effect generates topographical contrast (figure 10.14, left panel). In figure 10.14 (right panel) is shown an example of an image acquired by using SEs, illustrating iron nanoparticles obtained by pulsed laser deposition. A contribution to topographic contrast is also given by the backscattered electrons that directly strike the scintillator. Since these electrons have high energy, they are not deflected by the

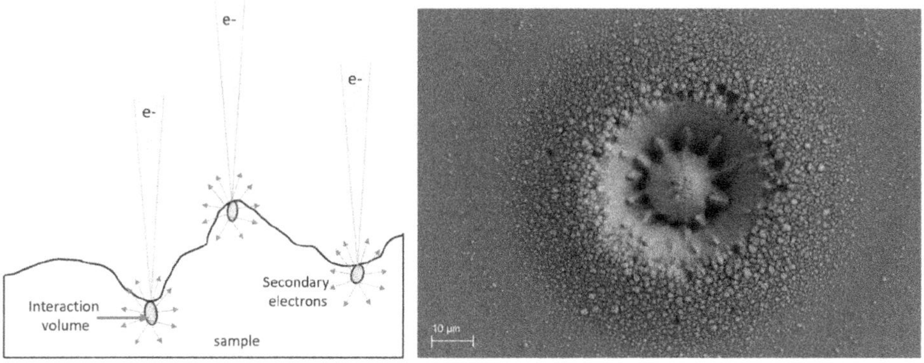

Figure 10.14. Left: an edge effect scheme. Right: Si surface structuring obtained using a pulsed laser deposition technique.

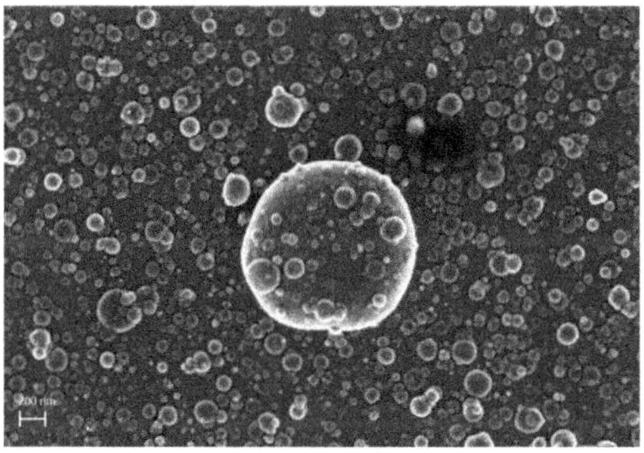

Figure 10.15. An example of an image acquired using an in-lens detector illustrating iron nanoparticles obtained by pulsed laser deposition.

potential of the scintillator, therefore only those emitted in the direction of the scintillator are able to reach it. With a detector located to one side of the specimen, a strong shadowing effect therefore occurs.

In-lens detector. In many cases, SEs are detected through in-lens detectors. They consist of a scintillator detector inserted inside the objective and the SEs pass upwards through the objective and arrive at the detector (see figure 10.13, right panel). As we can see from the image in figure 10.15, this system is very efficient thanks to the very small operating distances due to its geometric position in the beam path. Moreover, the combination with the electromagnetic lens also affects efficiency because its electrostatic field reduces the noise due to other electrons. The main advantage of using the in-lens detector is being able to obtain images determined only by SEs and therefore richer in surface information. In fact, the ET detector, which also collects many BSEs, provides images with greater contrast but with less structural detail (see figure 10.16).

BSE imaging. In addition to the contribution to topographic contrast, BSEs can provide information on atomic number contrast. In fact, the backscattering yield increases monotonically with Z (figure 10.17, left panel) and areas with a high

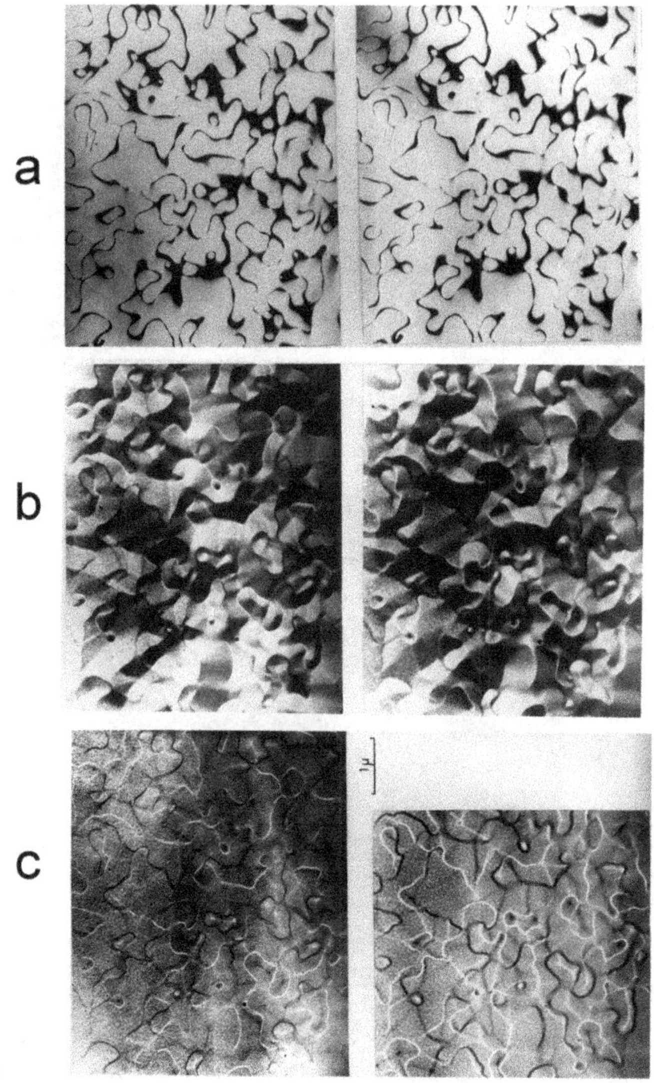

Figure 10.16. Electron microscope images of Cu–Mn–Al Heusler compound showing magnetic domain walls: (a) antiphase boundary dark-field imaging, (b) magnetic domains obtained through Foucault imaging and (c) magnetic domain walls obtained through Fresnel imaging [8] Wikimedia Common. Author: Alapper.

average atomic number appear brighter in the contrast [9]. Figure 10.17 (right panel) shows an example of a BSE image taken of a polished crystal showing a lamellar pattern. The light stripes have a composition of the $Sr_3Ru_3O_7$ phase while the dark stripes are of the Sr_2RuO_4 phase.

BSE detection takes place using a *solid-state detector* usually in the form of a thin flat plate, which is mounted under the lens poles (see figure 10.18).

This detector takes advantage of the fact that when a high-energy electron hits a semiconductor, it produces many electron–hole pairs. Normally these recombine

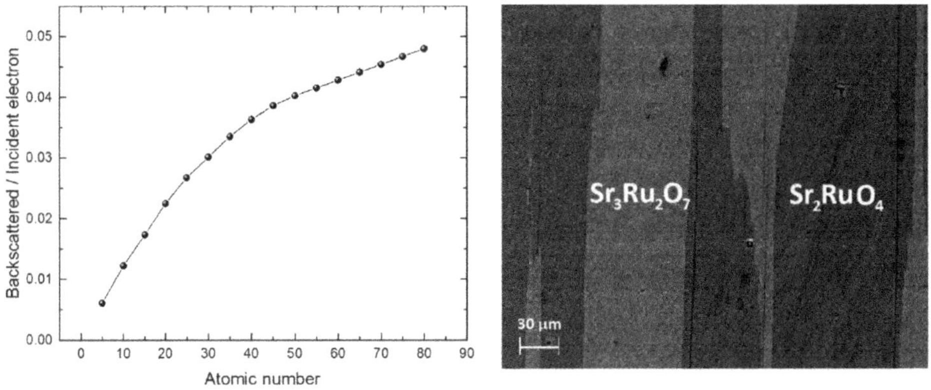

Figure 10.17. Left: The ratio of BSEs to the incident electron number (also called the yield η) as a function of the atomic number (interpolation of data from [13]). Right: a BSE image of different phases of strontium ruthenate oxide which result in the contrasting light and dark areas.

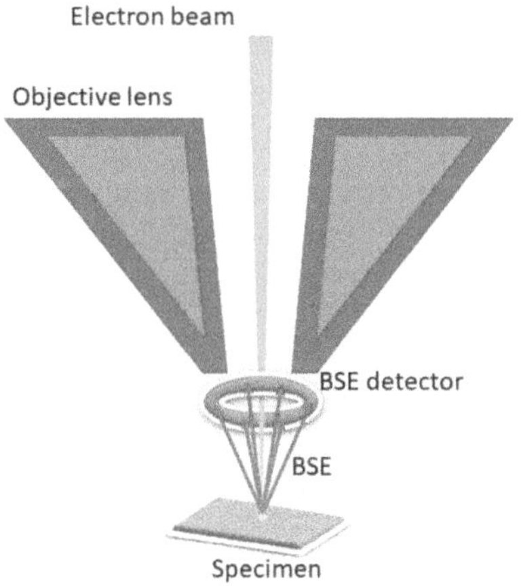

Figure 10.18. BSEs are detected by a solid-state detector whose position relative to the sample allows material differences in the samples to be displayed very efficiently.

quickly, but if a voltage is applied to the semiconductor then they can be separated, thus producing a current, which can subsequently be amplified. The detector is usually in the form of a thin flat plate, which is mounted under the lens poles. However, this detector is characterized by a relatively slow response time and is not suitable for rapid scans.

10.3 Compositional analysis in an electron microscope

X-rays can be used to gain chemical information about the specimen. From the characteristic x-ray spectrum emitted, the energy or the wavelength of each x-ray can be measured, and we can find out which elements are present in the sample. This analysis which has more qualitative aspects can be accompanied by a quantitative analysis which, always obtaining information from the x-ray spectrum, measures how many atoms are present of each type. X-ray spectrometers are generally of two types, The energy dispersive (ED) type simultaneously records the energy of all x-rays emitted and provides us with the intensity as a function of the energy of the photon. The wavelength dispersion (WD) type uses Bragg reflection by the crystal lattice of the sample, with the spectrometer 'tuned' to only one wavelength at a time. The use of one technique rather than the other is dictated by the needs of the study conducted. In fact, for example, the spectral resolution is better for the WD type than for the ED type, but the latter is faster and easier to use.

10.3.1 X-ray spectroscopy

As we have seen in section 10.2.2, if the primary electron beam hits a solid sample with a kinetic energy of some kiloelectronvolts, electrons in the inner shells of an atom are ejected and the atom is ionized. The excited atom returns to its stable state through an electron in one of its outer shells releasing its excess energy in the form of x-rays and filling the hole. The kinetic energy of the beam electrons depends on the accelerating voltage between the filament and the anode and we have

$$E_{\mathrm{k}} = \frac{1}{2} m_{\mathrm{e}} v^2 = eV,$$

where E_{k} is the kinetic energy, V is the applied voltage, m_{e} is the electron mass and v is the electron velocity. Once the internal shell has been ionized, the ionized shell, without its electron, is available to be filled again. Therefore, when an outermost electron transits the innermost shell, it emits an x-ray photon whose energy is equal to the difference in energy between the initial and final levels:

$$E_{\text{x-ray}} = E_{\text{final}} - E_{\text{initial}}.$$

The emitted x-rays' intensity as a function of wavelengths is shown in figure 10.19. The spectrum shows a mixture of all wavelengths emitted by the electrons braked by interaction with the atom as a continuous spectrum. As we can see from the figure, the continuous spectrum has a lower limit for wavelengths. This value corresponds to an upper limit for the energy emitted by an electron braked by the nucleus, which

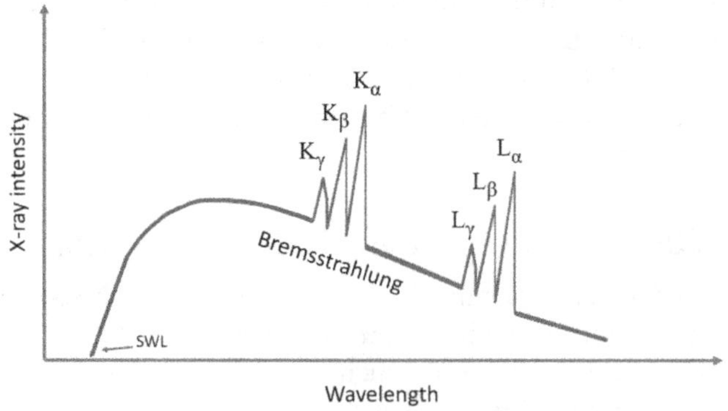

Figure 10.19. Schematic x-ray spectrum as a function of wavelength.

is when the electron stops completely, releasing all the kinetic energy to the emitted photon. It is called the shorter wavelength limit λ_{SWL} and it can be obtained considering

$$E = eV = hf,$$

$$\lambda = c/f,$$

$$\lambda_{SWL} = hc/eV,$$

where c is the light velocity, f is the frequency of the emitted photon and h is Planck's constant. The intensity of the continuous x-ray spectrum I_{cs} is given by

$$I_{cs} = AiZV^m,$$

where Z is the target atomic number, i is the tube current, A is a proportionality constant and m is a constant with a value of about 2 [10]. Superimposed on a continuous spectrum, if the accelerating voltage is raised above a certain critical value, depending on the target material we can observe several sharp intensity maxima that appear at certain wavelengths which are characteristic of the atomic species. These narrow peaks (or characteristic lines) constitute the so-called characteristic x-ray spectrum. Characteristic lines are due to the transitions of electrons between the internal orbits of bound electrons, whose energies are mainly governed by the main quantum number n. For $n = 1$ there is the shell K, occupied by a maximum of two electrons, for $n = 2$ there is the shell L, occupied by a maximum of eight electrons, for $n = 3$ there is the shell M, occupied by a maximum of 18 electrons and so on. Excluding shell K, the other shells are divided into sub-shells with slightly different energies, for example, shell L is divided into L_1, L_2 and L_3, while M is divided into five sub-shells. As the atomic number increases, these orbits are all filled and form a core around which the other electrons orbit. For a characteristic x-ray line to be produced, the kinetic energy of the incident electrons must be greater than the energy required to ionize one of these innermost shells. This condition occurs if the voltage V, which accelerates the electron beam, is greater than the excitation voltage, i.e. the voltage at which ionization energy is reached for a given shell. The intensity of any characteristic line, measured above the continuous spectrum for a given shell, is given by

$$I_{\text{Shell-line}} = B(V - V_{\text{Shell}})^n,$$

where B is a proportionality constant, V_{Shell} is the excitation voltage for the given shell and n is a constant with a value of about 1.5 [10]. The wavelength of any particular line decreases as the atomic number of the emitter increases. In particular, if we consider the characteristic line energies, these are proportional to the square of the atomic number Z (Moseley's law). In the spectrum of x-ray radiation, in correspondence to the energy and therefore of the wavelength at which the transition occurs, we observe a peak designated by the shell that initially contained the vacancy. Commonly, each group of characteristic lines are labelled with Greek characters from the most intense characteristic line to the least intense one

Table 10.3. Some commonly used x-ray energies in electronvolts [11].

Element	K_{α_1}	K_{α_2}	K_{β_1}	L_{α_1}	L_{α_2}	L_{β_1}
Fe	6403.84	6390.84	7057.98	705.0	705.0	718.5
Co	6930.32	6915.30	7649.43	776.2	776.2	791.4
Cu	8047.78	8027.83	8905.29	929.7	929.7	949.8

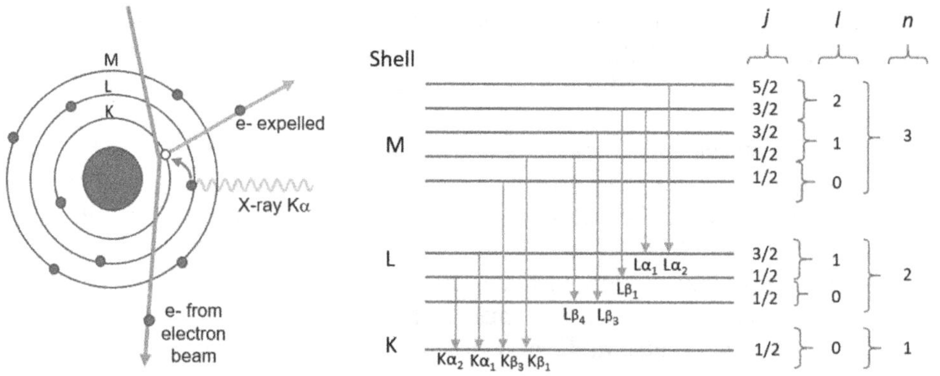

Figure 10.20. Left: sketch of characteristic x-rays K_α generated by an electron transition to an inner free energy level evacuated by a primary electron. Right: a scheme of the most important x-ray emission transitions, the K, L_1, L_2, L_3, etc levels, with their quantum numbers n, l and j (see [14]).

(table 10.3). As we also see from figure 10.20, not all transitions are allowed. In fact, the transitions are constrained by the dipole selection rules that establish the allowed transitions:

$$\Delta l = \pm 1$$

$$\Delta j = 0,\ 1.$$

As can be seen from the data in table 10.3, the components K_{α_1} and K_{α_2} have energies, and therefore wavelengths, that are very similar and often cannot be resolved as separate lines.

10.3.2 Energy dispersive x-ray spectroscopy (EDS)

In EDS systems an x-ray semiconductor detector is used, usually consisting of silicon or germanium. In these elements, the valence band of the electrons is entirely occupied, but the conduction band is separated by a very small potential barrier (1.1 eV for Si and 0.7 eV for Ge [8]). X-rays reach the detector thanks to the use of a beryllium window that is capable of resisting atmospheric pressure. However, this material absorbs all x-rays with energy less than 1 keV, but lower energy rays can be detected using a suitable polymer (see figure 10.21). Liquid nitrogen is used for cooling the detector to reduce noise caused by thermal excitation. When an x-ray photon is absorbed by these semiconductors, it provides the energy necessary to

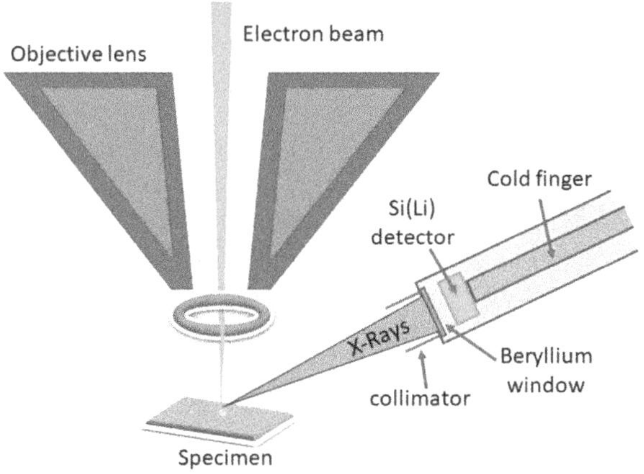

Figure 10.21. Mounting arrangements for a solid-state detector used for energy dispersive spectrometry.

move some valence electrons into the conduction band. Therefore, an incident photon generates a short current impulse caused by electrons in the conduction band and by holes in the valence band which move under the influence of a voltage bias applied to the detector. The pulse amplitudes are measured and then stored in a multichannel analyser where each channel corresponds to a certain interval of pulse heights. A spectrum is then plotted showing the number of counts per energy channel. Then, peaks forming at certain energy levels in the spectrum can be identified and their intensities are proportional to the amount of elements present within the specimen with that x-ray energy. The energy resolution range in the EDS technique is from 70 to 130 eV.

The concentration of a given element can be determined by carrying out a quantitative analysis. For example, the so-called *ZAF* method uses the ratio between the peak and a reference intensity (typically of pure elements such as Al, Si, Cr, Fe, Ni, etc [12]) as the primary input in order to be completely independent of scaling factors such as the beam current, acquisition time and solid angle of the detector. The weight fraction of the *i*th element is calculated as

$$C_i = (ZAF)\frac{I_i}{I_{(i)}},$$

where the Z factor is the atomic number correction accounting for the fraction of electrons backscattered by the sample and the volume of x-ray generation, A is a correction for the absorption of the generated x-rays in the sample, F is a correction for secondary x-ray fluorescence caused by x-rays of heavier elements in the sample exciting lower elements to give false intensity and, finally, $I_i/I_{(i)}$ is the ratio of the peak to the reference intensity of the *i*th element. Figure 10.22 presents an example of an EDS spectrum (black line) taken from a $Sr_2RuO_4/Sr_3Ru_2O_7$ eutectic crystal, showing x-ray intensity versus energy. Identification lines for the emission energies for strontium (Sr), are displayed and these correspond to peaks in the spectrum. Similarly, there are peaks indexed as ruthenium (Ru) and oxygen (O).

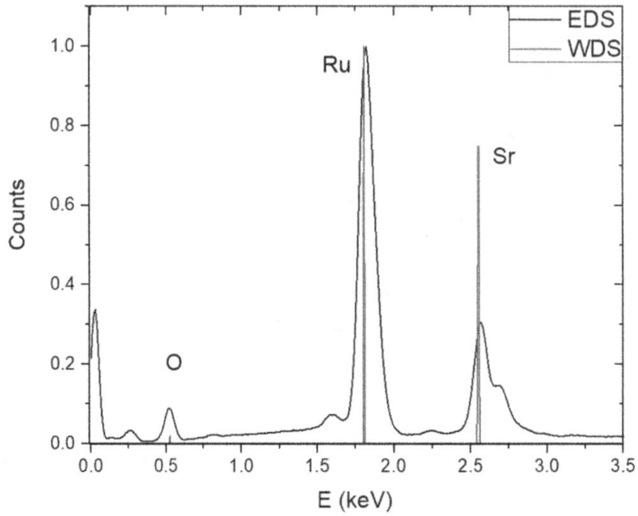

Figure 10.22. In this graph, the EDS spectrum (black line) and WDS spectrum (red line) of a $Sr_2RuO_4/Sr_3Ru_2O_7$ eutectic crystal are shown.

10.3.3 Bragg reflection

In 1913, for the first time, Lawrence Bragg found that crystalline materials give rise to characteristic patterns if impacted by x-ray radiation. Intense peaks of diffracted radiation were observed at well-defined wavelengths and directions of incidence. To explain this behaviour, Bragg considered the crystal to consist of parallel planes of ions placed at a distance d. The conditions for a peak intensity to take place are:

- Specular reflection of the x-rays by the individual atoms of the plane.
- Constructive interference between the rays reflected by consecutive planes.
- Elastic reflection.

In particular, according to Bragg's law, constructive interference occurs if the difference in path length between points A, B and C and A′, B′ and C′ (see figure 10.23) is an integral multiple of the wavelength λ, then

$$n\lambda = 2d \sin \theta,$$

where θ is the angle that the incident beam forms with the crystalline plane, λ is the wavelength of the radiation, d is the distance between two adjacent planes and n is a positive integer.

10.3.4 Wavelength dispersive x-ray spectroscopy (WDS)

Similarly to EDS, WDS uses the characteristic x-rays generated from a sample bombarded by electrons to identify the elemental constituents comprising the sample. Of the two methods, EDS is more commonly employed as data collection and analysis is a relatively quick and simple process because the complete spectrum of energies is acquired simultaneously. However, using WDS, the spectrum is

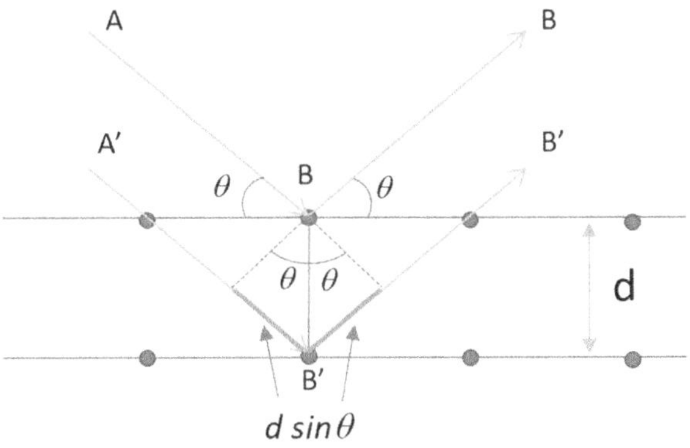

Figure 10.23. Bragg reflection by a family of lattice planes, separated by a distance d. The incident and reflected rays are shown for two adjacent planes.

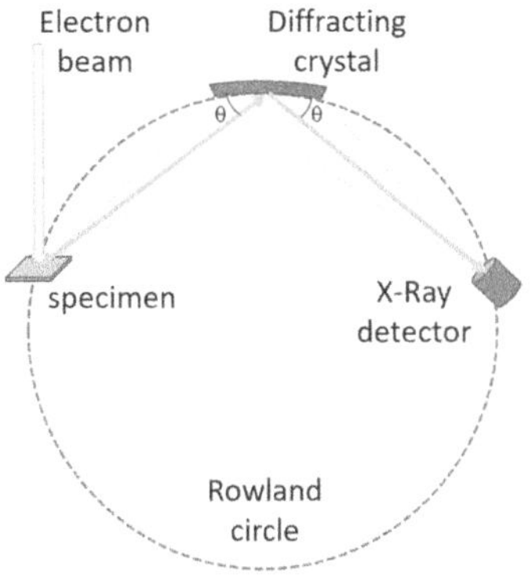

Figure 10.24. Rowland circle geometry: a crystal with fixed d is moved along a circle to vary θ so that x-rays of different λ are recorded.

acquired sequentially as the full wavelength range is scanned. Although it takes longer to acquire a full spectrum, the WDS technique has a much improved resolution compared to EDS, having peak widths in the range between 2 and 20 eV (see figure 10.22). Moreover, since no windows are needed, the x-rays are not absorbed and therefore light elements can be detected. Inside the spectrometer, analysing crystals of specific lattice spacings are used to diffract the characteristic x-rays from the sample into the detector. This is because the wavelength range is limited for a given value of $2d$, and several crystals of different spacings are therefore needed in order to cover the whole range of wavelengths of interest.

The wavelength of the x-rays diffracted into the detector may be selected by varying the position of the analysing crystal with respect to the sample, according to Bragg's law. A diffracted beam occurs only when this condition is met and therefore interference from peaks from other elements in the sample is inherently reduced. The most intense reflections are the first order ones ($n = 1$) while higher orders, adding unwanted peaks, have a relatively low intensity. However, x-rays from only one element at a time may be measured on the spectrometer and the position of the crystal must be changed to tune to another element. The final spectra are obtained through precise geometry between the sample and the reference crystal. The most precise geometry, called Johansson geometry, is shown in figure 10.24. The diffracting crystals are bent to a radius of $2R$ and then grounded to a radius R which is the radius of the focusing circle, called the Rowland circle. According to this configuration, all the points of reflection lie on the Rowland circle.

References

[1] Homann J 2006 Spielsand bei ca. 200 fach im Dunkelfeld *Wikimedia commons* https://commons.wikimedia.org/wiki/File:Sand_bei_200_fach_2.jpg

[2] Kallerna 2016 Thin section scan crossed polarizers Siilinjärvi apatite ore *Wikimedia Commons* https://en.wikipedia.org/wiki/File:Thin_section_scan_crossed_polarizers_Siilinj%c3%a4rvi_R636-105.90.jpg#filelinks

[3] Goodhew P J, Humphreys J and Beanland R 2001 *Electron Microscopy and Analysis* (London: Taylor and Francis)

[4] De Broglie L 1925 Recherches sur la théorie des quanta *Ann. Phys.* **10** 22–128

[5] McMullan G *et al* 2009 Experimental observation of the improvement in MTF from backthinning a CMOS direct electron detector *Ultramicroscopy* **109** 1144–7

[6] Goldstein J, Newbury D E, Joy D C, Lyman C E, Echlin P, Lifshin E, Sawyer L and Michael J R 2003 *Scanning Electron Microscopy and X-ray Microanalysis* (New York: Plenum)

[7] Reed S J B 2005 *Electron Microprobe Analysis and Scanning Electron Microscopy in Geology* (Cambridge: Cambridge University Press)

[8] Alapper 1970 Magnetic domains on antiphase boundaries in Heusler alloy *Wikimedia Commons* https://commons.wikimedia.org/wiki/File:Magnetic_Domains_on_Antiphase_Boundaries_in_Heusler_Alloy.jpg

[9] Böngeler R, Golla U, Kässens M, Reimer L, Schindler B, Senkel R and Spranck M 1993 Electron–specimen interaction in low voltage scanning electron microscopy *Scanning* **15** 1–18

[10] Cullity B D and Stock S R 2001 *Elements of X-ray Diffraction* 3rd edn (New York: Prentice-Hall)

[11] Thompson A C *et al* 2001 *X-Ray Data Booklet* (Berkeley, CA: Lawrence Berkeley National Laboratory, University of California)

[12] Ritchie N W M and Newbury D E 2015 Performing elemental microanalysis with high accuracy and high precision by scanning electron microscopy/silicon drift detector energy-dispersive x-ray spectrometry (SEM/SDD-EDS) *J. Mater. Sci.* **50** 493–518

[13] Sylvester P 2012 Use of the mineral liberation analyzer (MLA) for mineralogical studies of sediments and sedimentary rocks *Mineralogical Association of Canada Short Course Series* **42** 1–16

[14] Latośińska J N and Dolenko G N 1999 X-ray emission spectroscopy, methods *Encyclopedia of Spectroscopy and Spectrometry* (Oxford: Elsevier) pp 2463–7

Chapter 11

The second revolution of quantum mechanics: a path for beginners from superconductivity to quantum computers

Mario Cuoco and Marcello Sette

11.1 Introduction: the quantum world in a nutshell

In this chapter we will undertake a journey from the world of quantum phases of matter to the world where those phases are employed to set out specific physical states, which are then prepared, controlled and manipulated to perform operations such as information processing and computation. As mentioned we deal with quantum phases, thus implying that the physical configurations are subject to the laws of quantum mechanics. Although this process is not straightforward, we attempt to establish a short and accessible path through the knowledge that took around one hundred years to consolidate and implement our current view of the way electrons, protons and other fundamental atomic and subatomic constituents behave within the microscopic world. The complexity of this world is undoubted, as underlined by the famous quote by the Nobel Laureate Richard Feynman: 'I think I can safely say that nobody understands quantum mechanics'.

Although the applications of quantum mechanics are not simple, this branch of physics is opening up an entirely new world of technological opportunities, setting up what is currently dubbed the *second quantum revolution*, in contrast to the first quantum revolution that 'just' gave us the new rules that govern the physical world. One of the most celebrated examples of the second revolution is represented by the quantum computer. The recent announcement by Google of the demonstration that a programmable quantum chip can solve a problem that classical computers are practically not able to address, i.e. so-called quantum supremacy, indicates how far and fast research is moving in this area. The development of a quantum computer is reaching a level of maturity where one can foresee true technological breakthroughs and a profound conceptual change in the way we think about computation.

doi:10.1088/978-0-7503-2678-0ch11

A quantum computer is designed to carry out quantum operations, namely it manipulates quantum states in a controlled way such as to perform specific algorithms. We are familiar with bits and how a classical computer works and processes data. Let us then start from this perspective of classical states to smoothly get into quantum states. Here, it helps to draw a parallel about how physicists and computer scientists think in terms of states. A physical state can be encoded in observables, such as the position and velocity of an object, while the laws of physics describe their evolution in time. For example, if we throw a stone, then its position and velocity change at each time instant and Newton's law describes how the state varies in time, its trajectory, etc. On the other hand, a computer state is given by the value of its memory bits and a computer programme sets out how the computer processes the states and how it goes from one state to another. A computer state can be encoded in a string of bits and if a bit value is only 0 or 1 for a string with N elements one can have 2^N configurations. However, since the computer state is classical it can be in one configuration at each time and the computer programme determines the way you connect and vary the states. For each step of a classical algorithm we go from one state to another.

For quantum states the scenario is generally much richer. The typical way to visualize how to operate a quantum state is to introduce a qubit (or quantum bit). The qubit is the quantum mechanical counterpart of the classical bit. In classical computing the information is encoded in bits, with each bit having either the value zero or one. In quantum computing the information is built on qubits. A qubit is the most elementary unit that can be employed for quantum manipulation because it is based on two quantum states which are usually expressed as $|0\rangle$ and $|1\rangle$ (the use of $|\ldots\rangle$ is to distinguish them from classical encoding with the integers 0 and 1), but unlike the classical bit can also manifest as a combination of the two states, thus potentially introducing an *infinite* set of physical possibilities. This is one of the most important properties of the quantum world and it is usually called the *superposition principle*. One pictorial way to visualize the striking difference between bits and qubits is to use a vectorial representation for the states and construct a so-called Bloch sphere (see figure 11.1) in a three-dimensional space. When we think about classical states there are only two possible configurations, either $|0\rangle$ or $|1\rangle$ which on the Bloch sphere correspond to the vectors pointing at the north and south poles. A qubit can instead be in a generic superposition of $|0\rangle$ and $|1\rangle$ which means, within the mathematical correspondence, that it can point anywhere within the Bloch sphere. The superposition principle implies that the qubit can be somehow *simultaneously* in the states $|0\rangle$ and $|1\rangle$. The bizarre aspect of the quantum world is that when a qubit is measured (or more precisely when the corresponding observables are measured), the qubit will collapse into either $|0\rangle$ and $|1\rangle$ and the measured value will reflect that state. For example, when a qubit is in a superposition state of equal weights (e.g. the vectors within the equator plane of the Bloch sphere), a measurement will make it collapse into $|0\rangle$ and $|1\rangle$ with equal probability.

One may wonder why we need a three-dimensional vector on a unit sphere to describe a qubit. The answer is related to the fact that a qubit is the natural way to describe the *spin* of an electron. An oversimplified view of the electron is to imagine

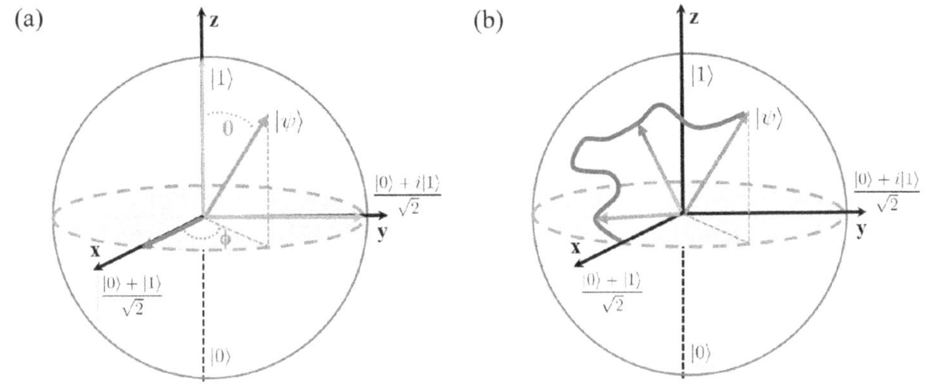

Figure 11.1. (a) Illustration of the Bloch sphere to describe all possible qubit configurations by means of a vector of unit length that can have a generic orientation that is set by the polar θ and azimuthal ϕ angles. The classical bits correspond to the vectors indicating the north (grey vector) and south poles, while a qubit ($|\psi\rangle$) can be represented as a unit vector (green) pointing anywhere on the sphere. The red (blue) vector along the x-(y-)axis within the equator plane indicates a special case of superposition of $|0\rangle$ and $|1\rangle$, having coefficients of equal amplitude in absolute value. Since the values of the angles θ and ϕ are in principle infinite, a qubit can be in an infinite number of configurations. This is the quantum power of obtaining infinitely many states by means of superposition. For the readers who are not familiar with complex numbers we note that 'i' in mathematics is the so-called imaginary unit such as $i^2 = -1$. (b) The operation on a qubit creates a vector evolving on the Bloch sphere along a given trajectory (the red line).

it as a sphere that can rotate around a given axis. The spin can be viewed as the intrinsic angular momentum of the electron associated with its rotation and thus it has a vectorial nature with a direction, related to the sense of rotation, and an intensity. The strange aspect of the quantum nature of the electron, however, implies that the spin projection along any given direction is *quantized* in the sense that it can assume only two discrete values that are indicated as +1 and −1 (in units of $\hbar/2$, i.e. half the reduced Planck constant $\hbar$). The states $|1\rangle$ and $|0\rangle$ correspond to the values +1 and −1 of the z-projection of the spin angular momentum which is usually denoted by S_z. In a similar way, if the spin is pointing along the x- or y-direction (i.e. the electron *rotates* around the x- or y-axis) one can again identify two quantum states with angular momentum +1 and −1 which are constructed by making a suitable superposition of $|1\rangle$ and $|0\rangle$ (see figure 11.1). In a quantum process with a quantum gate the qubit can evolve in time by changing its configuration on the Bloch sphere. If the qubit is the spin of an electron, then the quantum transformation can be operated, for example, by applying a magnetic field that varies its direction in time. By doing this, in principle, one can move continuously from the north pole (state $|1\rangle$) to the south pole (state $|0\rangle$) along a suitable trajectory on the sphere such as to match the requirements of the quantum algorithms.

The quantum world has other bizarre principles which one will always encounter, and one needs to get used to them if one aims to exploit them. This is, distinct from superposition, *entanglement* (see figure 11.2). Such concepts are crucial for understanding the quantum world.

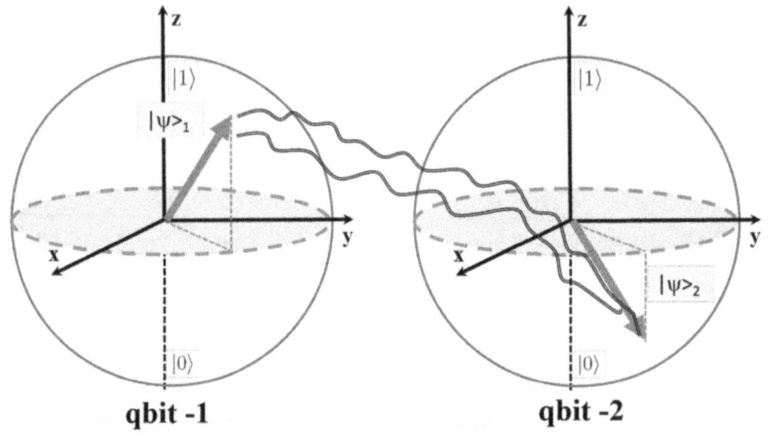

Figure 11.2. Sketch of two entangled qubits $|\psi_1\rangle$ and $|\psi_2\rangle$. The wavy double lines mimic the special 'connection' between them. Basically, the entanglement emerges from the results of the measurements. The outcome of the measurements on the single qubits could be 0 or 1. However, the output of the measurement on one qubit will always be correlated to that of the other qubit.

Later we will elaborate further on these aspects and specifically present how a quantum algorithm operates. Before that, within our presentation we will take a special trajectory by first selecting and describing a distinct quantum phase of matter, thus moving through aspects of materials that manifest extraordinary properties due to the collective action of their constituents and, in this context, we also briefly look at the microscopic scenario behind the emergent quantum configurations. Here, we aim to provide basic descriptive elements of how materials can be employed to yield a quantum bit.

The phase of matter that we are going to employ is the superconducting state because it is a paradigmatic case and it has genuine quantum properties that can even manifest at the macroscale. We will discuss what makes it special for exploitation towards the setting out of quantum bits and, potentially, the engineering of a quantum computer. There are many problems and challenges to be faced along this path. The goal here is to provide a simplified view of the main concepts as well as a prospective vision of the ultimate applications. Once we have touched on the physical states which can yield quantum bits, we will briefly look at how quantum computing can be accomplished and present how a quantum algorithm works.

The overall plan is as follows. The starting point of the targeted phase of matter refers to the basic phenomenology of superconductivity and the description of how superconducting materials can be designed to achieve elemental quantum bits (qubits) by realizing specific superconducting circuits. Hence, we provide a few examples of operations with qubits, considering the way they can be prepared and manipulated. This part focuses on one specific quantum algorithm and uses various examples in order to make the flow of the presentation less abstract.

11.2 Superconductivity: symmetry and quantum mechanics at the macroscopic scale

The word superconductivity explicitly states that we are dealing with an extraordinary type of conductivity and, in particular, refers to electrical conductivity. Thus, to understand what superconductivity is in a simple fashion, we need to begin by recalling a few basic properties of another state of matter which is more familiar and common in our day life, i.e. the metallic state or the state of a conventional electric conductor. Ohm's law expresses the key nature of the metallic state. The electrons can flow in a metallic wire only when a non-vanishing voltage V (or electric field) is applied and the resulting electron current I is such that $V = IR$, with R being the resistance of the wire with a dependence on the material's properties and the geometry. Furthermore, the canonical metallic state does not have a special magnetic response and, from a microscopic point of view, it can be 'imagined' as a quantum electron liquid, that is a liquid of electrons which is subject to the law of quantum mechanics.

When dealing with superconductivity one enters into a completely new realm of electronic quantum liquids. In short, avoiding technicalities, if somebody is asking for a direct and short description of the superconducting phase there is a series of fundamental properties which have to be listed in order to uniquely single out a superconductor:

1. A superconductor is separated by a well-defined temperature from a metal. There is a demarcating, or so-called critical temperature, T_c below which we have a superconductor and above which a metal. T_c can range from 10 K (–263 °C) to ~180 K (or even closer to the temperature of a conventional freezer if pressure is used) [1, 2].
2. Superconductors can sustain persistent currents in a loop for an *infinitely* long time even without an applied voltage. This property is an alternative way to state that the electric resistance in the superconducting state is zero, i.e. the current injected in the loop in practice does not decay in time.
3. Superconductors expel the magnetic field, according to the Meissner effect (figure 11.3). A magnetic field cannot penetrate into a superconductor due to its so-called *gauge* or *phase rigidity*, which is a manifestation of the symmetry that is broken when passing from a superconductor to a metal (we will expand a bit more on this aspect later). This property is also indicated as perfect diamagnetism (in the case of ideal screening), because the magnetization inside the superconducting material is equal and opposite to the applied external field, so that the resulting magnetic field is vanishing.
4. Superconductors with a not-simply connected shape (e.g. a ring or a doughnut) exhibit magnetic flux quantization, namely the flux has values which are multiples of a fundamental magnetic flux unit, $\Phi_0 = \frac{h}{2e}$, where h and e are the Planck constant and electron charge, respectively. This quantization aspect is typical of the quantum world and it has a topological character because the magnetic flux inside the superconducting ring can only

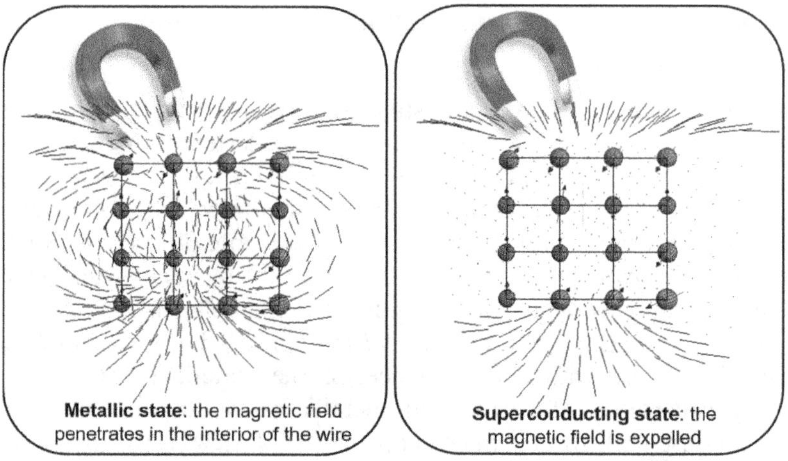

Figure 11.3. Schematic description of the Meissner effect. We have sketched the crystalline wire as a periodic arrangement of atoms (coloured circles) with a square geometry and included the spin of the electrons (arrows) orbiting close to the atoms. The magnetic field is described by the line forces through segments that indicate the orientation and the distribution. In the metallic state the field is everywhere within the crystal (left). For the superconducting state (right), the magnetic field is expelled due to its rigidity (see the main text).

have values which are an integer multiple of Φ_0, independent of the shape of the superconductor.

5. Putting two superconductors into contact weakly, you can observe the so-called Josephson effect, i.e. a coherent tunnelling of electron pairs (i.e. supercurrent flow) without any voltage drop or dissipation.
6. Superconductivity is a phase of matter where cooperative quantum effects lead to an effective *attraction* among electrons.

For the sake of completeness, we point out that some of the above characteristics can appear in a slightly different form in most superconducting materials. For example, concerning property 3 above, the magnetic field can partially penetrate inside the superconductor forming a so-called mixed phase where the normal metal and superconducting domains coexist. In other words, the demarcation between the superconductor and normal metal phase can be less sharp, both regarding the temperature and the magnetic field variations.

Comparing the superconducting and the metallic states, striking differences can be noted immediately: the superconductor can allow a flow of current without dissipation and it has a remarkable diamagnetic response. For the aims of this chapter, we would like to bring the attention of the reader to the Josephson effect, because it plays a role in building the superconducting devices which can eventually be employed as quantum bits. Before moving on to superconducting devices, however, we briefly look at what happens within the microscopic world (i.e. at length scales of the order of 10^{-9} m) of the materials that become superconductors. In particular, we touch upon the nature of the electronic organization that produces this remarkable phenomenon. This problem has a compelling history with many proposals before the BCS theory was formulated thanks to the pioneering work of John Bardeen, Leon N Cooper and John Robert Schrieffer [3].

There is one general and foundational concept that needs to be mentioned to construct a basic view of superconducting phenomenology, namely the idea of spontaneous symmetry breaking. Any time we deal with a physical configuration, it is instructive and useful to search for symmetries behind that state. This statement applies to our general understanding of nature and goes beyond the properties of materials. Symmetry is extremely crucial in understanding the physical world around us. Spontaneous symmetry breaking in this context is a fundamental organizing principle. Typically, the laws of physics are marked by specific symmetries, i.e. they do not change under some transformations. However, a particular state of matter that is realized may not exhibit all those symmetries. In this sense, superconductivity is a spontaneously symmetry broken phase of matter, but the symmetry that is involved has a genuine *quantum character*, thus it is not immediately understood and intuitively imagined. For this reason, before touching upon the case of superconductivity, let us move by means of analogies and consider few examples to better visualize and become familiar with the concepts of spontaneous symmetry breaking.

The laws of physics are translationally and rotationally invariant, however, the states which we often experience are not translationally or rotationally symmetric. A paradigmatic example is that of a material with a crystalline structure. We know that a solid with a crystalline structure becomes a liquid above the melting temperature. In the liquid phase it preserves translational symmetry, while if we look closely at a crystal the atoms occupy specific positions in space with a regular pattern and thus translational symmetry is broken. This can be experimentally observed by means of x-ray diffraction or scanning tunnelling microscopy, which allow the patterns of atoms to be seen. An important consequence of symmetry breaking is that the two phases have a different type of *rigidity*. In a word, a crystal is rigid and tends to keep its shape while the liquid is not. The differences in rigidity can be understood by looking at the nature of the forces which enter into play for the liquid and crystalline phases. Typically, the forces for the case of a liquid are viscous forces and thus they are proportional to the velocity of the liquid, while in a crystal a mechanical stress or an applied force is typically proportional to the displacement of the atoms and thus there can be restoring mechanisms if the atoms deviate from the equilibrium position, contrary to the liquid case. Another relevant concept which emerges from symmetry breaking is that of having a violation of conservation laws (in physics this is formally accounted for by Noether's theorem). Typically, symmetry means that nothing changes under a given transformation, which intuitively implies that something is conserved although we are making modifications to the physical configuration. Indeed, for a translationally symmetric system momentum is conserved, while for a crystal, which does not have translational symmetry, the momentum is not conserved. Hence, to wrap up the key concepts, if we have a broken symmetry state one should expect a change in the *rigidity* when going from the symmetric to asymmetric configuration, and a *violation of the conservation law*.

Going back to the case of the superconductor one can ask which symmetry is broken and what the associated rigidity is. One can start by saying what is not conserved when the symmetry is broken, rather than the symmetry aspect behind it. In short, the superconducting state or, more precisely, the superfluid is a state where the constituents (e.g. electrons or atoms) are paired and their number is not conserved. To understand how it is possible to have a physical state with a non-conserved number we have to recall the principle of superposition which we introduced before for the qubit. For a given energy, the electrons can only be in three possible configurations, which are typically labelled by counting the corresponding occupation. For example, we can have states with zero, one and two electrons as $|0\rangle$, $|1\rangle$ and $|2\rangle$, respectively (there cannot be more than two electrons for each energy due to the Pauli principle). Now, a generic state with a number of electrons that is not conserved can be constructed as a superposition of $|0\rangle$ and $|2\rangle$. However, since we want to set out a configuration that can never reduce to states $|0\rangle$ and $|2\rangle$, having instead a well-defined number, one needs to construct a special superposition through a complex phase factor $e^{i\phi}$, with 'i' being the imaginary unit fulfiling the relation $i^2 = -1$. The complex exponential function $e^{i\phi}$ can be expressed in terms of the trigonometric functions via Euler's relation as $e^{i\phi} = \cos[\phi] + i\sin[\phi]$. This formula is remarkable and tells us that $e^{i\phi}$ never vanishes and it is the unit complex number because it traces a unit circle in the complex plane as ϕ varies from 0 to 2π.

On the basis of the previous statements, it is intuitive to observe that the state $|\psi\rangle = \frac{1}{\sqrt{2}}(e^{i\phi/2}|0\rangle + e^{-i\phi/2}|2\rangle)$ never maintains the number of electrons because it is a superposition of states with inequivalent occupation numbers built on coefficients (i.e. $e^{\pm i\phi/2}$) that are never zero. The state $|\psi\rangle$ can be visualized in qubit language as a vector that is lying within the equator plane with ϕ being the azimuthal angle (figure 11.1). Here, the correspondence is that the state with zero electrons is the vector at the north pole while the state with a pair of electrons is that at the south pole. The rules of the quantum world allow us to construct a configuration that is, as much as possible, far from the classical states with a fixed number of electrons. Those configurations are the set of vectors lying within the equator of the Bloch sphere.

At this stage, we have seen that the quantum state $|\psi\rangle$ does not retain the number of electrons for a given energy. We did not say why, but in a superconductor an effective attraction between electrons can occur and states with pairs of electrons become energetically favourable. Above the critical temperature, the net interaction between two electrons is repulsive (Coulomb interaction). Below the critical temperature, the resulting interaction between two electrons can turn into being slightly attractive, a result of the electrons' interaction with the vibrations of the atoms in the crystal lattice (see figure 11.4). We know that in a material there are many electrons and thus if we aim to construct a physical state with a large number of electrons (let us say N pairs of electrons) we have to devise a sort of multi-qubit state comprising many configurations akin to $|\psi\rangle$. In this way one can obtain a qualitative view of the meaning of broken symmetry and the rigidity for the superconducting state. Indeed,

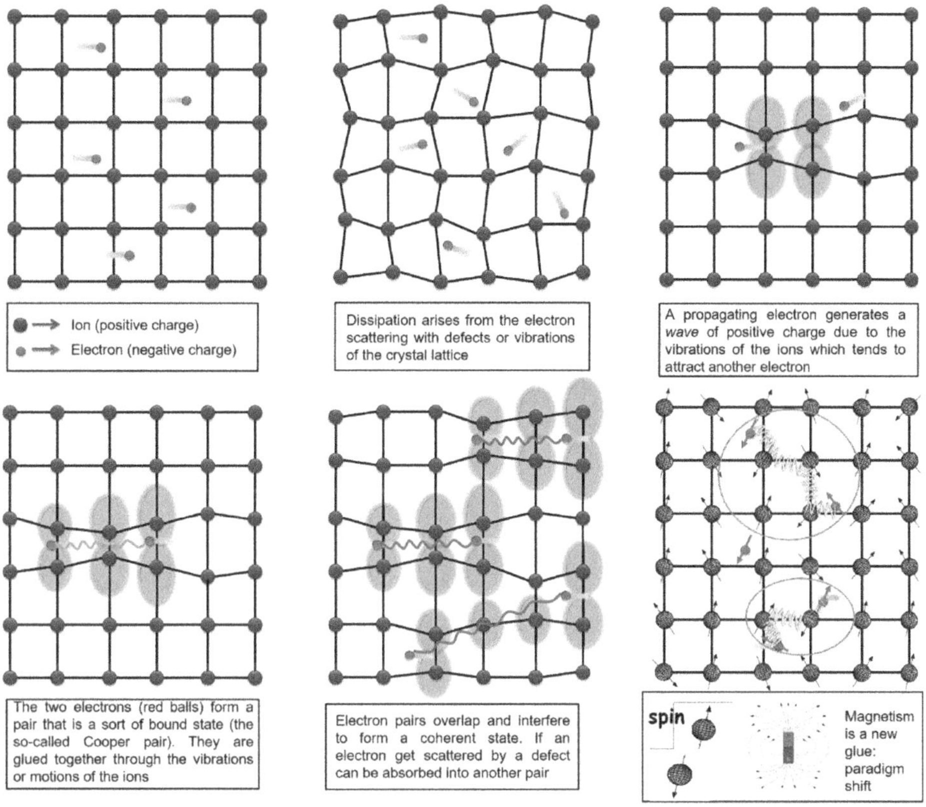

Figure 11.4. Schematic view of the electron pair formation in a representative lattice with square geometry where the ions (blue circles) can vibrate or there are positions which are not periodic. The deformations are amplified for graphical clarity but, of course, they do not correspond in scale to realistic atomic displacements in the material. The electrons (red circles) are scattered by the atom vibrations and can become trapped by a cooperative action of vibrating modes. In the last panel on the right we also sketch the possibility of forming pairs of electrons by means of magnetic waves connecting the spin of the electrons.

in principle each pair of electrons can choose a different value for ϕ and for a single pair of electrons all the ϕ are equivalent in the sense that they yield a state with a non-conserved number of particle. Using again the vectorial representation, for each pair of electrons we have a vector along the unit circle that can be at a generic angle ϕ. On the other hand, if we have N pairs of electrons there is a set $\{\phi_i; i = 1, 2, \ldots, N\}$ of phases that one needs to introduce for describing the multi-particle configuration where, in principle, each pair can choose a different value for the variable ϕ_i. In a superconductor, the interaction between the pairs yields a correlation between the phase and it turns out that the most energetically favourable configuration is that where all the phases are identical. We can now visualize the change-over from the disordered to the ordered phase. In the disordered case all the phases ϕ_i are random and there is no correlation between the electron pairs. On the other hand, in the ordered (superconducting) phase all the phases become locked and assume only one value. The emerging state resembles a large atom where all the electron pairs are in the same $|\psi\rangle$-like configuration with the same angle ϕ providing a quantum coherence that is macroscopic. In a cartoon picture we can put all the quantum states of the electron pairs in a row and describe them with the usual vector in a unit circle forming an angle which is completely arbitrary in the disordered phase, like having many clocks (unit circles) pointing to different hours. For the ordered superconducting phase, instead, all the clocks indicate the same hour. The choice of the same angle ϕ for all the electron paired states is the manifestation of the broken symmetry and the symmetry that is broken is related to the rotation of the quantum state $|\psi\rangle$ that in physics is called gauge symmetry. One can now also ask about the rigidity of the broken symmetry state. The rigidity of the superconductor arises when all the electron pairs are coherently locked together (or entangled), by realizing equal values for ϕ_i, then a variation of the angle turns into an energy cost. In this respect, the superconductor is rigid because it is not convenient to alter the relative angles between the coherent electron paired quantum states. The application of a magnetic field is a physical way to affect the angles ϕ. The fact that the superconductor is rigid to a change of the phase ϕ implies that the magnetic field cannot penetrate into the interior of the material. This is similar to the case of a crystal that is rigid with respect to deformations whose shape cannot change without applying a sufficiently strong mechanical strain. Indeed, for the superconductor one needs to exceed a critical amplitude of the magnetic field to destroy the ordering due to the coherence of the electron pairs or, in other words, to break the phase rigidity. This is a symmetry perspective for understanding the Meissner effect and the weird magnetic response of a superconductor.

11.3 Engineering quantum bits with superconductors

In this section we aim to provide brief basic elements of how a qubit can be constructed by employing superconductors and, in particular, superconducting circuits. First, let us recall the central challenge of quantum processing. We need qubits that one can control, for example with a quantum gate, and one has to build many qubits that should interact between each other in such a way that they are entangled and can operate coherently. Finally, at the end of any quantum process or operation one has to read the final state, thus one needs a quantum readout system. To make the system of qubits function correctly, the gate for each qubit should be fast and very efficient and also the readout has to operate with high speed so that one can read the state of the qubits before they lose the quantum coherence. Hence, there are important requirements to fulfil. On one hand, the qubits have to keep the coherence on a long time scale and for this they should be quite isolated, while on the other hand they should be strongly interacting between each other to be fast operating. For the above reasons, there is a subtle balance between the qubit interaction and the way one tries to implement processing and reading of the qubits. Let us see how superconductivity can be used to find a way to address the challenges of quantum processing.

As we discussed in the previous section, a qubit is based on two energy levels. Ideally, we would like to have only two states involved in the quantum processing.

For this purpose, it is crucial that the choice of the quantum system is such that the energy spectrum is made of energy levels which are not equally distributed. In this way, by suitably selecting the energy of the quantum control knob there will be a unique transition and coupling comprising only the selected energy levels. Indeed, by employing a harmonic time-dependent external field with frequency ω_{12}, one can induce a transition between two levels having energies E_1 and E_2 if the condition $|E_2 - E_1| = \hbar\omega_{12}$. Now, assuming that one can control the frequency of the external field and the amplitude of the energy separation with large precision, then one can avoid having a leakage and drive only a dynamics for the targeted levels to be engineered as qubits.

Why can superconductors match this requirement well? The answer is that the superconductors are remarkable elements in circuitry because they can build up Josephson junctions—an electrical element which is non-existent in normal conductors. If you fabricate a circuit which is based on two superconductors that are separated by a tiny insulating barrier then there will be a current made of electron pairs that will flow through the barrier. These devices are named after Brian Josephson, who predicted in 1962 that pairs of superconducting electrons could *tunnel* through the non-conducting separating layers from one superconductor to another. He also predicted the exact form of the current and voltage relations for the junction. Experimental evidence demonstrated that he was right, and Josephson was awarded the 1973 Nobel Prize in Physics for his great discoveries.

The Josephson effect is marked by a special relation between the current and the applied voltage. Indeed, the current that flows across the Josephson junction is expressed as $I = I_0 \sin[\Phi/\Phi_0]$, with I_0 being the maximal critical current that can flow, while $\Phi = \phi_1 - \phi_2$ is the phase difference between the two superconducting coherent states (here ϕ can be somehow related to the locking phase of the electron pairs within each superconductor). Here, Φ_0 is the quantum flux unit. Furthermore, the flux across the junction depends on the applied voltage through the following relation, $\Phi(t) = \int_{-\infty}^{t} V(t')\mathrm{d}t'$ or equivalently $\frac{\mathrm{d}\Phi}{\mathrm{d}t} = V(t)$. We will not explicitly derive the Josephson equations but we invite the reader to consult the famous Feynman lectures [4] for a simplified and instructive derivation.

With the Josephson relations, one can immediately obtain the expression of the energy of this element by simply recalling that the energy is the time integral of the power that in turn is given by the product of the voltage and the current. Hence, one has that

$$E_\mathrm{J} = \int V(t)I(t)\mathrm{d}t = \int \frac{\mathrm{d}\Phi}{\mathrm{d}t} I(t)\mathrm{d}t = \int I_0 \sin[\Phi/\Phi_0]\mathrm{d}\Phi = E_\mathrm{J}(1 - \cos[\Phi/\Phi_0]),$$

where $E_\mathrm{J} = I_0\Phi_0$ is the Josephson energy. An electric current flowing through a conductor generates a magnetic field surrounding it whose magnetic flux, Φ, is related to the current through the relation $\Phi = LI$, with L being the inductance. With similar arguments as those used for the superconductor, the energy of an inductor is given by $E_I = \frac{\Phi^2}{2L}$.

Comparing the energy of a superconductor with that of a conventional inductor, we observe that the non-linear inductance of the Josephson junction is essential for qubit design, as it allows the design of quantum anharmonic oscillators that have energy levels which are not uniformly distributed. A quantum harmonic oscillator cannot be used as a qubit, as there is no way to address only two of its states. To deepen this analysis one needs to construct the complete energy of the superconducting junction, including the charging part associated with the fact that a variation of the number of electron pairs costs energy due to electric interaction. The total energy of the superconducting Josephson circuit is given by $E_{sc} = \frac{(2e)^2}{2C_J}q^2 - E_J \cos(\Phi)$, where q is the effective charge of the Josephson junction and C_J its capacitance. Here, the charge and the phase are analogous to momentum and position in quantum mechanics. There are ways to functionalize the superconducting element as a qubit which are different in the form of the logical quantum states which are employed. In the superconducting charge qubit, for instance, different energy levels correspond to an integer number of Cooper pairs in the superconducting element. On the other hand, in the phase qubit, the energy levels are due to different quantum charge oscillation amplitudes across the Josephson junction. A complete quantum mechanical description of the Josephson circuit that makes a quantum bit goes beyond the scope of this chapter. Here, it is relevant to underline the role of the Josephson junction to design optimal two-level quantum systems that can be controlled and manipulated by electric fields.

11.4 The quantum world and quantum computers

Quantum computers will be capable of new modes of computation which classical computers are incapable of, even in principle.

In classical physics we follow a modelling of physical states in which they are independent (i.e. they do not affect each other). In the same way one can observe that in classical (Turing) computation the computer states do not interact among each other. The quantum theory of computation is a completely new arena because, as stated in the introduction, due to the superposition and entanglement principles the physical states are strongly interconnected.

One way to put the states in communication is via quantum interference, which is a phenomenon that is quite familiar when dealing with waves. Indeed, these principles permit new modes of information processing and communication which we call *quantum computation* and *quantum communication*. In table 11.1, we report a schematic view of the relationship between information processing and (quantum) physics.

Continuing by analogy, while a bit in classical physics is a degree of freedom which can assume one of two possible values, in quantum mechanics the simplest observable, beyond the trivial one with only one (eigen)value, is a Boolean observable, which has exactly two (eigen)values. For every physical system we can build Boolean observables, i.e. we can intuitively think in terms of *questions* which

Table 11.1. Relationship between information processing and (quantum or classical) physics.

Information processing	Physics
Computer	Physical system
Computation	Motion
Input	Initial state
Rules	Laws of motion
Output	Final state

have yes/no answers. Naively, a qubit can be imagined as a minimal quantum state exhibiting non-trivial Boolean observables.

In classical computation a single bit has a definite value. Its value may evolve, but the evolution rules are sequential and deterministic. In quantum computation a single qubit is a state (this is the crucial difference) where the two possible configurations can manifest together. The evolution of the state is sequential and deterministic as well (according to the rules of quantum mechanics), but it happens simultaneously for the two configurations. This naturally constitutes a sort of new parallel computation paradigm in which, however, we use no parallel computers and in which the parallel computation happens in different subspaces of the same single computational system.

To appreciate the power of such a paradigm, let us think for example about a system of 32 qubits, a limited system if we compare it to the analogue classical one made of 32 bits. However, as each qubit lives in a space spanned by two independent configurations, according to the quantum mechanics rules 32 qubits would evolve in a space of dimension 2^{32}, so the 32 qubit system allows the evolution of 2^{32} configurations at the same time. Hence, one can generate a parallel computation in a single quantum computer with 32 qubits which is equivalent to that performed in a parallel computer made of $2^{32} \approx 4 \times 10^9$ cores. This is very impressive if one considers that one of the best performing computers that is currently available, according to http://www.top500.org, has 10^7 cores.

Quantum states. Let us now have a closer look at the qubits and use a simplified representation where they are constructed with coefficients that are uniquely real numbers. For the following discussion we need to define a subset of the plane where the single-qubit lives. Consider a unit circle (a circle with radius one) and two orthogonal unit vectors constructed with respect to the centre of the circle. Let us call them $|1\rangle$ and $|-1\rangle$.

They correspond to the logical values False and True, respectively[1]. A qubit state is a linear combination of them (see figure 11.5),

$$|q\rangle = a|1\rangle + b|-1\rangle, \tag{11.1}$$

[1] Note that here we use a reverse logic for the correspondence.

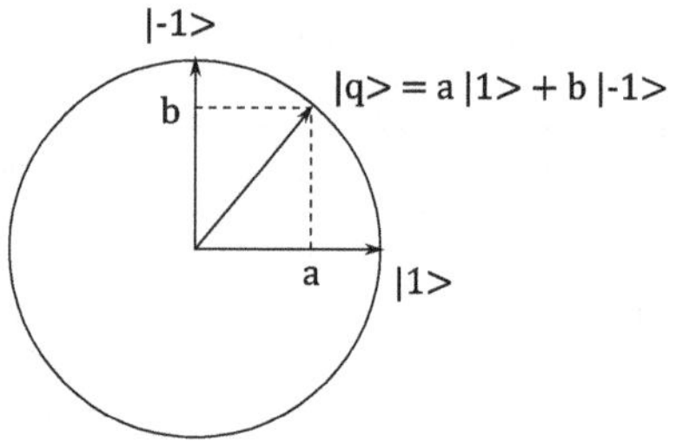

Figure 11.5. A qubit state can be represented as a vector in a plane built as a linear combination of two orthogonal unit vectors with coefficients a and b which are real numbers.

where a and b fulfill the relation[2]

$$|a|^2 + |b|^2 = 1. \tag{11.2}$$

We also explicitly note that

$$|-1\rangle \neq -|1\rangle.$$

In fact the two vectors are orthogonal, not opposite.

Let us now discuss the meaning of the state $|q\rangle$. If $a = 0$ (and $b = 1$) $|q\rangle$ represents the classical True value. If $b = 0$ (and $a = 1$) $|q\rangle$ represents the classical False value. In these cases we refer to these states with well-defined value as *sharp* configurations.

In contrast, a *not-sharp* $|q\rangle$ represents a state which comprises simultaneously the two orthogonal configurations. If we repeat a measurement of the state $|q\rangle$, according to the quantum rules the probability of observing the value True is $|b|^2$ and the probability of measuring the value False is $|a|^2$.

What about a system with two qubits? We extend the configurational space by considering four independent vectors[3]

$$|\psi\rangle = a|1, 1\rangle + b|1, -1\rangle + c|-1, 1\rangle + d|-1, -1\rangle \tag{11.3}$$

with the constraint

$$|a|^2 + |b|^2 + |c|^2 + |d|^2 = 1. \tag{11.4}$$

Here, the quantum interference will begin to play its relevant role.

More precisely, we will show that quantum computation consists of starting from an easily measurable observable which is first sharp (input), and then it becomes not-sharp (with the parallel computation spreading across the multiple configurations) and finally becomes sharp again (output and readout) (see figure 11.6).

Single-qubit state evolution. The individual qubit evolution is represented by a lifeline which, at each step of computation, may come across one single *gate*. There are many possible gates, but here we describe only two of them that we need for our final discussion of the quantum algorithm. To describe the general operating

[2] The normalisation constraint.

[3] Here $|1, -1\rangle$ is a shorthand of the tensor product $|1\rangle|-1\rangle$. Note that the order matters in the sequence.

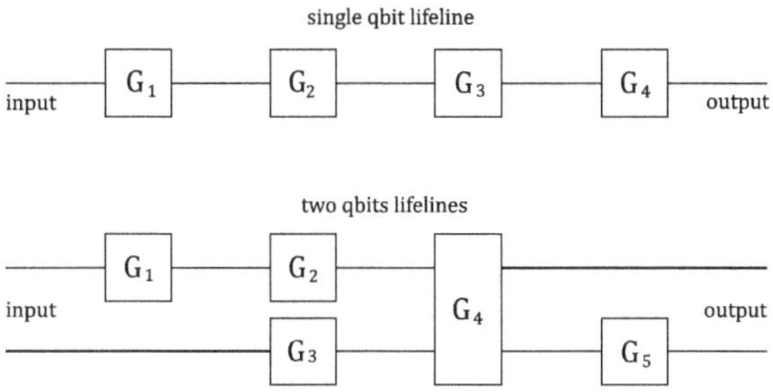

Figure 11.6. Single- and two-qubit example lifelines. Quantum gates (G) may act at each step of computation.

principle of a quantum gate on a state it is sufficient to demonstrate how it acts on *sharp* states[4].

The first is the Not gate, N for short. Below, we report how it works:

$$\begin{aligned} N(|1\rangle) &= |-1\rangle \\ N(|-1\rangle) &= |1\rangle. \end{aligned} \tag{11.5}$$

Since the output is *sharp* when the input is *sharp*, this gate is equivalent to the classical one.

Exercise 1. Show that, for any qubit $|q\rangle$, $N(N(|q\rangle)) = N^2(|q\rangle) = |q\rangle$. (Hint: it is sufficient to prove that $N^2(|1\rangle) = |1\rangle$ and $N^2(|-1\rangle) = |-1\rangle$.)

The second is the Hadamard gate, H for short. Here is how it works:

$$\begin{aligned} H(|1\rangle) &= \frac{1}{\sqrt{2}}(|1\rangle + |-1\rangle) \\ H(|-1\rangle) &= \frac{1}{\sqrt{2}}(|1\rangle - |-1\rangle). \end{aligned} \tag{11.6}$$

In this case, since the output is not-sharp when the input is sharp, this gate has no classical equivalent.

Exercise 2. Show that, for any qubit $|q\rangle$, $H(H(|q\rangle)) = H^2(|q\rangle) = |q\rangle$. (Hint: it is sufficient to prove that $H^2(|1\rangle) = |1\rangle$ and $H^2(|-1\rangle) = |-1\rangle$.)

Two-qubit state evolution. The most common two-qubit gate, and the only one we will use in this manuscript, is the Controlled Not gate (see figure 11.7), C_N for short. It operates on sharp states in the following way:

[4] Namely on the eigenvectors of the Boolean observable.

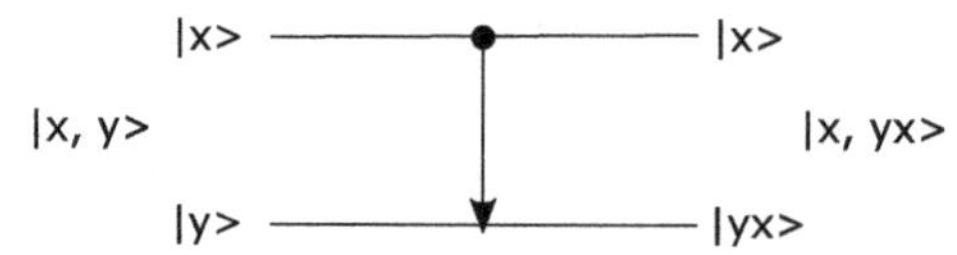

Figure 11.7. The Controlled Not gate. Here $x = \pm 1$ and $y = \pm 1$.

$$\begin{aligned} C_N(|1, 1\rangle) &= |1, 1\rangle \\ C_N(|1, -1\rangle) &= |1, -1\rangle \\ C_N(|-1, 1\rangle) &= |-1, -1\rangle \\ C_N(|-1, -1\rangle) &= |-1, 1\rangle. \end{aligned} \tag{11.7}$$

The first qubit state, the control state, is left unchanged by the gate. The second qubit (recall here that order matters) is changed to its orthogonal state only if the control state is True ($|-1\rangle$).

We can summarize the above equations into a single equation:

$$C_N(|x, y\rangle) = |x, yx\rangle \text{ for } x \in \{1, -1\} \text{ and } y \in \{1, -1\}. \tag{11.8}$$

As the order of input and output qubits matters, we will represent the gate as in figure 11.7.

Again, as the output is sharp when the input is sharp, this gate is equivalent to the classical one.

Exercise 3. Show that the Controlled Not gate is a reversible gate. (Hint: prove that $C_N(C_N(|x, y\rangle)) = C_N^2(|x, y\rangle) = |x, y\rangle$, when $x \in \{1, -1\}$ and $y \in \{1, -1\}$).

Exercise 4. Put three Controlled Not gates in a sequence in which the final effect is to swap the sharp states between two input qubits: $|x, y\rangle \longrightarrow |y, x\rangle$.

11.5 A quantum algorithm

An algorithm is a hardware independent specification of a computation. By hardware independent we mean that we do not specify how the computer is made, but only the effect that it should have on the information that it is provided with in its input.

A programme is a way of preparing a computer to perform a computational task. In general, to specify a computational task we specify the properties that the output information should have as a function of the input information. So a computational task is a problem to which an algorithm is the solution.

The Deutsch–Jozsa algorithm is a quantum algorithm, proposed by David Deutsch and Richard Jozsa in 1992 [5] with improvements by Richard Cleve, Artur Ekert, Chiara Macchiavello and Michele Mosca in 1998 [6]. Although of little practical use, it is one of the first examples of a quantum algorithm that is exponentially faster than any possible deterministic classical algorithm. It is also a

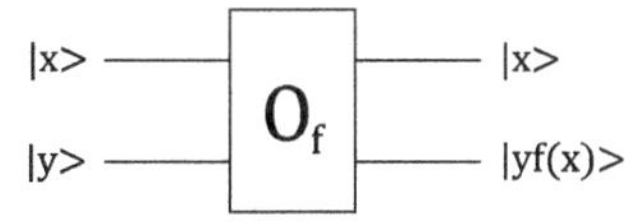

Figure 11.8. The Oracle hiding the f function. Here $x = \pm 1$ and $y = \pm 1$.

deterministic algorithm, meaning that it always produces an answer, and the answer is always correct.

Here we present a simplified version of the algorithm, using only two qubits (the original version involves $n + 1$ qubits).

It is instructive to first describe the problem that it solves. Consider an unknown Boolean function f of a single Boolean value

$$f\colon \{1, -1\} \to \{1, -1\}.$$

There are only four possible different functions of this type. Here are their truth tables:

x	$f(x)$
1	1
-1	1

x	$f(x)$
1	1
-1	-1

x	$f(x)$
1	-1
-1	1

x	$f(x)$
1	-1
-1	-1

Exercise 5. Check that $f^2(x) = 1$ for any $x \in \{1, -1\}$.

Suppose now we have a (perhaps quite complex) network hidden in a black box, called the Oracle, O_f for short, which evaluates $f(x)$ internally and performs the following computation (see figure 11.8):

$$O_f(|x, y\rangle) = |x, yf(x)\rangle \text{ for any } x \in \{1, -1\} \quad y \in \{1, -1\}. \tag{11.9}$$

Exercise 6. Prove that the Oracle performs a reversible computation, i.e. that $O_f^2(|x, y\rangle) = |x, y\rangle$. (Hint: look at the previous exercise.)

Exercise 7. Prove that the Controlled Not gate is equivalent to an Oracle which works with a function $f(x) = x$ (the second one in the previous list).

Exercise 8. Prove that to obtain from the Oracle the truth value of $f(x)$ it is sufficient to feed it with $|x, 1\rangle$.

As the Oracle produces sharp output states when is fed with sharp input states, it has a classical counterpart.

Furthermore, as it can directly produce the value of f (see the previous exercise), we need exactly two invocations of the Oracle with different values of x to determine what function f it hides. This is true both in classical and in quantum computation.

This is the proof that quantum computation cannot speed up any classical algorithm, only some of them.

But the problem we are going to solve with a quantum algorithm is different.

The statement of the problem. The problem we want to solve here is not to determine which function f is hidden in the Oracle, only a property of that function, namely the Boolean value

$$f(-1)f(1)$$

(i.e. if $f(-1) \neq f(1)$, the XOR of the two values[5]).

There is no classical way to compute

$$f(-1)f(1)$$

using only one invocation of the Oracle. In figure 11.9 we show an example of a classical network computing $f(-1)f(1)$.

It completes its task by invoking two parallel copies of the classical Oracle, so it has a time complexity of 1 and a space complexity of 2. Another equivalent classical way to have the same output is to invoke the same single classical Oracle twice, this time giving us a time complexity of 2 and a space complexity of 1.

What the quantum algorithm (see figure 11.10) does in short is to evaluate $f(-1)f(1)$ using only a single run of a single quantum Oracle. But the trick here is to invoke it with different inputs in different subspaces. In this way the Oracle, with its single run, performs different computations in (two) different universes or different subspaces yielding a linear superposition of our outputs.

Finally we use the reverse trick to recombine the different outputs in the superposition using in an interference phenomenon and to obtain the desired result.

Let us examine the quantum computation step by step.

At the beginning we feed the network with two qubits in the sharp state $|1\rangle$, the state corresponding classically to False or Empty:

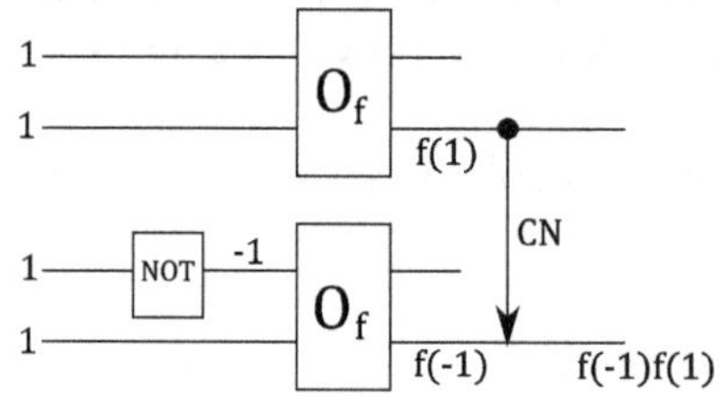

Figure 11.9. The classic parallel circuit which evaluates $f(-1)f(1)$. Note that it contains instances of the classical counterparts of the NOT, Oracle and Controlled Not circuits and that the lines hold Boolean values not states.

[5] Recall that for us the value 1 means False and the value -1 means True.

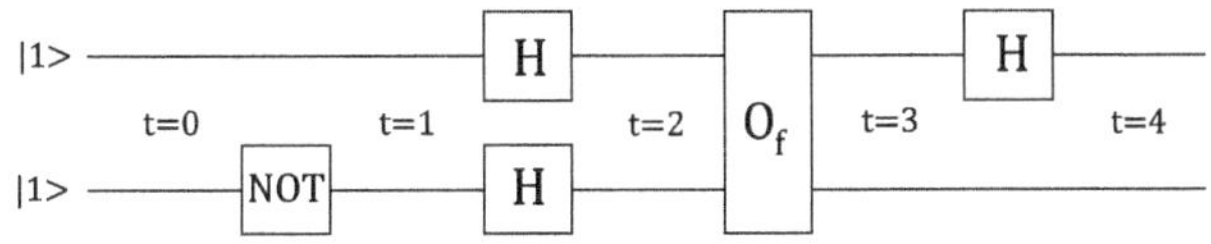

Figure 11.10. The quantum circuit which solves the problem of evaluating $f(-1)f(1)$. It contains a single instance of the Oracle, one Not and three Hadamard gates.

$$|\psi(0)\rangle = |1,\ 1\rangle. \tag{11.10}$$

After the first step a Not gate acts on the second qubit; the states of both the qubits are still sharp:

$$|\psi(1)\rangle = |1,\ -1\rangle. \tag{11.11}$$

At the second step the two qubits go through two parallel Hadamard gates:

$$\begin{aligned} |\psi(2)\rangle &= \left(\frac{1}{\sqrt{2}}(|1\rangle + |-1\rangle)\right)\left(\frac{1}{\sqrt{2}}(|1\rangle - |-1\rangle)\right) \\ &= \frac{|1,\ 1\rangle - |1,\ -1\rangle + |-1,\ 1\rangle - |-1,\ -1\rangle}{2}. \end{aligned} \tag{11.12}$$

Note here how each single Hadamard gate is unsharp in its input qubit state into two universes, then the tensor product mixes them into four universes, by virtue of quantum interference.

At the third step the qubits go through the Oracle:

$$\begin{aligned} |\psi(3)\rangle &= \frac{|1,\ 1f(1)\rangle - |1,\ -1f(1)\rangle + |-1,\ 1f(-1)\rangle - |-1,\ -1f(-1)\rangle}{2} \\ &= \frac{|1, f(1)\rangle - |1,\ -f(1)\rangle + |-1, f(-1)\rangle - |-1,\ -f(-1)\rangle}{2}. \end{aligned} \tag{11.13}$$

At this point it is better and simpler to split our discussion, making some hypotheses on the possible exclusive outcomes.

1. If $f(1) = f(-1) = f$, i.e. if $f(1)f(-1) = 1$, then we can rewrite and factorize $|\psi(3)\rangle$:

$$\begin{aligned} |\psi(3)\rangle &= \frac{|1, f\rangle - |1,\ -f\rangle + |-1, f\rangle - |-1,\ -f\rangle}{2} \\ &= \frac{|1\rangle(|f\rangle - |-f\rangle) + |-1\rangle(|f\rangle - |-f\rangle)}{2} \\ &= \left(\frac{|1\rangle + |-1\rangle}{\sqrt{2}}\right)\left(\frac{|f\rangle - |-f\rangle}{\sqrt{2}}\right). \end{aligned} \tag{11.14}$$

When the final Hadamard gate acts on the first qubit, recalling the solution of exercise 2, we obtain

$$|\psi(4)\rangle = |1\rangle\left(\frac{|f\rangle - |-f\rangle}{\sqrt{2}}\right). \tag{11.15}$$

2. On the other hand, if $f(1) = -f(-1) = f$, i.e. if $f(1)f(-1) = -1$, then again we can rewrite and factorize $|\psi(3)\rangle$:

$$\begin{aligned} |\psi(3)\rangle &= \frac{|1, f\rangle - |1, -f\rangle + |-1, -f\rangle - |-1, f\rangle}{2} \\ &= \frac{|1\rangle(|f\rangle - |-f\rangle) - |-1\rangle(|f\rangle - |-f\rangle)}{2} \\ &= \left(\frac{|1\rangle - |-1\rangle}{\sqrt{2}}\right)\left(\frac{|f\rangle - |-f\rangle}{\sqrt{2}}\right). \end{aligned} \tag{11.16}$$

When the final Hadamard gate acts on the first qubit, recalling again the solution of exercise 2, we obtain

$$|\psi(4)\rangle = |-1\rangle\left(\frac{|f\rangle - |-f\rangle}{\sqrt{2}}\right). \tag{11.17}$$

Taking into account our two hypotheses, the equations (11.15) and (11.16) become the single equation

$$|\psi(4)\rangle = |f(1)f(-1)\rangle\left(\frac{|f\rangle - |-f\rangle}{\sqrt{2}}\right). \tag{11.18}$$

The final output on the first qubit is just the solution to our problem and it has been obtained by a *single execution* of the Oracle block.

11.6 Exercise solutions

Solution 1.

$$\begin{aligned} N^2(|1\rangle) &= N(|-1\rangle) = |1\rangle, \\ N^2(|-1\rangle) &= N(|1\rangle) = |-1\rangle. \end{aligned} \tag{11.19}$$

Solution 2.

$$\begin{aligned}
H^2(|1\rangle) &= H\left(\frac{|1\rangle + |-1\rangle}{\sqrt{2}}\right) \\
&= \frac{(|1\rangle + |-1\rangle) + (|1\rangle - |-1\rangle)}{2} \\
&= \frac{2|1\rangle}{2} = |1\rangle. \\
H^2(|-1\rangle) &= H\left(\frac{|1\rangle - |-1\rangle}{\sqrt{2}}\right) \\
&= \frac{(|1\rangle + |-1\rangle) - (|1\rangle - |-1\rangle)}{2} \\
&= \frac{2|-1\rangle}{2} = |-1\rangle.
\end{aligned} \tag{11.20}$$

Solution 3. For any $x \in \{1, -1\}$ and $y \in \{1, -1\}$:

$$C_N^2(|x, y\rangle) = C_N(|x, yx\rangle) = |x, yx^2\rangle = |x, y\rangle. \tag{11.21}$$

Solution 4. Let us consider the three Controlled Not gates in the sequence depicted in figure 11.11:

$$\begin{aligned}
|\psi(0)\rangle &= |x, y\rangle \\
|\psi(1)\rangle &= |x, yx\rangle \\
|\psi(2)\rangle &= |yx^2, yx\rangle = |y, yx\rangle \\
|\psi(3)\rangle &= |y, xy^2\rangle = |y, x\rangle.
\end{aligned} \tag{11.22}$$

Solution 5. As $f(x) \in \{1, -1\}$, then $f^2(x) = 1$ for any input $x \in \{1, -1\}$.

Solution 6. For any $x \in \{1, -1\}$ and $y \in \{1, -1\}$,

$$\begin{aligned}
O_f^2(|x, y\rangle) &= O_f(|x, yf(x)\rangle) \\
&= |x, yf(x)f(x)\rangle = |x, yf^2(x)\rangle = |x, y\rangle,
\end{aligned} \tag{11.23}$$

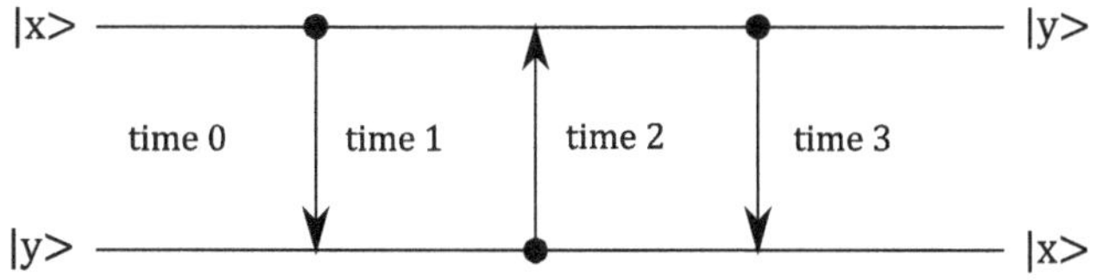

Figure 11.11. The quantum circuit which swaps the input states.

by virtue of exercise 5.

Solution 7. Let us recall that

$$C_N(|x, y\rangle = |x, yx\rangle.$$

Then, if $f(x) = x$,

$$O_f(|x, y\rangle) = |x, yf(x)\rangle = |x, yx\rangle. \tag{11.24}$$

Solution 8.

$$O_f(|x, 1\rangle) = |x, 1f(x)\rangle = |x, f(x)\rangle. \tag{11.25}$$

The previous computation produces a sharp state output in which the second qubit state represents the $f(x)$ value.

References

[1] Hamlin J J 2019 Superconductivity near room temperature *Nature* **569** 491–2

[2] Drozdov A P *et al* 2019 Superconductivity at 250 K in lanthanum hydride under high pressures *Nature* **569** 528–31

[3] Bardeen J, Cooper L N and Schrieffer J R 1957 Microscopic theory of superconductivity *Phys. Rev.* **106** 162

[4] Feynman R P, Leighton R B and Sands M 1965 *The Feynman Lectures on Physics, Volume III, Quantum Mechanics* (Reading, MA: Addison-Wesley)

[5] Deutsch D and Jozsa R 1992 Rapid solutions of problems by quantum computation *Proc. R. Soc. Lond.* A **439** 553–8

[6] Cleve R, Ekert A, Macchiavello C and Mosca M 1998 Quantum algorithms revisited *Proc. R. Soc. Lond.* A **454** 339–54

IOP Publishing

Modern Physics
A critical approach
Canio Noce

Chapter 12

A new quantum era: from quantum optics to quantum technologies

Antonio Capolupo and Canio Noce

12.1 Introduction

During the first 30 years of the twentieth century, quantum mechanics radically modified the framework of physics. Analogously, the last 30 years have been characterized by a similar revolution. Indeed, the birth and rapid growth of the new science of quantum information have changed the paradigms of information theory and computational science and engineering. The consequent ideas, both theoretical and technological, have focused the attention of scientists on the foundations of quantum mechanics, and how one may test and probe these fundamental issues using experimental means. Many astounding outcomes have already been achieved, such as the experimental demonstration of non-local quantum entanglement and quantum teleportation, and the control of matter and light at the level of individual particles. Interestingly, quantum optics is strictly tied to the historical debate on the nature of light, the wave–particle duality. Among the supporters of the particle nature of light one finds Isaac Newton, who adopted the particle interpretation in his work *Opticks*, as well as Pierre de Fermat, who considered light as being made of particles travelling along geometric rays. On the other hand, the wave nature of light was supported by Christian Huygens and later formalized by Augustin-Jean Fresnel, who developed a rigorous theory of wave diffraction. Moreover, for the electromagnetic field, the theory of radiation by James Clerk Maxwell unified electricity and magnetism, predicting wave solutions travelling at the speed of light in a vacuum. On this basis, Maxwell introduced the idea of light as an electromagnetic field, thus reinforcing the wave interpretation. At the beginning of the twentieth century, the wave theory was the most accredited interpretation of the fundamental nature of light. Even the introduction of the energy quantization hypothesis, due to Planck, to explain the nature of black-body radiation did not change the wave

doi:10.1088/978-0-7503-2678-0ch12

theory on the nature of light, since the energy quantization hypothesis was applied only to the walls of the cavity and not to the field.

A decisive step in favour of the particle interpretation of light came with the 1905 paper on the photoelectric effect by Albert Einstein [1]. To explain the emission of electrons from a metallic surface irradiated by ultraviolet rays, Einstein proposed that light was made up of discrete packets, or quanta, of energy. Each quantum carries an energy $E = h\nu$, where ν is the frequency and h is Planck's constant, in agreement with Planck's hypothesis. Thus through the photoelectric effect and energy quantization, the particle interpretation of light reacquired consensus in the scientific community.

A decade later, the quantum theory of the atom was being developed, with major contributions from Niels Bohr and Louis de Broglie. Energy quantization was essential to Bohr's atomic model, in which electrons can only occupy discrete orbitals at fixed energies E_i. In this context, the transition between two orbitals with energies E_i and E_j is possible only by means of a photon with frequency $\nu = |E_i - E_j|/h$, which is absorbed or emitted according to whether the final energy lies above or below the initial energy. Together with the particle nature of light, the wave nature of matter was becoming apparent, in particular thanks to the theory of matter waves by de Broglie. This insight would later lead Erwin Schrödinger to formulate his wave equation. Quantum mechanics then gave new impetus to the particle–wave debate.

Nevertheless, the quantum nature of light was long debated by the scientific community, since a large class of phenomena regarding radiation–matter interaction could be explained within the semiclassical theory, in which light is treated classically and only matter is quantized. Not even the photoelectric effect requires the quantization of the radiation field [2], as its experimental aspects can be completely understood from a semiclassical point of view[1]. As a consequence, despite the introduction of a quantized view of the electromagnetic field by Paul Dirac in 1927, later followed by the full theory of quantum electrodynamics (QED), the need for a quantum theory of light was still debated as late as the 1960s. The necessity of photon field quantization was understood with the discovery of Lamb shift and the need for quantum optics, in which quantum mechanics is applied to light, was finally understood with the discovery of the Hanbury Brown and Twiss (HBT) effect.

Here, we introduce some fundamental concepts of quantum optics and their applications to modern quantum technologies. In section 12.2 we show the phenomena of Lamb shift and scattering off the vacuum, which are explainable only with field quantization. Moreover, we present the HBT effect and we introduce the quantum theory of coherence. In section 12.3 we briefly report the studies that led to the birth of quantum computation and quantum information theory. The role of quantum optics in quantum information and in modern quantum technologies is

[1] A subtle exception regards the time delay between the incidence of light and the photoelectric emission.

discussed in sections 12.4 and 12.5, respectively. Section 12.6 is devoted to our conclusions.

12.2 Quantum optics and the quantum theory of coherence

In the 1960s, Ed Jaynes, who supported the semiclassical theory, had partially success in predicting the Lamb shift effect qualitatively without using field quantization. Nevertheless, he failed to provide an accurate numerical prediction. In fact, the electromagnetic field is intrinsically a quantum entity. The Lamb shift, together with scattering off the vacuum, represented a crucial example showing the need for field quantization. These phenomena, which will be briefly illustrated in the following, are completely explained by QED.

Lamb shift. Experimentally, a frequency splitting of about 1 GHz is observed between the $2s_{1/2}$ and energy levels of the hydrogen atom (figure 12.1), in contrast to the results of relativistic quantum mechanics, which predict that these two levels should be degenerate. This phenomenon is called Lamb shift [3]. It is due to the fluctuations of the vacuum energy which perturb the electric potential of the atomic nucleus and generate a fluctuation in the position of the electron. Then, the electron feels a Coulomb interaction from the positive nucleus which is slightly different with respect to the normal interaction. This effect is more significant near the nucleus, thus it affects the s orbital more than the p orbital, causing the shift of the levels.

Micromaser—scattering off the vacuum. Let us consider excited atoms in the vicinity of a cavity (a micromaser) (figure 12.2) [4]. When the motion of the atomic centre-of-mass is quantized and the atoms approach the cavity at a sufficiently low speed (their kinetic energy has to be smaller than the energy of the atom–field interaction), they are reflected from the cavity, even when it is initially empty. This reflection is due to the fast change in the coupling strength between the atoms and the vacuum fluctuation between the inside and the outside of the cavity.

Although QED allowed accurate descriptions of the Lamb shift and other phenomena, such as the anomalous magnetic moment of the electron, it was not clear if light quantization was necessary in optics. The need for quantum optics to study the nature and effects of light as quantized photons was definitively understood with the discovery of the HBT effect in 1956 [5]. Hanbury Brown and Twiss collected light in two different detectors and observed an interference effect between the two intensities which showed a positive correlation between the two signals. A similar effect was found for beams composed of other bosons, while anti-correlation

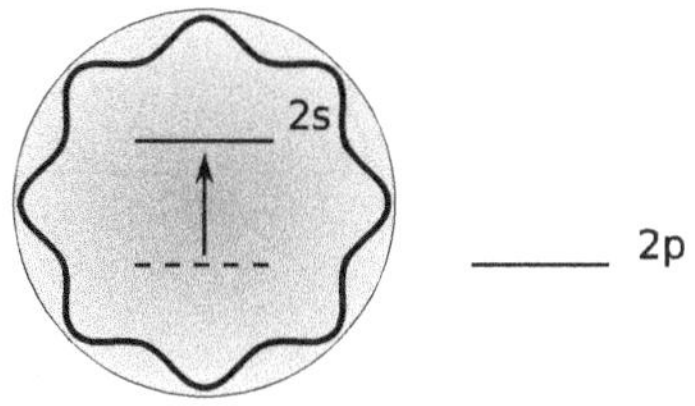

Figure 12.1. Schematic representation of the Lamb shift phenomenon for a hydrogen atom.

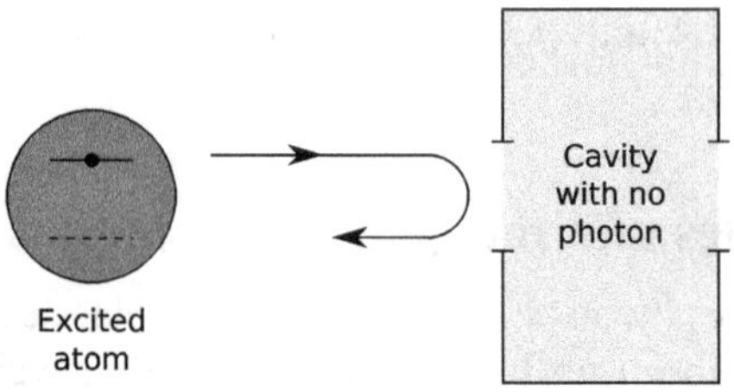

Figure 12.2. Scattering off the vacuum. A sufficiently slow excited atom approaching a cavity can be reflected by the cavity vacuum field, which behaves as an effective potential barrier.

effects were detected for an initial beam of fermions. These phenomena were interpreted by Roy J Glauber as genuine quantum effects.

A systematic study of the quantum properties of light occurred in 1963 when Glauber developed the quantum theory of optical coherence [6], which is essential for the understanding of laser physics[2]. The theory is based on the concept of coherent states of laser beams. These states were introduced originally by Schrödinger and John von Neumann in 1927.

Let us now define such states. We consider the wave function of an optical state in the position and momentum representations, $\phi(\hat{x})$ and $\phi(\hat{p})$, $\hat{x}$ and $\hat{p}$ being the amplitude and phase light quadratures, respectively[3]. The wave functions for the coherent state are Gaussian with identical widths, so that the variance associated with the measurement of any quadrature will be the same. The coherent states are characterized by the fact that they minimize the Heisenberg uncertainty relation,

$$\Delta\hat{x}^2\, \Delta\hat{p}^2 = \frac{1}{4}, \qquad \text{with} \quad \Delta\hat{x} = \Delta\hat{p}. \tag{12.1}$$

For such states the probability of finding n photons follows a Poissonian distribution, $P_n = \frac{\bar{n}^{-n}}{n!} e^{-\bar{n}}$, where $\bar{n}$ is the average photon number which is equal to the variance $\Delta n^2 = \bar{n}$ [7].

The coherent quantum states reproduce known results from classical optics, moreover they describe the quantum behaviour of light. These aspects of light were analysed in the 1970s and 1980s, in which many studies were focused on the quantum aspects of photon-counting statistics, and on the wave behaviour of light, paying attention to its phase-dependent properties. The analyses were developed through a non-linear optical process called spontaneous parametric down-conversion (SPDC) (figure 12.3) [8]. In the typical scenario, SPDC is realized by means of a laser pump directed at a non-linear crystal that transforms one photon of higher energy (called the pump photon), into a pair of photons of lower energy (called the signal photon and idler photon). The conservation of energy and momentum is satisfied and the two output photons are correlated in momentum and energy.

[2] Glauber was awarded the 2005 Nobel Prize for his contribution to the quantum theory of optical coherence.
[3] Quadrature operators represent observable quantities associated with the field mode. They correspond to the real and imaginary parts of the complex amplitude of the wave.

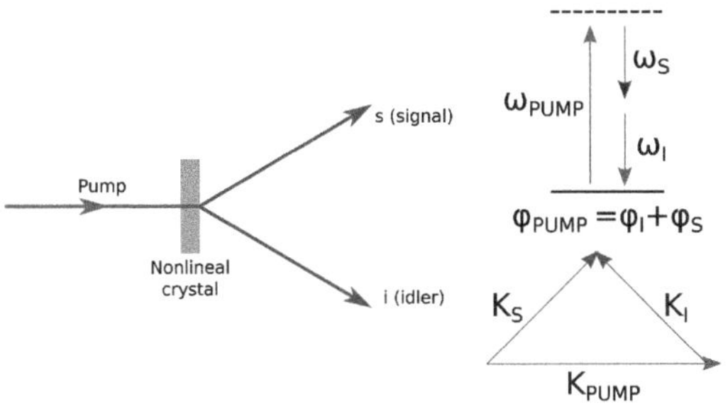

Figure 12.3. Schematic representation of the SPDC optical process.

Different types of SPDCs are obtained by varying the polarization of the input and output photons and the properties of the non-linear crystal. Through the SPDC process single photons (known as Fock states), entangled photon pairs and squeezed states are generated, which are relevant in modern quantum technologies.

Briefly, a quantum state is squeezed if it satisfies the condition equation (12.1) and has the property of either $\Delta\hat{x}^2 \leqslant 1/2$ or $\Delta\hat{p}^2 \leqslant 1/2$. In squeezed states the counting statistic shows a sub-Poissonian behaviour ($\Delta n^2 < \bar{n}$). This means that the photons are more equally spaced than in a coherent laser field.

12.3 Quantum computing and quantum information

Before the 1970s, the technology used in quantum mechanical experiments was not sufficiently developed to allow for fine control over the system. In particular, while global information on the system was accessible, it was impossible to probe its individual constituents experimentally. Such a situation is reminiscent of superconductivity, where the system (the superconductor) can only be studied as a whole. This state of affairs changed with the advent of new experimental techniques in the 1970s and the 1980s, which allowed for the manipulation of individual quantum systems.

The renewed experimental impetus also fed the debate on the foundations of quantum theory, resulting in the discovery of new phenomena and the establishment of fundamental theoretical limitations. The no-clone theorem, asserting the impossibility of cloning an unknown single quantum state, can be considered as one of the first achievements in quantum information [9].

Quantum information is concerned with the possibility of processing information by means of quantum mechanical systems. It is based on the quantum analogue of the bit, the qubit. While the classical bit can only be in either of the two states $|0\rangle$ and $|1\rangle$, the qubit can be in any linear superposition of the two [10],

$$|\psi\rangle = \alpha|0\rangle + \beta|1\rangle, \tag{12.2}$$

where α and β are complex coefficients satisfying $|\alpha|^2 + |\beta|^2 = 1$. In accordance with the laws of quantum mechanics, the outcome of a measurement on a qubit in the state equation (12.2) is $|0\rangle$ with probability $|\alpha|^2$ and $|1\rangle$ with probability $|\beta|^2$.

What is the advantage of quantum over classical information? Part of the answer lies in the quantity of information that can be stored in a qubit *before* the measurement process [11]. After a measurement is performed, the qubit state equation (12.2) collapses to either $|0\rangle$ or $|1\rangle$, so that a single measurement on a qubit corresponds to a single (classical) bit of information. However, prior to the measurement, a qubit can be in a continuum of states, as opposed to the discrete states (0, 1) of a classical bit. Thus, full knowledge of the state equation (12.2) requires a virtually infinite amount of classical information.

The storage capability of a qubit register grows exponentially with the number n of qubits. In general, an n-qubit state is a linear superposition of the states $\{|x_1x_2...x_n\rangle\}$, where the ith qubit is in the state $|x_i\rangle$ and $x_i = 0, 1$. Since there are 2^n such states, a general n-qubit superposition involves 2^n coefficients. To give an idea of how huge this number can become, consider that for 300 qubits one has $2^{300} \simeq 2 \times 10^{90}$, which is larger than the estimated number of atoms in the Universe $\sim 10^{82}$. These considerations would lead Richard Feynman to suggest that *quantum* systems be used to simulate nature [12].

The possibility for n-qubit states to exist in linear superposition is the key to quantum parallelism. The term refers to the ability to perform several operations at once (in parallel), whereas a classical computer would require each operation to be performed independently. To be concrete, suppose that one wishes to evaluate a function $f(x)$, which takes as an input a numerical value x encoded in a given qubit. Then one could prepare an n-qubit superposition to be passed as an input to f. The resulting state would be a superposition of the states $|f(x_i)\rangle$, where x_i is associated with the ith qubit. In a single operation one is able to store what would be the result of n computations performed on classical bits[4]. As a consequence, certain computations, such as the factorization of integers, are significantly more efficient on quantum computers[5]. Quantum algorithms have been developed to exploit quantum parallelism. The works of David Deutsch, Richard Jozsa and Peter Shor provided faster alternatives to classical algorithms in the solution of various problems.

The differences with respect to classical information theory, of course, are not limited to the information content of the qubits. There is a fundamental qualitative difference, in that multi-qubit states can be entangled. Entanglement is a purely quantum mechanical correlation, which has no classical counterpart. To explore this concept, consider a system made of two qubits. The possible states of the system are of the form

$$|\psi\rangle = \alpha_{00}|00\rangle + \alpha_{01}|01\rangle + \alpha_{10}|10\rangle + \alpha_{11}|11\rangle \quad \text{with}$$
$$|\alpha_{00}|^2 + |\alpha_{01}|^2 + |\alpha_{10}|^2 + |\alpha_{11}|^2 = 1.$$

In particular the Bell states [13]

[4] It should be remarked that, as n grows, the probability of extracting a specific result $f(x_i)$ generally decreases as $\frac{1}{n}$. This is irrelevant when one is only interested in features common to all the inputs, in which case quantum parallelism provides a substantial improvement with respect to classical strategies.

[5] Roughly speaking, in computer science a computational problem has an efficient solution if it can be solved quickly, compared to the inherent difficulty of the problem.

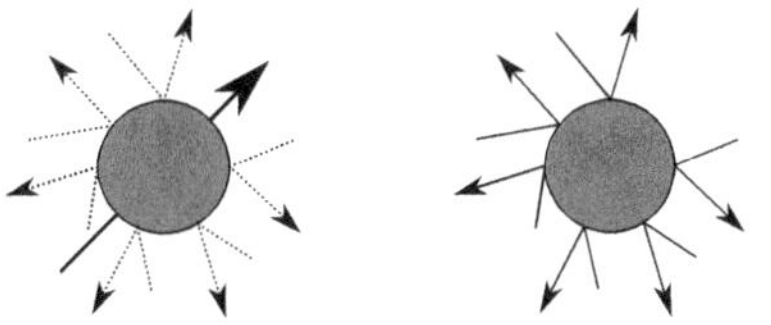

Figure 12.4. In the classical case the motion of the system is unaffected by incident photons. In the quantum case the interaction between the body and the scattered photons will cause them to be entangled, causing a loss of the quantum properties of the system under scrutiny, a process known as quantum decoherence.

$$|\psi_{\pm}\rangle = \frac{|00\rangle \pm |11\rangle}{\sqrt{2}}, \quad |\phi_{\pm}\rangle = \frac{|10\rangle \pm |01\rangle}{\sqrt{2}} \tag{12.3}$$

are of great importance in quantum information. For the states (12.3), the outcomes of a measurement performed on the first and the second qubit are correlated. Suppose, for instance, that the two qubits are initially prepared in the state $|\psi_{+}\rangle$. A measurement is then performed on the first qubit, yielding 0 as a result. The state collapses to $|00\rangle$, so that any following measurement necessarily finds the second qubit in the state $|0\rangle$. Similarly for when the first qubit is found in the state $|1\rangle$.

Note that neither $|\psi_{\pm}\rangle$ or $|\phi_{\pm}\rangle$ can be factorized as $|A_1\rangle \otimes |B_2\rangle$, with $|A_1\rangle$ and $|B_2\rangle$ belonging to the subspaces of the corresponding qubits $\mathcal{H}_1$ and $\mathcal{H}_2$. The state equations (12.3) are said to be *non-separable*. This property is common to all entangled states [13–15]. The correlations exhibited by entangled states persist even when the constituents are very far away from each other. This phenomenon was first pointed out by Einstein, Boris Podolsky and Nathan Rosen (EPR) in their seminal paper of 1935 [16].

12.4 The role of quantum optics in quantum information

In the previous section we discussed how quantum mechanics can be used to perform some tasks more efficiently with respect to the classical counterpart. We have assumed an ideal setting, neglecting all the interactions of the quantum systems with the environment. In reality, quantum systems are never isolated from the surrounding environment, and this continuous interaction leads to the emergence of quantum correlations between them (figure 12.4). The result is a loss of coherence within the system, a process known as quantum decoherence [17, 18]. It constitutes the major practical limitation in the realization of quantum devices.

In most cases, the primary source of decoherence is constituted by thermal excitations. However, these can be neglected at optical frequencies, making quantum optics the optimal framework to implement the ideas from quantum information theory. In this respect, the development of quantum optics was crucial for quantum information. Also, entangled photon states are essential to the execution of quantum protocols and a crucial role is played by the squeezed states. These states are generated by means of SPDC. Indeed, the principles of momentum and energy conservation lead to two output photons correlated in momentum and energy. The resulting state is the two-mode squeezed state:

$$|\psi_{\text{TMS}}\rangle = \sqrt{1 - |\lambda|^2}\sum_{n=0}^{\infty}\lambda^n|n\rangle|n\rangle, \tag{12.4}$$

where λ is the squeezing parameter. When $\lambda \to 1$ the state becomes the optical equivalent of the maximally entangled EPR state[6]. In realistic scenarios $\lambda < 1$ and the state is not maximally entangled.

The states of equation (12.4) are determinant for most quantum information tasks in the quantum optical setting, such as quantum teleportation [19] and quantum key distribution [13].

12.5 Quantum technologies

We have seen how the optical world happens to be the perfect playground to understand quantum mechanics and its applications. We now illustrate some of the classical and current trends in quantum optics and quantum technologies [20, 21].

12.5.1 The quantum teleportation protocol

The basic operation in quantum information is the quantum teleportation protocol. It plays a central role in both quantum computation and quantum communication, which makes it a privileged object of study in the quantum optical domain [22]. The typical setting in quantum communication is known as LOCC (local operations and classical communication). In this paradigm, the two parties (Alice and Bob) are spatially separated. Each of them can perform local generalized measurements in their own laboratory, and they can communicate classically. However, no exchange of quantum particles is allowed. This circumstance is sketched in (figure 12.5). Suppose that, under the LOCC conditions, Alice wishes to transmit an unknown quantum state to Bob. The accomplishment of this task constitutes the *quantum teleportation problem*. Classically, it is impossible to achieve such a transmission, yet it can be obtained in quantum information by a clever use of entanglement.

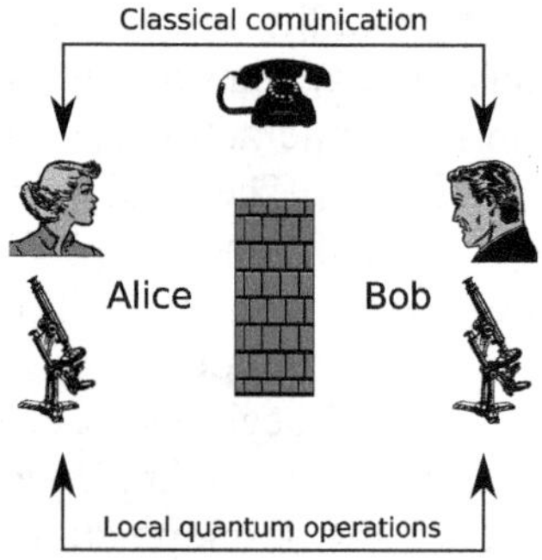

Figure 12.5. In the standard setting Alice and Bob can perform any generalized local measurement and can communicate classically. The wall indicates that they cannot exchange quantum particles.

[6] We note that such a state is unphysical as its main energy is infinite.

Let $|\psi\rangle = \alpha|0\rangle + \beta|1\rangle \in \mathcal{H}$ be the state to be teleported, and suppose that a Bell state of the form

$$|\phi\rangle = \frac{|0\rangle_{\text{Alice}}|0\rangle_{\text{Bob}} + |1\rangle_{\text{Alice}}|1\rangle_{\text{Bob}}}{\sqrt{2}}$$

is shared between Alice and Bob. Here the subscripts refer to the corresponding subsystems, localized within the respective laboratories. Denoting by $\mathcal{H}_{\text{Alice}}$ and $\mathcal{H}_{\text{Bob}}$ the relative Hilbert spaces, the total state is the product $|\psi\rangle|\phi\rangle \in \mathcal{H} \otimes \mathcal{H}_{\text{Alice}} \otimes \mathcal{H}_{\text{Bob}}$, which can be factorized as

$$|\psi\rangle|\phi\rangle = \frac{1}{2}\sum_{i=1}^{4}|A_i\rangle|B_i\rangle, \tag{12.5}$$

where $|A_i\rangle$ are entangled states in $\mathcal{H} \otimes \mathcal{H}_{\text{Alice}}$ [10], and $|B_1\rangle = \alpha\,|0\rangle_{\text{Bob}} + \beta\,|1\rangle_{\text{Bob}}$, $|B_2\rangle = \alpha\,|0\rangle_{\text{Bob}} - \beta\,|1\rangle_{\text{Bob}}$, $|B_3\rangle = \beta\,|0\rangle_{\text{Bob}} + \alpha\,|1\rangle_{\text{Bob}}$ and $|B_4\rangle = \beta\,|0\rangle_{\text{Bob}} - \alpha\,|1\rangle_{\text{Bob}}$. In practice, the states $|A_i\rangle$ are produced via beam splitters. Alice can perform local measurements on her subsystem[7], collapsing the state equation (12.5) to one of the $|A_i\rangle$. She can then communicate the result to Bob classically, who learns that his subsystem has to be in the corresponding state $|B_i\rangle$. Finally, Bob is able to replicate the original state by means of elementary transformations. For example, if Alice obtains $|A_2\rangle$, Bob simply operates the transformation $|0\rangle_{\text{Bob}} \to |0\rangle_{\text{Bob}}$, $|1\rangle_{\text{Bob}} \to -|1\rangle_{\text{Bob}}$ on the state of his subsystem $|B_2\rangle$ to reconstruct $|\psi\rangle$. A schematic illustration of quantum teleportation is shown in figure 12.6.

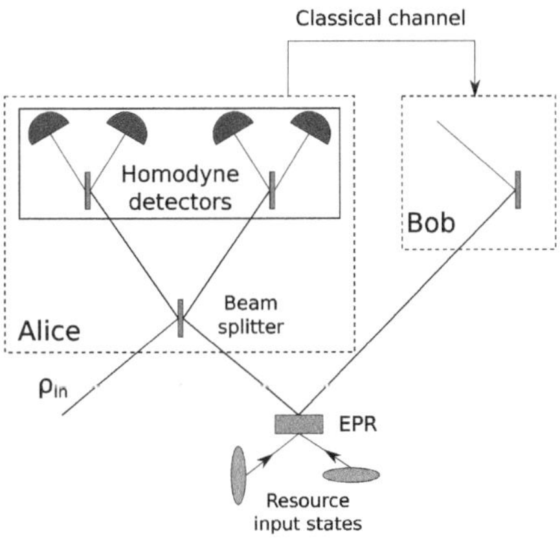

Figure 12.6. Quantum teleportation of a photon state. Alice and Bob share an entangled state. Alice combines the input state $\rho_i n$ with her half of the Bell state at a beam splitter. She then measures the quadrature of the field through a homodyne detector and communicates the result to Bob. He performs appropriate local operations on his half of the Bell state depending on Alice's classical results, reconstructing the input state.

[7] If the states are constructed from photons, one possibility is to use homodyne detectors to measure the field quadratures.

12.5.2 Quantum metrology and quantum state engineering

Quantum metrology uses quantum theory to obtain high sensitivity measurements of physical quantities [23]. In turn, a high measurement precision is of fundamental importance for the most advanced metrological optical experiments, such as those involved in the detection of gravitational waves.

A common problem in quantum metrology is quantum estimation, which aims to recover the value of a continuous parameter φ, encoded in the state ρ_φ of a quantum system. In order to retrieve the value, one first performs a measurement on the system, and then, by data-processing of the results, extrapolates the value of the parameter. Suppose θ is the value obtained through the extrapolation process. It is desirable that θ be as close as possible to the actual value φ. However, it is often the case that the extrapolated value θ carries some residual uncertainty. One can define the root mean square error (RMSE) $\delta\varphi_n$, where n is the number of copies of ρ_φ used in the process, as a measure of the estimation accuracy. It can be shown that, under reasonable conditions [23],

$$\delta\varphi_n \sim \frac{1}{\sqrt{n}}. \tag{12.6}$$

This follows from the central-limit theorem: if n is a large number of independent measurement outcomes, each having standard deviation $\Delta\sigma$, their average will converge to a Gaussian distribution with standard deviation $\Delta\sigma/\sqrt{n}$. This behaviour is referred to as the *standard quantum limit* (SQL).

The SQL can be circumvented by resorting to quantum strategies, which might involve squeezing or entanglement among the instruments employed for the measurements. By adopting adequate strategies, a far better limit can be reached [23]:

$$\delta\varphi_n \sim \frac{1}{n}. \tag{12.7}$$

Equation (12.7) is referred to as the Heisenberg limit, and represents the optimal rate at which the accuracy of a measurement can scale.

Due to the highly non-classical nature of the states required to achieve the limit equation (12.7), considerable effort has been put in the design and the realization of non-classical, non-Gaussian states of the radiation field[8]. Quantum state engineering (QSE) deals with the preparation, detection and reconstruction of such quantum states. Among the classes of non-Gaussian states used in the applications, the photon-subtracted and photon-added states, obtained through the addition or subtraction of photons from the radiation field, are of particular importance. They can be generated by means of beam splitters (figure 12.7).

[8] The coherent and squeezed states belong to the class of Gaussian states, as their wave functions are of Gaussian form.

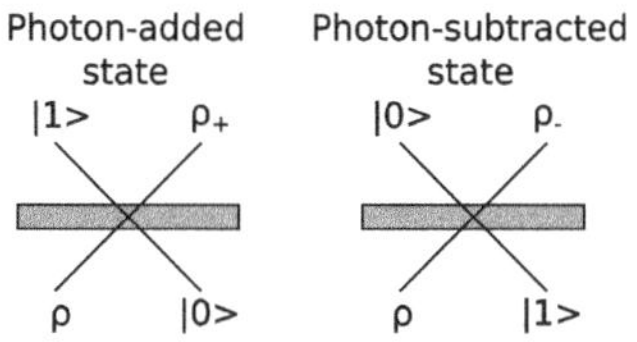

Figure 12.7. Beam splitter model for the photon-added and for the photon-subtracted states. In the first case a photon is added to the initial state. In the second case a photon is subtracted.

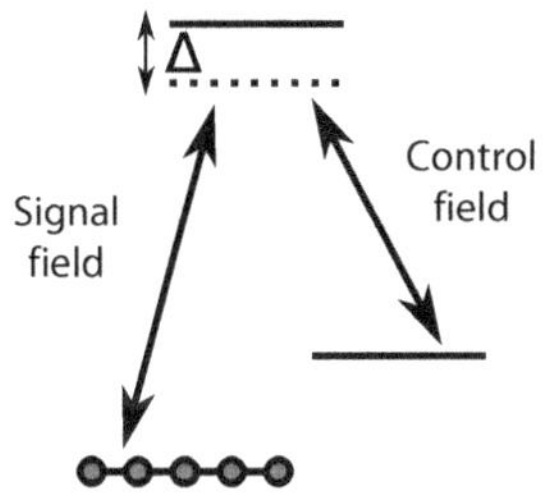

Figure 12.8. Electromagnetically induced transparency description. EIT is largely insensitive to the detuning Δ, which is the tuning of a laser to a frequency that is slightly off from the system's resonant frequency.

12.5.3 Quantum memory

Just like classical memory is essential for classical computers, quantum memory is fundamental for quantum information processing. It serves both as a synchronization device to coordinate several process in a quantum computer, and as a converter of heralded photons to photons-on-demand [24]. Quantum memory is also required in quantum repeaters, which are used to achieve long-range quantum communications. The necessity of quantum repeaters for long-range communications can be easily understood in view of our discussion of decoherence (section 12.4), that inevitably affects any quantum communication.

At a fundamental level quantum memory is required to store qubits faithfully and to release them on demand. Consequently it has to meet some performance criteria, which can be quantified through some specific parameters; for example, the efficiency η, defined as the ratio between the energies of the stored and retrieved pulses, and the multi-mode capacity, defined as the number of optical modes that can be stored in the memory cell satisfying the performance threshold.

A number of approaches have been proposed for quantum memory. Here we briefly present two of them.

Electromagnetically induced transparency and slow light

One approach to quantum memory is based on electromagnetically induced transparency (EIT) [25]. This non-linear optical phenomenon takes place in atoms with a peculiar energy level structure, as shown in figure 12.8. Two distinct ground states are coupled to an excited state by means of two optical fields: the weak signal, which is responsible for carrying the information content, and the strong control field, which is used to manipulate the atom. The control field acts as a switch for the signal absorption. When it is present the absorption of the signal is strongly inhibited and when it is removed the signal field is partially or completely absorbed. It is possible to show that an anomaly in the absorption spectrum always comes together with an anomaly in dispersion. In EIT the signal field is subject to a large positive dispersion, which results in a reduction of its group velocity. The control field can be used to modify the group velocity of the signal. In particular, lower velocities correspond to lower intensities of the control field. On this basis, it is possible to engineer quantum memories.

To illustrate how an EIT-based memory works, consider a light pulse, with EIT resonant frequency[9], entering the EIT medium (figure 12.9). The pulse is slowed down and compressed in order to fit inside the medium. Once the pulse is inside, its group velocity is brought down to zero by removing the control field. Then the pulse is stored in the EIT medium, and one can eventually turn the control field back on when the signal must be retrieved. In this way, the pulse can resume its propagation and leave the medium.

Off-resonant Faraday interaction

Another quantum memory model is based on the off-resonant Faraday interaction [26]. When an optical wave propagates through an ensemble of two-level atoms, if its frequency is off-resonant the absorption by the atoms can be neglected. Indeed, for sufficiently large detuning from the atomic transition frequency, the pulse is not able to excite any atoms and the amplitude remains substantially unchanged during the propagation in the medium. On the other hand, the Faraday interaction between the atoms and the field might induce a significant shift in the phase of the latter. This can in turn influence the phase of the atoms, an effect which can be exploited to design quantum memory (figure 12.10).

For definiteness, consider a pulse with macroscopic linear polarization along the y-axis, propagating along the z-direction through an off-resonant atomic gas, whose

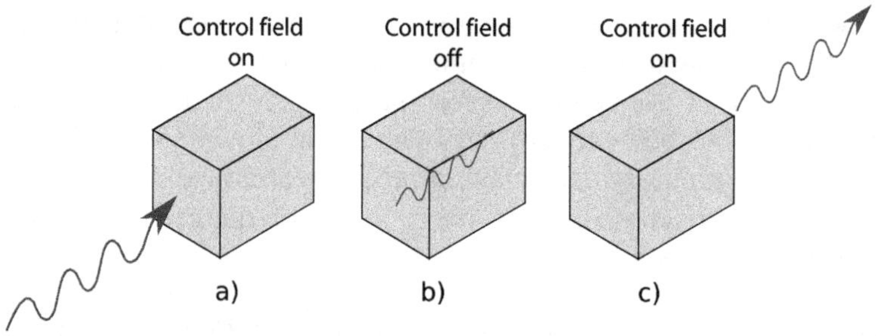

Figure 12.9. The signal enters the cell under the EIT conditions. During this first step the control field is on (a). While the spatially compressed pulse propagates inside the cell, the control field is turned off (b) and the information carried by the pulse is stored as a collective excitation of the ground states. To retrieve the information the control field is switched back on (c).

[9] This means that the frequency ω should be close to the frequency ω_{Raman} of the Raman transition between the two ground states.

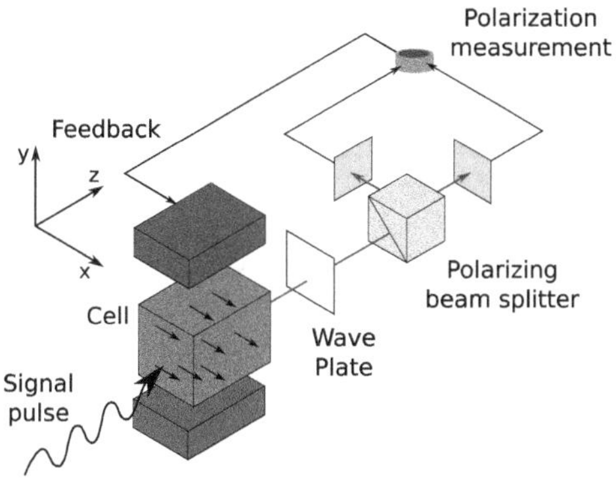

Figure 12.10. A quantum memory scheme based on off-resonant Faraday interaction. The quantum information is encoded in the polarization of the incident light. After the propagation through the cell the field is subjected to a polarimetric measurement. The result is fed back to the atoms by applying a magnetic field pulse.

atoms are initially prepared with the angular momentum oriented along the x-axis. Because of the interaction between the atoms and light, angular momentum is transferred from the photons to the atoms, putting the latter in rotation around the z-axis. Conversely, the wave polarization is affected by the z-component of the atomic angular momentum, which induces Faraday rotation. When the pulse leaves the medium, one determines its polarization and shifts the angular momentum of the atoms accordingly. In this way, one can store both the components of the wave polarization in the angular momentum of the atoms.

12.6 Conclusions and outlook

We have illustrated the main results achieved in quantum optics and shown its application in quantum technologies. We have presented the Lamb shift and the scattering off the vacuum, which are two phenomena showing the need for field quantization. We have introduced the quantum theory of coherence and briefly described the studies leading to the birth of quantum computation and quantum information theory. Moreover, we have described the role of quantum optics in modern quantum technologies and in quantum information by illustrating the quantum teleportation protocol, quantum metrology, quantum state engineering and quantum memory.

The progress in the understanding of quantum mechanics and in the ability to manipulate single atoms and photons has made it possible to observe entanglement even in systems containing billions of atoms. This fact turns out to be useful in various branches of science. Quantum optics and quantum information not only promise to be of fundamental importance in the development of new technologies, but they also represent a change of paradigm in interpreting many phenomena.

References

[1] Einstein A 1905 Concerning an heuristic point of view toward the emission and transformation of light *Ann. Phys.* **17** 132–48

[2] Muthukrishnan A, Scully M O and Zubairy M 2003 The concept of the photon—revisited *Opt. Photonics News* **14** 18–27

[3] Lamb W E and Retherford R C 1947 Fine structure of the hydrogen atom by a microwave method *Phys. Rev.* **72** 241–3

[4] Scully M O, Meyer G M and Walther H 1996 Induced emission due to the quantized motion of ultracold atoms passing through a micromaser cavity *Phys. Rev. Lett.* **76** 4144–7

[5] Brown R H and Twiss R Q 1957 Interferometry of the intensity fluctuations in light. I. basic theory: the correlation between photons in coherent beams of radiation *Proc. R. Soc.* A **31** 300–24

[6] Glauber R J 2007 *Quantum Theory of Optical Coherence* (New York: Wiley)

[7] Radmore P M and Barnett S M 1997 *Methods in Theoretical Quantum Optics* (Oxford Series in Optical and Imaging Sciences) (New York: Oxford University Press)

[8] Walborn S P, Monken C H, Pádua S and Souto Ribeiro P H 2010 Spatial correlations in parametric down-conversion *Phys. Rep.* **495** 87–139

[9] Wootters W K and Zurek W H A single quantum cannot be cloned *Nature* **299** 802–3

[10] Gerry C and Knight P 2004 *Introductory Quantum Optics* (Cambridge: Cambridge University Press)

[11] Ferraioli A G and Noce C 2019 The measurement problem in quantum mechanics *Sci. Philos.* **7** 41–58

[12] Feynman R P 1982 Simulating physics with computers *Int. J. Theor. Phys.* **21** 467–88

[13] Nielsen M A and Chuang I L 2011 *Quantum Computation and Quantum Information: 10th Anniversary Edition* (New York: Cambridge University Press)

[14] Horodecki R, Horodecki P, Horodecki M and Horodecki K 2009 Quantum entanglement *Rev. Mod. Phys.* **81** 865–942

[15] Plenio M B and Virmani S 2007 An introduction to entanglement measures *Quant. Inf. Comput.* **7** 1–51

[16] Einstein A, Podolsky B and Rosen N 1935 Can quantum-mechanical description of physical reality be considered complete? *Phys. Rev.* **47** 777–80

[17] Zurek W H 2003 Decoherence, einselection, and the quantum origins of the classical *Rev. Mod. Phys.* **75** 715–75

[18] Schlosshauer M 2005 Decoherence, the measurement problem, and interpretations of quantum mechanics *Rev. Mod. Phys.* **76** 1267–305

[19] Bennett C H, Brassard G, Crépeau C, Jozsa R, Peres A and Wootters W K 1993 Teleporting an unknown quantum state via dual classical and Einstein–Podolsky–Rosen channels *Phys. Rev. Lett.* **70** 1895–9

[20] Braunstein S L and van Loock P 2005 Quantum information with continuous variables *Rev. Mod. Phys.* **77** 513–77

[21] Browne D, Bose S, Mintert F and Kim M S 2017 From quantum optics to quantum technologies *Prog. Quant. Electron.* **54** 2–18

[22] Braunstein S L and Kimble H J 1998 Teleportation of continuous quantum variables *Phys. Rev. Lett.* **80** 869–72

[23] Giovannetti V, Lloyd S and Maccone L 2011 Advances in quantum metrology *Nat. Photonics* **5** 222–9

[24] Lvovsky A I, Sanders B C and Tittel W 2009 Optical quantum memory *Nat. Photonics* **3** 706–14

[25] Fleischhauer M, Imamoglu A and Marangos J P 2005 Electromagnetically induced transparency: optics in coherent media *Rev. Mod. Phys.* **77** 633–73

[26] Yousif M E 2017 The Faraday effect explained *IOSR J. Appl. Phys.* **9** 74–84

Part II

The modern physics behind experiments

IOP Publishing

Modern Physics
A critical approach
Canio Noce

Chapter 13

The Thomson experiment: cathode rays are still hot

Antonio Leo

13.1 Introduction

It is very likely that you are reading these words on a modern e-book reader or looking at an ultra-flat computer monitor. This is because cathode ray tube (CRT) monitor technology has been considered a dead technology since the first decade of the twenty-first century. Still, without CRTs, broadcasts would probably still be limited to radio programmes and computer outputs would even now be read on print-outs. Not to mention all the modern day applications which use cathode rays, such as industrial welders, lithographic processes and ultra-high resolution microscopes, just to name a few. That is why we can affirm that cathode rays are still hot.

But what are these cathode rays? How were they discovered? We will answer these (and other) questions in the following section, where a brief history of the discovery and understanding of cathode rays will be reported. Then, we will show how to realize one of the best known experiments on cathode rays, analysing in detail the physics behind it, as well as the apparatus and the procedure to realize the experiment.

13.2 History of cathode rays

The first experiments showing the existence of radiation between two electrical conductors (called electrodes) in a glass tube with rarefied air were performed by Michael Faraday in 1838 [1]. In particular, Faraday observed that once a high voltage is applied to the electrodes an arc is formed, originating from one electrode, which we will call the cathode, and ending on the other electrode, which we will call the anode. Moreover, Faraday also observed the existence of a dark space, i.e. an area where no luminescence was present, just in front of the cathode.

In the following years, many experiments were carried out with different gases and at lower and lower gas pressure values, leading to the observation of different

doi:10.1088/978-0-7503-2678-0ch13

phenomena [2]. Among these experiments, we mention those by Heinrich Geissler, which lead to the observation of the glow discharge phenomenon, which is the working principle of modern neon lights, and those by William Crookes, who succeeded in realizing high vacuum glass tubes (called Crookes tubes) where the remaining gas pressure was below 10^{-6} atm. Repeating the Faraday experiment in one of his high vacuum tubes, Crookes observed that the cathode dark space (often referred to as the Faraday dark space or Crookes dark space) spread across the whole tube, from the cathode to the anode. Moreover, he observed a glow at the end of the glass tube on the anode side.

At this point, a debate arose about what the nature of this radiation that emanated from the cathode was, soon to be baptized as cathode rays. There were two schools of thought—the one by Crookes himself, Arthur Schuster and others supported the idea of particle radiation [2], while the German school, led by Heinrich Hertz, supposed that cathode rays were a new form of æther waves, i.e. a type of electromagnetic radiation [3]. The debate was resolved by Joseph John Thomson, who in his seminal paper on cathode rays stated that [4]

> … we have in the cathode rays matter in a new state, a state in which the subdivision of matter is carried very much further than in the ordinary gaseous state: a state in which all matter—that is, matter derived from different sources such as hydrogen, oxygen, & c.— is of one and the same kind; this matter being the substances from which all the chemical elements are built up.

We have to note that, at the time Thomson wrote these words, the atom was the smallest known particle. Although it had already been suggested that atoms could be built up from smaller 'bricks', these were assumed to be the size of the hydrogen atom, the smallest atom. Following his experimental results, Thomson suggested that the mass of these new 'bricks', which he named corpuscles, had to be about 1000 times lighter than the hydrogen atom. Thus, he was advancing the idea of the existence of subatomic particles, which was a revolutionary idea at that time.

Thomson came to the cited conclusion through a series of three experiments. The first experiment was a variant of the experiment carried out in 1895 by Jean Baptiste Perrin, who demonstrated that cathode rays have negative charge. Thomson modified a Crookes tube by adding a pair of coaxial metallic cylinders with slits in them (see figure 13.1(a)), with an electrometer connected to the inner cylinder. The rays originating from cathode A passed through the slit in anode B so that the rays followed a quite linear path. The coaxial cylinders were arranged such that their slits were aligned together but not with the path of the rays. To let the cathode rays fall on the electrometer slit, a magnetic field had to be applied, which was demonstrated to modify the path of the rays. What Thomson found was that when the cathode ray path did not hit the electrometer, a small and irregular electrical charge was detected. If the cathode rays were bent so that they fell on the slit, then a well defined and significant negative charge could be measured. Bending the rays so much that they overshot the slits, the measured charge was a small fraction of the case of perfect aim. From these observations, Thomson concluded that 'however we

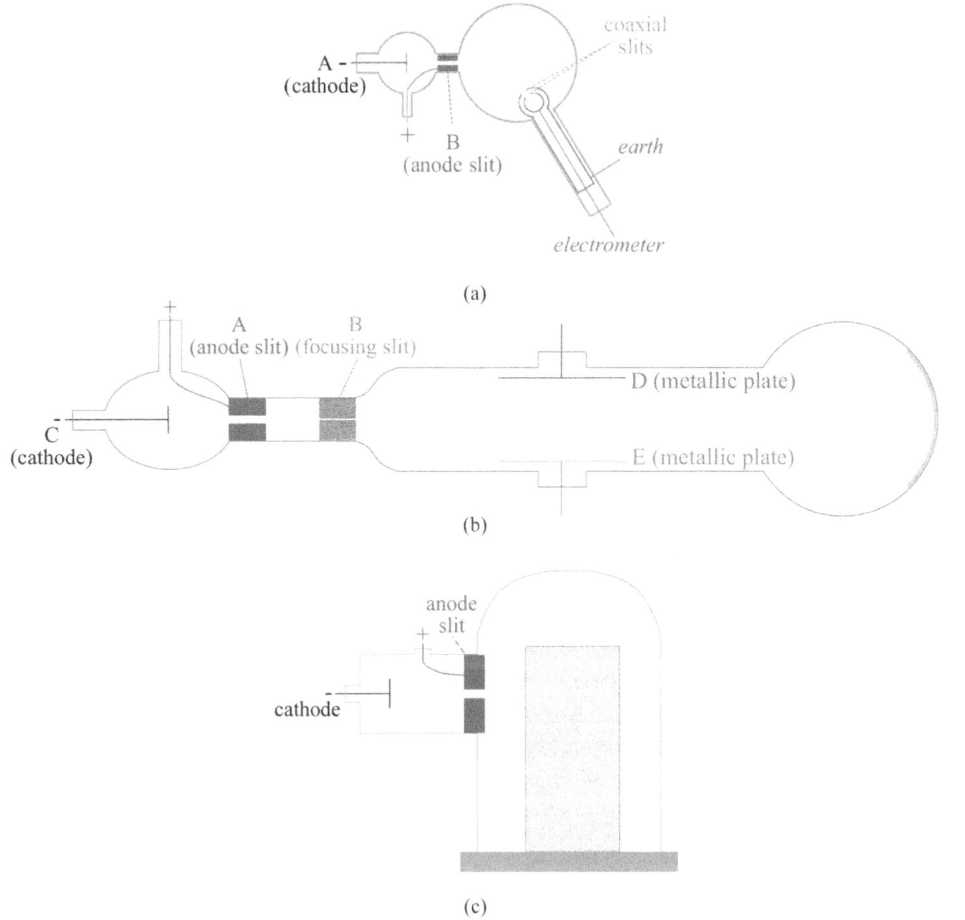

Figure 13.1. The experimental set-up diagrams of the three original Thomson's experiments on electrification and cathode rays connection (a), on the cathode rays charge sign (b), and on the influence of gas nature on the cathode rays bending (c). Reprinted from [4].

twist and deflect the cathode rays by magnetic forces, the negative electrification follows the same path as the rays, and that this negative electrification is indissolubly connected with the cathode rays' [4].

The second experiment performed by Thomson was also a variation of a previous experiment performed by Hertz. The experiment was carried out with the apparatus in figure 13.1(b): the cathode rays originate from a cathode, C, pass through a slit in the anode, A, and then through a second, thinner metallic slit, B, so that at the end of the glass tube a very narrow glowing spot can be observed. What Hertz tried to do was to modify the rays' path by applying an electrostatic force, thus he placed two metallic plates (D and E) parallel to the rays' path and applied a potential difference to them. Hertz' observations were that a small electrostatic force does not bend the rays, while they can be deviated with a large potential difference. These results were interpreted as proof that the rays were propagating through an ætherial medium.

Thomson correctly argued that the observed effects have to be ascribed to interference by the rarefied gas in the tube, which shows a non-zero conductivity as the D–E plates are electrified. Thus, Thomson spent a great deal of effort on obtaining a vacuum as high as possible in the tube. Once the right conditions were achieved, Thomson observed that 'the rays were depressed when the upper plate was connected with the negative pole of the battery, the lower with the positive, and raised when the upper plate was connected with the positive, the lower with the negative pole' [4]. This can only be a case of negatively charged cathode rays. Moreover, Thomson observed that the deflection was proportional to the potential difference between the plates.

Finally, in his third experiment, Thomson analysed how different gases could modify the path of the cathode rays once they were subject to the action of an applied magnetic field. He used a bell-jar with a cathode ray tube fastened on one side (see figure 13.1(c)). The jar was placed in a Helmholtz coil (not shown in the diagram of figure 13.1(c)). A vertical plate of glass divided into small squares was placed in the cathode ray path in order to evaluate the bending by the magnetic field. Surprisingly, he found that 'the path of the rays is independent of the nature of the gas' [4]. Other experiments were carried out by other researchers which demonstrated that the nature of the corpuscles was independent of the cathode material. Thus, Thomson's final conclusion was 'I can see no escape from the conclusion that they are charges of negative electricity carried by particles of matter'. To answer the questions he himself asked 'What are these particles? Are they atoms or molecules, or matter in a still finer state of subdivision?' [4]. Thomson carried out a series of measurements of the charge-to-mass ratio of the corpuscles, finding a value which is not too far from the CODATA recommended value $-e/m_e = -1.758820\,010\,76(53) \times 10^{11}$ C kg^{-1}.[1]

As mentioned, Thomson suggested that corpuscles have a much lighter mass than that of the hydrogen atom, but he went even further than that, and in 1904 he proposed his own model of atomic structure. In Thomson's model 'the atoms of the elements consist of a number of negatively electrified corpuscles enclosed in a sphere of uniform positive electrification' [5]. In other words, the corpuscles are like negative 'plums' embedded in a positive 'pudding', which is why Thomson's model is often referred to as the plum pudding model.

Currently, we know that Thomson's corpuscles are the fundamental subatomic negatively charged particles commonly known as electrons (a portmanteau of the words *electr*ic and *ion* by George Johnstone Stoney [6]) and that in the atomic structure they 'orbit' around a positively charged nucleus. Despite various efforts, Thomson failed to evaluate the charge of the electron. It was Robert Andrews Millikan who succeeded in providing definitive proof of the discreteness of electric charge (see chapter 14). Curiously, Thomson's son, George Paget Thomson, was one of the two physicists awarded the 1937 Nobel Prize in Physics for the observation of the wave properties of the electron, the other being Clinton Joseph Davisson (see chapter 15). The senior Thomson had himself been awarded the Nobel Prize in

[1] According to the NIST Database: https://physics.nist.gov/cgi-bin/cuu/Value?esme.

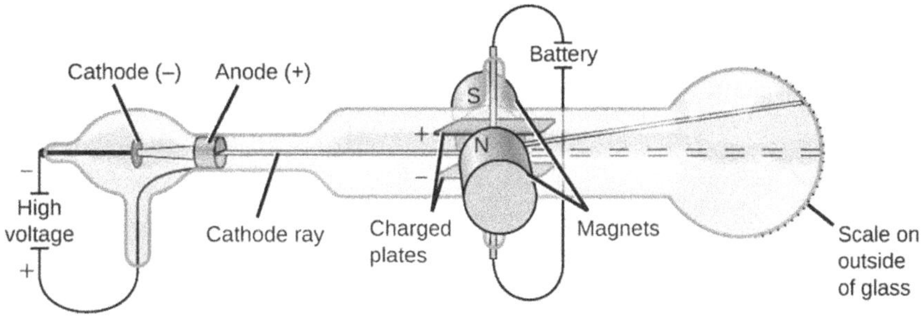

Figure 13.2. A diagram of the cathode ray tube for Thomson's experiment on the electron charge-to-mass ratio. Image reproduced from Openstax, under a CC BY 4.0 license.

Physics in 1906 as his great intuitions and experiments had been crucial toward the understanding of electromagnetism and matter physics. Even if you probably are not reading this work on a CRT monitor, it is also thanks to Thomson's work that we have the use of all the kinds of advanced and compact electronic devices available tday.

13.3 The physics behind the experiments

In the previous section, we have briefly discussed the three experiments which Thomson performed in order to clarify the connection between the charge and the cathode rays, the sign of this charge and the influence of the nature of the gas on the bending of cathode rays by an applied magnetic field. Through the results of these experiments, Thomson was also able to evaluate the electron's charge-to-mass ratio. In order to perform this evaluation, we can apply simple concepts of classical mechanics and electromagnetism.

Let us consider an experimental set-up such as the one drawn in figure 13.2, which is very similar to that used by Thomson in his second experiment. In this apparatus, a cathode is placed on one end of a vacuum tube. The electrons generated at the cathode are accelerated by the electric field between the cathode and the anode and they pass through the anode slit which helps to focus the beam. The focused electron beam can be deflected by applying an electrical potential between horizontal charged plates or by the magnetic field from a pair of magnets. Electrons will hit the opposite end of the tube, which has been treated in order to enhance the fluorescence of the spot on the glass. A scale helps to evaluate the beam deflection.

We will analyse the case of a beam bent by the magnetic field. Indeed, in this case, the possible presence of residual gas in the tube does not introduce disturbances in the experiment. In contrast, in the case of deflection by charged plates, these can induce ionization in the gas, thus altering the electric field bending effect (as Thomson himself observed [4]). Supposing that the magnetic field density $\vec{B}$ is uniformly directed along the magnets' common axis, a moving electron experiences a Lorentz force whose absolute value is

$$|F| = evB,$$

where v is the electron velocity. In an uniform magnetic field, the electron path is bent in a circular trajectory. If we define r as the bending radius of the trajectory, the equivalence between the Lorentz force and the centripetal force lead to the equation

$$m\frac{v^2}{r} = evB,$$

where m is the electron mass, and it follows that

$$\frac{e}{m} = \frac{v}{rB}. \tag{13.1}$$

In equation (13.1), r can be evaluated thanks to the graduated scale on the tube. The electron velocity v can be obtained by knowing the potential difference U_A between the cathode and the anode. Indeed, v is the velocity of the electron at the anode position, thus the electron kinetic energy $\frac{1}{2}mv^2$ has to be equal to the potential energy eU_A for a charge e at the anode position:

$$\frac{1}{2}mv^2 = eU_A \Rightarrow v = \sqrt{2\frac{e}{m}U_A}.$$

Finally, in order to obtain an explicit expression for the magnetic field in terms of measurable physical quantities, we can assume that the magnets form an Helmholtz coil. In this case, it is (see appendix A, equation (A.2))

$$B = \left(\frac{4}{5}\right)^{\frac{3}{2}}\frac{\mu_0 N}{R}I_H = KI_H, \tag{13.2}$$

where N is the number of coils, R is the coil radius and I_H is the bias current.

In terms of equation (13.2), equation (13.1) can be rewritten as

$$\frac{e}{m} = \frac{2U_A}{(KI_H r)^2}, \tag{13.3}$$

which will be the reference equation in our Thomson's experiment.

13.4 The experimental set-up

Figure 13.3(a) shows an apparatus that can be used to perform Thomson's experiment. This is a table-top apparatus[2] which includes (see figure 13.3(b)) a Thomson tube, four voltage power supplies, an ammeter, a voltmeter and a Helmholtz coil assembly.

In the Thomson tube there is a very low pressure of neon gas. Since the experiment will be carried out using a magnetic field to bend the cathode rays, the presence of the gas will not introduce disturbances. On the contrary, due to gas ionization by the travelling electrons, we will be able to visualize the otherwise

[2] The apparatus was manufactured by 3B Scientific.

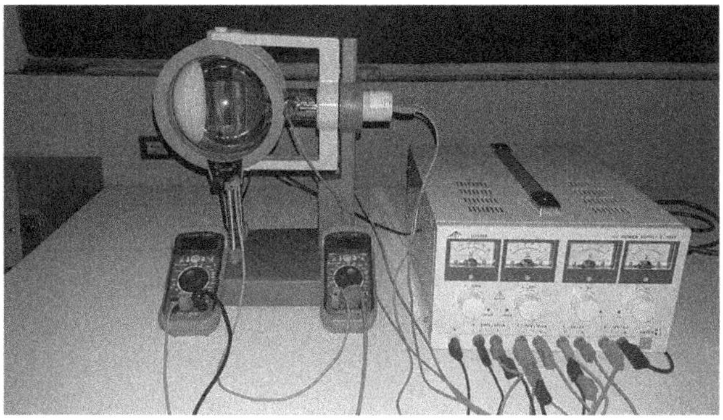

(a)

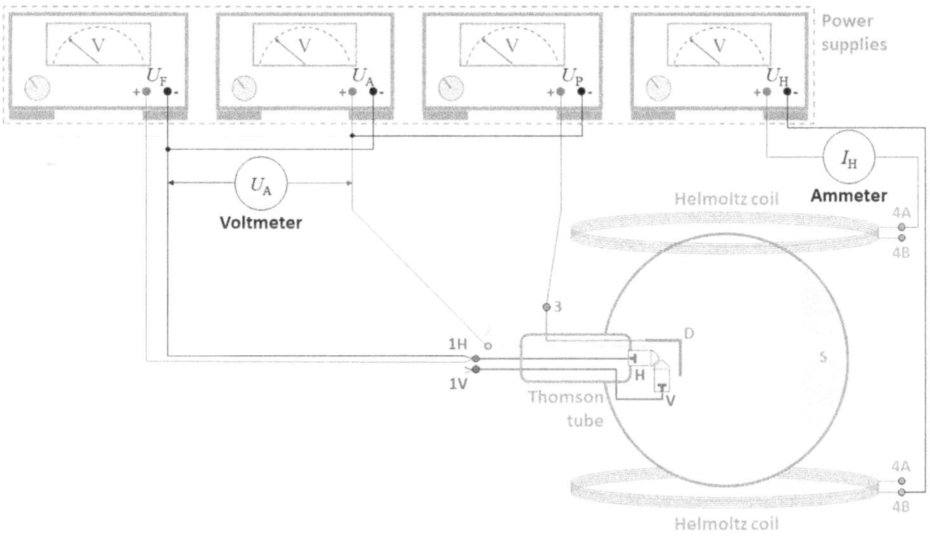

(b)

Figure 13.3. (a) A photograph of the apparatus used to perform the variants of Thomson's experiment described previously and (b) a sketch of this experimental set-up.

invisible cathode rays. Two electron guns are housed in the tube, i.e. the assembly of the cathode and the anode, one of which generates an axial beam (marked as H in figure 13.3(b)), while the other generates a vertical beam (V). The guns are of the hot cathode type, thus the cathode (a metallic filament) has to be heated in order to release electrons. The two filaments can be connected to a power supply through plug-in connectors (1H and 1V). The voltage bias to the anode (2) can be measured carefully using the voltmeter. The tube end opposite the gun side (marked as S in figure 13.3(b)) is treated with a substance which enhances the fluorescent glow from the electrons hitting the glass. A deflection plate (D) conveniently connected (3) to a power supply allows the beam focusing to be to tuned.

The solenoids in the Helmholtz coil assembly are made of copper wires wound around plastic supports. The number of turns in each solenoid is 320 and their radius is 67 mm. Due to its design, it can be assumed that the magnetic field generated by the Helmholtz coil is homogeneous over the whole beam deflection area. The two solenoids are connected to a power supply through plugs 4A and 4B (see figure 13.3(b)) in such a way that the current flows in each solenoid in the same direction. The bias current can be measured by the ammeter.

Once all the components of the apparatus are connected as illustrated in figure 13.3(b) and the Helmholtz coil assembly is aligned as shown in figure 13.3(a), the set-up is ready to perform the experiment.

13.5 How to determine the electron charge-to-mass ratio

In this section we will show two variants of Thomson's experiment: the first (which we will call the classical variant) is close to the original experiment and the second (the circular variant) shows how extreme the electron beam bending by a magnetic field can be.

Classical variant. To perform this variant of the experiment, we have to prepare the apparatus using the following steps:

1. Connect the horizontal electron gun to the power supply (see figure 13.3(b)), set the cathode heating voltage U_F and then wait some minutes in order for the the filament to begin producing electrons.
2. Set the anode voltage to the minimum value $U_{A,0}$ which produces a proper electron beam.
3. Focus the electron beam by tuning the deflection plate voltage U_P.

Once these steps have been performed, we should have a situation like that shown in figure 13.4(a). At this point, we can start the actual experiment, which can be performed using the following steps:

1. Set the Helmholtz coil current to $I_{H,0}$ so that the beam hits the border of the fluorescent area (see figures 13.4(b) and (c)).
2. Take note of $U_{A,0}$ and $I_{H,0}$ (together with their measurement error[3]) in a table such as table 13.1.
3. Set a new value for U_A and tune I_H so that the beam hits the border of the fluorescent area again and take note of the values in your table.
4. Repeat operations at point 3 in order to acquire a set of data suitable for a linear fit analysis.

An example of (U_A, I_H) data acquired in a classical variant experiment are presented in table 13.1. In order to perform a linear fit on this set of data, we need to linearize equation (13.3), thus we will consider the equation

[3] The estimation of the error associated with a measurement depends on different factors, first of all the characteristics of the instruments used to perform the measurement itself. For a more detailed description of the theory of error, see for example [7]. In our case, we are using digital instruments and we will assume that the error on a single measurement is one unit on the last digit of the reading.

(a)

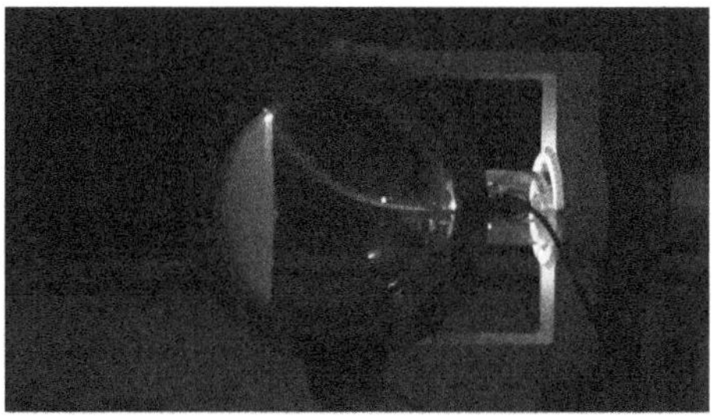

(b)

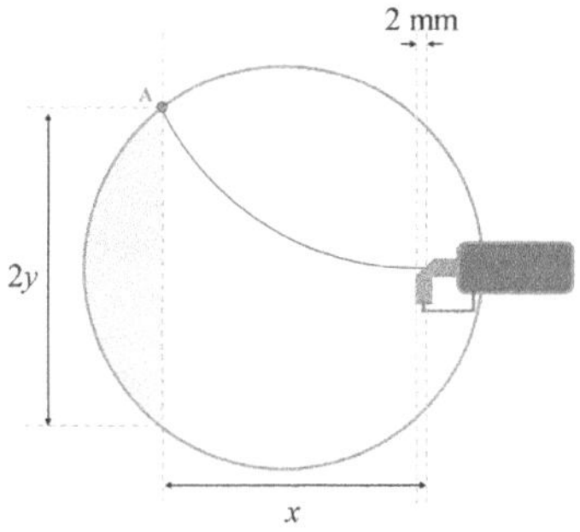

(c)

Figure 13.4. Photographs of the apparatus (a) while performing the initial alignment of the horizontal beam and (b) while performing the measurement in the classical variant. (c) A diagram of the beam configuration during the measurement.

Table 13.1. Data acquired in the presented classical variant of Thomson's experiment.

Measurement	U_A (V)	I_H (A)	I_H^2 (A^2)
0	174 ± 1	0.112 ± 0.001	0.0125 ± 0.0002
1	179 ± 1	0.115 ± 0.001	0.0132 ± 0.0002
2	184 ± 1	0.116 ± 0.001	0.0135 ± 0.0002
3	195 ± 1	0.120 ± 0.001	0.0144 ± 0.0002
4	213 ± 1	0.125 ± 0.001	0.0156 ± 0.0002
5	227 ± 1	0.129 ± 0.001	0.0166 ± 0.0003

$$I_H^2 = \frac{m}{e}\frac{2}{(Kr)^2}U_A = \alpha U_A, \tag{13.4}$$

where r can be evaluated by the relation (see figure 13.4(c) and appendix B, equation (B.3))

$$r = \frac{x^2 + y^2}{2y},$$

with $x = 82.44$ mm and $y = 51.84$ mm.

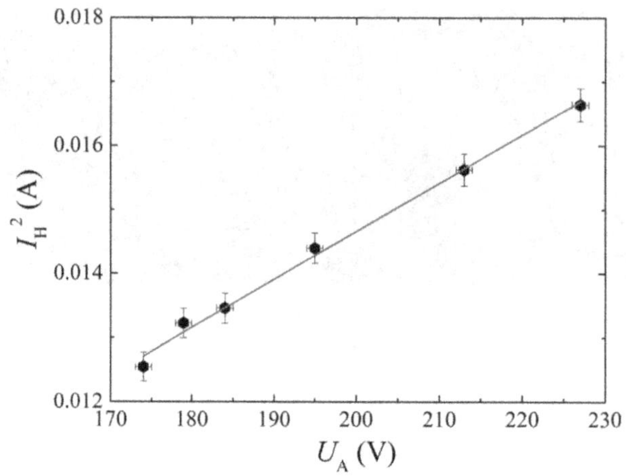

Figure 13.5. Data acquired in the presented first variant of Thomson's experiment. The red line is the result of a linear fit.

The I_H^2 data are included in table 13.1 and reported in figure 13.5. A linear fit has been performed on the (U_A, I_H^2) data resulting in $\alpha = (7.5 \pm 0.3) \times 10^{-5}$ A^2 V^{-1}. Thus, from equation (13.4), we evaluate

$$\frac{e}{m} = 1.72 \pm 0.06 \times 10^{11} \text{ C kg}^{-1},$$

a value in good agreement with the expected one.

Circular variant. The variant of Thomson's experiment we call the circular variant is performed in a very similar way to the classical variant. The preliminary steps of the experiment are the following:

1. Connect the vertical electron gun to the power supply (see figure 13.3(b)), set the cathode heating voltage U_F, and then wait some minutes until the filament is producing electrons.
2. Set the anode voltage to the minimum value $U_{A,0}$ which produce a proper vertical electron beam (figure 13.6(a)).
3. Eventually, focus the electron beam by tuning the deflection plate voltage U_P.

The actual experiment is carried out as follows:

1. Set the Helmholtz coil current to $I_{H,0}$ in order to obtain a circular beam whose diameter is equal to the distance between the beam origin and the border of the fluorescent area (see figure 13.6(c)).
2. Take note of $U_{A,0}$ and $I_{H,0}$ (together with their measurement error) in a table such as table 13.2.

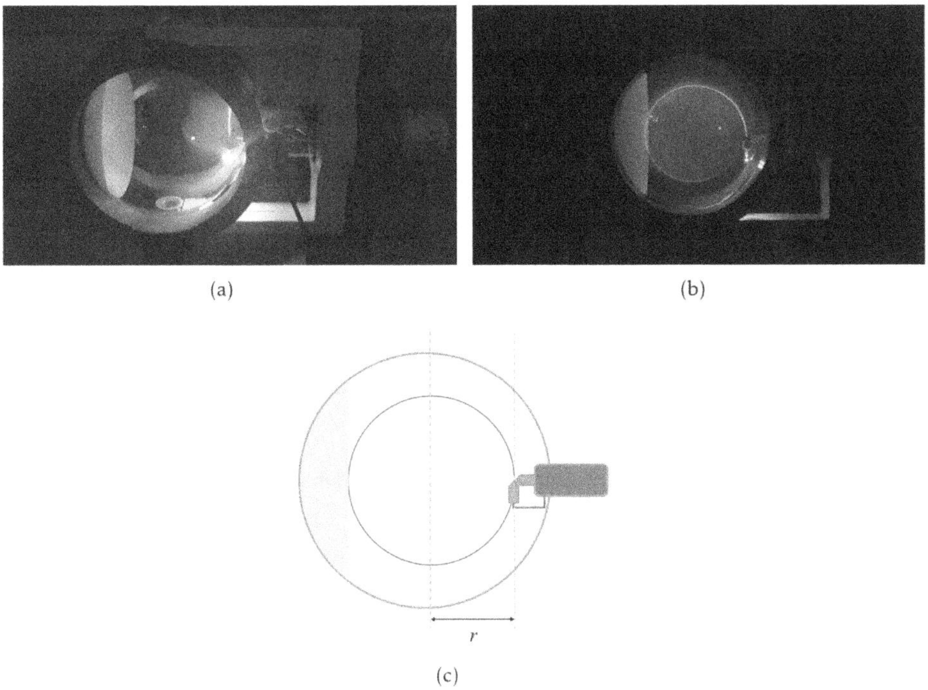

Figure 13.6. Photographs of the apparatus (a) while performing the initial alignment of the vertical beam and (b) while performing the measurement in the circular variant. (c) A diagram of the beam configuration during the measurement.

Table 13.2. Data acquired in the presented circular variant of Thomson's experiment.

Measurement	U_A (V)	I_H (A)	I_H^2 (A^2)
0	189 ± 1	0.25 ± 0.01	0.062 ± 0.005
1	209 ± 1	0.26 ± 0.01	0.068 ± 0.005
2	243 ± 1	0.27 ± 0.01	0.073 ± 0.005
3	267 ± 1	0.28 ± 0.01	0.078 ± 0.006

3. Set a new value for U_A and tune I_H in order to maintain the beam diameter as fixed and take note of the values in the table.
4. Repeat the operations of step 3 in order to acquire a set of data suitable for a linear fit analysis.

An example of (U_A, I_H) data acquired in a circular variant experiment is presented in table 13.2. Following the same procedure as described for the classical variant, a

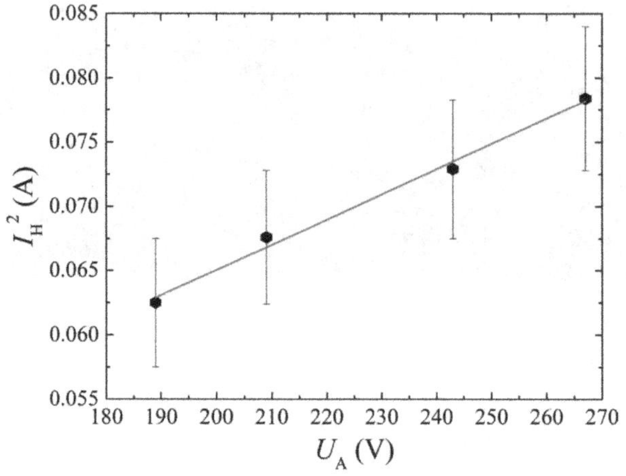

Figure 13.7. Data acquired in the presented circular variant of Thomson's experiment. The red line is the result of a linear fit.

linear fit has been performed on the (U_A, I_H^2) data based on equation (13.4). In this case it is $\alpha = (1.96 \pm 0.13) \times 10^{-4}$ A^2 V^{-1}. Since in this variant it is $r = 45$ mm, the charge-to-mass estimated value is

$$\frac{e}{m} = 1.85 \pm 0.12 \times 10^{11}\ \mathrm{C\,kg^{-1}},$$

a value which is again in good agreement with the expected one (figure 13.7).

Acknowledgements

The author would like to thank R De Luca and O Faella for useful discussions about the variants of Thomson's experiment presented here and their realization.

Appendix

A Helmholtz coils

A Helmholtz coil is composed of a pair of identical solenoids whose radius R is wider than their width w, with the two solenoids aligned on the same axis at a distance R and electrically connected in series (see figure A1(a)).

To evaluate the magnetic field density B at the centre of an Helmholtz coil, we start evaluating B at a point P at a distance x from the centre of a single coil (see figure A1(b)) through the Laplace equations. In our case, the magnetic field density $\vec{B}$ generated by an infinitesimal element $\mathrm{d}\vec{l}$ of a coil biased by a current $\vec{i}$ at a point P at a distance $\vec{r}$ from $\mathrm{d}\vec{l}$ can be written as

$$\mathrm{d}\vec{B} = \frac{\mu_0 i \mathrm{d}\vec{l} \times \vec{r}}{4\pi r^3}.$$

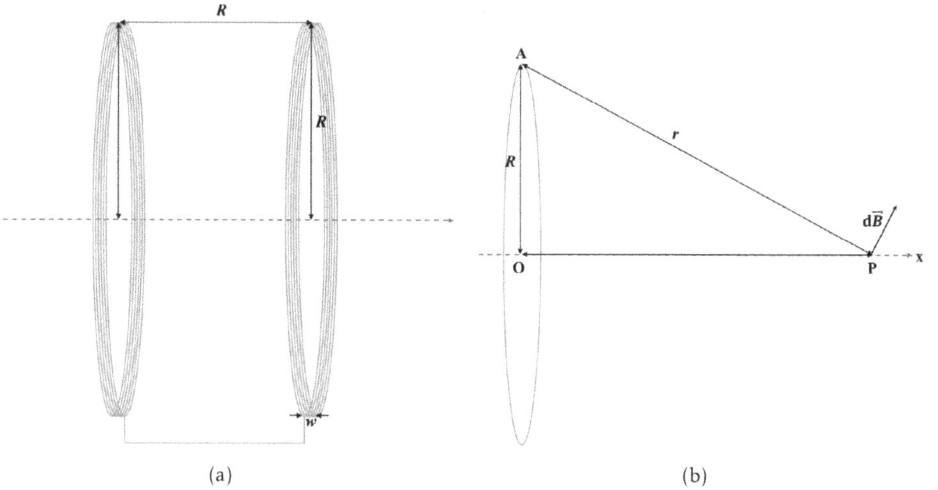

Figure A1. Drawings of (a) the usual arrangement of a Helmholtz coil and (b) the scheme used to evaluate the field generated.

In the case of a point P lying on the coil symmetry axis x, we can consider only the component of $\vec{B}$ along the axis, since the sum of the other components would be zero due to the symmetry. Thus

$$dB_x = \frac{\mu_0 i R dl}{4\pi r^3}. \tag{A.1}$$

Integrating equation (A.1) over dl we obtain

$$B_x = \frac{\mu_0 i R}{4\pi r^3} \int_0^{2\pi R} dl = \frac{\mu_0 i R^2}{2r^3} = \frac{\mu_0 i R^2}{2(R^2 + x^2)^{\frac{3}{2}}}.$$

For a solenoid with N coils the total magnetic field density in P is NB_x. Considering that in the centre of the Helmholtz coil it is $x = R/2$ and that we have an equal contribution from the two solenoids, finally it is

$$B = \left(\frac{4}{5}\right)^{\frac{3}{2}} \frac{\mu_0 i N}{R}. \tag{A.2}$$

B Evaluation of the bending radius for the classical variant of the experiment

In order to evaluate the bending radius r in the case of what we call the classical variant of Thomson's experiment, we start by defining the beam exit point of the axial electron gun (see figure 13.3(b)) as the origin O of a two-dimensional Cartesian coordinate system as the abscissa axis along the unbended beam direction (see figure A2). We will consider the two triangles AOC and AOB.

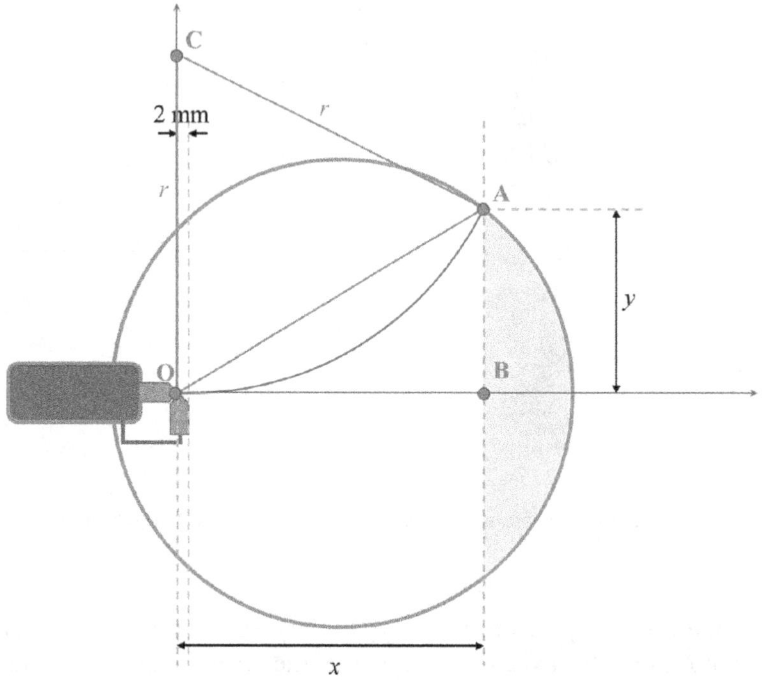

Figure A2. The diagram used to evaluate the bending radius of the beam in the classical variant of Thomson's experiment in our apparatus.

We start considering the isosceles triangle AOC. From the chord theorems it is

$$\overline{\mathrm{OA}} = 2r \sin\left(\frac{\mathrm{A\hat{C}O}}{2}\right),$$

which we can rewrite as

$$\overline{\mathrm{OA}} = 2r \sin(\mathrm{A\hat{O}B}) \tag{B.1}$$

since

$$\mathrm{A\hat{C}O} = \pi - 2\left(\frac{\pi}{2} - \mathrm{A\hat{O}B}\right).$$

Considering the right-angled triangle AOB, we have that

$$y = \overline{\mathrm{OA}} \sin(\mathrm{A\hat{O}B}), \tag{B.2}$$

thus, combining equation (B.1) with equation (B.2)

$$\frac{2r}{\overline{\mathrm{OA}}} = \frac{\overline{\mathrm{OA}}}{y} \Rightarrow r = \frac{\overline{\mathrm{OA}}^2}{2y}.$$

Finally, from the Pythagorean theorem $\overline{\mathrm{OA}}^2 = x^2 + y^2$ it follows that

$$r = \frac{x^2 + y^2}{2y}. \tag{B.3}$$

References

[1] Faraday M 1838 VIII. Experimental researches in electricity.—thirteenth series *Philos. Trans. R. Soc. Lond.* **128** 125–68

[2] Schuster A 1884 The Bakerian Lecture. Experiments on the discharge of electricity through gases. Sketch of a theory *Proc. R. Soc. Lond.* **37** 317–39

[3] Hertz H 1893 *Electric Waves: Being Researches on the Propagation of Electric Action with Finite Velocity through Space* (London: MacMillan)

[4] Thomson J J 1897 XL. Cathode rays *Philos. Mag. Ser. 5* **44** 293–316

[5] Thomson J J 1904 XXIV. On the structure of the atom: an investigation of the stability and periods of oscillation of a number of corpuscles arranged at equal intervals around the circumference of a circle; with application of the results to the theory of atomic structure *Philos. Mag. Ser. 6* **7** 237–65

[6] Stoney G J 1894 Of the 'electron,' or atom of electricity *Philos. Mag. Ser. 5* **38** 418–20

[7] Taylor J R 1982 *An Introduction to Error Analysis* 2nd edn (Sausalito, CA: University Science)

IOP Publishing

Modern Physics
A critical approach
Canio Noce

Chapter 14

The Millikan oil drop experiment

Marco Di Mauro

14.1 Introduction

The Millikan oil drop experiment is a landmark of modern physics, since it allowed it to be proved for the first time that electric charge is discrete, and to measure the elementary unit of charge. In this chapter, after a brief historical introduction, we describe this experiment in detail, with a careful treatment of the physics involved. After that, we briefly describe the table-top apparatus based on the Millikan experiment that we used in our lab to check the discretization of electric charge, the experimental procedure and the data analysis, closing with some comments about the results we obtained.

14.2 Historical introduction

In this section we summarize the historical background of Robert A Millikan's experiment, going from the atomic hypothesis to the discovery of the electron. This has the purpose of setting the experiment into context and explaining its fundamental importance for atomic and subatomic physics. Due to space constraints we can only provide a sketch of this complex and fascinating story. Good sources for further exploration are [1–5], where references to the original papers can also be found.

The idea that matter is not continuous, but made up of discrete entities, the so-called *atomic hypothesis*, has been around since at least Greek antiquity. This hypothesis has been invoked occasionally in physics since Isaac Newton's times, in particular in connection with the properties of gases. For example, Daniel Bernoulli managed to interpret gas pressure in what may be considered the first modern work on kinetic theory. However, factual evidence in support of the atomic hypothesis only started appearing in the nineteenth century, not from physics, but from chemistry (not that the two disciplines were so sharply separated at the time). In particular, in 1803, John Dalton hypothesized that matter is made up of atoms of different elements to explain Proust's law of definite proportions and his own law of

doi:10.1088/978-0-7503-2678-0ch14

multiple proportions. Over the course of many years, physicists' atoms and chemists' atoms, which were conceptually very different, came to be recognized as the same entities. The great successes of kinetic theory and statistical mechanics, in particular Albert Einstein's explanation of Brownian motion, definitely moved atoms from the realm of hypotheses to that of facts.

Starting from roughly the later half of the nineteenth century, different lines of research in physics began to shed more light on the intimate structure of matter. The results of Michael Faraday's experiments on electrolysis started to put forward the idea that electric charge may come in discrete units, in some way related to atoms of matter. The first mention of the word *electron* referred to such a unit of charge, and was coined in 1891 by George J Stoney, who was the first to attempt an estimate of it. Subsequent research in electromagnetism began to take into account this hypothesis, assuming electrical currents to be made up of moving charged particles, generally called 'atoms of electricity' or ions, and later electrons themselves. Meanwhile, research in spectroscopy, pioneered by Gustav R Kirchhoff and Robert W Bunsen, established among other results that heated rarefied gases emit light whose wavelength is distributed according to a well-defined pattern of lines, which is characteristic of the specific substance (see chapter 4). Some people, most notably James Clerk Maxwell, began to consider this as evidence of the fact that atoms should have some kind of internal structure, thus being more complex than point particles or rigid bodies. The recognition that Maxwell's theory implied that electromagnetic radiation should come from accelerated electric charges implied that this structure should involve something that is electrically charged. All this culminated in the masterful synthesis by Hendrik A Lorentz, who in particular was able to explain the Zeeman splitting of spectral lines by assuming that atoms contained charged particles. The extremely important idea that the electromagnetic and optical properties of substances are due to the fact that they are made of electrically charged constituents started to take a definite form and to give quantitative predictions.

The last strand of research was constituted by the study of electric conduction in rarefied gases, which led to the discovery of cathode rays. These rays are emitted by negative metallic electrodes under a very strong potential difference, and were soon recognized to be negatively charged.

All these separate lines of development were unified in 1897, when Joseph John Thomson measured the charge-to-mass ratio of the 'corpuscles' that made up cathode rays (see chapter 13) and discovered that it is always the same, regardless of the metal that composes the cathode. Therefore they were made up of some universal constituent of matter. However, immediately after this the problem emerged of explaining the size of the measured charge-to-mass ratio. In fact, such a ratio was thousands of times larger than that of the lightest known ion, the hydrogen ion (later to be identified with the proton). The two extreme hypotheses were that the constituents of cathode rays were as heavy as hydrogen ions, but had a much larger charge, or that they had the same charge as hydrogen ions, but were much lighter. The second hypothesis was revolutionary, since it implied that atoms (whose existence, we remark, was still not universally accepted) are not irreducible entities (despite their name), but that smaller, *subatomic* entities existed inside matter.

Many scientists, including Thomson himself, favoured the second hypothesis, which immediately began to spur on a great deal of activity in physics. Just to mention two of the most famous models, the Thomson plum pudding model of the atom and the Drude theory of electric and thermal conduction in metals were proposed only a few years later. The 'corpuscles' were baptized as electrons, thus giving a new meaning to the word originally coined by Stoney. However, a definitive proof of the fact that they really were subatomic entities required a separate measurement of their electric charge.

In 1899 Thomson himself attempted such a measurement[1], by employing a method invented by his student Charles Thomas Rees Wilson, who had discovered that supersaturated vapour can condense around charged particles. This is nothing other than the basic principle of a cloud chamber. By determining the number of charged particles by counting the droplets, and measuring the overall charge by electrometric methods, Thomson managed to estimate the charge of electrons with a 50 per cent accuracy. The first subatomic particle, the electron, thus became a fact of science, and atoms were finally proved to not be indivisible. Also, this was one of the first applications ever of the basic idea of the cloud chamber, which subsequently played a prominent role in elementary particle physics.

It is at this point that Millikan[2] came into the picture. In an impressive series of experiments performed starting from 1908 [7–10], in which he greatly improved Thomson's and Wilson's methods, Millikan provided the definitive proof of the fundamental discreteness of electric charge, of the subatomic nature of the electron and of the ionization mechanism. His experiments allowed Millikan to directly measure the charge of electrons with unrivalled precision (to within 0.6% of the modern value[3]), showing that it had the same modulus as the charge of the hydrogen ion[4].

As a by-product, Millikan's experiments provided fundamental evidence of the correctness of kinetic theory. In fact, Millikan managed to measure the order of magnitude of the kinetic energy of molecules due to thermal motion, showing that it

[1] In the same year he managed to prove that the particles produced in the photoelectric effect were the same that constituted cathode rays. At about the same time, the same was shown for the particles emitted in the β decay of radioactive substances and in thermionic emission, thus concluding that all those particles are identical, and that they were constituents of all atoms; atoms could in turn become ions by absorbing or releasing electrons.

[2] Robert Andrews Millikan was born in 1868. After his doctorate in 1895 he went to the University of Chicago, were he became full professor in 1910. At Chicago he undertook skilful experiments on the determination of the electric charge and the verification of Einstein's equation for the photoelectric effect. For this pioneering work he was awarded the Nobel Prize in 1923 [6]. In 1921 he moved to Caltech, were he stayed until his retirement, playing a major role in its becoming one of the world's leading institutions. There he began a massive study of cosmic rays, definitively proving their extraterrestrial origin, and giving them their name. He died in 1953.

[3] The error was later recognized to be due to a discrepancy in the values of the viscosity of air used by Millikan.

[4] For completeness we mention that Millikan's data analysis has raised some controversies [11] which, however, have been recently settled [12, 13].

matched the theoretical predictions, and to find and quantify the deviations from Stokes' law due to the granular nature of matter. His results agreed with the theoretical predictions put forward by Ebenezer Cunningham, based on kinetic theory.

Millikan's determination of the elementary unit of charge, combined with those for the charge-to-mass ratio of cathode rays, also allowed the electron mass to be accurately determined and provided independent estimates of many other relevant atomic quantities, such as all atomic and molecular weights, the Avogadro number, the Boltzmann constant, the Planck constant, etc, which constituted an improvement over existing estimates.

The great relevance of Millikan's work is well attested by the fact that his techniques are still widely used in the fields of condensed matter physics, nanotechnology and chemical physics. Examples include the search for fractionally charged excitations in condensed matter [14, 15], the measurement of the net charge of mobile nano-particles [16], the electrophoretic mobility of solutes (with the aim of monitoring kinetic interactions at the single particle level and intermediate steps of chemical reactions) [17] and cold atom analogues [18].

14.3 Description of the experiment

The essence of Millikan's improvement of previous methods was in performing the measurements on an individual charge carrier, rather than averaging over many of them [7], and in using oil droplets (he also considered other non-volatile liquids) in place of water droplets [8, 9], thus preventing them from evaporating, and hence ensuring that their mass remained constant for arbitrary amounts of time. Millikan's oil drop experiment consisted essentially of putting tiny charged oil droplets in a known electric field and studying their response to it. The best choice for the electric field is of course a uniform field, hence Millikan employed a condenser with planar and parallel plates. The apparatus is shown schematically in figure 14.1.

Oil droplets were sprayed into a chamber containing a condenser, and some of them could reach the region between the plates through a hole pierced in the centre

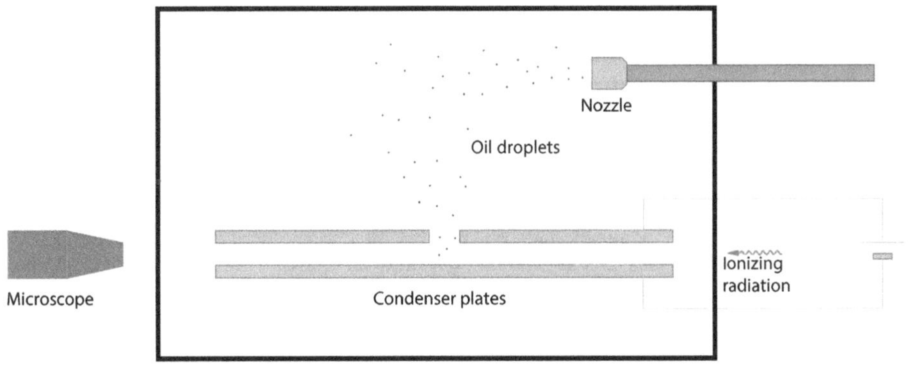

Figure 14.1. Millikan's oil drop experiment.

of one of them. The chamber was enclosed in a thermostat in order to avoid convection currents in the air, which could affect the motion of the oil droplets.

The oil droplets were charged by friction, and by ionizing the surrounding air by means of x-rays or by radiation emitted by a radioactive source; the droplets acquired charge by absorbing ions from the air. In the same way the charges of the droplets could vary and the variation could be studied as well[5]. Being very small, the droplets acquired very small charges, hence the discrete nature of electric charge could be seen more easily.

The charge was measured by observing the motion of the oil droplets with a microscope. Oil droplets could be held under observation for arbitrary amounts of time, hence very accurate measurements could be performed.

In Millikan's experiment, the condenser plates had a diameter of 22 cm and were separated by a distance of 15 mm.

Variants of the oil drop experiment were performed over the years. In particular, some of them (such as the very famous one performed by the Physical Science Study Committee [19]), used tiny plastic spheres. However, as Millikan himself emphasized [9], the best performance is attained with fluid spheres, for which a spherical shape and uniform density, both crucial as we shall see in the next section, are achieved with superior accuracy.

14.4 The dynamics of an oil droplet in a condenser

In Millikan's original experiment, and in our reproduction of it, the charge of an oil droplet is measured by studying its reaction to the uniform electrostatic field generated by a planar parallel plate condenser. In this section we study the dynamics of the droplet in the apparatus, to understand how this comes about. In addition to the importance of understanding Millikan's experiment, this constitutes an interesting problem in mechanics in its own right. In particular, it is a situation in which 'Aristotelian' dynamics can be seen in action and played with in the laboratory, with considerable control.

The droplet is acted on by the gravitational force $\vec{F}_g = m\vec{g}$. Here m is the mass of the droplet. Since a very small liquid droplet is spherical due to the surface tension, we can write $m = \frac{4}{3}\pi r^3 \rho_o$, where r is its radius, and ρ_o is the oil density. The radius cannot be measured directly, however, we shall see later how it is possible to compute it. The second force is the buoyancy due to air itself. By definition this force lays in the upwards vertical direction, and reads $F_b = \frac{4}{3}\pi r^3 \rho_A g$, where ρ_A is the air density, by Archimedes' law. Since the droplet is charged, it will also feel the force due to the electrostatic field, $\vec{F}_e = q\vec{E}$, where $E = \frac{U}{d}$, in which U is the potential

[5] As an interesting by-product, Millikan could observe that air ions could be absorbed by droplets which carried charges of the same sign as the ions, and that this process occurred with the electric field off. He then explained the capability of the ions to overcome the potential barrier as being due to their kinetic energy associated with thermal motion. He could also observe that this process did not occur (or better, that its probability became very low) when the charge on the droplet was so large that the potential barrier was higher than the average thermal energy predicted by the kinetic theory of gases. In this way Millikan's observations gave the first direct measurement of the predictions of kinetic theory for the average kinetic energy.

difference between the plates and d is the distance between them. If the plates of the condenser are kept perfectly horizontal, this force is vertical, that is, parallel to the gravitational force. Therefore, the motion will occur along a vertical axis, which we assume to be directed upwards. The electrostatic force will be directed upwards if $qU > 0$, and downwards if $qU < 0$. In the actual experiment, the electrostatic force is tuned so that it is directed upwards, and strong enough to make the particle move upwards. Therefore, in the following, we shall assume we are in this situation.

The oil droplet will also feel the friction due to the viscosity of air. Now, the typical orders of magnitude for air density and viscosity are $\rho_A \sim 1$ kg m^{-3} and $\eta \sim 10^{-5}$ kg m^{-1} s^{-1}. Moreover, typical velocities measured for the droplets are $v \sim 10^{-3}$ m s^{-1}. Since $\rho_o \sim 10^3$ kg m^{-3} (see below, section 14.6), we have $r \simeq L \sim 10^{-3}$ mm. Hence the Reynolds' number is of the order $R \sim 10^{-4}$, therefore we are definitely in the regime where this force is linear in velocity and, since the droplets are spherical, we can use Stokes' law (see appendix A). Thus

$$F_v = -bv = -6\pi\eta rv.$$

The friction force has a minus sign in front, i.e. it is directed downwards, since by assumption the droplet is moving upwards. Since all forces but friction are constant, the dynamics is analogous to that of a falling body in a viscous fluid, where the only constant force is weight. In that case, as discussed in appendix A, the forces balance themselves after a very rapid transient, so that we can safely assume that the motion has a constant speed. In the present case the constant force acting on the droplet is $\vec{F} = \vec{F}_e + \vec{F}_g + \vec{F}_b$. In fact, we can give a rough estimate of the duration of the transient: since $\tau = \frac{m}{b} = \frac{2r^2\rho_o}{\eta}$, with the above numbers we obtain $\tau \sim 10^{-4}$ s, which is abundantly below our capacity of measurement.

The force equilibrium equation, taking all ingredients into account, is

$$q\frac{U}{d} - \frac{4}{3}\pi r^3(\rho_o - \rho_A)g - 6\pi\eta rv_1 = 0.$$

We have called the limit speed v_1 because later on we shall consider a second limit speed. The above equation can be solved with respect to q to give

$$q = \frac{d}{U}\left[\frac{4}{3}\pi r^3(\rho_o - \rho_A)g + 6\pi\eta rv_1\right]. \tag{14.1}$$

The quantity in square brackets is positive, hence q has the same sign as U, consistent with our requirement. All the quantities appearing in equation (14.1) are known or measurable, with the exception of the radius of the droplet. This quantity can, however, be determined in terms of known quantities in the following way. It is sufficient to turn the electric field off and to observe the resulting motion of the oil droplet. As there is no electric field to contrast the weight, the droplet will begin falling with a constant speed v_2. The equilibrium equation is now

$$-\frac{4}{3}\pi r^3(\rho_o - \rho_A)g + 6\pi\eta r v_2 = 0. \tag{14.2}$$

Note that friction is always directed in the opposite direction with respect to the motion. This is an algebraic equation which can be easily solved with respect to r. The only physically relevant solution is

$$r = \sqrt{\frac{9\eta v_2}{2g(\rho_o - \rho_A)}}. \tag{14.3}$$

We can now substitute this into equation (14.1) to obtain a formula for the charge:

$$q = 9\sqrt{2}\pi d\sqrt{\frac{\eta^3}{(\rho_o - \rho_A)g}}\frac{(v_1 + v_2)v_2^{1/2}}{U}. \tag{14.4}$$

This is the formula that is used in the Millikan experiment to measure the electric charges of oil droplets. As can be seen, it only involves measurable and known quantities.

14.5 Description of the experimental apparatus

The apparatus that was actually used for our experiments is the one shown in figure 14.2(1).

It is a compact, table-top apparatus which allows charged oil droplets (the droplets are charged by the friction with the nozzle) to be sprayed in a chamber, to make them move between the plates of a condenser, and to perform all the

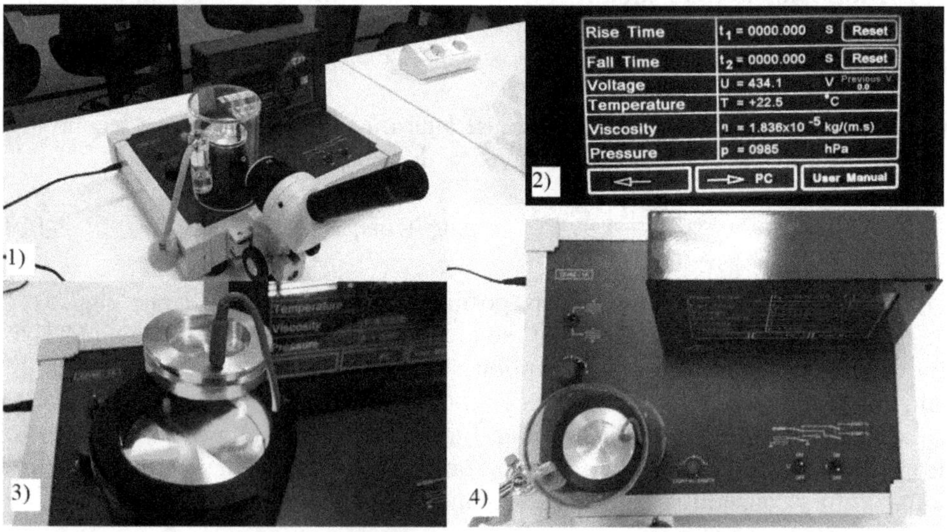

Figure 14.2. (1) The experimental apparatus. (2) The display. (3) The condenser plates with the green LED. (4) The apparatus from above, so that all the controllers can be seen.

measurements which are needed to measure their charge. However, it does not include (fortunately!) any source of ionizing radiation.

The display shows the value of the modulus of the potential difference, the temperature of the air inside the chamber (together with its pressure and viscosity at that temperature) and the readings of two chronometers (figure 14.2(2)). The purpose of the chronometers is to measure the time taken by a droplet to travel a given distance upwards or downwards, respectively.

The plates of the condenser have a diameter of 50 mm, and the inter-space is 3 mm (figure 14.2(3)).

The condenser potential difference can be regulated (figure 14.2(4)), with a maximum value of about 630 V, and the polarity can be chosen as well. In fact, the droplets usually acquire a negative charge when sprayed, but they can also acquire a positive charge, and as we discussed above we need the product qU to be positive.

The dynamics of the droplets in the chamber may be observed using a microscope with 30× magnification, whose focus can be regulated. The chamber is illuminated by means of a green LED. The eyepiece of the microscope includes a grading whose purpose is to measure the distance travelled vertically by a droplet. The spacing is 0.1 mm. Figure 14.3 shows the typical image that can be seen in the eyepiece.

More details concerning the apparatus can be found in the manufacturer's manual [20].

14.6 Measurement of the electric charge

Since we have an expression for the electric charge of the droplet, we can measure it. An important point to stress is that the quantity

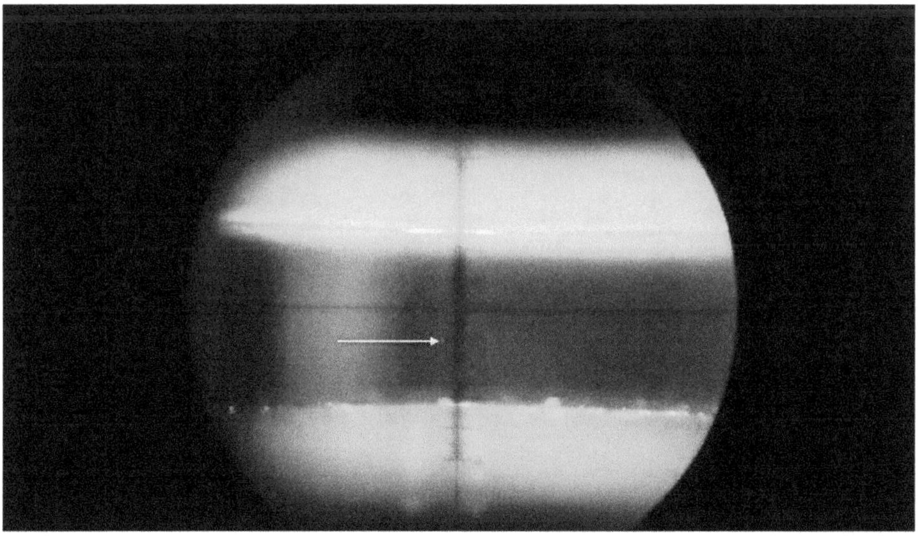

Figure 14.3. The typical picture seen in the eyepiece. The yellow arrow indicates a typical oil droplet (the light may be regulated so that only the larger, more suitable droplets are visible). The grading can be seen as well. The horizontal green lines (the curvature is just an optical effect in the photo) are the plates of the condenser.

$$\frac{9\sqrt{2}\pi d}{\sqrt{(\rho_o - \rho_A)g}}\eta^{3/2} \equiv P(T) \tag{14.5}$$

is not a constant, since the fluid densities and the air viscosity are functions of the temperature T. The viscosity at room temperature ($T \simeq 20$ °C) and atmospheric pressure varies linearly with the temperature. From the tabulated values of the density of the oil we used, which are given in [20], we find the following relation:

$$\rho_o(T) = (886 - 0.6T)\ \mathrm{kg\ m^{-3}},$$

where T is the numerical value of the temperature in degrees celsius (°C). Since this is much larger than ρ_A, independently of the value of T, we neglect the latter quantity. Concerning the viscosity of air, we recall that, as is typical for gases, it grows with temperature. Moreover, for a large range of temperatures, which amply includes room temperature, the viscosity of air grows linearly with temperature. We considered the following linear relation, which is obtained by fitting the values for the viscosity coefficient of dry air between 15 °C and 30 °C (which can be found for example in [21]):

$$\eta(T) = (1.73 + 0.0047T) \times 10^{-5}\ \mathrm{kg\ m^{-1}\ s^{-1}}. \tag{14.6}$$

Putting this all together, we see that in the temperature and pressure ranges of interest the function $P(T)$ behaves linearly with a very good approximation, as shown in figure 14.4 (where the function $p(T) = 10^{-10}P(T)$, which has numerical values of the order of unity, is plotted).

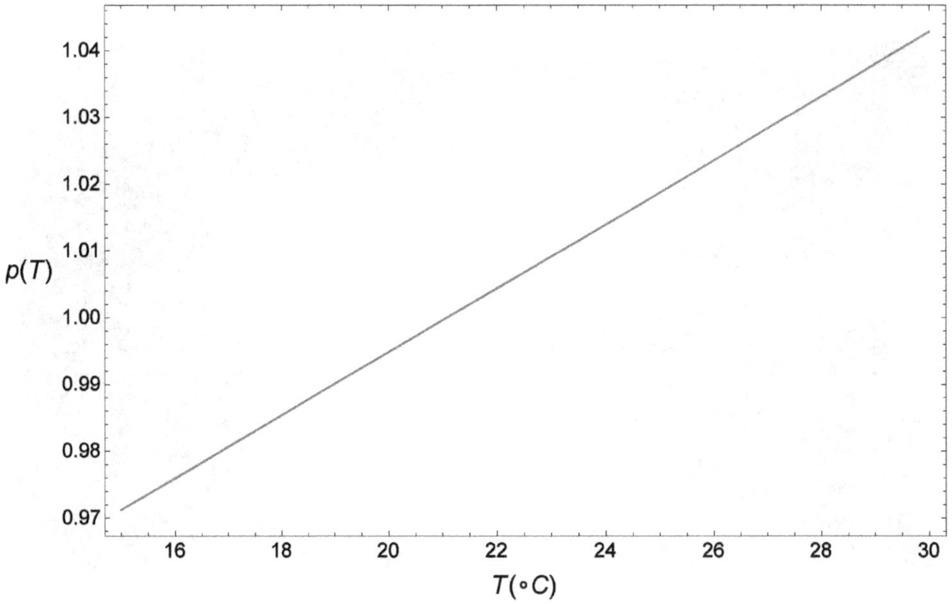

Figure 14.4. Plot of the function $p(T) = 10^{-10}P(T)$ as a function of temperature, in the range of the temperatures of interest.

With the numerical values of this function, the charge q of an oil droplet can be obtained by measuring only the temperature T, the limit velocities in the presence and absence of an electric field, and the potential difference which generates the electric field. In the original Millikan experiment, as said, the charge could be varied by ionizing the air around the droplet, by means of some kind of ionizing radiation. In that case, the above formula allows charge variations to be measured as well.

14.7 The experimental procedure

As stated at the end of the previous section, the measurement of the charge involves measuring the temperature T, the speeds v_1 and v_2, and the potential difference U.

The first step for doing so, after preparing the apparatus, is to choose a droplet on which the measurements should be performed. The chosen droplet should not be too big, otherwise it falls down too quickly and not even the highest electric field attainable can make it go up (however, these droplets usually fall very fast, so that after a few seconds no more of them can be seen), or too small, since in that case it will be more subject to possible air movements and other sources of error, and in addition it will be less clearly visible (larger droplets look brighter under the microscope). Then the potential difference must be chosen so that the droplet starts moving upwards. This means that the polarity of the former must be appropriately chosen, and the modulus must be large enough to overcome the gravitational force.

Temperature can be promptly read off from the display, which constantly shows its value, with an accuracy of half a degree. In normal conditions, the temperature of the air inside the chamber tends to increase (about 1 °C per half an hour) because of natural warming of the apparatus. Note that, since the variation of viscosity with temperature is taken into account by the use of equation (14.6), there is no need to separately read off the viscosity from the display. An alternative way to perform the experiment involves reading the viscosity directly from the display, and substituting it into equation (14.4).

Also the potential difference is reported on the display, with an accuracy of 0.1 V. Speeds are obtained by exploiting the fact that they remain constant during the motion. The distance travelled by the droplet is measured by comparing the path against the scale which can be seen inside the eyepiece of the microscope (particular care should be taken to keep this scale vertical since the eyepiece can be rotated), and then it is divided by the time taken, which is measured using the chronometers. The circuitry allows us to automatically switch from the ascending time to descending time in the moment when the electric field is turned off. Typical speeds are of the order of 1 mm s^{-1}, hence they are easily measurable.

The display also shows pressure, however, pressure data are only needed if the corrections to Stokes' law described in appendix B are taken into account, which we do not.

By putting all these data together, the charge can be computed by hand (the value of $p(T)$ can be obtained graphically from figure 14.4 or by some computer programme). The sign of the charge (which is not relevant, however) can be deduced from the polarization that has been used to make the droplet go upwards.

As explained in the next section, in order to have a small error on the charge, it is vital to perform as many measurements of the charge of a single droplet as possible. This can be achieved by making the droplet go up and down, by repeatedly turning the electric field on and off, and measuring the times taken to go up and down[6]. Of course, the values of the potential difference and of the temperature must be recorded each time.

An alternative procedure for measuring the charge of a droplet is described in appendix C.

14.8 Data analysis

Real proof of the discretization of electric charge along the steps of Millikan would require a very large number of measurements on many different droplets, in order to be able to say with confidence that all droplets have charges which are integer multiples of a fundamental unit, that is the greatest common divisor, which is identified with the charge of a single electron. As described in the introduction, Millikan could expose his chamber to ionizing radiation, so he was also able to measure the variations of the droplet charge due to the ionization of air, and to check that the charge always varied by an integer multiple of this fundamental unit. In our laboratory there was no possibility to do the latter[7], nor to perform a very large number of measurements. This was due not only to time constraints and to the difficulty of keeping the droplet between the plates of the condenser for a very long time, but also to the fact that our apparatus is not shielded against heating, and in fact, as also remarked above, its temperature keeps rising constantly while working. The latter effect invariably started disturbing the motion of the droplets after some time.

For these reasons we limited ourselves to check that the measured values of the droplet charge were integer multiples of the electron charge.

What was done in practice was to perform as many separate measurements on the same oil droplet as possible, so to have a set of many values for its charge. Then the average value $\bar{q}$ was compared with the electron charge, by computing the number

$$n = \frac{\bar{q}}{e},$$

where the value $e = 1.6022 \times 10^{-19}$ C was used[8]. This number should of course be close to an integer, but for the result to be reliable the error should also be estimated. Assuming that the number of measurements on the single droplet is large enough, the error associated with their mean value $\bar{q}$ is the standard deviation of the mean, defined as

[6] The fact that the droplet must go up when the electric field is on was required from the beginning specifically for the purpose of performing repeated measurements in this fashion.

[7] There is actually the possibility that the charge of a droplet varies by absorption of an atmospheric ion without any exposure of the chamber to ionizing radiation. However, this is a very rare event, and cannot be controlled.

[8] The value accepted to date is $e = 1.602\,176\,634 \times 10^{-19}$ C, see for example [22].

$$\Delta\bar{q} = \frac{\sigma_q}{\sqrt{N}},$$

where σ_q is the standard deviation of the values of q with respect to the average $\bar{q}$. The result was considered satisfactory if the error associated with n, i.e.

$$\Delta n = \frac{\Delta\bar{q}}{e},$$

was sensibly lower than 0.5. In that case, in fact, we can be reasonably sure that the droplet carries a number of electron charges given by the integer that is closest to n.

The obtained results have been satisfactory in many cases, with the values of n lying between a few units and a few tens. A few sets of measurements, on the other hand, gave unsatisfactory results, i.e. n was not close to an integer, and/or the error was too large. Apart from insufficient sets of data, such unsatisfactory results were probably due to inaccurate measurements of speeds (since the chronometers were started and stopped by hand, some error unavoidably arises), to uncontrolled temperature fluctuations and to imperfect screening against air turbulence (which could also be due to the temperature rising too much). Another possible source of error could be the lack of consideration of the correction to the Stokes' law described in appendix B. However, at least the last two of these should have been ruled out by our focusing on droplets that were not too small.

Acknowledgements

The author would like to thank R De Luca for useful discussions about the Millikan experiment and its realization.

Appendices

A. Moving in a viscous fluid

Consider a particle of mass m in a viscous fluid. It is well known that whenever the particle moves, the fluid will exert a force on it opposing its movement, and such a force, for not too high speeds, will be proportional to the velocity of the particle relative to the fluid[9]. Mathematically,

$$\vec{F}_{\mathrm{f}} = -b\vec{v},$$

where the coefficient b contains the information about the geometry of the moving particle and the viscosity of the fluid. Now suppose that the particle moves in the fluid under the action of a constant force $\vec{F}$. We can write the equation of motion of the particle as

[9] The relevant quantity is the so-called Reynolds number [23], which is defined as $R = \frac{\rho v L}{\eta}$, where ρ and η are, respectively, the density and the dynamic viscosity coefficient of the fluid, and L is the scale of the linear dimension of the moving object; the friction force is linear in the velocity if this number is much smaller than one, which is surely the case of the Millikan experiment, as shown in the main text (this is essentially due to the small size of the oil droplets).

$$\vec{F} - b\vec{v} = m\vec{a} = m\frac{\mathrm{d}\vec{v}}{\mathrm{d}t}.$$

Since the force will tend to move the particle in the same direction as the force, the velocity will be parallel to it. Hence the friction will also act on the same line. Thus the motion is rectilinear, and we can project the above equation on this line:

$$\frac{\mathrm{d}v}{\mathrm{d}t} = \frac{F}{m} - \frac{b}{m}v.$$

This equation can be easily integrated by separation of the variables. We assume the initial condition is $v(t = 0) = 0$, i.e. that the particle is initially at rest with respect to the fluid. The result is then

$$v(t) = \frac{F}{b}\left(1 - \mathrm{e}^{-\frac{b}{m}t}\right).$$

As is apparent, the speed tends asymptotically to a limit value $v_L = \frac{F}{b}$. In practice, we may safely assume that the particle moves with constant speed v_L after a transient, whose duration is of the order of $\tau = \frac{m}{b}$ (independent of the force $\vec{F}$). In fact, as shown in figure 14.5, after a time of $\sim 3\tau$, the speed of the particle is already $\sim 95\%$ of the limit speed.

Indeed, the limit speed can be obtained by simply imposing force equilibrium:

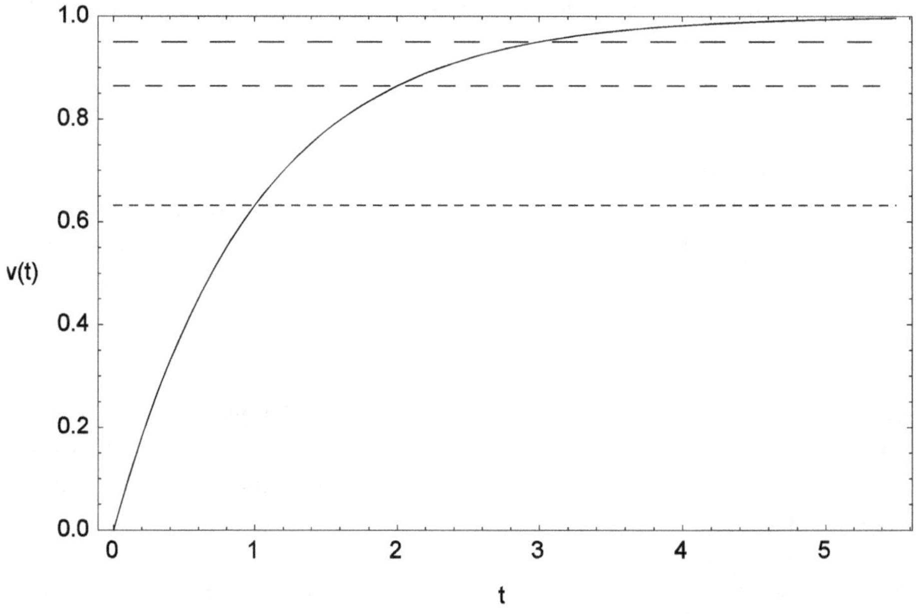

Figure 14.5. Plot of the speed $v(t)$ (solid line) with $F = b = m = 1$, for which the asymptote is at $v_L = 1$. The dashed lines denote the values of the velocity after $t = \tau$, $t = 2\tau$ and $t = 3\tau$, respectively, showing that after $t = 3\tau$, the speed of the particle is already above $\sim 95\%$ of the limit speed.

$$\frac{\mathrm{d}v}{\mathrm{d}t} = \frac{F}{m} - \frac{b}{m}v = 0 \qquad \Leftrightarrow \qquad F = bv.$$

The latter equation has sometimes been called 'Aristotle's equation of motion' in the literature [24]. In fact, it may be considered as a modern, quantitative version of Aristotle's empirical statement that to put an object in motion with a constant speed, a constant force is needed, while in the absence of this force the object stops [25]. From the above equation it is immediate to find the limit speed $v_L = F/b$.

The most common instance of this dynamics is given by a particle that falls in a viscous fluid under the effect of its weight. In that case the direction in which the motion proceeds is by definition the vertical one and $\vec{F} = m\vec{g}$. In that case we have (neglecting buoyancy which in this case is oriented along the same line as the other forces) $v_L = \frac{mg}{b}$.

Stokes' law

In the case of a spherical object, moving slowly enough in the fluid so the flow is laminar[10], the coefficient b, which we left unspecified above, has a simple expression in terms of the radius of the object and the viscosity of the fluid. This is relevant for the Millikan experiment since oil droplets are spherical due to surface tension, and typically move at speeds of the order of a few millimetres per second, so the occurrence of turbulence may be excluded. This expression is given by

$$b = 6\pi\eta r,$$

where η is the coefficient of viscosity of the fluid, and r is the radius of the spherical object. The proof of this relation rests on quite advanced analytical methods of fluid mechanics, hence it will not be reproduced here. The details may be found, for example, in [23].

B. Corrections to Stokes' law

While performing his experiments Millikan noticed that, when considering smaller droplets, the value of the elementary unit of electricity tended to disagree with the one found for larger droplets. In particular, it was larger the smaller the speed v_2 of falling under gravity was. He interpreted this effect as being due to the fact that Stokes' law ceases to be strictly valid when the diameter of the sphere becomes comparable to the mean free path of the molecules of the fluid [8–10]. It is to be emphasized that at that time there were no experimental data about the limits of validity of Stokes' law [10]. The idea is that a very small droplet does not see the air as a continuum, but starts to sense the dishomogeneities and voids in it, effectively increasing its speed. If this effect is not taken into account, the charge computed from equation (14.4) will be larger than the real value. By performing an ingenious series of experiments, and by referring to experiments performed by others, Millikan

[10] Again, this happens if the Reynolds number is very small.

managed to propose a corrected form of Stokes' law. In the following we briefly sketch his reasoning, referring to [10] for details.

If we assume that Stokes' law is valid, we can use equation (14.2), which we recall was derived under this hypothesis, to express the falling speed of the droplet as

$$v_2 = \frac{2}{9}\frac{gr^2(\rho_o - \rho_A)}{\eta}.$$

Under the assumption that corrections to the Stokes' law are small, following Millikan we can speculate that this expression can be corrected to first order as

$$v_{2,\,\mathrm{corr.}} = v_2\left(1 + A\frac{l}{r}\right),$$

where l is the mean free path of a molecule in the fluid (which gives the scale of the inhomogeneities), A is a parameter and r is the value of the radius calculated from the uncorrected Stokes' law, i.e. using equation (14.3)[11]. This amounts to substituting Stokes' expression of friction with

$$F_{f,\,\mathrm{corr.}} = -6\pi\eta r v\left(1 + A\frac{l}{r}\right)^{-1}.$$

This implies that the value of the droplet charge computed from equation (14.4) is shifted to

$$q_{\mathrm{corr.}} = q\left(1 + A\frac{l}{r}\right)^{-\frac{3}{2}}. \tag{14.7}$$

Millikan was also able to quantify this correction and to estimate the parameter A. In general, the product Al is inversely proportional to the pressure, i.e. $Al = \alpha/p$. In the case of air, $\alpha = 82\ \mu\mathrm{m\ hPa}$ [20].

This result can be used to increase the reliability of our experiment. If the corrected relation (14.7) is used in place of equation (14.4), we can in fact expect that the values of the charges obtained will be accurate also for smaller droplets. However, in view of the increased complexity of the computation, and of the fact that our apparatus presented other sources of error for smaller droplets anyway, we preferred to stick to the uncorrected formula (14.4), trying to avoid the use of droplets that are too small.

C. The static method

To find the charge carried by an oil droplet, an alternative experimental procedure to that described in section 14.7 may be employed. Namely, instead of making the droplet go upwards, one may make it float at a fixed height by tuning the potential difference in such a way that it balances the weight (hence the name *static method*). In this case the charge is given by equation (14.4), with $v_1 = 0$, i.e. (putting $v_2 \equiv v$)

[11] The use of the uncorrected rather than the corrected value introduces only negligible errors.

$$q = 9\sqrt{2}\pi d\sqrt{\frac{\eta^3}{(\rho_o - \rho_A)g}}\frac{v^{3/2}}{U_0} = P(T)\frac{v^{3/2}}{U_0},$$

where U_0 is the value of the potential difference that achieves the balance, and $P(T)$ is again given by equation (14.5). In this case the measurement procedure is similar to the above, but now only the downwards velocity must be measured. However, this simplicity is counterbalanced by the fact that tuning the differential potential to exactly stop the droplet can be quite challenging. In this case, if the number of measurement sets taken on a single droplet is not large enough to use the methods described in section 14.8, the data analysis is necessarily rougher. In fact, the standard formula for error propagation can be used:

$$\frac{\Delta n}{n} = \frac{\Delta U_0}{U_0} + \frac{3}{2}\frac{\Delta v}{v},$$

where we neglect the error on $P(T)$, and $\bar{A}$ and ΔA denote, respectively, the average value of the quantity A and the maximal error on A, obtained from the semi-dispersion of the measured values.

References

[1] Kaplan I 1962 *Nuclear Physics* (Reading, MA: Addison-Wesley)

[2] Bransden B H and Joachain C J 1983 *Physics of Atoms and Molecules* (New York: Longman)

[3] Pais A 1986 *Inward Bound. On Matter and Forces in the Physical World* (Oxford: Clarendon)

[4] Mehra J and Rechenberg H 1982 *The Historical Development of Quantum Theory* vol 1, part 1 (New York: Springer)

[5] Whittaker E T 1951 *A History of the Theories of Ether and Electricity* vol 1 (London: Nelson)

[6] Millikan R 1924 The electron and the light-quant from the experimental point of view *Nobel Lecture* http://www.nobelprize.org/prizes/physics/1923/millikan/lecture

[7] Millikan R 1910 A new modification of the cloud method of determining the elementary electrical charge and the most probable value of that charge *Philos. Mag.* **19** 209–28

[8] Millikan R 1911 The isolation of an ion, a precise measurement of its charge, and the correction of Stokes' law *Phys. Rev. Ser. I* **32** 349–97

[9] Millikan R 1913 On the elementary electrical charge and the Avogadro constant *Phys. Rev.* **2** 109–43

[10] Millikan R 1917 *The Electron* (Chicage, IL: University of Chicago Press)

[11] Holton G 1978 Subelectrons, presuppositions, and the Millikan–Ehrenhaft dispute *Hist. Stud. Phys. Sci.* **9** 161–224

[12] Franklin A 1997 Millikan's oil-drop experiments *Chem. Educat.* **2** 1–14

[13] Goodstein D 2000 In defense of Robert Andrews Millikan *Am. Sci.* **89** 54–60 https://www.jstor.org/stable/27857400

[14] Mar N M *et al* 1996 Improved search for elementary particles with fractional electric charge *Phys. Rev.* D **53** 6017–32

[15] Lee E R, Halyo V, Lee I T and Perl M L 2004 Automated electric charge measurements of fluid microdrops using the Millikan method *Metrologia* **41** S147–58

[16] Mojarad N and Krishnan M 2012 Measuring the size and charge of single nanoscale objects in solution using an electrostatic fluidic trap *Nat. Nanotechnol* **7** 448–52
[17] Faez S 2016 How to replace the oil droplet in Millikan's experiment with a single virus arXiv: 1601.01226
[18] Lembessis V E 2019 Cold atoms interacting with highly twisted laser beams mimic the forces involved in Millikan's experiment *Laser Phys.* **29** 035503
[19] The Physical Science Study Committee (PSSC) 1959 The Millikan Experiment https://www.youtube.com/watch?v=WSlS_aOc6Jc
[20] 3B Scientific Millikan Apparatus' Manual https://www.a3bs.com/product-manual/1018882_EN.pdf
[21] Çengel Y A 2008 *Introduction to Thermodynamics and Heat Transfer* 2nd edn (New York: McGraw-Hill)
[22] Bich W 2019 The third-millennium International System of Units *Riv. Nuovo Cim.* **42** 49–102
[23] Batchelor G K 1967 *An Introduction to Fluid Mechanics* (Cambridge: Cambridge University Press)
[24] Susskind L and Hrabovsky G 2014 *Classical Mechanics: The Theoretical Minimum* (London: Penguin)
[25] Aristotle 2018 *Physics* Translated, with an introduction and notes by C D C Reeve (Indianapolis/Cambridge: Hackett)

Chapter 15

The Davisson–Germer experiment

Veronica Granata

15.1 Introduction

The Davisson–Germer experiment demonstrated the wave nature of the electron, confirming the earlier hypothesis of Louis de Broglie [1]. By putting wave–particle duality on a firm experimental footing, it represented a major step forward in the development of quantum mechanics. In 1927, at the Bell Laboratories (Murray Hill, NJ), Clinton J Davisson and Lester H Germer fired electrons at reduced speed against a target of crystalline nickel, which could be rotated to observe the angular dependence of the scattered electrons. Their electron detector was mounted on an arc so that it could be rotated to observe electrons at different angles. It was a great surprise to them to find that at certain angles there was a peak in the intensity of the scattered electron beam. This peak indicated wave behaviour for the electrons, and could be interpreted by the Bragg law to give values for the lattice spacing in the nickel crystal. This experiment, similarly to Arthur H Compton's which highlighted the particle behaviour of light, confirmed de Broglie's hypothesis of the wave nature of electrons and completed the hypothesis of wave–particle duality.

15.2 Historical introduction

In this section the history of the experiment of Davisson and Germer, which was based on de Broglie's hypothesis of wave–particle duality, will be briefly introduced. This section aims to put the experiment in context and explain its fundamental importance for quantum physics. Due to space constraints, only a sketch of this complex and fascinating story will be provided. The best sources of further information are [2–5].

The Davisson–Germer experiment (1927) was the first measurement of the wavelengths of electrons. Davisson, who worked at the Bell Research Laboratories, received the Nobel Prize in Physics for the year 1937 together with George P Thomson from the University of Aberdeen in Scotland, who also found experimental indications of electron diffraction independently [6].

doi:10.1088/978-0-7503-2678-0ch15

Davisson (1881–1958) and Germer (1896–1971) investigated the reflection of electron beams on the surface of nickel crystals. When the beam strikes the crystal, the nickel atoms in the crystal scatter the electrons in all directions. Their detector measured the intensity of the scattered electrons with respect to the incident electron beam. Their normal polycrystalline samples exhibited a very smooth angular distribution of scattered electrons. In early 1925, one of their samples was inadvertently recrystallized in a laboratory accident that changed its structure into a nearly mono-crystalline form. As a result, the angular distribution manifested sharp peaks at certain angles. As Davisson and Germer soon found out, other mono-crystalline samples also exhibited such anomalous patterns, which differed with the chemical constitution, the angle of incidence and the orientation of the sample. Only in late 1926 they did understand what was going on, when Davisson attended a meeting of the British Association for the Advancement of Science in Oxford. There Born spoke about de Broglie's matter-waves and Schrödinger's wave mechanics. Their later measurements completely confirmed the quantum mechanical predictions for electron wavelength λ as a function of momentum p: $\lambda = h/p$. However, their initial experiments (unlike Thomson's) were conducted in the context of industrial materials research on filaments for vacuum tubes, not under any specific theoretical guidance.

The phenomenon of electron diffraction is quite general and can be explained by the wave nature of atomic particles. The planes of atoms in the crystal (Bragg planes) are regularly spaced and can produce a constructive interference pattern, if the so-called Bragg condition ($n\lambda = 2d \sin\theta = D \sin\varphi$, where d is the spacing of atomic planes and D is the spacing of the atoms in the crystal) is satisfied (for more details see the next section).

In their experiment, Davisson and Germer found that the intensity reached a maximum at $\varphi = 50°$ (for an initial kinetic energy of the electrons of 54 eV, normal incidence as indicated and φ as the scattering angle). From a philosophical point of view this experiment reveals a striking feature. It demonstrates the existence of de Broglie waves. Yet we can speak of causation, not in a deterministic but in a probabilistic sense. There is clearly, on the observational level, a conditional dependence of the intensity of the reflected beam on the set of antecedent conditions. These antecedent conditions are also conditionally prior to their respective effects. There is of course no local causal mechanism, as the causal situation covers a stream of particles. There is only a certain likelihood that one particular particle in these experiments will be scattered in a particular direction.

However, sufficiently much is known about the scattering of atomic particles to establish a causal dependence between the antecedent and consequent conditions. In the Davisson–Germer experiment the wavelength of the electron beam, scattered at 50°, is 0.165 nm. This is the effect to which specific antecedent conditions correspond—the electron beam has an initial kinetic energy of 54 eV, the lattice spacing of the nickel atoms is known, from which the spacing of the Bragg planes can be calculated, and the condition for constructive interference is also known. There is quite a general dependence of the interference effects on the regular spacing of the atom planes in the crystal. This is used regularly in the study of atomic properties and is completely analogous to the use of x-ray diffraction by Max von Laue, Paul Knipping and

Walter Friedrich in 1912. Under certain conditions, particles such as electrons thus exhibit wave-like characteristics such as electromagnetic radiation [7].

15.3 Description of instrumentation

The experimental set-up of Davisson and Germer has been reproduced with laboratory instrumentation that is easily usable by students. With this instrumentation it is possible to measure the wavelength of the electrons, thus confirming the de Broglie hypothesis. The apparatus that was used is shown in figure 15.1. The experimental set-up, called a diffraction tube, consists of an highly evacuated tube with an electron gun, which is a pure tungsten heater filament and a cylindrical anode, all contained in a clear glass bulb (the ampoule).

The apparatus is schematically shown in figure 15.2. The electrons emitted by the heated cathode are constrained to a narrow beam by an aperture and are then focussed by means of an electron optical system and accelerated by a voltage V. The resulting tight, monochromatic beam then passes through a micro-mesh nickel grating situated at the aperture of the gun. Onto this grid, a thin layer of polycrystalline graphitized carbon has been deposited by vaporization. This layer affects the electrons in the beam much like a diffraction grating.

The result of this diffraction is seen in the form of an image comprising two concentric rings that become visible on the fluorescent screen. A spot resulting from the undeflected electron beam continues to be visible at the centre of the rings, as shown in figure 15.3. Each ring corresponds to a Bragg reflection on the atoms of the graphite layer. A magnet is also supplied with the tube. This allows the direction of the electron beam to be changed, which may be necessary if the graphite target has slight damage as a result of the manufacturing process or due to later overheating.

The anode voltage can be regulated (figure 15.2) in the range of 0–5000 V. Regarding the geometric characteristics of the experimental set-up, the distance from the graphite target to the fluorescent screen is (125 ± 2) mm, the fluorescent screen and glass bulb have diameters of, respectively, 100 and 130 mm, and the total length is 260 mm, while the lattice constants of graphite are $d_{10} = 0.213$ nm and

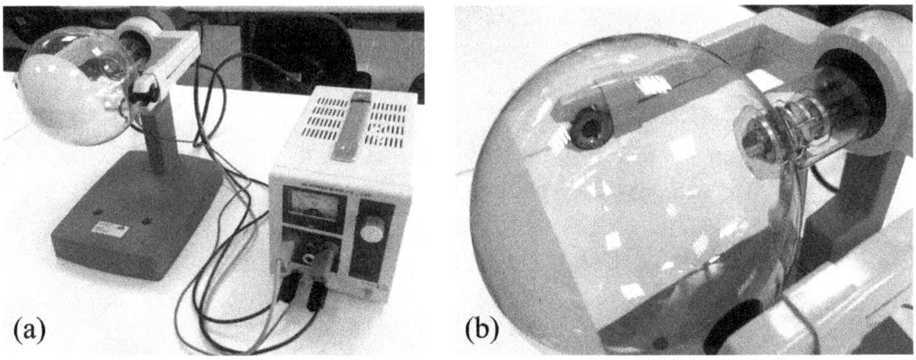

Figure 15.1. (a) The experimental apparatus. (b) Detail of the interior of the diffraction tube—a glass bulb with the electron gun.

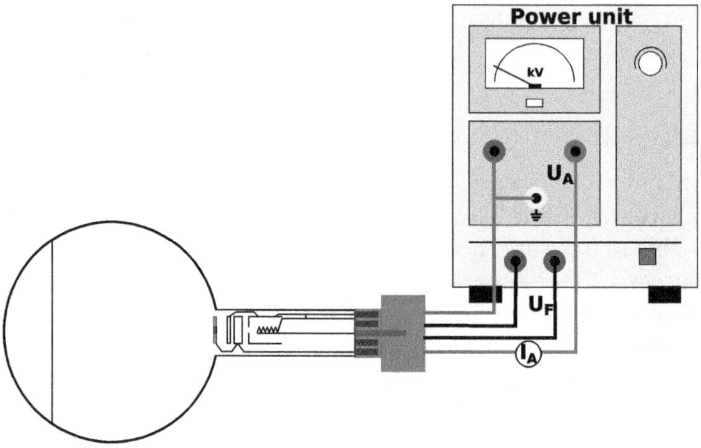

Figure 15.2. The experimental set-up of Davisson and Germer.

Figure 15.3. Diffraction rings around the non-diffusion electron beam visible on the fluorescent screen.

$d_{11} = 0.123$ nm. More details concerning the apparatus can be found in the manufacturer's manual [8].

The variation of the anode voltage V involves the modification of the radius r_i of the diffraction rings, i.e. the reduction of the voltage causes the increase of the radius r_i. This observation agrees with de Broglie's postulate, according to which the λ wavelength is differently proportional to the impulse p.

15.4 Measurement of the reticular step of graphite

15.4.1 Theoretical outline

Bragg's law

When x-rays of a fixed wavelength are incident on a crystal surface at certain incident angles, intense reflected x-rays are produced when the wavelengths of the

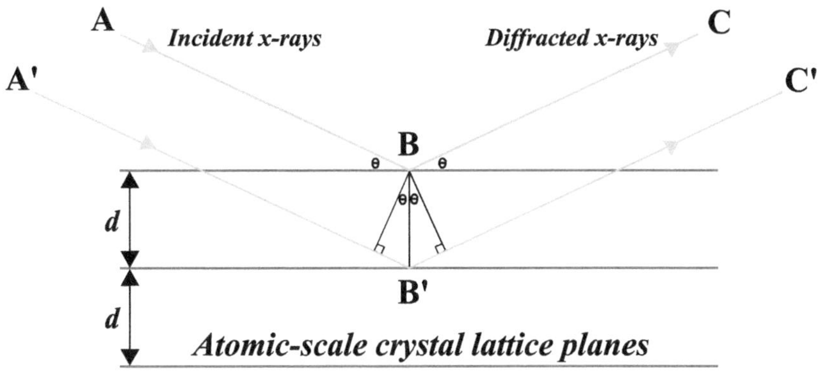

Figure 15.4. Bragg's law reflection. The diffracted x-rays exhibit constructive interference when the distance between paths ABC and A′B′C′ differs by an integer number of wavelengths (λ).

scattered x-rays interfere constructively. In order for the waves to interfere constructively, the differences in the travel path must be equal to integer multiples of the wavelength. When this constructive interference occurs, a diffracted beam of x-rays will leave the crystal at an angle equal to that of the incident beam.

Consider a single crystal [9] with aligned planes of lattice points separated by a distance d, see figure 15.4. Monochromatic x-rays A and A′ are incident upon the crystal at an angle θ and the travel path length difference between the ray paths ABC and A′B′C′ is an integer multiple of the wavelength. Constructive interference will occur for a combination of the specific wavelength (λ), crystal lattice planar spacing (d) and angle of incidence (θ). Each rational plane of atoms in a crystal will undergo refraction at a single, unique angle (for x-rays of a fixed wavelength). The relationship between the wavelength of the incident x-rays, angle of incidence and spacing between the crystal lattice planes of atoms is known as Bragg's law [10], expressed as

$$n\lambda = 2d \sin \theta,$$

where n (an integer) is the 'order' of reflection, λ is the wavelength of the incident x-rays, d is the inter-planar spacing of the crystal and θ is the angle of incidence. The formula is explained analytically by considering an optical path difference of $2d \sin \theta$.

The principle of Bragg's law is applied in the construction of instruments such as Bragg spectrometers, which are often used to study the structure of crystals and molecules.

Since graphite, which is the target of the electron beam in the experiment presented, has a hexagonal structure (as illustrated in the diagram of figure 15.5) with two distances of the lattice planes d_{10} and d_{11}, in the first order of diffraction ($n = 1$) two diffraction rings with diameters of, respectively, r_{10} and r_{11} will be observed (see figure 15.6).

Measurement of the distances of the lattice planes

In order to determine the dependence of the radius r of the diffraction rings as a function of the acceleration voltage of the electron beam, it is necessary to consider the geometry of the diffraction tube, shown in figure 15.7, where

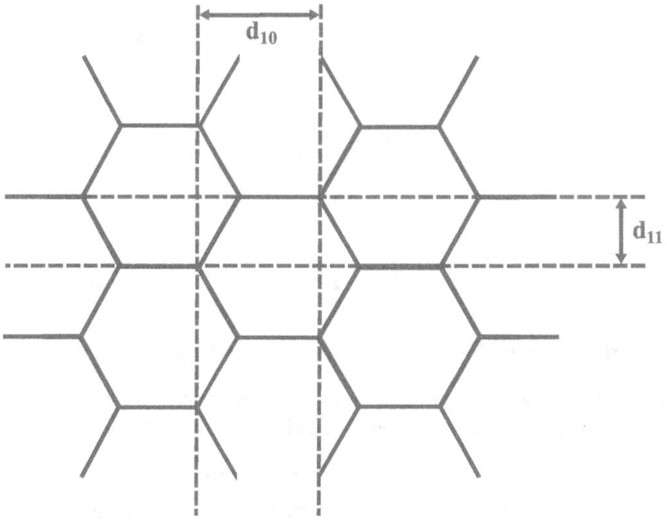

Figure 15.5. Hexagonal structure of graphite with distance characteristics d_{10} and d_{11} between the lattice planes.

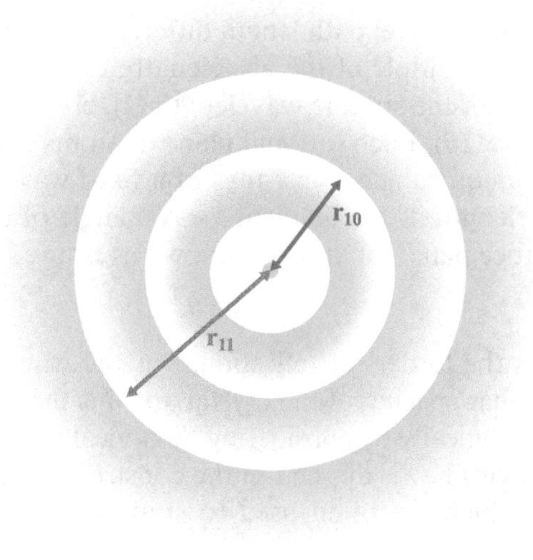

Figure 15.6. Diffraction rings with diameters of r_{10} and r_{11} observed on the fluorescence screen.

- θ is the angle of deviation of the electrons.
- r is the radius of the diffraction rings.
- L is the distance between the graphite sample and the plane containing the rings.
- R is the radius of the diffraction tube.

Using the trigonometric relations and the chord theorem, we obtain the following relations between the variables r, L, R and θ:

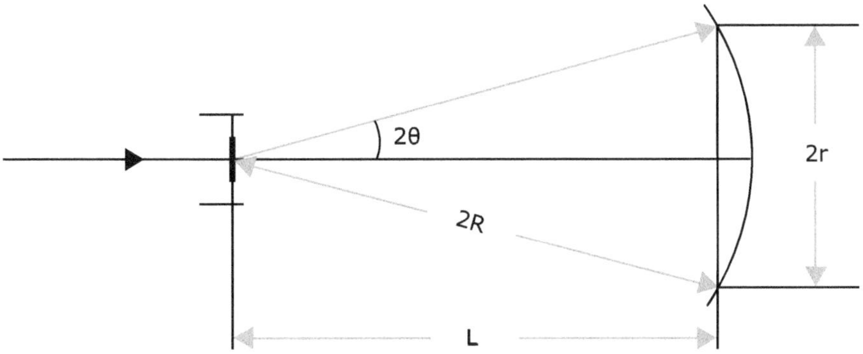

Figure 15.7. Schematic representation of Debye–Scherrer diffraction.

$$\begin{cases} r = L\tan 2\theta \\ r = R\sin 4\theta \end{cases}. \tag{15.1}$$

Since

$$\tan 2\theta = \frac{\sin 4\theta}{1 + \cos 4\theta},$$

and combining the two relationships (15.1), we obtain

$$L = \frac{r}{\tan 2\theta}$$

from the form of duplication for the tangent function, and replacing $r = R\sin 4\theta$, the following expression is obtained for the quantity L:

$$L = R(1 + \cos 4\theta).$$

Applying the duplication formula for the cosine function and simple mathematical steps, the relation that gives the distance L as a function of the variables R and θ is

$$L = 2R(1 - \sin^2 2\theta).$$

Using the trigonometry theorems it is possible to express the length L, the distance between the graphite sample and the plane containing the diffraction rings, as a function of the radius of the latter:

$$L = R\left[1 + \sqrt{1 - \frac{r^2}{R^2}}\right]. \tag{15.2}$$

Since the diffraction angles θ are small ($\theta \ll 1$) the following approximations can be used,

$$\tan\theta \approx \theta, \quad \sin\theta \approx \theta,$$

and if $n = 1$, Bragg's law takes the following simplified form:

$$\lambda = 2d\theta.$$

From the first equation of (15.1), the radius of the diffraction rings can be written as

$$r = L2\theta,$$

from which we obtain the angle 2θ,

$$2\theta = \frac{r}{L}.$$

The wavelength can thus be rewritten in terms of the distances r and L:

$$\lambda = d\frac{r}{L}. \tag{15.3}$$

Define this wavelength as λ_B (the *Bragg wavelength*) and, since r and L depend on the voltage V, equation (15.3) is rewritten as

$$\lambda_B(V) = d\frac{r(V)}{L(V)}, \tag{15.4}$$

where the dependence of L on r is given by equation (15.2).

The de Broglie hypothesis states that all matter exhibits wave-like properties and relates the observed wavelength of matter to its momentum, in the relationship

$$\lambda_D = \frac{h}{p};$$

recall that h is Planck's constant. This wavelength is called the *de Broglie wavelength*.

In the diffraction tube that reproduces the Davisson–Germer experiment, the electrons are accelerated by the voltage V and subsequently affect the target with velocity v. From mechanical energy conservation,

$$eV = \frac{1}{2}mv^2,$$

then the speed of the electrons is

$$v = \sqrt{\frac{2eV}{m}},$$

from which we obtain the momentum

$$p = mv = m\sqrt{\frac{2eV}{m}},$$

or

$$p = \sqrt{2meV}.$$

Replacing the momentum p in the de Broglie relation, we obtain that the wavelength λ_D is

$$\lambda_D(V) = \frac{h}{\sqrt{2meV}}. \tag{15.5}$$

According to the de Broglie hypothesis, the two wavelengths λ_B and λ_D must be equal, so they equate equation (15.4) with equation (15.5):

$$d\frac{r}{R\left[1 + \sqrt{1 - \left(\frac{r}{R}\right)^2}\right]} = \frac{h}{\sqrt{2meV}}. \tag{15.6}$$

Developing the Taylor series (15.6) for $\frac{r}{R} \ll 1$, the relation that links the radius of the diffraction rings r and the voltage V is obtained:

$$r(V) = \frac{2Rh}{d\sqrt{2me}}\frac{1}{\sqrt{V}}. \tag{15.7}$$

Since R, the radius of the diffraction tube, is provided by the constructor ($2R = (125 \pm 2)$ mm), the factor

$$\alpha = \frac{2Rh}{\sqrt{2me}}$$

is a constant, where m and e are, respectively, the mass and the charge of the electron. By replacing the numeric values of all the constants, α is equal to

$$\alpha = 5.04 \times 10^{-12}\,\text{m}^2\,\sqrt{\text{kV}},$$

for the purpose of the experiment it is more convenient to express the unit of measurement of α in units of 'nm cm':

$$\alpha = 0.504\frac{\text{nm cm}}{\frac{1}{\sqrt{\text{kV}}}}.$$

By replacing α in equation (15.7), it is possible to obtain the dependence of the radius of the diffraction rings r on the voltage V:

$$r(V) - \frac{\alpha}{d}\frac{1}{\sqrt{V}}. \tag{15.8}$$

As anticipated, the radius r is inversely proportional to the acceleration voltage of the electrons, with precision, r decreases as the square root of the acceleration voltage increase.

Through equation (15.8), the distances of the lattice planes can be obtained by measuring the rays of the diffraction rings as a function of applied voltage.

15.4.2 Experimental part

In order to demonstrate de Broglie wave–particle duality and verifies his equation, in the proposed experiment, which tries to reproduce the original experiment of

Table 15.1. A sample of experimental data collected for the Davisson and Germer experiment. The table contains the difference of potential V, its error and the radius r. (*Numeric values are merely used as an example.)

V (V)	$\left(\frac{V}{\mathrm{kV}}\right)^{-\frac{1}{2}}$	$\Delta\left(\frac{1}{\sqrt{\frac{V}{\mathrm{kV}}}}\right)$	r (cm)
	1.41*	0.01*	4.2*

Davisson and Germer, the difference of potential V (measured in kilovolts) is varied and is measured using the radius r (measured in centimeters) of a single diffraction ring. The data can be collected in a table such as the example shown in table 15.1. The errors in the voltage V and in the radius r are produced by the sensitivity of the instruments. In particular, in the instrumentation used in the experiment the ΔV sensitivity is equal to 0.1 kV, while the error on the quantity $\frac{1}{\sqrt{V}}$ is obtained from the formula of propagation of errors [11]:

$$\Delta\left(\frac{1}{\sqrt{V}}\right) = \frac{\partial\left(\frac{1}{\sqrt{V}}\right)}{\partial V}\Delta V = \frac{1}{2\sqrt{V}}\Delta V.$$

From the experimental data it is possible to estimate the reticular step d through a linear fit. In fact, if in equation (15.8)

$$\begin{cases} y = r \\ x = \frac{1}{\sqrt{V}} \end{cases}$$

arises, a linear relationship between x and y is obtained,

$$y = \frac{\alpha}{d}x, \tag{15.9}$$

where the quantity

$$m = \frac{\alpha}{d}$$

represents the angular coefficient of the line equation (15.9). From the linear interpolation the coefficient m is obtained and from it the reticular step d and the corresponding error are obtained through the relations

$$\begin{cases} d = \frac{\alpha}{m} \\ \Delta d = \frac{\Delta m}{m}d \end{cases}.$$

By comparing the experimental value obtained with that reported by the manufacturer, it can be concluded that the experimental results are in a close agreement with the de Broglie equation. The starting hypothesis, the equality of the λ_B and λ_D wavelengths, is verified, therefore the de Broglie hypothesis is validated.

Acknowledgements

The author would like to thank A Stabile for useful discussions about the Davisson and Germer experiment and its realization.

Appendix

A Relativistic approximation

The momentum of the electrons present in de Broglie's relation ($\lambda_D = \frac{h}{p}$) was obtained [1] by considering the electron as a non-relativistic particle. It is given by

$$p = \sqrt{2meV},$$

and for a non-relativistic particle the relation of the conservation of mechanical energy is

$$eV = \frac{1}{2}mv^2.$$

However, around 4 kV, the electrons in the diffraction tube of the experiment have a speed equal to about 12.5% of the speed of light in a vacuum, c.

To take account of any relativistic corrections [12], the classical relationship of conservation of mechanical energy must be replaced with the relation

$$eV = mc^2(\gamma - 1), \tag{A.1}$$

where γ is the Lorentz factor

$$\gamma = \frac{c}{\sqrt{c^2 - v^2}}$$

and m is the electron rest mass

$$m = 0.5110\frac{\text{MeV}}{c^2}.$$

Replacing γ in equation (A.1),

$$eV = mc^2\left(\frac{c}{\sqrt{c^2 - v^2}} - 1\right).$$

Performing simple calculations, it is possible to obtain the velocity of a particle in the relativistic case:

$$v^{\text{rel}} = c\sqrt{1 - \frac{1}{\left(1 + \frac{eV}{mc^2}\right)^2}}. \tag{A.2}$$

While the relativistic momentum is given by the expression

$$p^{\text{rel}} = mv^{\text{rel}}\gamma.$$

Replacing v^{rel} of equation (A.2), it is possible to obtain

$$p^{\text{rel}} = \sqrt{2meV + \left(\frac{eV}{c}\right)^2}.$$

Consequently the relativistic momentum is greater than the classical one, therefore it follows that the de Broglie relativistic wavelength $\lambda_{\text{D}}^{\text{rel}}$ is smaller than the classical one, that is,

$$\lambda_{\text{D}}^{\text{rel}} = \frac{h}{\sqrt{2meV + \left(\frac{eV}{c}\right)^2}}, \tag{A.3}$$

while the classic expression is

$$\lambda_{\text{D}} = \frac{h}{\sqrt{2meV}}. \tag{A.4}$$

It is interesting to evaluate the relative error that is committed considering the relativistic electron instead of the 'slow' one. By developing the relationship (A.3) in series, with $\frac{eV}{mc^2} \ll 1$, the estimate of this error is provided by

$$\left|\frac{\lambda_{\text{D}}^{\text{rel}} - \lambda_{\text{D}}}{\lambda_{\text{D}}}\right| \simeq \frac{eV}{4mc^2} = \frac{5}{10\,000}V, \tag{A.5}$$

where the voltage V is expressed in kilovolts.

Instead the relative measurement error in the determination of the wavelength (A.4) is provided by the expression [11]

$$\left|\frac{\Delta\lambda_{\text{D}}}{\lambda_{\text{D}}}\right| = \frac{1}{2}\frac{\Delta V}{V}.$$

In the case of the instrumentation used in the experiment described, the ΔV sensitivity is equal to 0.1 kV, therefore

$$\left|\frac{\Delta\lambda_{\text{D}}}{\lambda_{\text{D}}}\right| = \frac{2}{100}\frac{1}{V}. \tag{A.6}$$

In table A1 the values of the two relative errors, equations (A.5) and (A.6), have been calculated for some values of the applied voltage. It is possible to observe that for voltage values lower than 6.3 kV the relative wavelength measurement error is

Table A1. Comparison of relative errors in the classical case and in the relativistic case, at certain voltage values.

	1 kV	5 kV	6.3 kV	10 kV
$\left\vert \frac{\Delta\lambda_D}{\lambda_D} \right\vert$	$\frac{2}{100}$	$\frac{4}{1000}$	$\frac{3}{1000}$	$\frac{2}{1000}$
$\left\vert \frac{\lambda_D^{rel} - \lambda_D}{\lambda_D} \right\vert$	$\frac{5}{10\,000}$	$\frac{2.5}{1000}$	$\frac{3}{1000}$	$\frac{5}{1000}$

greater than the relativistic one, and therefore predominant. Therefore the relativistic effects are negligible, while for values higher than 6.3 kV the relativistic error prevails over the measurement error and consequently the approximation of a 'slow' electron is no longer valid.

In the case of the instrument used to reproduce the experiment of Davisson and Germer, the maximum voltage that can be reached with the generator supplied is 5 kV, for this reason the electron can be considered 'slow'.

References

[1] de Broglie L 1925 Recherches sur la théorie des quanta *Ann. Phys.* **10** 22–128

[2] Greenberger D *et al* 2009 *Compendium of Quantum Physics: Concepts, Experiments, History and Philosophy* (Berlin: Springer)

[3] Taylor E F and French A P 1979 *An Introduction to Quantum Physics* (Cheltenham: Stanley Thornes)

[4] Russo A 1981 Fundamental research at Bell Laboratories: the discovery of electron diffraction *Hist. Stud. Phys. Sci.* **12** 117–60

[5] Krane K 1983 *Modern Physics* (New York: Wiley)

[6] Davisson C and Germer L H 1927 Diffraction of electrons by a crystal of nickel *Phys. Rev.* **30** 705–40

[7] Weinert F 2009 Davisson–Germer experiment *Compendium of Quantum Physics* ed D Greenberger, K Hentschel and F Weinert (Berlin: Springer)

[8] 3B Scientific *Electron Diffraction Tube Manual* https://www.3bscientific.com/product-manual/1013889_EN.pdf.

[9] McQuarrie D A 1997 *Physical Chemistry: A Molecular Approach* (Sausalito, CA: University Science)

[10] Bragg W L 1934 *The Crystalline State* vol 1 (New York: Macmillan)

[11] Taylor J R 1997 *An Introduction to Error Analysis: The Study of Uncertainties in Physical Measurements* (Sausalito, CA: University Science)

[12] Stabile A Lecture notes held in the framework of *Progetto Lauree Scientifiche*, Dipartimento di fisica 'E.R. Caianiello', Università di Salerno, Italia

Chapter 16

Current transport and light emission in semiconductors: a simple way to determine the Planck constant

Sergio Pagano and Martina Moccaldi

16.1 Introduction

Semiconductors have contributed to the technological development of the last century in a fundamental way. To understand how semiconductors work and how they can be used, the knowledge of the basic concepts of condensed matter physics is essential. This branch of physics has its roots in the early investigations in the eighteenth and nineteenth centuries of the properties of matter (see chapter 9), and in the great progress made in the twentieth century in understanding atomic theory and the peculiar way in which microscopic particles interact to form macroscopic bodies, which goes under the name of quantum mechanics. In this chapter, we show how common semiconducting devices (namely light emitting diodes (LEDs)) can be used to carry out a simple experiment with a fundamental result—the measurement of Planck's constant (h). This constant, named after German physicist Max Karl Planck, is crucial for understanding how matter behaves at a microscopic scale. Despite its extremely small value ($h = 6.626\,070\,04 \times 10^{-34}$ J s), the experiment can be easily performed in any high school laboratory by students and it only requires some LEDs and simple electrical instruments.

We first present a brief overview of the structure of matter, devoting particular attention to materials exhibiting a crystalline configuration of atoms (section 16.1). Then, we focus on the electrical conductivity of materials (section 16.2) and on semiconductors (section 16.3), in particular highlighting how they can be used to realize LEDs. Finally, section 16.4 is devoted to illustrating the experiment for the computation of Planck's constant.

doi:10.1088/978-0-7503-2678-0ch16

16.2 The structure of matter

Solid state physics studies the fundamental properties of matter when it is solid, its capacity of transferring electrical current, heat and vibrations, and its response to changes in temperature and to external electric and magnetic fields.

It is known that all of the matter around us is made up of atoms, which are themselves composite structures made up of electrons, light elementary particles with negative charge, and a nucleus, which in turn is formed of other elementary particles, protons and neutrons, massive particles with a positive and a neutral electrical charge, respectively. The internal particles of an atomic nucleus are held together by means of nuclear forces, whereas the electrons are bonded to the nucleus by electromagnetic forces. The bonds between atoms are also electromagnetic in nature. In this microscopic world the laws of nature are no longer prescribed by classical physics, instead they are investigated through quantum mechanics.

Atoms are linked together through different kinds of chemical bonds, which differ according to the kinds of atoms involved. Chemical bonds have several names, depending on the bonding mechanism and its strength: covalent, ionic, van der Waals, metallic and hydrogen bonds. According to the bonding type, the melting point of a material changes radically, as shown in figure 16.1. In particular, it is higher in covalent and ionic bonds and lower in molecular and metal bonds. If the bonding strength is not large enough, the material may not be in its solid form at room temperature (e.g. water). Once atoms are joined together by means of chemical bonds, they can form different kinds of solid materials: crystalline, amorphous or polycrystalline. A solid is called *crystalline* if its atoms are regularly distributed in space, giving rise to a regular structure. Examples of crystals are graphite and

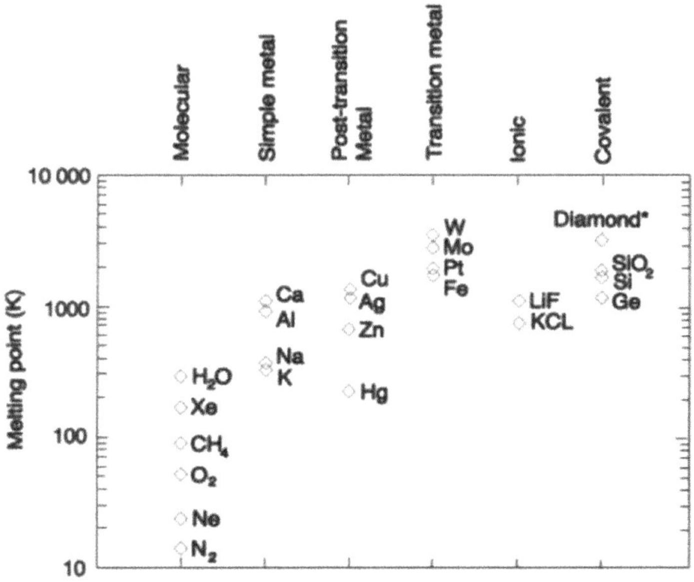

Figure 16.1. Correlation between the melting point of a material and the type of internal bonding.

quartz. In crystals inter-atomic forces keep the atoms at fixed distances. However, they can oscillate around their equilibrium positions, like masses connected by springs. These oscillations are responsible for sound propagation in solids and also affect the electrical conductivity, as we will see later. The crystalline regularity is sometimes broken by the presence of defects, inclusions of different materials, a change in the spatial orientation of the crystals and so on. In such cases, the material maintains a crystalline order on a small scale and appears as many differently oriented small crystals joined together, forming a so-called *polycrystalline* solid. When atoms forming a solid are organized in a random manner, much like in a liquid, but without the possibility of moving freely, the solid is called *amorphous*.

Solid state physics mainly investigates the structure and the properties of crystals, which are easier to describe due to their regular structure [1].

According to the position of atoms in the crystalline structure, it is possible to distinguish a wide range of different crystals, which were classified by Auguste Bravais in 1848 into 14 classes. It can indeed be proved that every crystal has a lattice belonging to one of these categories.

Even the most perfect crystals exhibit impurities, such as the absence of an atom in the crystalline structure (a vacancy), the presence of an atom in a position different to the regular positions of the crystalline lattice (an interstitial impurity) and the replacement of an atom with a different natured atom (a replacement impurity).

16.3 Electrical conductivity of materials

The capability of solids to carry electric current is described by the electrical conductivity σ and its inverse ρ, called resistivity. They relate the amount of applied electric field E to the current density J that flows in the material. In a typical material it takes the form of a simple linear relation, the so-called Ohm's law (see chapter 5):

$$\vec{J} = \sigma\vec{E} = \frac{1}{\rho}\vec{E}.$$

According to the value of resistivity, materials can be divided into four fundamental classes: conductors, insulators, semiconductors and superconductors (see table 16.1).

Table 16.1. Ranges of electrical conductivity for various materials.

Material	Examples	σ $(\Omega\ \mathrm{cm})^{-1}$
Superconductor	Niobium, lead, tin, aluminium	Infinite ($T < T_c$)
Metal	Copper, aluminium, iron, steel, tin	$\approx 10^4$–10^6
Semiconductor	Silicon, germanium, indium arsenide, indium phosphate, gallium arsenide	$\approx 10^{-6}$–10^3 (room temperature)
Insulator	Sulphur, diamond, silicon dioxide, nylon, Teflon, polystyrene	$\approx 10^{-20}$–10^{-10}

Conductors exhibit high conductivity. In such materials it appears that electrons can move easily when an electric field is applied. Most of the elements belonging to this class are metals, such as aluminium, silver, gold and copper, but there are also non-metal conductors such as carbon. However, the flow of electric current is not completely free and produces a loss of energy, which is converted into thermal energy, so conductors become hot when they carry large currents (the Joule effect).

Insulators are characterized by very low conductivity, because their electrons are strongly bonded to their atoms and cannot move freely, even if an intense field is applied. This category of materials includes marble, fused quartz, PVC plastics, rubber and so on.

Semiconductors have intermediate values of conductivity, because they have few free electrons to carry current, and their number strongly depends on the temperature. Examples of semiconductors are silicon, germanium and gallium arsenide. The crucial feature of these materials is that it is possible to profoundly change their ability to conduct electricity, as we will discuss in detail in section 16.3.

Superconductors constitute a special class of materials, made of simple elements and composites, such as niobium, lead, tin, aluminium, niobium nitride, $YBa_2Cu_3O_7$, etc, whose peculiar property is to have *infinite* conductivity when cooled below a certain temperature (called the critical temperature T_c) and for not too large currents. The understanding of their physical properties is deeply routed in quantum mechanics, took several decades to be explained and, to some extent, is still an open question and so is outside the scope of this chapter.

The assumption of free electrons in a conductor leads to a problem. Ohm's law relates the electric field $\vec{E}$, which exerts a force $-e\vec{E}$ on the electrons (Coulomb's law), to the current density $\vec{J}$, which is proportional to the velocity $\vec{v}$ of the electrons:

$$J = -en|\vec{v}|.$$

This is in contrast to Newton's second law of dynamics: a force should produce an acceleration, not a constant velocity. To solve this problem it is necessary to assume that electrons in a conductor are not completely free. They are accelerated by the electric field but then undergo random collisions which nullify the accumulated velocity. Such collisions occur randomly with an average time τ. The average velocity, in the direction of the field, is then given by the acceleration due to the electric field $-e\vec{E}/m$ times the time τ as

$$\vec{v} = \frac{-e\vec{E}\tau}{m}.$$

Taking into account all these aspects, Ohm's law is recovered:

$$\vec{J} = \frac{ne^2\tau}{m}\vec{E},$$

where a microscopic derivation of the conductivity σ is obtained (the Drude model). Here m is the effective electron mass in the crystal, n is the density of free electrons and e is the electron charge.

Now the question is: what causes the collisions? In a perfect crystalline metal, quantum mechanics tells us that electrons should be absolutely free to move. However, as we remarked above, no real crystal is perfect. The presence of defects is an obstacle to the free motion of electrons. Moreover, at finite temperatures, atoms oscillate somewhat around their equilibrium crystalline positions. This also acts as an obstacle to electron motion. Taken together, the defects and finite temperature produce the finite, although small, resistivity of metals. Moreover, it can be expected that, if the temperature rises, collisions increase and the metal resistivity $\rho = \sigma^{-1}$ should increase as well. This is exactly what happens (see figure 16.2).

In the case of semiconductors, as also shown in figure 16.2, experimental data reveal an opposite behaviour; the reason for this will be discussed in the next section.

16.4 Semiconductors

From the historical point of view, the term *semiconductor* was introduced by Alessandro Volta in 1782. In 1833 Michael Faraday observed that conductivity of certain materials increases with temperature. At the time there was no microscopic theory of electrical current. The discovery of the electron by Joseph John Thomson in 1897 gave raise to a fruitful period of crucial experiments. In 1901 Eduard Riecke discovered that the electrical current in metals is due to the motion of electrons and Johan Koenigsberger supposed that the resistivity of semiconductors depends on temperature in 1903. In the following 30 years, the evolution of quantum mechanics allowed this experimental evidence to be explained. Indeed, in 1931 Alan Wilson proposed a band model for solids and the presence of donors and acceptors. In 1936 the Bell Telephone Laboratories financed research projects on semiconductors and in 1947 John Bardeen, Walter Brattain and William Shockley realized the first transistor, for which they were awarded the Nobel Prize for Physics. Their invention has been essential for modern technological evolution.

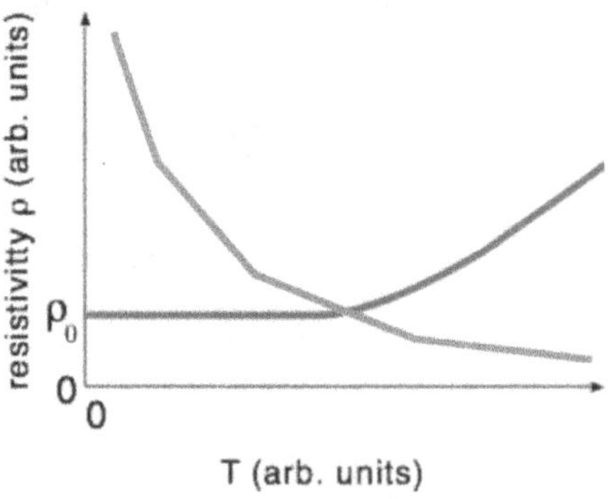

Figure 16.2. Dependence of resistivity of metals (red line) and semiconductors (green line) on temperature.

A key property of semiconductors is that their conductivity σ covers several orders of magnitude and strongly depends on the temperature and impurity content, as highlighted in table 16.1. These variations are due to their electronic configuration.

As already mentioned, quantum mechanics predicts that electrons are free to move in a perfect crystal. However, the same theory says that electrons cannot possess any energy, as free particles do, but they are located in certain energy bands, separated by gaps, whose size and shape strongly depend on the material. Moreover, electrons are quantum particles, they have to obey to Pauli's exclusion principle: they must have all different quantum numbers, which includes energy. As a result, all the electrons present in a solid will fill the various energy bands, starting from the lowest, much like people taking seats in theatre. Taking this analogy further, when a number of rows of seats are all filled, people seated there cannot move, as all the seats around them are filled. Only people on the edges of the filled rows have space to move, as they have empty seats nearby. When all the electrons have accommodated themselves in the allowed energy bands, different situations can occur.

In metals the last filled energy band, called the valence band, is not completely full, so electrons having the highest energy (the Fermi energy) have space to acquire extra energy. Consequently, they can move if an electric field (a force) is applied, much like the people in the theatre sitting at the edges of the occupied rows. For this reason, metals conduct electric current.

In insulators, the last filled energy band is full, so electrons can move if they first acquire enough energy to overcome the gap towards the next allowed band, called the conduction band. This can only happen for very high temperatures or applied electric fields. Thus, under normal conditions, electrons are blocked and the material cannot conduct electricity.

Semiconductors are essentially insulators in which the value of the energy gap to the next allowed band is quite small ($\approx$1 eV instead of several electronvolts). In this case, already at room temperature, a few electrons can leave the last energy level due to thermal activation, moving to the conduction band and therefore participating in electrical conduction. Moreover, the higher the temperature, the more electrons are promoted into the conduction band, and the lower the material resistivity is, as shown in figure 16.2.

Since an electron has moved from the valence to the conduction band, a *hole* is left in the original band. Such a hole behaves as a positively charged particle that is free to move under an electric field, so it contributes to conduction as well. Therefore, in semiconductors two types of electricity carriers are present: *electrons* and *holes*.

Because of the possibility of easily changing their electrical conductivity, it is evident why semiconductors have drawn a great deal of interest in both scientific research and technological applications.

16.4.1 Doped semiconductors

The electronic features of semiconductors can be deeply modified by introducing certain impurities in the lattice. This mechanism is known as *doping*. For example, we can replace some tetravalent atoms of silicon or gallium in the crystal lattice with the same number of pentavalent (such as phosphorus or arsenic) or trivalent atoms (such as aluminium or boron). It is worth highlighting that these impurities are inserted into the lattice structure of the crystal and not into the interstitial areas.

The most used semiconductor is silicon, which has four valence electrons shared with its neighbouring atoms, generating a very stable bond. Due to the low number of free electrons at room temperature (10^{10} electrons per cm^3), pure silicon is substantially an insulator. By doping this material with pentavalent atoms, extreme changes in its conductivity can be produced.

A pentavalent atom shares four electrons with the other tetravalent atoms of the crystal lattice, restoring the covalent bond (see figure 16.3). The remaining electron is loosely attached to the original atom so a little energy, such as the thermal energy at room temperature, can push it into the conduction band, adding one more carrier for electricity. For this reason, a pentavalent impurity is called a *donor*. Each donor atom donates one electron to the conduction band, thus drastically changing the conductivity of the material. For example, pure silicon contains almost 5×10^{22} atoms per cm^3 and its conduction band is composed of almost 10^{10} electrons per cm^3 at room temperature. By adding a tiny amount of doping, one phosphorus atom for

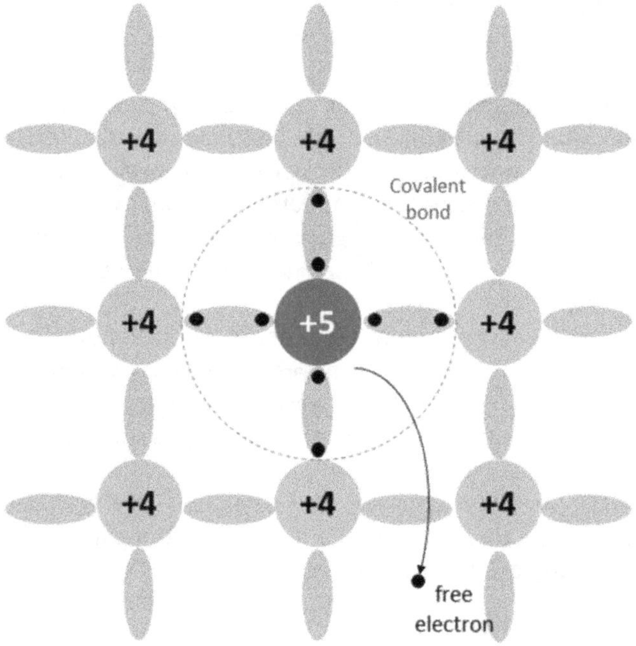

Figure 16.3. Representation of an n-doped semiconductor obtained by introducing a pentavalent atom in a crystal lattice composed of tetravalent atoms.

every 50 million silicon atoms, almost 10^{15} electrons per cm^3 are introduced into the conduction band [2], thus improving the conductivity of the material by *five orders of magnitude*. Therefore, it is clear how strong the effect of even a little doping in a semiconductor can be. These kinds of doped semiconductors are called *n-doped semiconductors* since negative charges are introduced in the conduction band, without a corresponding increase of electronic holes in the valence band. In terms of the energy band picture, the effect of the added donor impurities is realized by adding energy levels filled with electrons inside the forbidden gap and very close to the conduction band. Therefore, electrons at those levels require very little energy to jump to the conduction bands. At room temperature essentially all these jumps occur, thus realizing an n-doped semiconductor where the free electron density is equal to the density of added donor impurities.

Trivalent impurities (such as boron, gallium, indium, aluminium and so on) added in a tetravalent lattice share three electrons with the other atoms through covalent bonds, leaving an unattached bond which attracts an electron and therefore generates an electronic hole (see figure 16.4). Due to this mechanism, such a trivalent impurity is called an *acceptor*. This kind of doping produces an increase of mobile positive charges, so the resulting semiconductor is called a *p-doped semiconductor*. In terms of the energy band picture, the effect of the added acceptor impurities is to realize additional energy levels inside the forbidden gap and very close to the valence band. Therefore, electrons on the top of the valence band require very little energy to jump to the new available levels, leaving behind a hole. At room temperature

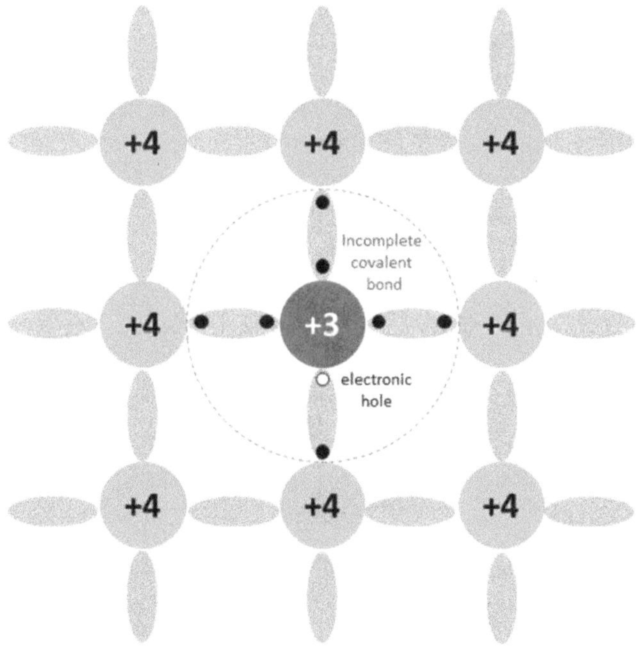

Figure 16.4. Representation of a p-doped semiconductor obtained by introducing a trivalent atom in a crystal lattice composed of tetravalent atoms.

essentially all these jumps occur, thus realizing a p-doped semiconductor where the holes exhibit the same density of added acceptor impurities. In summary, the doping modifies the conductivity properties of a semiconductor because it makes admissible energy levels that are forbidden in pure semiconductors. These levels are close to the conductivity band in the case of n-doping and the valence band in the case of p-doping. As a result, the doping produces an increase of electrons in the conductivity band (n-doping) or an increase of electronic holes in the valency band (p-doping). In both cases, the semiconductor becomes a conductor, whose conductivity can be determined by the amount of introduced impurities.

16.4.2 P–n junctions and diodes

P-doped semiconductors and n-doped semiconductors are electrically neutral. With a bit of technology, it is possible to dope two adjacent regions of the same semiconductor with different doping types, thus realizing a *p–n junction*, as shown in figure 16.5. In this configuration, on one side of the junction (on the left in figure 16.5) there are a large number of electrons and very few holes, and on the other side (on the right in figure 16.5) there are a large number of holes and very few electrons, and all of them are free to move. The difference in concentration between the left and right sides of the junction pushes electrons towards the right and holes towards the left (a *diffusion current*). However, as these charges move, they break the neutrality of the semiconductor, leaving fixed positive charges to the left and fixed negative charges to the right. Such a charged bi-layer generates a strong electric field which acts against the motion of electrons and holes. An equilibrium state is reached when there is no more flow of carriers across the p–n junction and a stable charged region is present at the junction position. This region, called the *depletion layer*, is very thin, having a thickness that depends on the doping levels and is of the order of 1 μm. An electrical field is developed across the depletion layer having a high intensity (almost 10^4–10^5 V cm^{-1}) and produces a potential difference ΔV_{pn} across it of the order of 1 V. In this circumstance, a free charge needs extra energy to cross the depletion region.

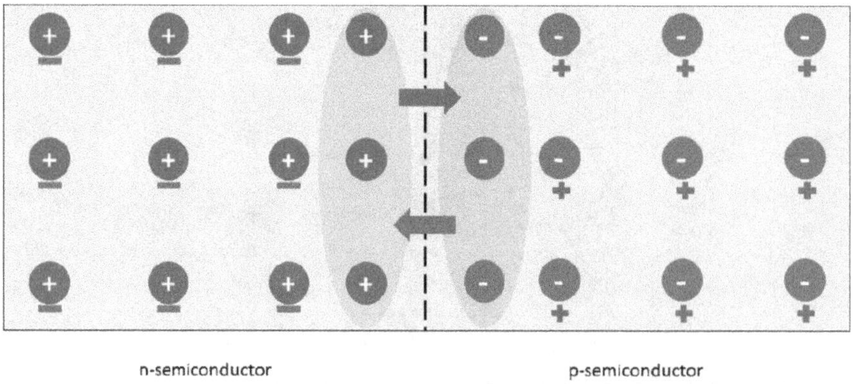

Figure 16.5. A p–n junction.

When a proper potential difference is applied at the sides of a p–n junction, the equilibrium is modified and the charges may receive the necessary energy to overcome the barrier. The resulting device can be arranged in two different manners, according to the sign of the potential. If the positive potential is applied on the p side (*forward bias*), ΔV_{pn} decreases and the motion of charges through the separation interface can occur (see figure 16.6). The intensity of electrical current quickly grows with the bias potential.

Instead, if the positive potential is applied to the n-side (*reverse bias*) ΔV_{pn} increases and the motion of charges is further inhibited (see figure 16.7).

The external voltage needed to reduce the potential barrier in order to have a significant current depends on the type of semiconductor used and the temperature. For example, this voltage at room temperature is about 0.6–0.7 V for silicon and 0.3–0.35 V for germanium. A device composed of a p–n junction where a potential difference is applied is called a *p–n junction diode* or simply a *diode*. This kind of device is widely employed in all electronic applications and is the basis for a number of specific realizations, such as solar cells, lasers, LEDs, etc.

The main feature of a p–n junction diode is that the current moves in only one direction. Therefore, differently to resistors, diodes are characterized by an non-linear, almost exponential, current–voltage (I–V) relation

$$I = I_s[e^{V/\eta V_T} - 1],$$

where I_s is the saturation current, of the order of 10^{-9}–10^{-18} A, η is a quality parameter whose value is between 1 and 2, $V_T = K_B T/e$ is the thermal voltage which is almost equal 0.25 mV at room temperature, with K_B being Boltzmann's constant, T is the absolute temperature and e is the electron charge.

When an electron crosses the diode junction in the p-direction it finds a sea of holes. As a consequence the electron, which is in the conduction band, jumps down in energy to fill the hole in the valence band and releases the energy difference, roughly equal to the semiconductor energy gap. A similar process occurs when a

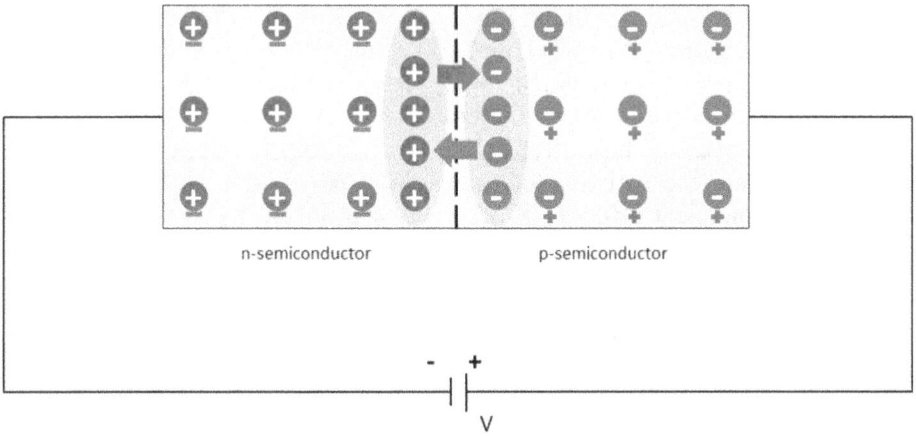

Figure 16.6. Direct bias of a p–n junction.

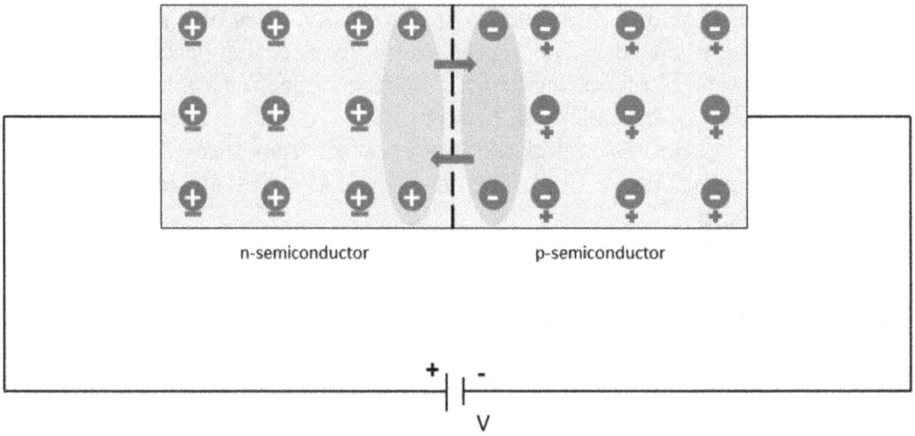

Figure 16.7. Inverse bias of a p–n junction.

hole crosses the junction in the n-direction. In both cases, an energy almost equal to the gap energy is released. Such energy can take the form of an electromagnetic wave (photons) or lattice vibration wave (phonons), or both, depending on the semiconductor properties. In silicon most of the energy is released as phonons and very little as photons. As a result, keeping in mind that the photon energy E is proportional to its frequency ν by the relation $E = h\nu$, where h is the Planck constant, forward biased a silicon diode emits electromagnetic waves falling in the infrared spectral region. Other semiconductors, such as gallium arsenide, emit most of the energy as photons, so they generate higher frequency (visible) light when forward biased. A device that employs such a mechanism to generate light when forward biased is called a light emitting diode (LED). The wavelength of the emitted radiation depends on the gap energy value and therefore on the kind of semiconductor used. The most common LEDs are built by employing alloys of gallium, arsenic, indium, phosphor and aluminium. By choosing the appropriate semiconductor and doping, it is possible to realize LEDs with emission wavelengths from 360 nm (ultraviolet) to 950 nm (infrared), covering the whole visible spectrum.

It is important to note that, when an LED is just forward biased, a small current flows in the device and is circulated by the external battery. The potential energy added by the battery compensates the energy loss in the circuit. The value of the potential energy given by the battery is 1 eV per electron. Each electron crossing the LED will generate the emission of one photon of energy $h\nu$. The low current guarantees the practical absence of other dissipation mechanisms, such as Joule heating. Therefore, energy conservation implies that

$$eV = h\nu. \tag{16.1}$$

If we measure the voltage of initial illumination of an LED and know its emitted wavelength then, from equation (16.1) and knowledge of the electron charge, we can compute the value of the Planck constant.

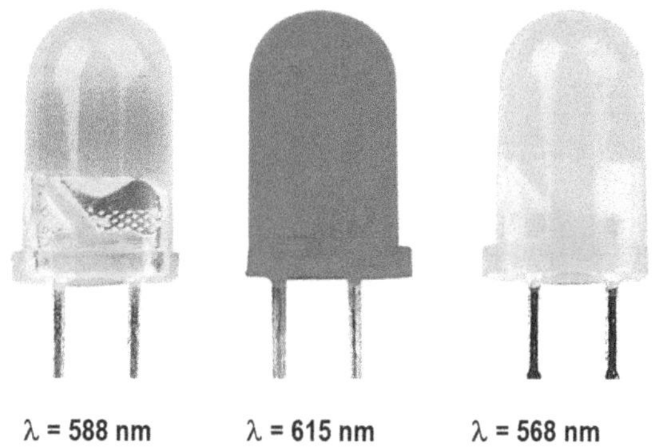

Figure 16.8. Examples of different LED types with the emitted wavelength.

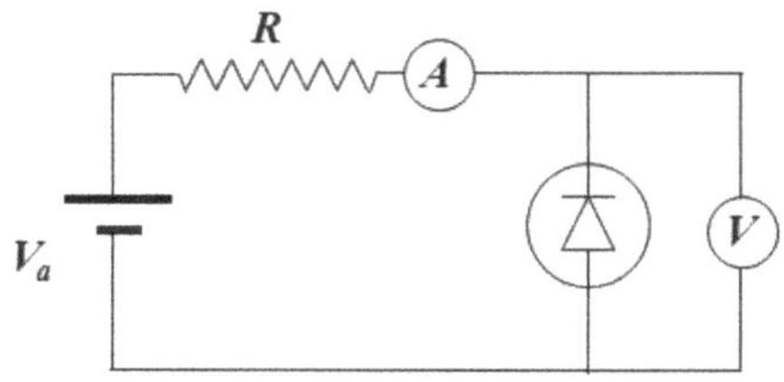

Figure 16.9. Schematic diagram of the experiment.

16.5 Experimental determination of the Planck constant

According to equation (16.1), there is a direct proportionality between the voltage V at which an LED just lights up and the frequency ν of the emitted radiation. Commercial LEDs of different colours, corresponding to different emitted wavelengths, can be purchased easily. Three common LEDs, with emitted radiation at 568, 588 and 615 nm, are shown in figure 16.8. It is easy to obtain the characteristic of emission for each LED by looking at the manufacturer's data sheet. In order to measure V, a variable voltage source, in the range 2–10 V, is connected to the chosen LED through a limiting resistor of value about 100 Ω. The diode bias voltage can be measured using a voltmeter connected in parallel to the LED. Every diode has a mark to indicate the two p and n sides. In the case of LEDs, typically the longer lead corresponds to the anode (p side) and must be connected to the positive side of the voltage source. In figure 16.9 a schematic diagram of the connections is presented. Starting from the lowest voltage generated by the power source, it has to be slowly increased until an emission of light from the LED is detected. The voltage drop across the LED, as read by the voltmeter, has to be written down, together with the LED type. This process can be repeated using different LEDs, for which the emission wavelength has been previously determined. At the end a table similar to table 16.2 can be obtained.

Table 16.2. Example of experimental results.

LED	λ	V	ν	E
Red	615 nm	1.92 V	487×10^{12} Hz	3.08×10^{-19} J
Yellow	588 nm	2.01 V	510×10^{12} Hz	3.22×10^{-19} J
Green	568 nm	2.10 V	528×10^{12} Hz	3.37×10^{-19} J

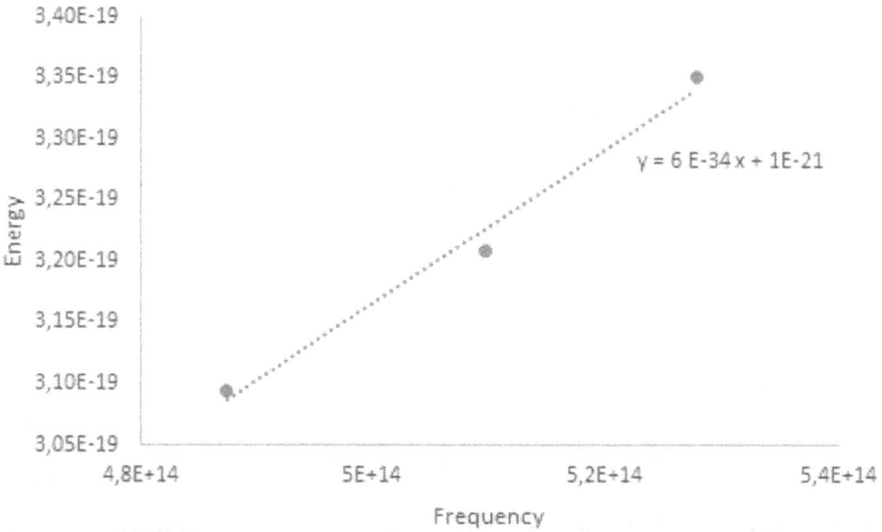

Figure 16.10. Example of linear regression between the energy E of electrons and the emission frequency ν of photons.

The last two columns are computed from the experimental data using

$$\nu = \frac{c}{\lambda},$$
$$E = eV,$$

where $c = 2.997\,924\,58 \times 10^9$ m s^{-1} is the speed of light in a vacuum and $e = 1.602\,176\,634 \times 10^{-19}$ C is the electron charge. By using a simple linear regression formula, available in any spreadsheet software, it is possible to compute the best value of the proportionality constant and also graph the data. As an example, in figure 16.10 the experimental data and the regression line are reported. The resulting value for h ($7.0 \pm 0.6 \times 10^{-34}$ J s) appears to be in good agreement with the generally accepted value ($6.626\,070\,04 \times 10^{-34}$ J s).

16.6 Conclusions

Solid state physics includes the essential knowledge that has given birth to modern electronics and has allowed the huge technological improvements of the last century. At the basis of this branch of physics is the unconventional and counterintuitive description of the microscopic world of quantum mechanics, where a fundamental role is played by the Planck constant. The simple experiment described in this chapter shows how quantum physics laws not only rule the microscopic world but are also crucial to explain the way common devices, such as LEDs, operate. The experiment can easily be performed with minimal equipment and by students with limited knowledge of electrical quantities. Nevertheless, it allows a fairly good determination of the Planck constant. It can also be used as a starting point for further investigations in solid state physics and electronics.

References

[1] Kittel C 2005 *Introduction to Solid State Physics* (New York: Wiley)

[2] Halliday D, Resnick R and Krane K S 2001 *Physics* vol 2 (New York: Wiley)

Part III

Learning modern physics through exercises and applets

Chapter 17

Graded exercises and problems in modern physics

Antonio Stabile and Canio Noce

This chapter contains a collection of exercises and problems in special relativity and quantum physics, i.e. modern physics, spanning just about the full scope of this book. The exercises and problems are of varying difficulty, and these are accompanied by comprehensive solutions. Indeed, the basic ideas and methods of modern physics are illustrated by means of a carefully chosen set of problems, accomplished with detailed step-by-step solutions.

Most of the exercises are straightforward and uncomplicated, so these may be accessible to general undergraduate students. However, some of them, mainly the problems, are at a higher level of complexity, with challenge and interest for a practising modern physicist. Accordingly, this chapter will be useful for a wide range of readers including both students studying special relativity and quantum mechanics, and science lecturers.

We would like to point out that all of these exercises and problems are based on the actual physical examples presented in the previous chapters of this book, even if sometimes the mathematical structure of the models under consideration is simplified intentionally in order to access the physics more rapidly. A variety of topics are covered, including the postulates of special relativity and quantum mechanics, the intriguing aspects of relativity as well as the solution of Schrödinger's equation in simple cases.

Importantly, these exercises and problems have all been assigned to undergraduate students, so that we are confident that an undergraduate science student or a researcher, who has had even minimal exposure to modern physics, may benefit from this material, certainly improving their problem-solving skills. Finally, we suggest the general reader consult chapters 3, 6 and 8 where the theory behind the exercises and problems is presented and discussed.

doi:10.1088/978-0-7503-2678-0ch17

17.1 Relativistic physics

Exercise 1

How fast must a rod be moving if its length is observed to shrink to 90% of its rest length?

Solution
From the expression for length contraction (see chapter 3), we obtain

$$L = \frac{L_0}{\gamma}, \quad \frac{L}{L_0} = \sqrt{1 - (V/c)^2} = 0.90,$$
$$V = c\sqrt{1 - (0.90)^2} = 0.44c.$$

Exercise 2

A train is in motion with respect to a railway station with a speed of 108 km h^{-1}. As determined by the station clocks, how long will it take for the train's clock to fall behind by five microseconds?

Solution
From the time dilation expression (see chapter 3) $\Delta t_{\text{station}} = \gamma \Delta t_{\text{train}}$:

$$\Delta t_{\text{station}} = \frac{\Delta t_{\text{train}}}{\sqrt{1-(V/c)^2}} = \frac{\Delta t_{\text{train}}}{\sqrt{1-\left(\frac{108 \text{ km h}^{-1}}{3\times 10^8 \text{ m s}^{-1}}\right)^2}}$$
$$= \frac{\Delta t_{\text{train}}}{\sqrt{1-\left(\frac{3\times 10 \text{ m s}^{-1}}{3\times 10^8 \text{ m s}^{-1}}\right)^2}} \approx \frac{\Delta t_{\text{train}}}{1-5\times 10^{-15}}.$$

It follows that

$$5 \times 10^{-15}\,\Delta t_{\text{station}} = \Delta t_{\text{station}} - \Delta t_{\text{train}} \equiv 5 \times 10^{-6} \text{ s},$$
$$\Delta t_{\text{station}} = 10^9 \text{ s} \approx 31.7 \text{ yr}.$$

Exercise 3

At what fraction of the speed of light must a particle move so that its kinetic energy is equal to its rest energy?

Solution
Using energy conservation in relativistic mechanics (see chapter 3), we get

$$m_0c^2\,(\gamma - 1) = m_0c^2 \Rightarrow \gamma = 2.$$

Solving, $v = \sqrt{3}/2c \approx 0.866c$.

Exercise 4

A spaceship goes on a mission to TRAPPIST-1, 40 light years from Earth. (a) What should the constant speed of the spaceship be, with respect to the Earth, to be able to reach its destination in 60 Earth years? (b) From the point of view of an observer in the spaceship how many years have passed?

Solution

(a) The speed of the spaceship is

$$V = \frac{40 \text{ light years}}{60 \text{ years}} = \frac{40 \text{ ly}}{60 \text{ yr}} = 0.67c.$$

(b) The time measured in the spaceship in which the observer is at rest is a proper time (see chapter 3):

$$\Delta t_{\text{spaceship}} = \frac{\Delta t_{\text{Earth}}}{\gamma} = \Delta t_{\text{Earth}}\sqrt{1 - (V/c)^2} = 60 \text{ yr } \sqrt{1 - (0.67)^2} = 44.5 \text{ yr}.$$

Exercise 5

An inertial observer O sees an ellipse with an eccentricity of 0.3. What must the speed of a observer O′, moving relative to O, be in order that the ellipse be seen as a circle?

Solution

The observer O sees an ellipse moving with speed V, while the observer O′ sees a circle at rest, with radius R_0, in its frame. The observer O measures only in the direction parallel to speed $\mathbf{V}$, $\mathbf{V} \equiv (V; 0)$, a length contraction (see chapter 3), therefore,

$$R_x = \frac{R_x'}{\gamma} \equiv \frac{R_0}{\gamma},$$
$$R_y = R_y' \equiv R_0.$$

The eccentricity of the ellipse[1] measured by the observer O is

$$e = \sqrt{1 - \left(\frac{R_x}{R_y}\right)^2} = \sqrt{1 - \left(\frac{1}{\gamma}\right)^2} = \sqrt{1 - \left[1 - (V/c)^2\right]^2} = V/c.$$

Therefore, $V = 0.3c$.

[1] Let a and b be, respectively, the semi-major and the semi-minor axes of an ellipse. Therefore, the eccentricity can be expressed as $e = \sqrt{1 - \left(\frac{b}{a}\right)^2}$.

Exercise 6

Two identical spaceships A and B have a length of 30 m. Spaceship A travels to the right and spaceship B travels to the left, with velocities $0.75c$ and $0.90c$, respectively, relative to the Earth (see figure 17.1(a)). What is the velocity and the length of spaceship B measured from spaceship A (see figure 17.1(b))?

Solution
Let $V_A = 0.75c$ and $V_B = -0.9c$ be the velocities relative to the Earth, respectively, of the spaceships A and B. The relative velocity V_{AB} is (see chapter 3)

$$V_{AB} = \frac{V_A - V_B}{1 - \frac{V_A V_B}{c^2}} = \frac{0.75c - (-0.90c)}{1 - \frac{(0.75c)(-0.90c)}{c^2}} = 0.98c.$$

From the point of view of A, the length of the spaceship B is contracted:

$$L_A = \frac{L_0}{\gamma_{AB}} = L_0\sqrt{1 - (V_{AB}/c)^2} = 30\ \text{m}\sqrt{1 - (0.98)^2} \approx 6.0\ \text{m}.$$

Exercise 7

A rocket moves with a speed of $0.60c$ and forms an angle of 45° with respect to the x-axis, as determined by an observer O. What is the velocity of the rocket as determined by a second observer O′ moving with a speed of $0.80c$ along the common x–x'-axis with respect to the observer O?

Solution
For observer O the rocket velocity is

$$\mathbf{v} = (0.60\cos 45°c,\ 0.60\sin 45°c) = (0.42c,\ 0.42c).$$

Using the velocity transformation relation (see chapter 3), we have for observer O′

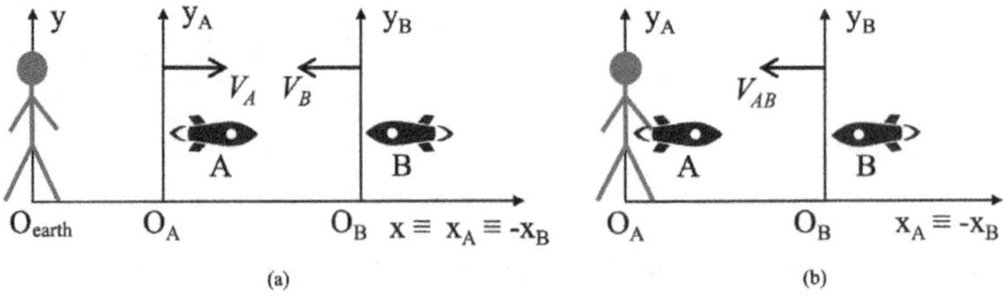

Figure 17.1. (a) Observer at rest with the Earth. (b) Observer at rest with the spaceship A.

$$v'_x = \frac{v_x - V}{1 - \frac{v_x V}{c^2}} = \frac{0.42c - 0.80c}{1 - \frac{(0.42c)(0.80c)}{c^2}} = 0.57c,$$

$$v'_y = \frac{v_y}{\gamma\left(1 - \frac{v_x V}{c^2}\right)} = \frac{0.42c}{1 - \frac{(0.42c)(0.80c)}{c^2}}\sqrt{1 - (0.80c/c)^2} = 0.38c.$$

The speed measured by observer O′ is $\mathbf{v}' = (0.57c,\ 0.38c)$:

$$v' = \sqrt{v_x'^2 + v_y'^2} = \sqrt{(0.57c)^2 + (0.38c)^2} = 0.69c,$$

and the angle θ' that velocity makes with the x–x'-axis is

$$\tan\theta' = \frac{v'_y}{v'_x} = \frac{0.38c}{0.57c} = 0.67.$$

Therefore, $\theta' = 33.7°$.

Exercise 8

What is the speed and the momentum of an electron at rest that is accelerated through a potential difference of 10^5 V?

Solution

From energy conservation, we have

$$\text{eV} = m_0c^2(\gamma - 1),$$

where $m_0 = 0.51 \times \text{MeV}/c^2$ is the electron rest mass. Then, the velocity is

$$v = c\sqrt{1 - \frac{1}{\left(1 + \frac{\text{eV}}{m_0c^2}\right)^2}} = c\sqrt{1 - \frac{1}{\left(1 + \frac{10^5\,\text{eV}}{0.51 \times 10^6\,\text{eV}}\right)^2}} = \frac{\sqrt{11}}{6}c \approx 0.55c.$$

The relativistic momentum is $P = \gamma m_0 v$ (see chapter 3). Using the following Lorentz factor $\gamma = 1 + \frac{\text{eV}}{m_0c^2}$, we have

$$P = \gamma m_0 v = \left(1 + \frac{\text{eV}}{m_0c^2}\right) m_0 c\sqrt{1 - \frac{1}{\left(1 + \frac{\text{eV}}{m_0c^2}\right)^2}} = \sqrt{2m_0\,\text{eV} + \left(\frac{\text{eV}}{c}\right)^2},$$

$$P = \sqrt{2 \times (0.51 \times 10^6\,\text{eV}/c^2) \times (10^5\,\text{eV}) + \left(\frac{10^5\,\text{eV}}{c}\right)^2} \approx 0.33\,\text{MeV}/c.$$

Exercise 9

An airplane moves with a constant speed $V = 900\ \text{km h}^{-1}$ relative to the Earth. Its length is 100 m, as measured from an observer O′ on the airplane. If an observer O on the Earth sees two lights turn on simultaneously at the ends of the airplane: (a) What is the spatial separation of the two lights as measured by O? (b) What would observer O′ see on the airplane?

Solution

(a) For the observer O the lights (events) are simultaneous, then $\Delta t = 0$. While the observer O′ sees the lights at two different points: $\Delta x' = L_0 = 100$ m. Let us consider the Lorentz transformations (see chapter 3) for finite intervals:

$$\Delta t' = \gamma\,(\Delta t - V/c^2 \Delta x),$$
$$\Delta x' = \gamma\,(\Delta x - V\Delta t).$$

Therefore,

$$\Delta t' = -\gamma \frac{\Delta x}{c^2} = -\frac{VL_0}{c^2},$$
$$\Delta x = \frac{\Delta x'}{\gamma} = L_0\sqrt{1 - (V/c)^2}.$$

In general, we note that observer O sees the two events closer together than O′, $\Delta x < \Delta x'$:

$$\Delta x = 100\text{ m}\sqrt{1 - \left(\frac{250\text{ m s}^{-1}}{3 \times 10^8\text{ m s}^{-1}}\right)^2} = 100\text{ m}\sqrt{1 - 6.9 \times 10^{-13}}$$
$$\approx 100\,(1 - 3.5 \times 10^{-13})\text{ m}.$$

(b) For the observer O′ the chronological order of events $\Delta t'$ depends on the relative speed V. Indeed, in the case $V > 0$ we have $\Delta t' < 0$ while $V < 0$, $\Delta t' > 0$. The simultaneity is recovered only if $V = 0$. This is an example in which the chronological order is not an invariant. Indeed, the value measured by observer O′ is

$$|\Delta t'| = \frac{(250\text{ m s}^{-1}) \times (100\text{ m})}{(3 \times 10^8\text{ m s}^{-1})^2} \approx 2.8 \times 10^{-13}\text{ s}.$$

Exercise 10

Observer O measures the area $\sqrt{3}$ m^2 of an equilateral triangle ABC at rest in his xy-plane. An observer O′ moving relative to O at $0.6c$ observes the triangle A′B′C′. The triangle symmetry axis forms an angle $\theta = 15°$ with respect to the direction x–x' as measured by observer O (see figure 17.2). What is the area measured by O′?

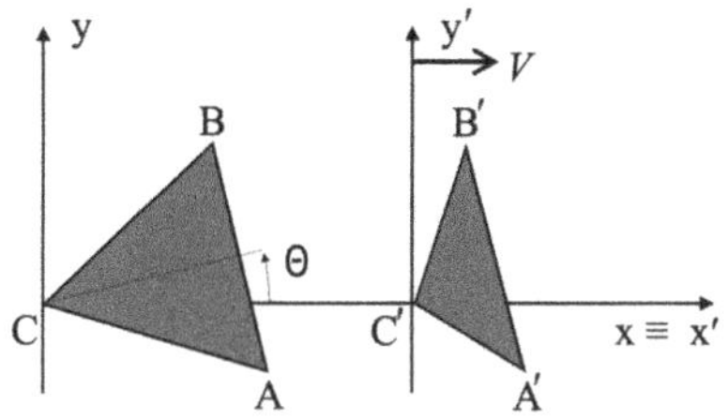

Figure 17.2. The equilateral triangle ABC is at rest with the observer O and forms an angle θ with the x-axis, while the scalene triangle A′B′C′ is at rest with the observer O′.

Solution

For observer O the area[2] $\mathcal{A}_0$ is

$$\mathcal{A}_0 = \frac{L_0^2 \sin \pi/3}{2} = \frac{L_0^2\sqrt{3}}{4} = \sqrt{3}\ \text{m}^2 \approx 1.73\ \text{m}^2.$$

Hence, the proper length of the side of the triangle is $L_0 = 2$ m. We indicate with

$$\begin{aligned} A &\equiv (L_0 \cos 15°,\ -L_0 \sin 15°) \approx (1.93\ \text{m},\ -0.518\ \text{m}),\\ B &\equiv (L_0 \cos 45°,\ L_0 \sin 45°) \approx (1.41\ \text{m},\ 1.41\ \text{m}),\\ C &\equiv (0,\ 0), \end{aligned}$$

the vertices of the equilateral triangle as measures by O.

Since there is length contraction only in the x–x' direction, with $\gamma = \frac{1}{\sqrt{1-(0.60)^2}} = 1.25$, the vertices measures by O′ are

$$\begin{aligned} A' &\equiv \left(\frac{L_0 \cos 15°}{\gamma},\ -L_0 \sin 15°\right) = \left(\frac{1.93}{1.25}\ \text{m},\ -0.518\ \text{m}\right) = (1.54\ \text{m},\ -0.518\ \text{m}),\\ B' &\equiv \left(\frac{L_0 \cos 45°}{\gamma},\ L_0 \sin 45°\right) = \left(\frac{1.41}{1.25}\ \text{m},\ 1.41\ \text{m}\right) = (1.12\ \text{m},\ 1.41\ \text{m}),\\ C' &\equiv (0,\ 0). \end{aligned}$$

The length of the sides and the angles measured by observer O′ are

$$\begin{aligned} L_{\text{C}'\text{A}'} &= \sqrt{(1.54\ \text{m})^2 + (0.518\ \text{m})^2} \approx 1.62\ \text{m},\\ L_{\text{C}'\text{B}'} &= \sqrt{(1.12\ \text{m})^2 + (1.41\ \text{m})^2} \approx 1.80\ \text{m},\\ \tan\theta'_{\text{C}'\text{A}'} &= -\frac{0.518}{1.54} \approx -0.336 \ \Rightarrow\ \theta'_{\text{C}'\text{B}'} = -18.59°,\\ \tan\theta'_{\text{C}'\text{B}'} &= \frac{1.41}{1.12} \approx 1.26 \ \Rightarrow\ \theta'_{\text{C}'\text{B}'} = 51.54°. \end{aligned}$$

The area $\mathcal{A}'_0$ measured by observer O′ is

[2] Let vectors **AB** and **AC** point, respectively, from A to B and from A to C; the area of the parallelogram ABDC is $\mathcal{A}_{\text{ABCD}} = |\mathbf{AB}||\mathbf{AC}|\sin\alpha$, where α is the angle between the vectors. Therefore, the area of triangle ABC is half of parallelogram area: $\mathcal{A}_{\text{ABC}} = \frac{1}{2}|\mathbf{AB}||\mathbf{AC}|\sin\alpha$.

$$\mathcal{A}' = \frac{L_{C'A'}L_{C'B'}\sin\left(\theta'_{C'B'} - \theta'_{C'A'}\right)}{2}$$
$$= \frac{(1.62\ \text{m}) \times (1.80\ \text{m}) \times \sin(18.59^\circ + 51.54^\circ)}{2} \approx 1.4\ \text{m}^2.$$

We note that the result $\mathcal{A}' = 1.4\ \text{m}^2$ could be obtained as $\mathcal{A}' = \mathcal{A}_0/\gamma$. Indeed,

$$\mathcal{A}' = \frac{\mathcal{A}_0}{\gamma} = \frac{\sqrt{3}\ \text{m}^2}{1.25} = 1.4\ \text{m}^2.$$

To prove this, we calculate the area in an alternative way. We rewrite the generic area for the observer O as follows[3]:

$$\mathcal{A}_0 = \frac{1}{2}\det\begin{bmatrix} A_x & A_y \\ B_x & B_y \end{bmatrix}.$$

Similarly for the observer O′ we have

$$\mathcal{A}' = \frac{1}{2}\det\begin{bmatrix} A'_x & A'_y \\ B'_x & B'_y \end{bmatrix}.$$

Only in the propagation direction do we have the contraction of lengths, then

$$\mathcal{A}' = \frac{1}{2}\det\begin{bmatrix} \frac{A_x}{\gamma} & A_y \\ \frac{B_x}{\gamma} & B_y \end{bmatrix} = \frac{1}{2\gamma}\det\begin{bmatrix} A_x & A_y \\ B_x & B_y \end{bmatrix} = \frac{\mathcal{A}_0}{\gamma}.$$

We note that in the limit $\gamma \to 1$ (i.e. $V \to 0$) the area $\mathcal{A}' \to \mathcal{A}_0$, while in limit $\gamma \to \infty$ (i.e. $V \to c$) the area $\mathcal{A}' \to 0$.

Exercise 11

The longest wavelength of light emitted by hydrogen in the Balmer series is $\lambda = 656$ nm, but this wavelength measured from a distant galaxy is $\lambda' = 1458$ nm. Find the speed at which the galaxy is receding from Earth, assuming the shift to be due to the Doppler effect.

Solution

From the equation of the transverse Doppler effect (see chapter 3) we obtain

[3] Let $A \equiv (A_x, A_y)$, $B \equiv (B_x, B_y)$ and $C \equiv (C_x, C_y)$ be three generic triangle vertices in a Cartesian coordinate system, the area can be computed as 1/2 times the absolute value of the determinant:

$$\mathcal{A}_{ABC} = \frac{1}{2}\det\begin{bmatrix} A_x & A_y & 1 \\ B_x & B_y & 1 \\ C_x & C_y & 1 \end{bmatrix}.$$

$$\lambda = \lambda' \sqrt{\frac{1+\beta}{1-\beta}}.$$

Placing $\kappa = \frac{\lambda}{\lambda'}$, we have

$$\beta = \frac{\kappa^2 - 1}{\kappa^2 + 1}.$$

Replacing the numerical values, $\kappa = 0.45$, we find $\beta = 0.66$ or $V = 0.66c$.

Exercise 12

Stars typically emit red light of atomic hydrogen with a wavelength of 656.3 nm (called the Hα spectral line). Compute the wavelength of that light observed on Earth from stars receding directly from us with relative speeds of $v_1 = 10^{-1}c$, $v_2 = 10^{-2}c$ and $v_3 = 10^{-3}c$.

Solution
Using the formula $\lambda = \lambda' \sqrt{\frac{1+\beta}{1-\beta}}$ we have

$$\begin{aligned} v_1 &= 10^{-1}c \longrightarrow \lambda = 725.5 \text{ nm}, \\ v_2 &= 10^{-2}c \longrightarrow \lambda = 6662.9 \text{ nm}, \\ v_3 &= 10^{-3}c \longrightarrow \lambda = 656.9 \text{ nm}. \end{aligned}$$

Exercise 13

The light from a nearby star is observed to be shifted toward blue by 2%, i.e. $v' = 1.02\nu_0$. (a) How fast is it moving? (b) Is the star approaching or receding from Earth? (Assume the motion is directly toward or away from Earth.)

Solution
We have

$$\frac{\Delta v}{v_0} \equiv \chi = \frac{v' - v_0}{v_0} = 0.02.$$

From the transverse Doppler effect (see chapter 3), i.e. $v' = v_0 \sqrt{\frac{1+\beta}{1-\beta}}$, we obtain

$$\beta = \frac{(\chi + 1)^2 - 1}{(\chi + 1)^2 + 1}.$$

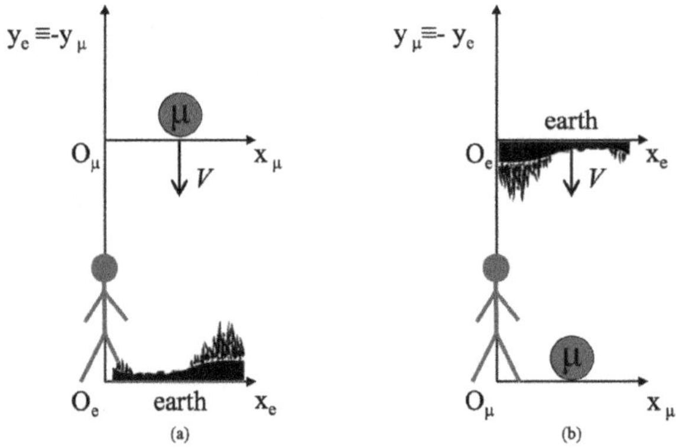

Figure 17.3. (a) From the point of view of an observer at rest with the Earth the muon moves with speed V. (b) From the point of view of an observer at rest with the muon the Earth moves with speed V.

(a)

$$\beta = \frac{(0.02 + 1)^2 - 1}{(0.02 + 1)^2 + 1} = 1.98 \times 10^{-2}.$$

Therefore, the velocity is $V = 1.98 \times 10^{-2}c$.

(b) The star is receding from Earth.

Problem 1

A muon with a lifetime of 2.2×10^{-6} s is created in the upper atmosphere at about 10 km from Earth. When it is created it has a velocity of 99.92% of the light in a vacuum in a direction towards Earth. (a) What is the average distance that it will travel before decaying, as determined by an observer on Earth (see figure 17.3(a))? (b) From the point of view of the muon, how far will it travel towards Earth before it disintegrates (see figure 17.3(b))?

Solution

(a) For an observer on Earth, classically, in $\Delta t = 2.2 \times 10^{-6}$ s and with a speed $0.9992c$, the distance d_{muon} that the muon can travel is

$$d_{\text{muon}} = v\Delta t = (0.9992 \times 2.998 \times 10^8 \text{ m s}^{-1}) \times (2.2 \times 10^{-6} \text{ s}) = 659.0 \text{ m},$$

so that the muon cannot, on average, reach the Earth. However, due to time dilation (see chapter 3) the muon lifetime $\Delta t_{\text{muon}}^{\text{rel}}$ is increased for the observer on Earth:

$$\Delta t_{\text{muon}}^{\text{rel}} = \gamma \Delta t_{\text{muon}} = \frac{\Delta t_{\text{muon}}}{\sqrt{1 - (V/c)^2}} = \frac{2.2 \times 10^{-6} \text{ s}}{\sqrt{1 - \left(\frac{0.9992c}{c}\right)^2}}$$

$$\approx 25 \times 2.2 \times 10^{-6} \text{ s} = 55 \times 10^{-6} \text{ s}.$$

The average distance travelled, as determined by an Earth observer $d^{\text{rel}}_{\text{muon}}$, is

$$d^{\text{rel}}_{\text{muon}} = v\Delta t^{\text{rel}}_{\text{muon}} = (0.9992 \times 2.998 \times 10^8 \text{ m s}^{-1}) \times (55 \times 10^{-6} \text{ s}) = 16.5 \text{ km}.$$

Thus, an observer on Earth determines that, on average, a muon will reach the Earth (16.5 km > 10 km).

(b) From the point of view of an observer at rest with respect to the muon, the distance classically travelled by the Earth d_{Earth}, is

$$d_{\text{Earth}} = v\Delta t = (0.9992 \times 2.998 \times 10^8 \text{ m s}^{-1}) \times (2.2 \times 10^{-6} \text{ s}) = 659.0 \text{ m},$$

so that Earth does not reach the muon. However, the initial distance of the Earth to muon, $d^{\text{rel}}_{\text{Earth}}$, is shortened because of the Lorentz contraction:

$$d^{\text{rel}}_{\text{Earth}} = \frac{h}{\gamma} = 10 \text{ km} \sqrt{1 - \left(\frac{0.9992c}{c}\right)^2} = \frac{10 \text{ km}}{25} = 400.0 \text{ m}.$$

Thus, an observer on the muon determines that, on average, the Earth will reach him/her (400 m < 659 m), in agreement with the observer on Earth.

Problem 2

A rocket ship $L_0 = 240$ m long, at rest with respect to an observer O′, travels at a constant velocity $V = 0.6c$ relative to an inertial observer O on the ground. A light source is placed midway in the rocket ship, at point B, between two mirrors, points A and C (see figure 17.4). The light source emits two flashes, one in the direction of the nose of the ship (point C) and one in the direction of the tail (point A). Subsequently, the two flashes return to the source where they are absorbed. Determine: (a) the time that the flashes take to reach the mirrors as measured by an observer O′ on the rocket ship; (b) the time that the flashes take to reach the mirrors as measured by the observer O; (c) the time that the flashes take to return to the source as measured from both O′ and O; and (d) the total time of the two flashes (the time interval between the events of emission and absorption) as measured by O′ and O.

Solution

(a) As determined by observer O′ at rest with respect to the rocket ship, the times are

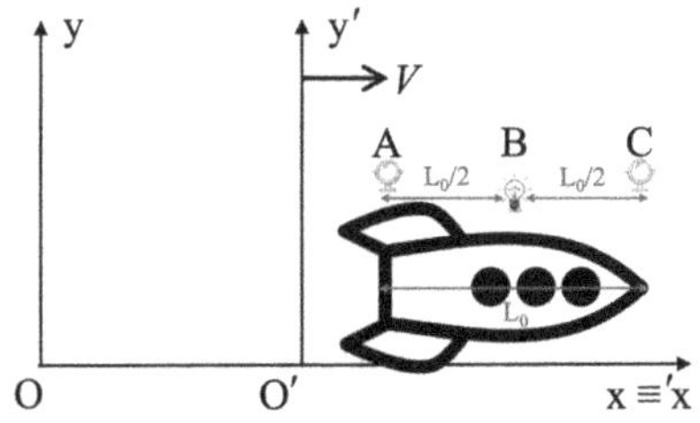

Figure 17.4. The rocket ship.

$$\Delta t'_{\mathrm{B}\to\mathrm{A}} = t'_{\mathrm{A}} - t'_{\mathrm{B}} = \frac{L_0}{2c} = 0.4\ \mu\mathrm{s},$$
$$\Delta t'_{\mathrm{B}\to\mathrm{C}} = t'_{\mathrm{C}} - t'_{\mathrm{B}} = \frac{L_0}{2c} = 0.4\ \mu\mathrm{s}.$$

The flashes arrive at A and C at the same time.

(b) In the rocket ship we have the following distances travelled:

$$\Delta x'_{\mathrm{B}\to\mathrm{A}} = x'_{\mathrm{A}} - x'_{\mathrm{B}} = 0 - \frac{L_0}{2} = -\frac{L_0}{2},$$
$$\Delta x'_{\mathrm{B}\to\mathrm{C}} = x'_{\mathrm{C}} - x'_{\mathrm{B}} = L_0 - \frac{L_0}{2} = \frac{L_0}{2}.$$

Using the inverse Lorentz transformation (see chapter 3), we find the times as measured by observer O:

$$\Delta t_{\mathrm{B}\to\mathrm{A}} = \gamma\left(\Delta t'_{\mathrm{B}\to\mathrm{A}} + \frac{V}{c^2}\Delta x'_{\mathrm{B}\to\mathrm{A}}\right) = \gamma\left(\frac{L_0}{2c} - \frac{V}{c^2}\frac{L_0}{2}\right) = \gamma\frac{L_0}{2c}\left[1 - \frac{V}{c}\right] = 0.2\ \mu\mathrm{s},$$
$$\Delta t_{\mathrm{B}\to\mathrm{C}} = \gamma\left(\Delta t'_{\mathrm{B}\to\mathrm{C}} + \frac{V}{c^2}\Delta x'_{\mathrm{B}\to\mathrm{C}}\right) = \gamma\left(\frac{L_0}{2c} + \frac{V}{c^2}\frac{L_0}{2}\right) = \gamma\frac{L_0}{2c}\left[1 + \frac{V}{c}\right] = 0.8\ \mu\mathrm{s}.$$

For the observer O the arrivals of the flashes on mirrors A and C are not simultaneous.

(c) For the observer O′ the time taken by the flash to return from the mirror to the source, obviously, is the same:

$$\Delta t'_{\mathrm{A}\to\mathrm{B}} = t'_{\mathrm{B}} - t'_{\mathrm{A}} = \frac{L_0}{2c} = 0.4\ \mu\mathrm{s},$$
$$\Delta t'_{\mathrm{C}\to\mathrm{B}} = t'_{\mathrm{B}} - t'_{\mathrm{C}} = \frac{L_0}{2c} = 0.4\ \mu\mathrm{s}.$$
$$\Delta x'_{\mathrm{A}\to\mathrm{B}} = x'_{\mathrm{B}} - x'_{\mathrm{A}} = \frac{L_0}{2} - 0 = \frac{L_0}{2},$$
$$\Delta x'_{\mathrm{C}\to\mathrm{B}} = x'_{\mathrm{B}} - x'_{\mathrm{C}} = \frac{L_0}{2} - L_0 = -\frac{L_0}{2}.$$

While for the observer O, we have

$$\Delta t_{\mathrm{A}\to\mathrm{B}} = \gamma\left(\Delta t'_{\mathrm{A}\to\mathrm{B}} + \frac{V}{c^2}\Delta x'_{\mathrm{A}\to\mathrm{B}}\right) = \gamma\left(\frac{L_0}{2c} + \frac{V}{c^2}\frac{L_0}{2}\right) = \gamma\frac{L_0}{2c}\left[1 + \frac{V}{c}\right] = 0.8\ \mu\mathrm{s},$$
$$\Delta t_{\mathrm{C}\to\mathrm{B}} = \gamma\left(\Delta t'_{\mathrm{C}\to\mathrm{B}} + \frac{V}{c^2}\Delta x'_{\mathrm{C}\to\mathrm{B}}\right) = \gamma\left(\frac{L_0}{2c} - \frac{V}{c^2}\frac{L_0}{2}\right) = \gamma\frac{L_0}{2c}\left[1 - \frac{V}{c}\right] = 0.2\ \mu\mathrm{s}.$$

(d) The total times as measured by observer O′ and O are

$$\Delta t'_{\mathrm{A}} = \Delta t'_{\mathrm{B}\to\mathrm{A}} + \Delta t'_{\mathrm{A}\to\mathrm{B}} = 0.8\ \mu\mathrm{s},$$
$$\Delta t'_{\mathrm{C}} = \Delta t'_{\mathrm{B}\to\mathrm{C}} + \Delta t'_{\mathrm{C}\to\mathrm{B}} = 0.8\ \mu\mathrm{s},$$
$$\Delta t_{\mathrm{A}} = \Delta t_{\mathrm{B}\to\mathrm{A}} + \Delta t_{\mathrm{A}\to\mathrm{B}} = 1.0\ \mu\mathrm{s},$$
$$\Delta t_{\mathrm{C}} = \Delta t_{\mathrm{B}\to\mathrm{C}} + \Delta t_{\mathrm{C}\to\mathrm{B}} = 1.0\ \mu\mathrm{s}.$$

It is important to note that the time separations, such as $\Delta t'_{B\to A}$, $\Delta t_{B\to A}$, $\Delta t'_{A\to B}$ etc, are not proper times. Indeed, it is easy to verify that

$$\Delta t_{B\to A} \neq \gamma\, \Delta t'_{B\to A} \Rightarrow 0.2\ \mu s \neq \frac{0.4\ \mu s}{\sqrt{1-(0.6)^2}} = 1.25 \times 0.4\ \mu s = 0.5\ \mu s$$

$$\Delta t'_{B\to A} \neq \gamma\, \Delta t_{B\to A} \Rightarrow 0.4\ \mu s \neq \frac{0.2\ \mu s}{\sqrt{1-(0.6)^2}} = 1.25 \times 0.2\ \mu s = 0.25\ \mu s.$$

If the observers O and O′ measure the time separation between two events that, for both observers, occur at different spatial locations, then the time separations are not related by simply dividing or multiplying by γ. Instead, the time separations $\Delta t'_A$ and $\Delta t'_C$ are proper times:

$$\Delta t_A = \gamma \Delta t'_A = \frac{0.8\ \mu s}{\sqrt{1-(0.6)^2}} = 1.25 \times 0.8\ \mu s = 1.0\ \mu s,$$

$$\Delta t_C = \gamma \Delta t'_C = \frac{0.8\ \mu s}{\sqrt{1-(0.6)^2}} = 1.25 \times 0.8\ \mu s = 1.0\ \mu s.$$

Problem 3

Determine the relativistic motion of a charge q in a constant uniform electric field $\vec{E}$ with initial speed orthogonal to the field. (a) What is the relativistically correct trajectory? (b) What is the classic limit of this trajectory? (c) What is the proper time of an observer at rest with the charge?

Solution

(a) Choose coordinates so that y is parallel to $\vec{E} = (0, E_0, 0)$ and choose an inertial reference frame so that there is no motion of charge q along the z-axis. Therefore, the motion of q is confined to the xy-plane. In this system the initial velocity is $\vec{v}_0 = (v_0, 0)$. The equation of motion, in the frame with respect to which the motion is observed, in terms of relativistic momentum $\vec{P} = (P_x, P_y)$ is

$$\frac{d}{dt}\vec{P} = q\vec{E},$$

explicitly

$$\frac{d}{dt}P_x = 0, \quad \frac{d}{dt}P_y = qE_0.$$

Integrating we obtain

$$P_x(t) = P_{0x}, \quad P_y(t) = qE_0 t + P_{0y}.$$

Using the relation between the relativistic moment $\vec{P}$, velocity $\mathbf{v}$ and energy $\mathcal{E}$,

$$\vec{P} = \frac{\mathcal{E}}{c^2}\vec{v} = \frac{\sqrt{m_0^2c^2 + P^2}}{c}\vec{v},$$

we obtain

$$v_x(t) = \frac{c^2}{\mathcal{E}}P_x(t) = \frac{cP_{0x}}{\sqrt{m_0^2c^2 + P_{0x}^2 + (qE_0t + P_{0y})^2}},$$

$$v_y(t) = \frac{c^2}{\mathcal{E}}P_y(t) = \frac{c(qE_0t + P_{0y})}{\sqrt{m_0^2c^2 + P_{0x}^2 + (qE_0t + P_{0y})^2}}.$$

Imposing the initial conditions of velocity $\vec{v}_0 = \vec{v}(0) = (v_0, 0)$, we obtain

$$P_{0x} = \frac{m_0v_0}{\sqrt{1 - (v_0/c)^2}} \equiv P_0, \quad P_{0y} = 0.$$

So,

$$v_x(t) = \frac{cP_0}{\sqrt{\mathcal{P}_0^2 + (qE_0t)^2}},$$

$$v_y(t) = \frac{cqE_0t}{\sqrt{\mathcal{P}_0^2 + (qE_0t)^2}},$$

where

$$\mathcal{P}_0^2 = m_0^2c^2 + P_0^2 = \frac{m_0^2c^2}{1 - (v_0/c)^2}.$$

Note that in the limit $t \to \infty$, $v_x \to 0$ and $v_y \to c$. Using relations of the velocity, we can derive the components of acceleration:

$$a_x(t) = \frac{dv_x}{dt} = -\frac{cq^2E_0^2P_0t}{[\mathcal{P}_0^2 + (qE_0t)^2]^{3/2}},$$

$$a_y(t) = \frac{dv_y}{dt} = \frac{cqE_0\mathcal{P}_0^2}{[\mathcal{P}_0^2 + (qE_0t)^2]^{3/2}}.$$

Note that in the limit $t \to \infty$ both components of acceleration go to zero. The coordinates as a function of time take the form

$$v_x(t) = \frac{dx}{dt} \Rightarrow x(t) - x_0 = \int \frac{cP_0}{\sqrt{\mathcal{P}_0^2 + (qE_0t)^2}}\, dt = \frac{cP_0}{qE_0}\sinh^{-1}\left(\frac{qE_0t}{\mathcal{P}_0}\right),$$

$$v_y(t) = \frac{dy}{dt} \Rightarrow y(t) - y_0 = \int \frac{cqE_0t}{\sqrt{\mathcal{P}_0^2 + (qE_0t)^2}}\, dt = \frac{c}{qE_0}\sqrt{\mathcal{P}_0^2 + (qE_0t)^2}.$$

Imposing the initial conditions of position $\mathbf{r}(0) = (0, 0)$, we have $x_0 = 0$ and $y_0 = \frac{c\mathcal{P}_0}{qE_0}$. Hence,

$$x(t) = \frac{c\mathcal{P}_0}{qE_0}\sinh^{-1}\left(\frac{qE_0t}{\mathcal{P}_0}\right),$$

$$y(t) = \frac{c\mathcal{P}_0}{qE_0}\left[\sqrt{1 + \left(\frac{qE_0t}{\mathcal{P}_0}\right)^2} - 1\right].$$

To derive the trajectory we eliminate time t. To do this, we rewrite the equation $x(t)$ in the following way:

$$\sinh\left(\frac{qE_0x}{c\mathcal{P}_0}\right) = \frac{qE_0t}{\mathcal{P}_0}.$$

Then, replacing this relation into $y(t)$, we find that the trajectory in the plane xy is a *catenary*:

$$y = \frac{c\mathcal{P}_0}{qE_0}\left[\cosh\left(\frac{qE_0x}{c\mathcal{P}_0}\right) - 1\right].$$

(b) The classic limit is recovered supposing $v \ll c$ and that the work done by the field qE_0x is small with respect to the energy $c\mathcal{P}_0$. Indeed, we have $\mathcal{P}_0 \approx m_0c$, $P_0 \approx m_0v_0$ and $\cosh(ax) \approx 1 + a^2x^2/2$, from which

$$y \approx \frac{m_0c^2}{qE_0}\left[1 + \frac{1}{2}\left(\frac{qE_0x}{m_0v_0c}\right)^2 - 1\right] = \frac{1}{2}\frac{qE_0}{m_0v_0^2}x^2 = \frac{1}{2}\frac{a_{\text{clas}}}{v_0^2}x^2,$$

where $a_{\text{clas}} = \frac{qE_0}{m_0}$ is the classic acceleration. Therefore, the trajectory obtained is the usual parabola (see figure 17.5).

(c) The proper time t_0 of an observer at rest with the charge is

$$t_0 = \int_0^t \frac{\mathrm{d}t}{\gamma} = \int_0^t \sqrt{1 - (v/c)^2}\,\mathrm{d}t = \int_0^t \sqrt{1 - (v_x/c)^2 - (v_y/c)^2}\,\mathrm{d}t$$
$$= \int_0^t \frac{m_0c\mathrm{d}t}{\sqrt{\mathcal{E}_0^2 + (qE_0t)^2}} = \frac{m_0c}{qE_0}\sinh^{-1}\left(\frac{qE_0t}{\mathcal{E}_0}\right),$$

from which

$$t_0 = \frac{m_0c}{qE_0}\ln\left[\frac{qE_0t}{\mathcal{E}_0} + \sqrt{1 + \left(\frac{qE_0t}{\mathcal{E}_0}\right)^2}\right].$$

As $t \to \infty$, t_0 increases much more slowly than t, according to the law

$$t_0 = \frac{m_0c}{qE_0}\ln\left[2\frac{qE_0t}{\mathcal{E}_0}\right].$$

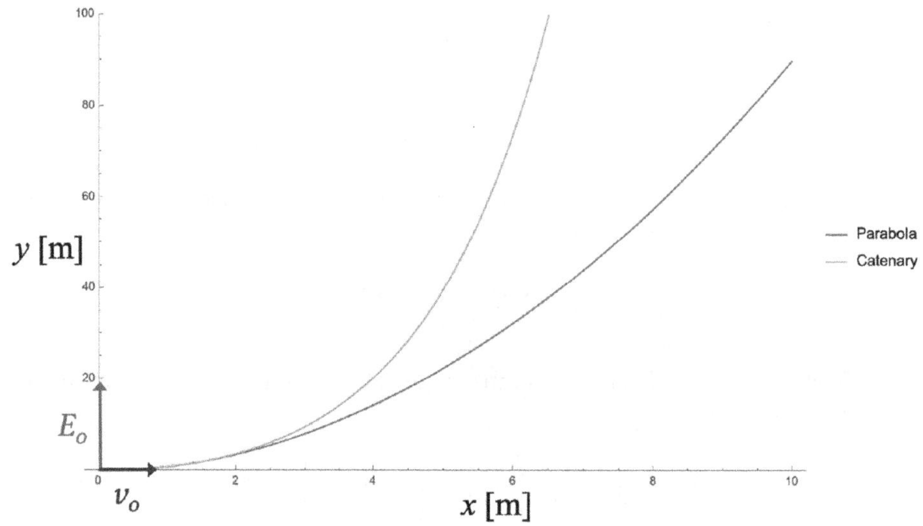

Figure 17.5. The blue line is the classical trajectory, while the orange line is the relativistic trajectory. The trajectories were obtained for a positron with initial speed $v_0 = c/3$ accelerated by an electric field $E_0 = 10^5$ N C^{-1}.

17.2 Quantum physics

Exercise 1

Determine the ratio of the energy emitted from a black-body (T = 3000 K) in wavelength bands of width 300 Å centred on λ_1 = 6000 Å (visible) and λ_2 = 30 000 Å (infrared).

Solution

From chapter 6 we have

$$u_\lambda \mathrm{d}\lambda = \frac{8\pi hc}{\lambda^5}\frac{\mathrm{d}\lambda}{(\mathrm{e}^{hc/\lambda k_\mathrm{B}T} - 1)} = G(\lambda)\mathrm{d}\lambda.$$

The bandwidth $\Delta\lambda$ = 300 Å is sufficiently small that we may use $G(\lambda)$ as constant over this interval, to obtain

$$\frac{\Delta u_{\lambda_2}}{\Delta u_{\lambda_1}} = \frac{G(\lambda_2)}{G(\lambda_1)} = \frac{\lambda_1^5}{\lambda_2^5}\frac{\mathrm{e}^{hc/\lambda_1 k_\mathrm{B}T} - 1}{\mathrm{e}^{hc/\lambda_2 k_\mathrm{B}T} - 1},$$

replacing the numerical values we have

$$\frac{\Delta u_{\lambda_2}}{\Delta u_{\lambda_1}} = 5.517.$$

This result indicates that only 19% of radiation, at that temperature, is emitted as visible light.

Exercise 2

Knowing that the maximum sensitivity of the human eye is in the region of yellow–green light of frequency 5.5×10^{14} Hz, at what temperature will a black-body be most visible?

Solution

Using Wien's law in the form of equation (6.21) (for details about this formula see chapter 6) we have

$$\frac{\nu_{\max}}{T} = \frac{2.82k_B}{h} = 5.87 \times 10^{10} \text{ Hz K},$$

from which

$$T = \frac{5.5 \times 10^{14} \text{ Hz}}{5.87 \times 10^{10} \text{ Hz K}^{-1}} = 9.3 \times 10^3 \text{ K}.$$

Exercise 3

What power does the human body emit (supposing we identify it with a black-body) at a temperature 30 °C with a surface $S = 2$ m^2? Why is it not visible in the dark?

Solution

From the Stefan–Boltzmann law, we write

$$E = \sigma T^4 = W/S \rightarrow W = S\sigma T^4$$
$$\frac{P}{S} = \sigma T^4 \Longrightarrow P = S\sigma T^4.$$

Recalling that 30 °C = 303 K we have

$$W = 5.67 \times 10^{-8} \text{ W m}^{-2} \text{ K}^{-4} \times (303 \text{ K}^{-1})^4 \times 2 \text{ m}^2 = 956 \text{ W}.$$

As for the second question, we apply the Wien displacement law, so that

$$\lambda_{\max} T = a = 2.89 \times 10^{-3} \text{ m K}^{-1} \rightarrow \lambda_{\max} = \frac{2.89 \times 10^{-3} \text{ m K}^{-1}}{303 \text{ K}^{-1}}$$
$$= 9.538 \times 10^{-11} \text{ m} \approx 950 \text{ nm}.$$

This wavelength falls into the infrared band, which is not visible to the human eye.

Exercise 4

What are the momentum, wavelength and frequency of a 10 keV photon?

Solution

$$p = \frac{E}{c} = \frac{10\ \text{keV}}{c} = 10^4\ \text{eV}/c,$$

$$\lambda = \frac{h}{p} = \frac{hc}{E} = \frac{(4.14 \times 10^{-15}\ \text{eV s})(2.99 \times 10^8\ \text{m s}^{-1})}{10\ \text{keV}} = \frac{12.4 \times 10^{-7}\ \text{eV m}}{10\ \text{keV}}$$

$$= \frac{12.4\ \text{keV Å}}{10\ \text{keV}} = 1.24\ \text{Å},$$

$$\nu = \frac{c}{\lambda} = \frac{2.99 \times 10^8\ \text{m s}^{-1}}{1.24\ \text{Å}} = \frac{2.99 \times 10^8\ \text{m s}^{-1}}{1.24 \times 10^{-10}\ \text{m}} = 2.41 \times 10^{19}\ \text{Hz}.$$

Exercise 5

The kinetic energies of photoelectrons range from zero to 2 eV when light of wavelength 3500 Å falls on a surface. (a) What is the stopping potential for this light? (b) What is the threshold wavelength for the material?

Solution

(a) From $eV_s = K_{max}$ (see chapter 7),

$$eV_s = 5eV \Rightarrow V_s = 5V.$$

(b) Using $eV_s = h\nu - eW_0$, we obtain the threshold wavelength

$$\frac{hc}{\lambda_{th}} = eW_0 = \frac{hc}{\lambda} - eV_s \Rightarrow \lambda_{th} = \frac{hc}{\frac{hc}{\lambda} - eV_s} = \frac{12.4\ \text{keV Å}}{\frac{12.4\ \text{keV Å}}{3.5\ \text{kÅ}} - 2\ \text{eV}} = 8037\ \text{Å}.$$

Exercise 6

Silicon is illuminated with ultraviolet light of wavelength 2500 Å. If the work function of silicon is 2.25 eV, what is the maximum kinetic energy of the emitted electrons?

Solution

$$K_{max} = \frac{hc}{\lambda} - eW_0 = \frac{12.4\ \text{keV Å}}{2.5\ \text{kÅ}} - 2.25\ \text{eV} = 2.71\ \text{eV}.$$

Exercise 7

What is the percentage change in the wavelength of an x-ray of wavelength 0.200 Å and 135° Compton scattering from an electron?

Solution
From the Compton relation

$$\lambda' = \lambda + \frac{h}{m_0 c}(1 - \cos\theta) = 0.200\ \text{Å} + \frac{h}{(0.511\ \text{MeV}/c^2)c}(1 - \cos 135°)$$
$$= 0.200\ \text{Å} + \frac{12.4\ \text{keV Å}}{0.511\ \text{MeV}}\left(1 + \frac{\sqrt{2}}{2}\right) = 0.241\ \text{Å},$$
$$\frac{\lambda' - \lambda}{\lambda} = \frac{0.241\ \text{Å} - 0.200\ \text{Å}}{0.200\ \text{Å}} = 0.0205.$$

Therefore the percentage change is 2.05%.

Exercise 8

In a Compton experiment an x-ray of wavelength 0.150 Å undergoes a 30° scattering. What is the wavelength of the scattered photon and the kinetic energy of the electron after the scattering event?

Solution
Using the Compton relation

$$\lambda' = \lambda + \frac{h}{m_0 c}(1 - \cos\theta) = 0.150\ \text{Å} + \frac{h}{(0.5\ \text{MeV}/c^2)c}(1 - \cos 30°)$$
$$= 0.150\ \text{Å} + \frac{12.4\ \text{keV Å}}{0.5\ \text{MeV}}\left(1 - \frac{\sqrt{3}}{2}\right) = 0.153\ \text{Å}. \quad (17.1)$$

From energy conservation $h\nu + m_0c^2 = h\nu' + E$:

$$\frac{hc}{\lambda} + m_0c^2 = \frac{hc}{\lambda'} + E \Rightarrow E - m_0c^2 = \frac{hc}{\lambda} - \frac{hc}{\lambda'}$$
$$= 12.4\ \text{keV Å}\left(\frac{1}{0.150\ \text{Å}} - \frac{1}{0.153\ \text{Å}}\right) = 1.62\ \text{keV}.$$

Hence, the kinetic energy is 1.62 keV.

Exercise 9

Estimate the de Broglie wavelength for: (a) a person of 70 kg walking in the park at a speed of 4 km h^{-1}; (b) a tennis ball of mass 60 g and with a speed of 150 km h^{-1} played in a tennis match; and (c) an electron if it is accelerated through a potential difference of 100 V.

Solution
(a) From the expression of the de Broglie wavelength, $\lambda = h/mv$, we obtain

$$\lambda_{\text{person}} = \frac{h}{mv} = \frac{6.63 \times 10^{-34}\ \text{J s}}{(70\ \text{kg}) \times (4/3.6\ \text{m s}^{-1})} = 8.52 \times 10^{-26}\ \text{Å}.$$

(b) For the tennis ball we have

$$\lambda_{\text{ball}} = \frac{h}{mv} = \frac{6.63 \times 10^{-34}\ \text{J s}}{(0.06\ \text{kg}) \times (150/3.6\ \text{m s}^{-1})} = 2.65 \times 10^{-24}\ \text{Å}.$$

(c) Using the energy conservation in classical mechanics $\frac{1}{2}mv^2 = eV$, we have

$$v = \sqrt{\frac{2eV}{m}},$$

while the momentum is

$$mv = m\sqrt{\frac{2eV}{m}} = \sqrt{2meV}.$$

The de Broglie wavelength for the electron is

$$\lambda_{\text{electron}} = \frac{h}{p} = \frac{h}{\sqrt{2meV}}$$

$$= \frac{6.63 \times 10^{-34}\ \text{J s}}{\sqrt{2 \times (9.11 \times 10^{-31}\ \text{kg}) \times (1.60 \times 10^{-19}\ \text{C}) \times (10^2\ \text{V})}} = 1.23\ \text{Å}.$$

Quantum phenomena are negligible in the macroscopic world (person, tennis ball), while they are crucial in the description of the microscopic world (electron).

Exercise 10

Suppose that the momentum can be measured to an accuracy of 0.01%. (a) What is the uncertainty in the position of a particle with a mass of 1 g and a speed of 2 m s^{-1}? What is the uncertainty in the position for an electron moving with a speed of $0.2c$.

Solution

(a) We have $\Delta p/p = 10^{-4}$:

$$\Delta p = 10^{-4}p = 10^{-4}mv.$$

Using Heisenberg's uncertainty principle $\Delta x \Delta p \geqslant h/4\pi$ (see chapter 7)

$$\Delta x \geqslant \frac{h}{4\pi\Delta p} = \frac{10^4 h}{4\pi mv} = \frac{10^4 \times (6.63 \times 10^{-34}\ \text{J s})}{4\pi \times (10^{-3}\ \text{kg}) \times (2\ \text{m s}^{-1})} = 2.64 \times 10^{-18}\ \text{Å}.$$

(b) For the electron we have to use the relativistic relation for the mass $m = \gamma m_0$:

$$\Delta x \geqslant \frac{h}{4\pi\Delta p} = \frac{10^4 h}{4\pi\gamma m v} = \frac{10^4 h\sqrt{1-(v/c)^2}}{4\pi \times (0.511\ \text{MeV}/c^2) \times (0.2c)}$$
$$= \frac{10^4 \times (12.4\ \text{keV Å})\sqrt{1-(0.2)^2}}{4\pi \times (0.511\ \text{MeV}) \times (0.2)} = 96.6\ \text{Å}.$$

Exercise 11

What is the frequency of the photon that is absorbed when a hydrogen atom undergoes a transition from $n = 3$ to $n = 7$?

Solution

From the Bohr model the energy levels are $E_n = -E_0/n^2$ (see chapter 7), where $E_0 = 13.6$ eV is the lowest energy level of hydrogen atom. Therefore,

$$E_3 = -\frac{E_0}{9} = -1.51\ \text{eV}, \quad E_7 = -\frac{E_0}{49} = -0.278\ \text{eV},$$

and from the Bohr postulate the energy of the emitted photon is $E_\gamma = E_7 - E_3 = 1.23$ eV. The frequency of the photon is given by

$$\nu_\gamma = \frac{E_\gamma}{h} = \frac{1.23\ \text{eV}}{4.14 \times 10^{-15}\ \text{eV s}} = 2.97 \times 10^{-16}\ \text{Hz}.$$

Exercise 12

Find the correction to frequency of an emitted photon when the recoil kinetic energy of the hydrogen nucleus is taken into account.

Solution

Supposing that the hydrogen is initially at rest, from conservation of energy *e* momentum gives

$$E_{\text{H}}^i = E_{\text{H}}^f + E_\gamma + E_{\text{N}},$$
$$0 = -P_{\text{N}} + P_\gamma,$$

where E_{H}^i and E_{H}^f are the initial and final energies of the hydrogen atom, E_γ and P_γ are the photon energy and the photon momentum, and E_{N} and P_{N} are the nuclear kinetic energy and the nuclear momentum. We indicate with λ_γ the correct wavelength and λ_0 the uncorrected wavelength, hence

$$\frac{1}{\lambda_0} = \frac{E_{\text{H}}^i - E_{\text{H}}^f}{hc} = \frac{1}{\lambda_\gamma} + \frac{E_{\text{N}}}{hc},$$
$$\sqrt{2m_{\text{N}}E_{\text{N}}} = \frac{h}{\lambda_\gamma} \Rightarrow E_{\text{N}} = \frac{h^2}{2m_{\text{N}}\lambda_\gamma^2},$$

where m_N is the nuclear mass (just a proton). From which

$$\frac{\lambda_\gamma - \lambda_0}{\lambda_0} = \frac{h}{2m_N c\lambda_\gamma} = \frac{hc}{2m_N c^2\lambda_\gamma} = \frac{12.4\ \text{keV Å}}{2\times(939\times MeV)\lambda_\gamma} = \frac{6.6\ \mu\text{Å}}{\lambda_\gamma}.$$

The typical values of the wavelengths are $\sim 10^3$ Å. Therefore, the fractional change is of the order of 10^{-9}.

Exercise 13

Show numerically the tunnelling probability: (a) for a ball of mass $m = 1$ kg moving at a speed $v = 30\ \text{m s}^{-1}$ against a gravity potential barrier of height $h = 46$ m and width $L = 10$ m; and (b) for an electron with energy $E = 1$ eV is incident upon a rectangular barrier of potential energy $V_0 = 2$ eV and width $L = 1$ Å. (c) What happens if the energy of a particle is greater than the potential barrier?

Solution

(a) The kinetic energy of the ball is $E = \frac{1}{2}mv^2 \sim 450$ J, while the gravitational potential energy of the barrier is $V_0 = mgh \sim 451$ J. Therefore,

$$V_0 - E = 1\ \text{J}.$$

Using the wave vector for a finite rectangular potential barrier (see section 8.3.4 of chapter 8) we have

$$\beta = \frac{\sqrt{2m(V_0 - E)}}{\hbar} = \frac{\sqrt{2\times(1\ \text{kg})\times(1\ \text{J})}}{1.1\times10^{-34}\ \text{J s}} \approx 10^{34}\ \text{m}^{-1},$$

then

$$\beta L \approx (10^{34}\ \text{m}^{-1})\times(10\ \text{m}) = 10^{35},$$

and

$$\sinh(\beta L) \approx \frac{1}{2}e^{\beta L} \approx \frac{1}{2}e^{10^{35}} \approx 10^{4\times10^{34}}.$$

Hence, the tunnelling probability T (see chapter 8) is

$$T = \left[1 + \frac{V_0^2\sinh^2(\beta L)}{4E(V_0 - E)}\right]^{-1} \approx \left[\frac{V_0\sinh^2(\beta L)}{4}\right]^{-1} \sim 10^{-2}\times10^{-8\times10^{34}},$$

$$T \sim 10^{-8\times10^{34}} \sim 0.$$

That is, the tunnelling probability for the ball is essentially zero. In other words, in agreement with the predictions of classical mechanics, the ball does not cross the gravity potential barrier:

$$R = \left[1 + \frac{4E(V_0 - E)}{V_0^2\sinh^2(\beta L)}\right]^{-1} \approx [1 + 10^{-4\times10^{34}}]^{-1} = 1.$$

(b) In the case of the electron we have

$$V_0 - E = 1\ \text{eV} = 1.6 \times 10^{-19}\ \text{J}$$

and

$$\beta = \frac{\sqrt{2m(V_0 - E)}}{\hbar} = \frac{\sqrt{2 \times (9.1 \times 10^{-31}\ \text{kg}) \times (1.6 \times 10^{-19}\ \text{J})}}{1.1 \times 10^{-34}\ \text{J s}} \approx 4.5 \times 10^9\ \text{m}^{-1},$$

then

$$\beta L = (4.5 \times 10^9\ \text{m}^{-1}) \times (10^{-10}\ \text{m}) = 0.45,$$
$$\sinh(\beta L) = \frac{e^{\beta L} - e^{-\beta L}}{2} = \frac{e^{0.45} - e^{-0.45}}{2} \approx 0.47.$$

The tunnelling probability T for the electron is

$$T = \left[1 + \frac{V_0^2 \sinh^2(\beta L)}{4E(V_0 - E)}\right]^{-1} = \left[1 + \frac{(2\ \text{eV})^2\ (0.47)^2}{4(1\ \text{eV})(1\ \text{eV})}\right]^{-1} = [1 + 0.22]^{-1} \sim 0.82.$$
$$T \sim 0.82.$$

Therefore, contrary to the classical prediction, the electron has a probability of 82% to cross the barrier of potential V_0.

(c) In this case we have $E - V_0 > 0$. To calculate the reflection and transmission coefficients R and T, it is unnecessary to perform all calculations (see chapter 8), instead all we need to do is replace in the equation

$$\beta \to \mathrm{i}\tilde{k},$$

where $\tilde{k} = \sqrt{\frac{2m}{\hbar^2}(E - V_0)}$. Using the relation between the hyperbolic-sine and sine, i.e. $\sinh(\mathrm{i}x) = \mathrm{i}\sin(x)$, we obtain the new expressions for the transmission and reflection coefficients:

$$T = \left[1 + \frac{V_0^2 \sin^2(\tilde{k}\ L)}{4E(E - V_0)}\right]^{-1},$$
$$R = \left[1 + \frac{4E(E - V_0)}{V_0^2 \sin^2(\tilde{k}L)}\right]^{-1}.$$

It is easy to verify that $R + T = 1$. In this case, R and T oscillate periodically between a minimum and maximum:

$$\frac{4E(E - V_0)}{(2E - V_0)^2} \leqslant T \leqslant 1,$$
$$0 \leqslant R \leqslant \frac{V_0^2}{(2E - V_0)^2}.$$

The complete transmission $T = 1$ ($R = 0$) occurs only for certain values of the energy $E = E_T$:

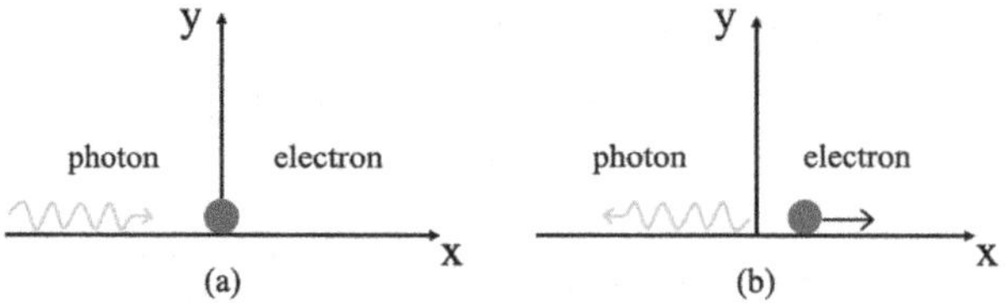

Figure 17.6. (a) Photon and electron before the collision. (b) Photon and electron after the collision.

$$\sin(\tilde{k}L) = 0 \Rightarrow \tilde{k}(E_T)L = n\pi,$$

$$E_T = \frac{h^2 n^2}{8m^2L^2} + V_0.$$

This result is in agreement with scattering phenomena at low energy and simulates a well-known phenomenon as the *Ramsauer–Townsend effect*[4].

While the maximum reflection (minimum transmission) occurs for the energy $E = E_R$:

$$\sin(\tilde{k}L) = 1 \Rightarrow \tilde{k}(E_R)L = (2n + 1)\frac{\pi}{2},$$

$$E_R = \frac{h^2(2n + 1)^2}{32m^2L^2} + V_0.$$

This result cannot be explained in classical mechanics, where the particle is always transmitted, but it is a typically quantum effect which requires the wave-like behaviour of matter.

Problem 1

An x-ray photon with energy of 100 keV experiences a 'head-on' collision with an electron at rest (see figure 17.6(a)). (a) Using conservation of momentum and energy, find the recoil velocity of the electron (see figure 17.6(b)). (b) Check that this result agrees with the value determined from the Compton equation.

Solution

(a) Before scattering the photon and the electron have energy and momentum of, respectively, $E_\gamma = h\nu$, $P_\gamma = \frac{h}{\lambda}$ and $E_e = m_0c^2$, $P_e = 0$. After scattering the photon has $E'_\gamma = h\nu'$, $P'_\gamma = \frac{h}{\lambda'}$ and the electron $E'_e = mc^2 = \gamma m_0 c^2$, $P'_e = mv = \gamma m_0 v$. From the conservation of energy and momentum (the momentum of a photon $\frac{h}{\lambda} = \frac{E_\gamma}{c}$), we obtain

[4] It is a physical phenomenon involving the scattering of low-energy electrons by atoms of a noble gas (argon, krypton, xenon, etc). The probability of collision between the electrons and gas atoms obtains a minimum value for electrons with a certain amount of kinetic energy.

$$E_\gamma + m_o c^2 = E'_\gamma + \frac{m_0 c^2}{\sqrt{1-(v/c)^2}},$$

$$\frac{E_\gamma}{c} + 0 = -\frac{E'_\gamma}{c} + \frac{m_0 v}{\sqrt{1-(v/c)^2}}.$$

The simultaneous solution of the energy and the momentum equations gives

$$\begin{aligned} v &= \frac{(2E_\gamma + m_0c^2)^2 - (m_0c^2)^2}{(2E_\gamma + m_0c^2)^2 + (m_0c^2)^2}c \\ &= \frac{(2\times 0.1\text{ MeV} + 0.5\text{ MeV})^2 - (0.5\text{ MeV})^2}{(2\times 0.1\text{ MeV} + 0.5\text{ MeV})^2 + (0.5\text{ MeV})^2}c \approx 0.32c. \end{aligned}$$

(b) Using the Compton relation, we find

$$\lambda' - \lambda = \frac{h}{m_0 c}(1 - \cos\theta) = \frac{h}{m_0 c}(1 - \cos\pi) = \frac{2h}{m_0 c} \Rightarrow \lambda' = \lambda + \frac{2h}{m_0 c},$$

from which

$$\frac{1}{E'_\gamma} = \frac{1}{E_\gamma} + \frac{2}{m_0 c^2} = \frac{1}{0.1\text{ MeV}} + \frac{2}{0.5\text{ MeV}} \Rightarrow E'_\gamma = \frac{1}{14}\text{ MeV}.$$

Substituting E'_γ into the energy equation and solving for v,

$$v = c\sqrt{1 - \frac{1}{\left(1 + \frac{E_\gamma - E'_\gamma}{m_0 c^2}\right)^2}} = c\sqrt{1 - \frac{1}{\left(1 + \frac{0.1\text{ MeV} - 0.071\text{ MeV}}{0.5\text{ MeV}}\right)^2}} \approx 0.32c.$$

Problem 2

Consider the one-dimensional wavefunction

$$\psi(x) = \psi_0 \frac{e^{-kx}}{x},$$

where ψ_0 and k are constants. Using Schrödinger's equation, find the potential $V(x)$ and energy E for which this wave function is an eigenfunction (assume that as $x \to \infty$, $V(x) \to 0$).

Solution

Differentiating the given wave function,

$$\frac{d}{dx}\psi(x) = -\frac{(kx+1)}{x}\psi(x),$$
$$\frac{d^2}{dx^2}\psi(x) = \frac{(k^2x^2+2kx+2)}{x^2}\psi(x).$$

Substituting it in the time-independent Schrödinger equation,

$$\left[-\frac{\hbar^2}{2m}\frac{d^2}{dx^2} + V(x)\right]\psi(x) = E\psi(x),$$

we have

$$\left[-\frac{\hbar^2}{2m}\frac{(k^2x^2+2kx+2)}{x^2} + V(x)\right]\psi(x) = E\psi(x),$$
$$\left[-\frac{\hbar^2}{2m}\frac{(k^2x^2+2kx+2)}{x^2}\right]\psi(x) = [E - V(x)]\,\psi(x).$$

Therefore,

$$[E - V(x)] = -\frac{\hbar^2}{2m}\frac{(k^2x^2+2kx+2)}{x^2}.$$

Imposing $V(x) \to 0$ when $x \to \infty$, we find

$$E = -\frac{\hbar^2 k^2}{2m},$$
$$V(x) = \frac{\hbar^2}{2m}\frac{(2kx+2)}{x^2}.$$

Problem 3

An electron moves in a one-dimensional box of length L with potential (see figure 17.7)

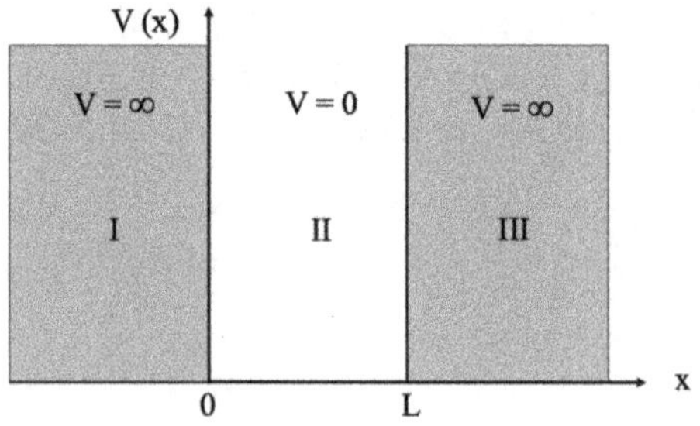

Figure 17.7. The one-dimensional infinite potential well of length L. It is divided into three regions, with regions I and III having $V(x) = \infty$ and region II having $V(x) = 0$.

$$V(x) = \infty, \ x < 0,$$
$$V(x) = 0, \ 0 < x < L,$$
$$V(x) = \infty, \ x > L.$$

(a) If the energy of the ground state is 10 eV, what is the dimension of the box? (b) What is the zero point energy predicted by the uncertainty principle? (c) Check that the Heisenberg principle is verified for a generic eigenfunction.

Solution

(a) It has been shown that the wave function and the energy eigenvalues have the form (see chapter 8)

$$\psi_n(x) = \sqrt{\frac{2}{L}} \sin\left(\frac{n\pi x}{L}\right),$$
$$E_n = \frac{h^2 n^2}{8mL^2}.$$

The energy of the ground state is $E_1 = \frac{h^2}{8mL^2}$, imposing that $E_1 = 10$ eV, we find

$$L = \frac{h}{\sqrt{8mE_1}} = \frac{6.63 \times 10^{-34} \text{ J s}}{\sqrt{8 \times (9.11 \times 10^{-31} \text{ kg}) \times (10 \times 1.60 \times 10^{-19} \text{ J})}} = 1.94 \text{ Å}.$$

(b) The electron is confined to a box of length L, which means that the uncertainty in the position is $\Delta x = L$—the particle has to be somewhere in the box. The minimum uncertainty in the momentum from the Heisenberg relation is

$$\Delta p = \frac{\hbar}{2L},$$

which corresponds to a minimum energy E_{min}:

$$E_{\text{min}} = \frac{(\Delta p)^2}{2m} = \frac{h^2}{32\pi^2 mL^2}.$$

This result is in agreement with the quantitative result derived in our case: $E_{\text{min}} < E_1$.

(c) The expectation values of the position x and momentum p in the nth state are

$$\langle x\rangle_{\psi_n} = \int_0^L \psi_n(x)^*[x\psi_n(x)]\mathrm{d}x = \frac{2}{L}\int_0^L x\sin^2\!\left(\frac{n\pi x}{L}\right)\mathrm{d}x = \frac{L}{2},\ \forall n \geqslant 1$$

$$\langle p\rangle_{\psi_n} = \int_0^L \psi_n(x)^*[p\psi_n(x)]\mathrm{d}x = \frac{2}{L}\int_0^L \sin\left(\frac{n\pi x}{L}\right)\left[-\mathrm{i}\hbar\frac{\mathrm{d}}{\mathrm{d}x}\sin\left(\frac{n\pi x}{L}\right)\right]\mathrm{d}x$$

$$= -\mathrm{i}\frac{2n\pi\hbar}{L^2}\int_0^L \sin\left(\frac{n\pi x}{L}\right)\cos\left(\frac{n\pi x}{L}\right)\mathrm{d}x = 0,\ \forall n \geqslant 1.$$

The average position is $L/2$ and the average momentum is zero. This last agrees with the classical result: the particle has equal probability of travelling in either the $+x$- or $-x$-direction. The expectation values of x^2 and p^2 are

$$\langle x^2\rangle_{\psi_n} = \int_0^L \psi_n(x)^*[x^2\psi_n(x)]\mathrm{d}x = \frac{2}{L}\int_0^L x^2\sin^2\!\left(\frac{n\pi x}{L}\right)\mathrm{d}x = L^2\!\left(\frac{1}{3} - \frac{1}{2n^2\pi^2}\right),$$

$$\langle p^2\rangle_{\psi_n} = \int_0^L \psi_n(x)^*[p^2\psi_n(x)]\mathrm{d}x = \frac{2}{L}\int_0^L \sin\left(\frac{n\pi x}{L}\right)\left[-\hbar^2\frac{\mathrm{d}^2}{\mathrm{d}x^2}\sin\left(\frac{n\pi x}{L}\right)\right]\mathrm{d}x$$

$$= \frac{2}{L}\int_0^L \sin\left(\frac{n\pi x}{L}\right)\left[\left(\frac{\hbar\pi n}{L}\right)^2\sin\left(\frac{n\pi x}{L}\right)\right]\mathrm{d}x$$

$$= \left(\frac{\hbar\pi n}{L}\right)^2\frac{2}{L}\int_0^L \sin^2\!\left(\frac{n\pi x}{L}\right)\mathrm{d}x = \left(\frac{\hbar\pi n}{L}\right)^2.$$

The uncertainties Δx_n and Δp_n take the form:

$$\Delta x_n = \sqrt{\langle x^2\rangle_{\psi_n} - \langle x\rangle_{\psi_n}^2} = L\sqrt{\frac{1}{12} - \frac{1}{2n^2\pi^2}},$$

$$\Delta p_n = \sqrt{\langle p^2\rangle_{\psi_n} - \langle p\rangle_{\psi_n}^2} = \sqrt{2mE_n} = \frac{\hbar\pi n}{L}.$$

Therefore, the uncertainty principle for a generic eigenfunction is

$$\Delta x_n \Delta p_n = \frac{\hbar}{2}\sqrt{\frac{\pi^2 n^2}{3} - 2}.$$

This product increases with increasing n, having a minimum value for $n = 1$. The value of this product for $n = 1$ is about equal to $0.568\hbar$ which obeys the Heisenberg uncertainty principle, which states that the product will be greater than or equal to $\hbar/2$.

IOP Publishing

Modern Physics
A critical approach
Canio Noce

Chapter 18

Using applets to learn modern physics

Marco Figliolia, Antonio Stabile and Canio Noce

Applets are certainly among the most successful resources used in physics education to understand theory and its application simultaneously. Indeed, applets may be considered computer experiments, and thus they allow students to recognize connections to the topic of study in a different way, using animations. Moreover, the use of an animation is not a passive activity, since students can learn real-time information about the physical systems that they experience difficulty in understanding, and more importantly they can actively interact with the system they are observing. Finally, on the teacher's side, instructors can use the applets in their didactic activities, helping in the presentation of topics with which their students could experience some difficultly.

In the following, we present three applets devoted to specific applications of modern physics. Namely, we propose an applet where the reader may apply the basic concepts of special relativity (see chapters 3 and 17), looking for the so-called Minkowski diagrams, and two applets related to quantum mechanics. In these latter applets, the reader can find the description of the eigenvalues and eigenvectors of the infinite-well of potential as well as an application for the tunnelling effect (see chapters 8 and 17).

18.1 Time dilation and length contraction shown in space–time frames

The Lorentz transformations (see chapter 3) are a family of linear transformations from a coordinate frame $(t, \mathbf{x})$ to another frame $(t', \mathbf{x}')$—each coordinate in one frame is a linear function of all the coordinates in the other frame—that moves at a constant velocity V relative to the former. Assuming that the relative motion is along the common x–x'-axis, the transformation takes the form

doi:10.1088/978-0-7503-2678-0ch18

$$
\begin{aligned}
t' &= \frac{t - \frac{V}{c^2}x}{\sqrt{1 - \frac{V^2}{c}}},\\
x' &= \frac{x - Vt}{\sqrt{1 - \frac{V^2}{c}}},\\
y' &= y,\\
z' &= z,
\end{aligned}
$$

where c is the speed of light. Since in our choice of relative motion the coordinates y and z do not vary, we can consider the transformation of the reference frames only with the coordinates t and x.

In general, the coordinates of a physical event are given by x and t. All possible values of space and time can be represented on a space–time diagram in which the space axis is horizontal and the time axis is vertical. It is appropriate to use the same dimensions for the coordinates x and t. To do this, multiplying the time t by the velocity of light c, the quantity ct represents the distance that light travels in vacuum at instant t. For convenience we indicate the latter quantity with x_0: $x_0 = ct$. Then, the Lorentz transformation equations can be written as follows:

$$x_0' = \frac{x_0 - \beta x}{\sqrt{1 - \beta^2}}, \tag{18.1}$$

$$x' = \frac{x - \beta x_0}{\sqrt{1 - \beta^2}}, \tag{18.2}$$

where $\beta = \frac{V}{c}$. The inverse transformations are

$$x_0 = \frac{x_0' + \beta x'}{\sqrt{1 - \beta^2}}, \tag{18.3}$$

$$x = \frac{x' + \beta x_0'}{\sqrt{1 - \beta^2}}. \tag{18.4}$$

Let the x and x_0 axes of the frame S be depicted by two mutually perpendicular straight lines. To draw the axes of the frame S' in this diagram, let us resort to equations (18.1) and (18.2), from which the origins of the reference frames S and S' are seen to coincide (when $x = 0$ and $x_0 = 0$, $x' = 0$ and $x_0' = 0$ as well). The equation of motion of S' relative to S can be obtained by setting $x_0' = 0$ (the space axis in the frame S') and $x' = 0$ (the time axis in the frame S'). From equation (18.1), imposing $x_0' = 0$, we obtain the following relation $x_0 = \beta x$, while from equation (18.2) imposing $x' = 0$ we obtain $x_0 = \frac{x}{\beta}$. We note that, since $V < c$ and then $\beta < 1$, the angle $\varphi = \arctan \beta$ is less than $\pi/4$. From simple geometric considerations, we note that φ is the angle between the axes x–x' and x_0–x_0' (see figure 18.1). Thus, the Lorentz transformation reduces to a conversion of the orthogonal reference frame

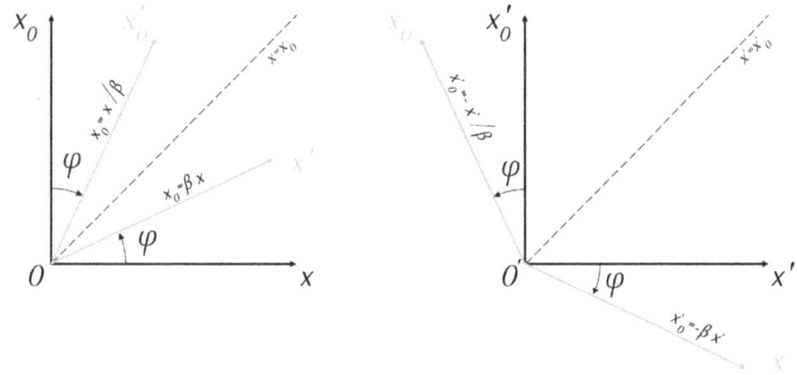

Figure 18.1. Geometric illustration of the Lorentz transformation. This transformation reduces to the rotation of the x'- and x_0'-axes through the angle $\varphi = \arctan\beta$ with respect to the x- and x_0-axes. (a) The transition from the frame S to S' corresponds to the convergence of the x'- and x_0'-axes. (b) The inverse transition to the divergence of the x- and x_0-axes.

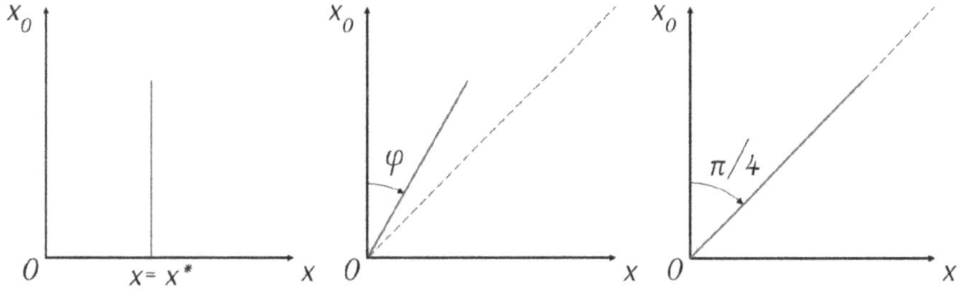

Figure 18.2. (a) The world line of an object at rest at the point $x = x^*$. (b) The world line of an object moving uniformly along the x-axis. (c) The world line of the light ray.

(x, x_0) to a non-orthogonal one (x', x_0'). The x'- and x_0'-axes rotate around the origin, in the direction of the bisector of the coordinate angle, through the same angle φ.

An event is a natural phenomenon that happens at a given moment of time t and a given point in space x. The set of all events is called *Minkowski space* [1]. The sequence of events happening with a particle in the Minkowski space yields a certain curve called the *world line* of the particle. This curve in Minkowski space corresponds to the motion of the particle. Let us analyse some simple cases. (a) A particle located at the point $x = x^*$ in the frame S. Its world line in the plane (x, x_0) of the Minkowski space is a straight line parallel to the x_0-axis (see figure 18.2(a)). (b) A particle moving uniformly along the x-axis in the frame S at velocity v. Its world line in this frame is a straight line inclined at the angle $\theta = \arctan\beta$, $(\theta < \pi/4)$, to the x_0-axis (see figure 18.2(b)). (c) A photon whose world line is a straight line making a $\pi/4$ angle with the x_0-axes (see figure 18.2(c)).

The world line of a particle with an arbitrary motion can be represented by $x = x(x_0)$ in the plane (x, x_0), as it is represented in figure 18.3. The inclination of the world line to the x_0-axis at each given point is determined by the derivative $\frac{dx}{dx_0}$ at that point:

$$\tan\theta = \beta.$$

Thus, the inclination angle is determined from the following equation:

$$\theta = \arctan\beta = \arctan\frac{v}{c},$$

where v is the instantaneous velocity of the point or the object. Since $\beta < 1$ always, the angle θ cannot exceed $\pi/4$ for any moving object. Let us consider the two pairs of equilateral hyperbolas in the plane (x, x_0) (see figure 18.4):

$$x_0^2 - x^2 = 1 \tag{18.5}$$

$$x_0^2 - x^2 = -1. \tag{18.6}$$

One can readily subdivide the plane (x, x_0) into four quadrants, each containing one hyperbola. We note that the light ray world lines are the asymptotes of the hyperbolas. Each of these hyperbolas (18.5) and (18.6) intersects only one of the

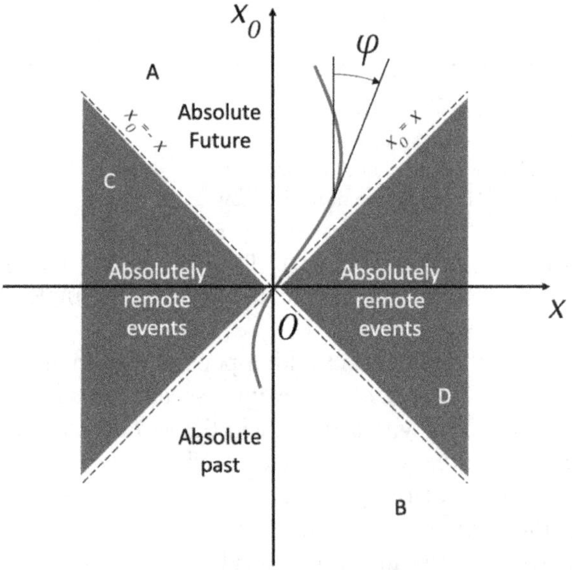

Figure 18.3. The intersection of the space–time cone by the plane (x, x_0). The events located in quadrant A represent the absolute future with respect to event O, while the events located in quadrant B represent the absolute past. All events located in quadrants C and D represent absolutely remote events with respect to event O. In red is shown the world line of an arbitrary motion. In this case, the angle formed by the tangent line to the world line and the x_0-axis is defined from the relation $\theta = \arctan(v/c)$, where v is the instantaneous velocity of a particle.

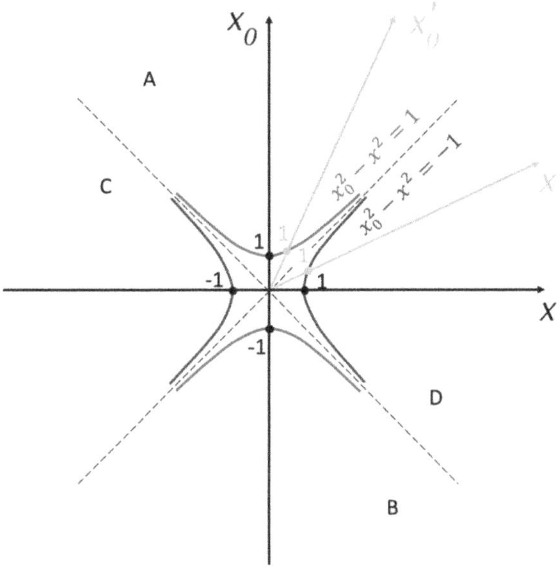

Figure 18.4. The four equilateral hyperbolas $x_0^2 - x^2 = 1$ and $x_0^2 - x^2 = -1$ are plotted in the coordinate system (x, x_0). Since the Lorentz transformation leaves the expression $x_0^2 - x^2 = x_0'^2 - x'^2$ invariant, we shall also obtain the hyperbolas $x_0'^2 - x'^2 = 1$ and $x_0'^2 - x'^2 = -1$ in the new oblique-angled coordinate system. This means that these four equilateral hyperbolas cross the axes x, x_0, x' and x_0' at distances from the origin equal to unity.

axes, x or x_0. The intersection points of hyperbola (18.5) with the x_0-axis are determined from the condition $x = 0$. We see that the hyperbolas (18.5) intersect the x_0-axis at the points $x_0 = \pm 1$. In a similar way one can find that the hyperbolas (18.6) intersect the x-axis at the points $x = \pm 1$. The hyperbolas are often referred to as *calibration curves*.

Since the expression $x_0^2 - x^2 = x_0'^2 - x'^2$ (see chapter 3) is the invariant of the Lorentz transformation equations (18.3) and (18.4), equations (18.5) and (18.6) are valid in the frame S'. It follows directly that the same hyperbolas cut off the unitary sections on the new oblique-angled axes x' and x_0' as well. We note that the unitary sections of the x and x' axes are far from being equal.

Now, using the Minkowski space we can show the relativity of the time intervals and the geometrical interpretation of the space contraction. Let us show the x- and x_0-axes and x'- and x_0'-axes and plot of the hyperbola that passes through the quadrant A and D of the coordinate systems S and S' (see figure 18.5). Let a clock be at rest at the origin of the coordinate system S' (see figure 18.5(a)). Its world line is the x_0'-axis. At the moment of time $x_0 = 0$ a moving clock was at the origin of the coordinate system S where its reading was compared against one of the clocks of the system S' located at the same point. The point A′ in the frame S is simultaneous with the point A″ (events located at the straining line $x_0 =$ cost). The sections OA″ and OA′ correspond to the time reading of the clocks of the systems S and S'. This means that if the clocks of S' measures the *proper time* interval OA′, the time interval by two clocks of S, located at the point O and x^*, is equal to OA″. It is possible to note

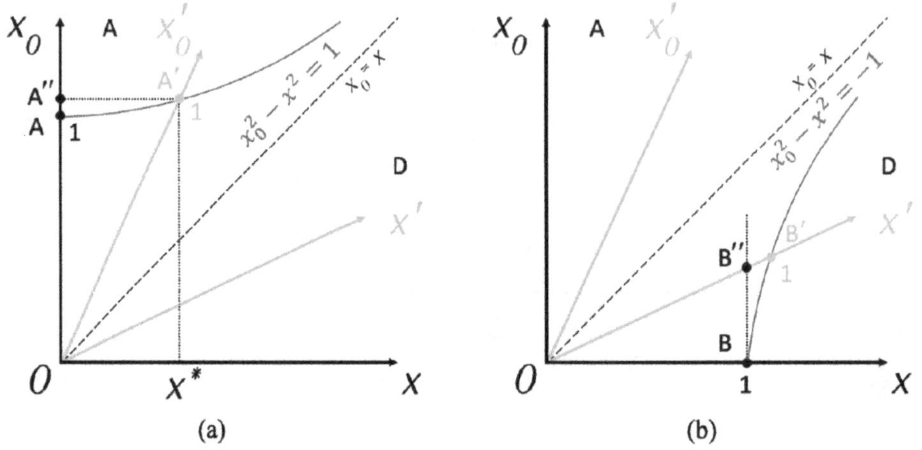

Figure 18.5. (a) A geometric illustration of the relativity of the time intervals between two events. The hyperbola $x_0^2 - x^2 = 1$ intersects the x_0'-axis at point A′ and the x_0-axis at point A. Thus, OA = 1 and OA′ = 1. Let a clock be at rest in the frame S', its world line coincides with the x_0'-axis. In the frame S the point A′ is simultaneous with the point A″ at which the clock will indicate the time determined by the section OA″ relative to the reading of another clock from S located at the point O. This implies that the time interval, during which the clock in the frame S' moves, is less in terms of S' than in terms of S. (b) A geometric illustration of the relativity of ruler lengths. OB is a ruler at rest in the frame S. The world lines of its ends are the x_0-axis and BB″. The hyperbola $x_0^2 - x^2 = -1$ intersects the x'-axis at point B′ and the x-axis at point B. Thus, OB = 1 and OB′ = 1. To find simultaneously the positions of the ruler's ends in the frame S', the world lines of the ruler's ends should intersect with the x'-axis. Then the ruler's length in the frame S' turns out to be equal to OB″. This implies that the ruler length is less in terms of S than in terms of S'.

from figure 18.5(a) that the time interval registered by the clock of S' is smaller, because OA′ < OA″ (if OA′ = 1 then OA = 1, therefore OA″ > 1): we have the *time dilation*. Analogously for the case with the clock at rest in S.

Now, let us consider the geometric illustration of the contraction of the moving ruler. In figure 18.5(b) the section OB represents the length of a ruler at rest in S, the *proper length*. Its world lines in the frame S are straight lines parallel to the x_0-axis and passing through the points O and B. In terms of the frame S', the simultaneous positions of the ends of the section OB at the moment $x_0' = 0$ correspond to the intersection of its world lines with x', i.e. to the points O and B″. From figure 18.5(b), we note that the ruler's length in S is smaller, because OB″ < OB′ (if OB = 1 then OB′ = 1, therefore OB″ < 1): we have the *length contraction*. Analogously for the case with the ruler at rest in S'.

The video clip presented in figure 18.6 summarizes some of the consequences of special relativity. Specifically, the animation describes the simultaneity concept, time dilation and length contraction, as depicted using Lorentz transformations (see chapter 3) or the superimposed Minkowski diagrams for the two observers presented above. We give a simple representation of the time dilation and length contraction equations, derived from the Lorentz transformations, Minkowski diagrams and the (1 + 1)-dimensional increment form of the Minkowski line element. Fixing the proper length ($\Delta x = -6.24$) and the proper time ($\Delta x_0 = 2.6$), the animation shows

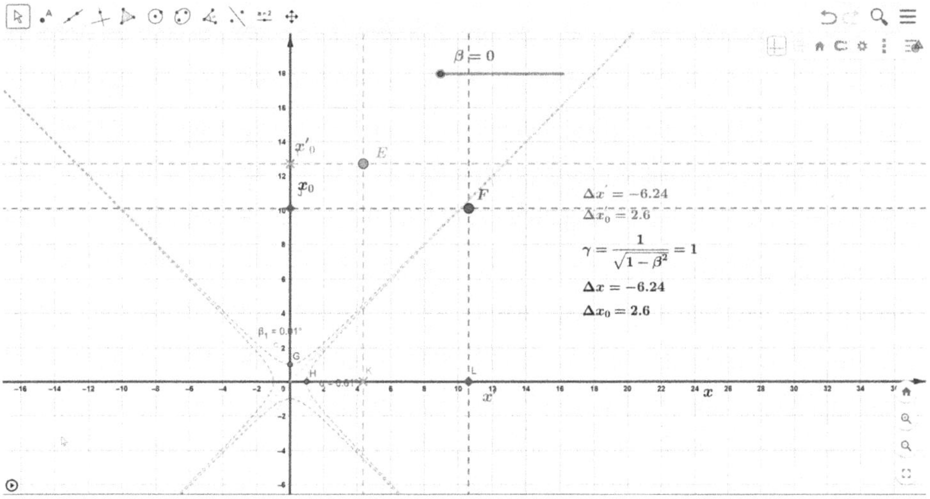

Figure 18.6. Video clip showing how the Minkowski diagrams can help to understand some consequences of special relativity. Animation available online at https://iopscience.iop.org/book//978-0-7503-1665-1.

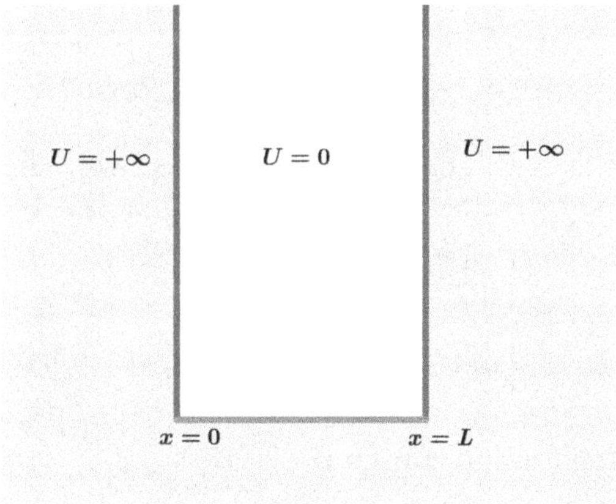

Figure 18.7. Infinite square-well potential energy.

that these quantities vary in the moving frame when its velocity β changes. According to the description presented previously, the superimposed Minkowski diagrams for two observers are also shown.

18.2 Finite potential well

Let us recall briefly the basic features of a free particle confined in a limited spatial region (see chapter 8). The potential energy, in the case considered, is zero inside the region and infinite outside (see figure 18.7):

$$U(x) = \begin{cases} 0, & \text{for } 0 \leqslant x \leqslant L \\ +\infty, & \text{for } \left| x - \dfrac{L}{2} \right| > \dfrac{L}{2} \end{cases}.$$

The infinite potential energy outside the region, $x < 0 \cup x > L$, means that there is zero probability of ever finding the particle there: the wavefunction of the particle is exactly zero, $\phi(x) = 0$, at $x < 0 \cup x > L$. While inside the region, $0 \leqslant x \leqslant L$ with $U(x) = 0$, the particle is free and the time-independent Schrödinger equation is written as follows:

$$-\frac{\hbar^2}{2m}\frac{\mathrm{d}^2}{\mathrm{d}x^2}\phi(x) = E\phi(x) \Longrightarrow \phi''(x) = -k^2\phi(x),$$

where $k^2 = \frac{2mE}{\hbar^2}$, $\phi''(x) = \frac{\mathrm{d}^2}{\mathrm{d}x^2}\phi(x)$ and the general solution for the wavefunctions $\phi(x)$ is

$$\phi(x) = A\sin(kx + \theta_0).$$

Imposing the boundary conditions, i.e. $\phi(0) = \phi(L) = 0$, we find the quantization condition for the energy E_n of the particle:

$$E_n = \frac{n^2h^2}{8mL^2}.$$

We have a series of discrete allowed energy values, with intermediate values not allowed. The stationary solution for the wavefunction $\phi(x)$ is therefore

$$\phi_n(x) = \begin{cases} \sqrt{\frac{2}{L}}\sin\left(\frac{n\pi x}{L}\right), & \text{for } 0 \leqslant x \leqslant L \\ 0, & \text{for } \left| x - \frac{L}{2} \right| > \frac{L}{2} \end{cases},$$

where $\sqrt{\frac{2}{L}}$ is the normalization constant.

The video clip presented in figure 18.8 sketches the relevant characteristics of this important quantum example. Specifically, the animation describes how the eigenfunction and the corresponding eigenenergy change when the quantum number n varies. To obtain this result we fix the width of the quantum well ($a = 8$ Å) and the mass of the particle contained within the well (we are assuming that this particle is indeed an electron). We also report how the probability of finding the electron in the half part of the potential well is modified as a function of the quantum number n.

18.3 Walking through the wall: the quantum tunnelling effect

In this section we briefly recall the behaviour of a quantum particle facing a finite potential barrier (see chapter 8):

$$U(x) = \begin{cases} U_0, & \text{for } 0 \leqslant x \leqslant a \\ 0, & \text{for } x < 0 \ \vee \ x > a. \end{cases}$$

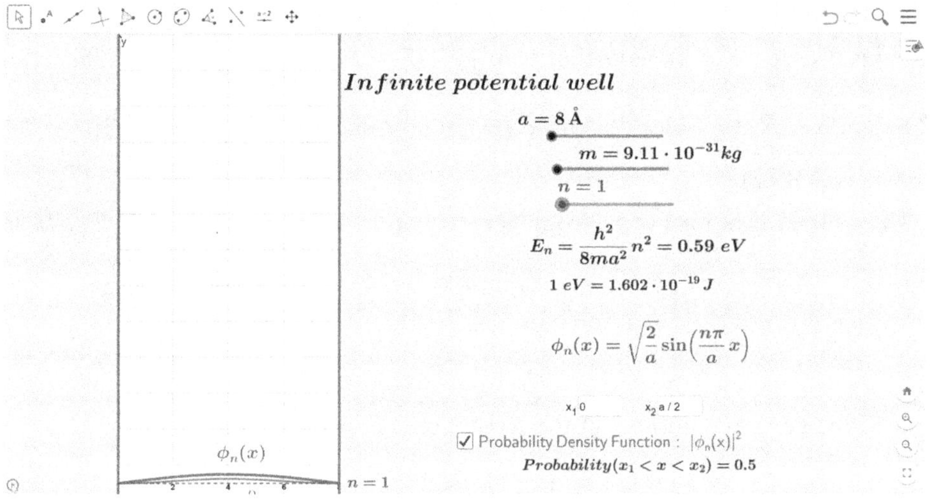

Figure 18.8. Video clip showing the relevant quantum aspect of a particle within an infinite quantum well. Animation available online at https://iopscience.iop.org/book//978-0-7503-1665-1.

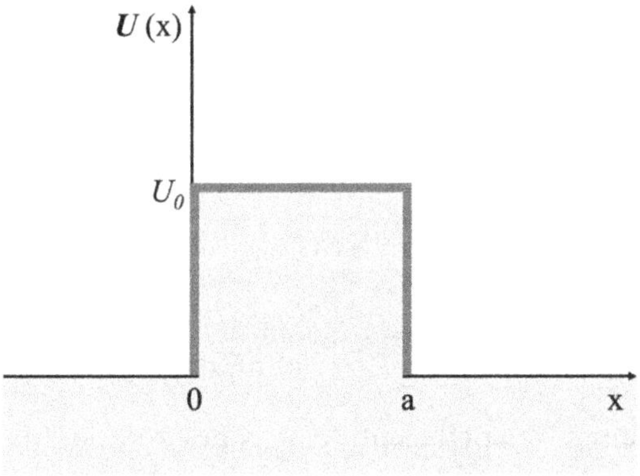

Figure 18.9. Finite rectangular potential barrier.

In figure 18.9, we can distinguish three regions: regions I and III with $U(x) = 0$, and region II with $U(x) = U_0 > 0$. In regions I and III the time-independent Schrödinger's equation becomes

$$\phi'' = -\frac{2mE}{\hbar^2}\phi = -k^2\phi,$$

with $k = \sqrt{\frac{2\,mE}{\hbar^2}}$. In region II Schrödinger's equation can be written as follows:

$$\phi'' = \frac{2m}{\hbar^2}(U_0 - E)\phi = \beta^2\phi.$$

Now, we can consider two distinct cases: $0 < E < U_0$ and $E > U_0 > 0$. In the first case, i.e. $0 < E < U_0$, we have that $\beta = \sqrt{\frac{2m}{\hbar^2}(U_0 - E)}$ is a real number. While in the second case, i.e. $E > U_0 > 0$, $\beta = \sqrt{\frac{2m}{\hbar^2}(U_0 - E)}$ is an imaginary number: $\beta = i\tilde{k}$, $\tilde{k} = \sqrt{\frac{2m}{\hbar^2}(E - U_0)}$.

The general solutions of Schrödinger's equation in the first case $0 < E < U_0$ are

$$\phi(x) = \begin{cases} \phi_{\rm I}(x) = A e^{ikx} + B e^{-ikx}, & \text{for } x < 0 \\ \phi_{\rm II}(x) = C e^{-\beta x} + D e^{\beta x}, & \text{for } 0 \leqslant x \leqslant a \\ \phi_{\rm III}(x) = F e^{ikx} + G e^{-ikx}, & \text{for } x > a. \end{cases}$$

Instead the general solutions in the second case $E > U_0 > 0$ are

$$\phi(x) = \begin{cases} \phi_{\rm I}(x) = A e^{ikx} + B e^{-ikx}, & \text{for } x < 0 \\ \phi_{\rm II}(x) = C e^{-i\tilde{k}x} + D e^{ix}, & \text{for } 0 \leqslant x \leqslant a \\ \phi_{\rm III}(x) = F e^{ikx} + G e^{-ikx}, & \text{for } x > a, \end{cases}$$

where A, B, C, D, F and G are complex constants to be determined.

Finally, after some calculations (see chapter 8), we find the transmission ($T = |F/A|^2$) and reflection coefficients ($R = |B/A|^2$). In the case $0 < E < U_0$ the coefficients takes the form

$$\begin{aligned} T_{E<U_0} &= \left[1 + \frac{U_0^2 \sinh^2(\beta a)}{4E(U_0 - E)}\right]^{-1} \\ R_{E<U_0} &= \left[1 + \frac{4E(U_0 - E)}{U_0^2 \sinh^2(\beta a)}\right]^{-1}. \end{aligned} \tag{18.7}$$

In general, the transmission coefficient $T_{E<U_0}$ is not zero and is used to describe the probability of a particle moving toward a finite potential barrier—the *tunnelling effect*. While in the case $E > U_0 > 0$ the coefficients are

$$\begin{aligned} T_{E>U_0} &= \left[1 + \frac{V_0^2 \sin^2(\tilde{k}a)}{4E(E - V_0)}\right]^{-1}, \\ R_{E>U_0} &= \left[1 + \frac{4E(E - V_0)}{V_0^2 \sin^2(\tilde{k}a)}\right]^{-1}. \end{aligned} \tag{18.8}$$

In this case the transmission and reflection coefficients oscillate periodically between a minimum and maximum. The complete transmission $T_{E>U_0} = 1$ ($R = 0$) occurs only for certain values of the energy $E_T = \frac{h^2 n^2}{8m^2 L^2} + U_0$ (see chapter 17). This result is in agreement with scattering phenomena at low-energy and simulates the well-known

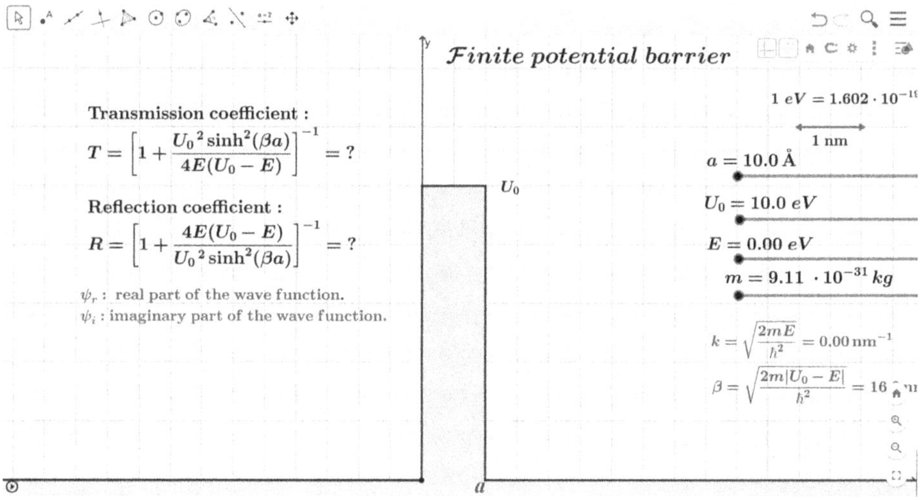

Figure 18.10. Video clip showing what happens when a particle hits a barrier. Animation available online at https://iopscience.iop.org/book//978-0-7503-1665-1.

phenomenon the *Ramsauer–Townsend effect*, while the maximum reflection (minimum transmission) occurs for the energy $E_R = \frac{h^2(2n+1)^2}{32m^2L^2} + U_0$.

We note that in both cases we have $R + T = 1$, and equations (18.8) can be recovered by equations (18.7), replacing β with $i\tilde{k}$.

The video clip presented in figure 18.10 describes the pertinent characteristics of the tunnelling effect. Namely, the animation describes how the transmission (T) and reflection (R) coefficients change when the energy of the incoming particle (we are assuming that this particle is an electron), hitting the barrier from the left, varies. We also show how the wavefunction is modified as a function of the energy of the incident particle, and we report the corresponding variations of the wave number of the wavefunction in the barrier (β) and outside the barrier (k). We assume that this energy can be as large as three times the barrier height U_0. We remark that the picture of motion has been obtained by fixing the width (a = 10 Å) and the height (U_0 = 10 eV) of the barrier.

Reference

[1] Resnick R 1968 *Introduction to Special Relativity* (New York: Wiley)